ANALOG CIRCUITS and DEVICES

ANALOG CIRCUITS and DEVICES

Editor-in-Chief
Wai-Kai Chen

CRC PRESS

Boca Raton   London   New York   Washington, D.C.

Library of Congress Cataloging-in-Publication Data

Catalog record is available from the Library of Congress

Visit the CRC Press Web site at www.crcpress.com

The material included here first appeared in *The VLSI Handbook* (CRC Press, 2000), Wai-Kai Chen, editor.

© 2003 by CRC Press LLC

No claim to original U.S. Government works
International Standard Book Number 0-8493-1736-3
Printed in the United States of America 1 2 3 4 5 6 7 8 9 0
Printed on acid-free paper

Preface

The purpose of *Analog Circuits and Devices* is to provide, in a single volume, a comprehensive reference covering the broad spectrum of devices and their models, amplifiers, analog circuits and filters, and compound semiconductor digital integrated circuit technology. The book has been written and developed for practicing electrical engineers in industry, government, and academia. The goal is to provide the most up-to-date information in the field.

Over the years, the fundamentals of the field have evolved to include a wide range of topics and a broad range of practice. To encompass such a wide range of knowledge, the book focuses on the key concepts, models, and equations that enable the design engineer to analyze, design, and predict the behavior of large-scale systems. While design formulas and tables are listed, emphasis is placed on the key concepts and theories underlying the processes.

The book stresses fundamental theory behind professional applications. In order to do so, the text is reinforced with frequent examples. Extensive development of theory and details of proofs have been omitted. The reader is assumed to have a certain degree of sophistication and experience. However, brief reviews of theories, principles, and mathematics of some subject areas are given.

The compilation of this book would not have been possible without the dedication and efforts of John Choma, Jr., Rolf Schaumann, Bang-Sup Song, Stephen I. Long, and, most of all, the contributing authors. I wish to thank them all.

Wai-Kai Chen
Editor-in-Chief

Editor-in-Chief

Wai-Kai Chen is Professor and Head Emeritus of the Department of Electrical Engineering and Computer Science at the University of Illinois at Chicago. He is now serving as Academic Vice President at International Technological University. He received his B.S. and M.S. in electrical engineering at Ohio University, where he was later recognized as a Distinguished Professor. He earned his Ph.D. in electrical engineering at University of Illinois at Urbana/Champaign.

Professor Chen has extensive experience in education and industry and is very active professionally in the fields of circuits and systems. He has served as visiting professor at Purdue University, University of Hawaii at Manoa, and Chuo University in Tokyo, Japan. He was editor of the *IEEE Transactions on Circuits and Systems, Series I and II*, president of the IEEE Circuits and Systems Society, and is the founding editor and editor-in-chief of the *Journal of Circuits, Systems and Computers*. He received the *Lester R. Ford Award* from the Mathematical Association of America, the *Alexander von Humboldt Award* from Germany, the *JSPS Fellowship Award* from Japan Society for the Promotion of Science, the *Ohio University Alumni Medal of Merit for Distinguished Achievement in Engineering Education*, the *Senior University Scholar Award* and the *2000 Faculty Research Award* from the University of Illinois at Chicago, and the *Distinguished Alumnus Award* from the University of Illinois at Urbana/Champaign. He is the recipient of the *Golden Jubilee Medal*, the *Education Award*, and the *Meritorious Service Award* from IEEE Circuits and Systems Society, and the *Third Millennium Medal* from the IEEE. He has also received more than a dozen honorary professorship awards from major institutions in China.

A fellow of the Institute of Electrical and Electronics Engineers and the American Association for the Advancement of Science, Professor Chen is widely known in the profession for his *Applied Graph Theory* (North-Holland), *Theory and Design of Broadband Matching Networks* (Pergamon Press), *Active Network and Feedback Amplifier Theory* (McGraw-Hill), *Linear Networks and Systems* (Brooks/Cole), *Passive and Active Filters: Theory and Implements* (John Wiley), *Theory of Nets: Flows in Networks* (Wiley-Interscience), and *The VLSI Handbook* and *The Circuits and Filters Handbook* (CRC Press).

Contributors

R. Jacob Baker
University of Idaho
Boise, Idaho

Andrea Baschirotto
Università di Pavia
Pavia, Italy

Marc Borremans
Katholieke Universiteit Leuven
Leuven-Heverlee, Belgium

Charles E. Chang
Conexant Systems, Inc.
Newbury Park, California

David J. Comer
Brigham Young University
Provo, Utah

Donald T. Comer
Brigham Young University
Provo, Utah

Bram De Muer
Katholieke Universiteit Leuven
Leuven-Heverlee, Belgium

Geert A. De Veirman
Silicon Systems, Inc.
Tustin, California

Maria del MarHershenson
Stanford University
Stanford, California

Donald B. Estreich
Hewlett-Parkard Company
Santa Rosa, California

John W. Fattaruso
Texas Instruments, Incorporated
Dallas, Texas

Mohammed Ismail
The Ohio State University
Columbus, Ohio

Johan Janssens
Katholieke Universiteit Leuven
Leuven-Heverlee, Belgium

John M. Khoury
Lucent Technologies
Murray Hill, New Jersey

Thomas H. Lee
Stanford University
Stanford, California

Harry W. Li
University of Idaho
Moscow, Idaho

Chi-Hung Lin
The Ohio State University
Columbus, Ohio

Stephen I. Long
University of California
Santa Barbara, California

Sunderarajan S. Mohan
Stanford University
Stanford, California

Alison Payne
Imperial College
University of London
London, England

Hirad Samavati
Stanford University
Stanford, California

Bang-Sup Song
University of California
La Jolla, California

Michiel Steyaert
Katholieke Universiteit Leuven
Leuven-Heverlee, Belgium

Donald C. Thelen
Analog Interfaces
Bozeman, Montana

Chris Toumazou
Imperial College
University of London
London, England

Meera Venkataraman
Troika Networks, Inc.
Calabasas Hills, California

Chorng-kuang Wang
National Taiwan University
Taipei, Taiwan

R.F. Wassenaar
University of Twente
Enschede, The Netherlands

Louis A. Williams, III
Texas Instruments, Inc.
Dallas, Texas

Min-shueh Yuan
National Taiwan University
Taipei, Taiwan

C. Patrick Yue
Stanford University
Stanford, California

Contents

1

Bipolar Junction Transistor (BJT) Circuits

David J. Comer
Donald T. Comer
Brigham Young University

1.1   Introduction

The *bipolar junction transistor* (or BJT) was the workhorse of the electronics industry from the 1950s through the 1990s. This device was responsible for enabling the computer age as well as the modern era of communications. Although early systems that demonstrated the feasibility of electronic computers used the vacuum tube, the element was too unreliable for dependable, long-lasting computers. The invention of the BJT in 1947[1] and the rapid improvement in this device led to the development of highly reliable electronic computers and modern communication systems.

Integrated circuits, based on the BJT, became commercially available in the mid-1960s and further improved the dependability of the computer and other electronic systems while reducing the size and cost of the overall system. Ultimately, the microprocessor chip was developed in the early 1970s and the age of small, capable, personal computers was ushered in. While the metal-oxide-semiconductor (or MOS) device is now more prominent than the BJT in the personal computer arena, the BJT is still important in larger high-speed computers. This device also continues to be important in communication systems and power control systems.

0-8493-1736-3/03/$0.00+$1.50
© 2003 by CRC Press LLC

Because of the continued improvement in BJT performance and the development of the heterojunction BJT, this device remains very important in the electronics field, even as the MOS device becomes more significant.

1.2   Physical Characteristics and Properties of the BJT

Although present BJT technology is used to make both discrete component devices as well as integrated circuit chips, the basic construction techniques are similar in both cases, with primary differences arising in size and packaging. The following description is provided for the BJT constructed as integrated circuit devices on a silicon substrate. These devices are referred to as "junction-isolated" devices.

The cross-sectional view of a BJT is shown in Fig. 1.1.[2]

This device can occupy a surface area of less than $1000\ \mu m^2$. There are three physical regions comprising the BJT. These are the emitter, the base, and the collector. The thickness of the base region between emitter and collector can be a small fraction of a micron, while the overall vertical dimension of a device may be a few microns.

Thousands of such devices can be fabricated within a silicon wafer. They may be interconnected on the wafer using metal deposition techniques to form a system such as a microprocessor chip or they may be separated into thousands of individual BJTs, each mounted in its own case. The photolithographic methods that make it possible to simultaneously construct thousands of BJTs have led to continually decreasing size and cost of the BJT.

Electronic devices, such as the BJT, are governed by current–voltage relationships that are typically nonlinear and rather complex. In general, it is difficult to analyze devices that obey nonlinear equations, much less develop design methods for circuits that include these devices. The basic concept of modeling an electronic device is to replace the device in the circuit with linear components that approximate the voltage–current characteristics of the device. A model can then be defined as a collection of simple components or elements used to represent a more complex electronic device. Once the device is replaced in the circuit by the model, well-known circuit analysis methods can be applied.

There are generally several different models for a given device. One may be more accurate than others, another may be simpler than others, another may model the dc voltage–current characteristics of the device, while still another may model the ac characteristics of the device.

Models are developed to be used for manual analysis or to be used by a computer. In general, the models for manual analysis are simpler and less accurate, while the computer models are more complex and more accurate. Essentially, all models for manual analysis and most models for the computer include only linear elements. Nonlinear elements are included in some computer models, but increase the computation times involved in circuit simulation over the times in simulation of linear models.

1.3   Basic Operation of the BJT

In order to understand the origin of the elements used to model the BJT, we will discuss a simplified version of the device as shown in Fig. 1.2. The device shown is an npn device that consists of a p-doped

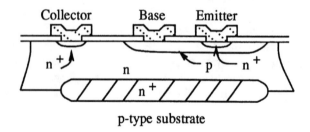

p-type substrate

FIGURE 1.1   An integrated npn BJT.

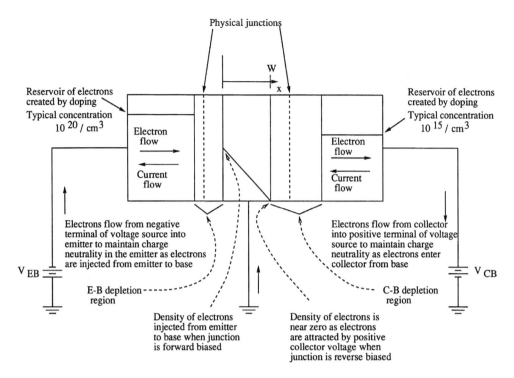

FIGURE 1.2 Distribution of electrons in the active region.

material interfacing on opposite sides to n-doped material. A pnp device can be created using an n-doped central region with p-doped interfacing regions. Since the npn type of BJT is more popular in present construction processes, the following discussion will center on this device.

The geometry of the device implied in Fig. 1.2 is physically more like the earlier alloy transistor. This geometry is also capable of modeling the modern BJT (Fig. 1.1) as the theory applies almost equally well to both geometries. Normally, some sort of load would appear in either the collector or emitter circuit; however, this is not important to the initial discussion of BJT operation.

The circuit of Fig. 1.2 is in the active region, that is, the emitter–base junction is forward-biased, while the collector–base junction is reverse-biased. The current flow is controlled by the profile of electrons in the p-type base region. It is proportional to the slope or gradient of the free electron density in the base region. The well-known diffusion equation can be expressed as:[3]

$$I = qD_nA\frac{dn}{dx} = -\frac{qD_nAn(0)}{W} \tag{1.1}$$

where q is the electronic charge, D_n is the diffusion constant for electrons, A is the cross-sectional area of the base region, W is the width or thickness of the base region, and $n(0)$ is the density of electrons at the left edge of the base region. The negative sign reflects the fact that conventional current flow is opposite to the flow of the electrons.

The concentration of electrons at the left edge of the base region is given by:

$$n(0) = n_{bo}e^{qV_{BE}/kT} \tag{1.2}$$

where q is the charge on an electron, k is Boltzmann's constant, T is the absolute temperature, and n_{bo} is the equilibrium concentration of electrons in the base region. While n_{bo} is a small number, $n(0)$ can be large for values of applied base to emitter voltages of 0.6 to 0.7 V. At room temperature, this equation can be written as:

$$n(0) = n_{bo}e^{V_{BE}/0.026} \tag{1.3}$$

In Fig. 1.2, the voltage $V_{EB} = -V_{BE}$.

A component of hole current also flows across the base–emitter junction from base to emitter. This component is rendered negligible compared to the electron component by doping the emitter region much more heavily than the base region.

As the concentration of electrons at the left edge of the base region increases, the gradient increases and the current flow across the base region increases. The density of electrons at $x = 0$ can be controlled by the voltage applied from emitter to base. Thus, this voltage controls the current flowing through the base region. In fact, the density of electrons varies exponentially with the applied voltage from emitter to base, resulting in an exponential variation of current with voltage.

The reservoir of electrons in the emitter region is unaffected by the applied emitter-to-base voltage as this voltage drops across the emitter–base depletion region. This applied voltage lowers the junction voltage as it opposes the built-in barrier voltage of the junction. This leads to the increase in electrons flowing from emitter to base.

The electrons injected into the base region represent electrons that were originally in the emitter. As these electrons leave the emitter, they are replaced by electrons from the voltage source, V_{EB}. This current is called emitter current and its value is determined by the voltage applied to the junction. Of course, conventional current flows in the opposite direction to the electron flow.

The emitter electrons flow through the emitter, across the emitter–base depletion region, and into the base region. These electrons continue across the base region, across the collector–base depletion region, and through the collector. If no electrons were "lost" in the base region and if the hole flow from base to emitter were negligible, the current flow through the emitter would equal that through the collector. Unfortunately, there is some recombination of carriers in the base region. When electrons are injected into the base region from the emitter, space charge neutrality is upset, pulling holes into the base region from the base terminal. These holes restore space charge neutrality if they take on the same density throughout the base as the electrons. Some of these holes recombine with the free electrons in the base and the net flow of recombined holes into the base region leads to a small, but finite, value of base current. The electrons that recombine in the base region reduce the total electron flow to the collector. Because the base region is very narrow, only a small percentage of electrons traversing the base region recombine and the emitter current is reduced by a small percentage as it becomes collector current.

In a typical low-power BJT, the collector current might be $0.995 I_E$. The current gain from emitter to collector, I_C/I_E, is called α and is a function of the construction process for the BJT. Using Kirchhoff's current law, the base current is found to equal the emitter current minus the collector current. This gives:

$$I_B = I_E - I_C = (1 - \alpha)I_E \tag{1.4}$$

If $\alpha = 0.995$, then $I_B = 0.005 I_E$. Base current is very small compared to emitter or collector current. A parameter β is defined as the ratio of collector current to base current resulting in:

$$\beta = \frac{\alpha}{1 - \alpha} \tag{1.5}$$

This parameter represents the current gain from base to collector and can be quite high. For the value of α cited earlier, the value of β is 199.

1.4   Use of the BJT as an Amplifier

Figure 1.3 shows a simple configuration of a BJT amplifier. This circuit is known as the *common emitter configuration.*

A voltage source is not typically used to forward-bias the base–emitter junction in an actual circuit, but we will assume that V_{BB} is used for this purpose. A value of V_{BB} or V_{BE} near 0.6 to 0.7 V would be appropriate for this situation. The collector supply would be a large voltage, such as 12 V. We will assume that the value of V_{BB} sets the dc emitter current to a value of 1 mA for this circuit. The collector current entering the BJT will be slightly less than 1 mA, but we will ignore this difference and assume that $I_C = $ 1 mA also. With a 4-kΩ collector resistance, a 4-V drop will appear across R_C, leading to a dc output voltage of 8 V. The distribution of electrons across the base region for the steady-state or quiescent conditions is shown by the solid line of Fig. 1.3(a).

If a small ac voltage now appears in series with V_{BB}, the injected electron density at the left side of the base region will be modulated. Since this density varies exponentially with the applied voltage (see Eq. 1.2), a small ac voltage can cause considerable changes in density. The dashed lines in Fig. 1.3(a) show the distributions at the positive and negative peak voltages. The collector current may change from its quiescent level of 1 mA to a maximum of 1.1 mA as e_{in} reaches its positive peak, and to a minimum of 0.9 mA when e_{in} reaches its negative peak. The output collector voltage will drop to a minimum value of 7.6 V as the collector current peaks at 1.1 mA, and will reach a maximum voltage of 8.4 V as the collector current drops to 0.9 mA. The peak-to-peak ac output voltage is then 0.8 V. The peak-to-peak value of e_{in} to cause this change might be 5 mV, giving a voltage gain of $A = -0.8/0.005 = -160$. The negative sign occurs because when e_{in} increases, the collector current increases, but the collector voltage decreases. This represents a phase inversion in the amplifier of Fig. 1.3.

In summary, a small change in base-to-emitter voltage causes a large change in emitter current. This current is channeled across the collector, through the load resistance, and can develop a larger incremental voltage across this resistance.

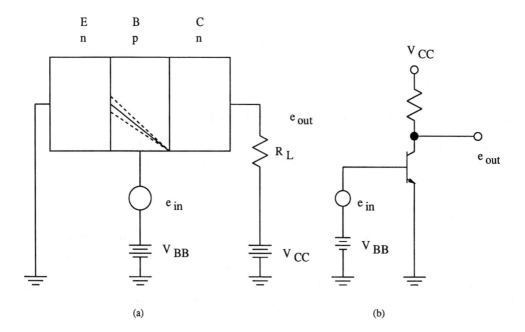

(a)   (b)

FIGURE 1.3   A BJT amplifier.

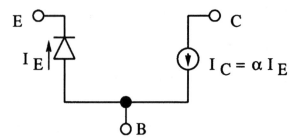

FIGURE 1.4   Large-signal model of the BJT.

1.5   Representing the Major BJT Effects by an Electronic Model

The two major effects of the BJT in the active region are the diode characteristics of the base–emitter junction and the collector current that is proportional to the emitter current. These effects can be modeled by the circuit of Fig. 1.4.

The simple diode equation represents the relationship between applied emitter-to-base voltage and emitter current. This equation can be written as

$$I_E = I_1(e^{qV_{BE}/kT} - 1) \tag{1.6}$$

where q is the charge on an electron, k is Boltzmann's constant, T is the absolute temperature of the diode, and I_1 is a constant at a given temperature that depends on the doping and geometry of the emitter-base junction.

The collector current is generated by a dependent current source of value $I_C = \alpha I_E$.

A small-signal model based on the large-signal model of Fig. 1.4 is shown in Fig. 1.5. In this case, the resistance, r_d, is the dynamic resistance of the emitter-base diode and is given by:

$$r_d = \frac{kT}{qI_E} \tag{1.7}$$

where I_E is the dc emitter current.

1.6   Other Physical Effects in the BJT

The preceding section pertains to the basic operation of the BJT in the dc and midband frequency range. Several other effects must be included to model the BJT with more accuracy. These effects will now be described.

Ohmic Effects

The metal connections to the semiconductor regions exhibit some ohmic resistance. The emitter contact resistance and collector contact resistance is often in the ohm range and does not affect the BJT operation in most applications. The base region is very narrow and offers little area for a metal contact. Furthermore, because this region is narrow and only lightly doped compared to the emitter, the ohmic resistance of the base region itself is rather high. The total resistance between the contact and the intrinsic base region can be 100 to 200 Ω. This resistance can become significant in determining the behavior of the BJT, especially at higher frequencies.

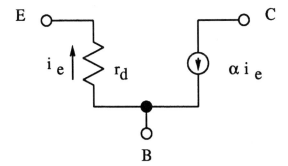

FIGURE 1.5 A small-signal model of the BJT.

Base-Width Modulation (Early Effect)

The widths of the depletion regions are functions of the applied voltages. The collector voltage generally exhibits the largest voltage change and, as this voltage changes, so also does the collector–base depletion region width. As the depletion layer extends further into the base region, the slope of the electron distribution in the base region becomes greater since the width of the base region is decreased. A slightly steeper slope leads to slightly more collector current. As reverse-bias decreases, the base width becomes greater and the current decreases. This effect is called *base-width modulation* and can be expressed in terms of the Early voltage,[4] V_A, by the expression:

$$I_C = \beta I_B\left(1 + \frac{V_{CE}}{V_A}\right) \tag{1.8}$$

The Early voltage will be constant for a given device and is typically in the range of 60 to 100 V.

Reactive Effects

Changing the voltages across the depletion regions results in a corresponding change in charge. This leads to an effective capacitance since

$$C = \frac{dQ}{dV} \tag{1.9}$$

This depletion region capacitance is a function of voltage applied to the junction and can be written as:[4]

$$C_{dr} = \frac{C_{Jo}}{(\phi - V_{app})^m} \tag{1.10}$$

where C_{Jo} is the junction capacitance at zero bias, ϕ is the built-in junction barrier voltage, V_{app} is the applied junction voltage, and m is a constant. For modern BJTs, m is near 0.33. The applied junction voltage has a positive sign for a forward-bias and a negative sign for a reverse-bias. The depletion region capacitance is often called the *junction capacitance*.

An increase in forward base–emitter voltage results in a higher density of electrons injected into the base region. The charge distribution in the base region changes with this voltage change, and this leads to a capacitance called the *diffusion capacitance*. This capacitance is a function of the emitter current and can be written as:

$$C_D = k_2 I_E \qquad (1.11)$$

where k_2 is a constant for a given device.

1.7   More Accurate BJT Models

Figure 1.6 shows a large-signal BJT model used in some versions of the popular simulation program known as SPICE.[5] The equations for the parameters are listed in other texts[5] and will not be given here. Figure 1.7 shows a small-signal SPICE model[5] often called the hybrid-π equivalent circuit. The capacitance, C_π, accounts for the diffusion capacitance and the emitter–base junction capacitance. The collector–base junction capacitance is designated C_μ. The resistance, r_π, is equal to $(\beta + 1)r_d$. The transductance, g_m, is given by:

$$g_m = \frac{\alpha}{r_d} \qquad (1.12)$$

The impedance, r_o, is related to the Early voltage by:

$$r_o = \frac{V_A}{I_C} \qquad (1.13)$$

R_B, R_E, and R_C are the base, emitter, and collector resistances, respectively. For hand analysis, the ohmic resistances R_E and R_C are neglected along with C_{CS}, the collector-to-substrate capacitance.

1.8   Heterojunction Bipolar Junction Transistors

In an npn device, all electrons injected from emitter to base are collected by the collector, except for a small number that recombine in the base region. The holes injected from base to emitter contribute to

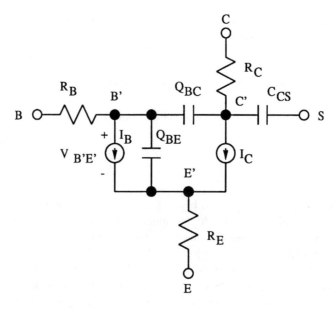

FIGURE 1.6 A more accurate large-signal model of the BJT.

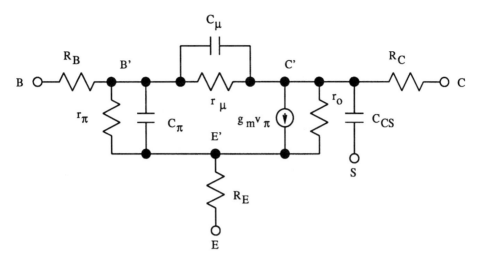

FIGURE 1.7   The hybrid-π small-signal model for the BJT.

emitter junction current, but do not contribute to collector current. This hole component of the emitter current must be minimized to achieve a near-unity current gain from emitter to collector. As α approaches unity, the current gain from base to collector, β, becomes larger.

In order to produce high-β BJTs, the emitter region must be doped much more heavily than the base region, as explained earlier. While this approach allows the value of β to reach several hundred, it also leads to some effects that limit the frequency of operation of the BJT. The lightly doped base region causes higher values of base resistance, as well as emitter–base junction capacitance. Both of these effects are minimized in the *heterojunction BJT* (or HBJT). This device uses a different material for the base region than that used for the emitter and collector regions. One popular choice of materials is silicon for the emitter and collector regions, and a silicon/germanium material for the base region.[6] The difference in energy gap between the silicon emitter material and the silicon/germanium base material results in an asymmetric barrier to current flow across the junction. The barrier for electron injection from emitter to base is smaller than the barrier for hole injection from base to emitter. The base can then be doped more heavily than a conventional BJT to achieve lower base resistance, but the hole flow across the junction remains negligible due to the higher barrier voltage. The emitter of the HBJT can be doped more lightly to lower the junction capacitance. Large values of β are still possible in the HBJT while minimizing frequency limitations. Current gain-bandwidth figures exceeding 60 GHz have been achieved with present industrial HBJTs.

From the standpoint of analysis, the SPICE models for the HBJT are structurally identical to those of the BJT. The difference is in the parameter values.

1.9   Integrated Circuit Biasing Using Current Mirrors

Differential stages are very important in integrated circuit amplifier design. These stages require a constant dc current for proper bias. A simple bias scheme for differential BJT stages will now be discussed.

The diode-biased current sink or *current mirror* of Fig. 1.8 is a popular method of creating a constant-current bias for differential stages.

The concept of the current mirror was developed specifically for analog integrated circuit biasing and is a good example of a circuit that takes advantage of the excellent matching characteristics that are possible in integrated circuits. In the circuit of Fig. 1.8, the current I_2 is intended to be equal to or "mirror" the value of I_1. Current mirrors can be designed to serve as sinks or sources.

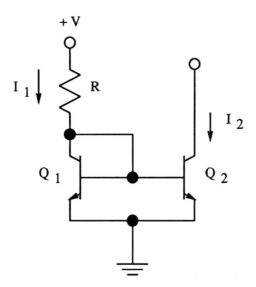

FIGURE 1.8   Current mirror bias stage.

The general function of the current mirror is to reproduce or mirror the input or reference current to the output, while allowing the output voltage to assume any value within some specified range. The current mirror can also be designed to generate an output current that equals the input current multiplied by a scale factor K. The output current can be expressed as a function of input current as:

$$I_O = KI_{IN} \qquad (1.14)$$

where K can be equal to, less than, or greater than unity. This constant can be established accurately by relative device sizes and will not vary with temperature.

Figure 1.9 shows a multiple output current source where all of the output currents are referenced to the input current. Several amplifier stages can be biased with this multiple output current mirror.

Current Source Operating Voltage Range

Figure 1.10 shows an ideal or theoretical current sink in (a) and a practical sink in (b). The voltage at node A in the theoretical sink can be tied to any potential above or below ground without affecting the value of I. On the other hand, the practical circuit of Fig. 1.10(b) requires that the transistor remain in the active region to provide a current of:

$$I = \alpha \frac{V_B - V_{BE}}{R} \qquad (1.15)$$

This requires that the collector voltage exceed the voltage V_B at all times. The upper limit on this voltage is determined by the breakdown voltage of the transistor. The output voltage must then satisfy:

$$V_B < V_C < (V_B + BV_{CE}) \qquad (1.16)$$

where BV_{CE} is the breakdown voltage from collector to emitter of the transistor. This voltage range over which the current source operates is called the *output voltage compliance range* or the *output compliance*.

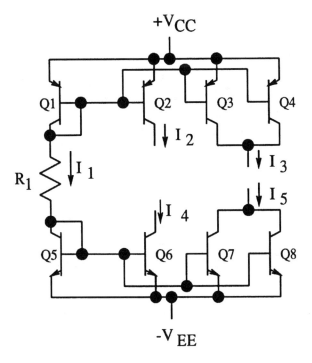

FIGURE 1.9   Multiple output current mirror.

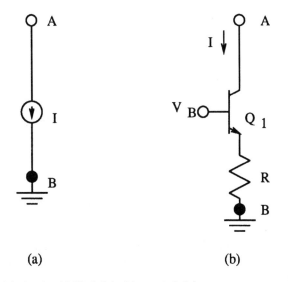

FIGURE 1.10   Current sink circuits: (a) ideal sink, (b) practical sink.

Current Mirror Analysis

The current mirror is again shown in Fig. 1.11. If devices Q_1 and Q_2 are assumed to be matched devices, we can write:

$$I_{E1} = I_{E2} = I_{EO}e^{V_{BE}/V_T} \tag{1.17}$$

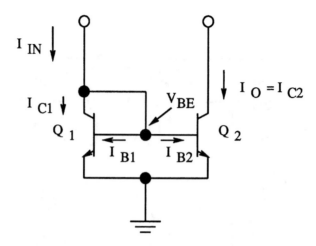

FIGURE 1.11   Circuit for current mirror analysis.

where $V_T = kT/q$, $I_{EO} = AJ_{EO}$, A is the emitter area of the two devices, and J_{EO} is the current density of the emitters. The base currents for each device will also be identical and can be expressed as:

$$I_{B1} = I_{B2} = \frac{I_{EO}}{\beta + 1} e^{V_{BE}/V_T} \tag{1.18}$$

Device Q_1 operates in the active region, but near saturation by virtue of the collector–base connection. This configuration is called a *diode-connected transistor*. Since the collector-to-emitter voltage is very small, the collector current for device Q_1 is given by Eq. 1.8, assuming $V_{CE} = 0$. This gives:

$$I_{C1} = \beta I_{B1} \approx \frac{\beta}{\beta + 1} I_{EO} e^{V_{BE}/V_T} \tag{1.19}$$

The device Q_2 does not have the constraint that $V_{CE} \approx 0$ as device Q_1 has. The collector voltage for Q_2 will be determined by the external circuit that connects to this collector. Thus, the collector current for this device is:

$$I_{C2} = \beta I_{B2}\left(1 + \frac{V_{C2}}{V_A}\right) \tag{1.20}$$

where V_A is the Early voltage. In effect, the output stage has an output impedance given by Eq. 1.13. The current mirror more closely approximates a current source as the output impedance becomes larger.

If we limit the voltage V_{C2} to small values relative to the Early voltage, I_{C2} is approximately equal to I_{C1}. For integrated circuit designs, the voltage required at the output of the current mirror is generally small, making this approximation valid.

The input current to the mirror is larger than the collector current and is:

$$I_{IN} = I_{C1} + 2I_B \tag{1.21}$$

Since $I_{OUT} = I_{C2} = I_{C1} = \beta I_B$, we can write Eq. 1.21 as:

$$I_{IN} = \beta I_B + 2I_B = (\beta + 2)I_B \tag{1.22}$$

Relating I_{IN} to I_{OUT} results in:

$$I_{OUT} = \frac{\beta}{\beta + 2} I_{IN} = \frac{I_{IN}}{1 + 2/\beta} \qquad (1.23)$$

For typical values of β, these two currents are essentially equal. Thus, a desired bias current, I_{OUT}, is generated by creating the desired value of I_{IN}. The current I_{IN} is normally established by connecting a resistance R_1 to a voltage source V_{CC} to set I_{IN} to:

$$I_{IN} = \frac{V_{CC} - V_{BE}}{R_1} \qquad (1.24)$$

Control of collector/bias current for Q_2 is then accomplished by choosing proper values of V_{CC} and R_1.

Figure 1.12 shows a multiple-output current mirror.

It can be shown that the output current for each identical device in Fig. 1.12 is:

$$I_O = \frac{I_{IN}}{1 + \dfrac{N+1}{\beta}} \qquad (1.25)$$

where N is the number of output devices.

The current sinks can be turned into current sources by using pnp transistors and a power supply of opposite polarity. The output devices can also be scaled in area to make I_{OUT} larger or smaller than I_{IN}.

Current Mirror with Reduced Error

The difference between output current in a multiple-output current mirror and the input current can become quite large if N is large. One simple method of avoiding this problem is to use an emitter follower to drive the bases of all devices in the mirror, as shown in Fig. 1.13.

The emitter follower, Q_0, has a current gain from base to collector of $\beta + 1$, reducing the difference between I_O and I_{IN} to:

$$I_{IN} - I_O = \frac{N+1}{\beta + 1} I_B \qquad (1.26)$$

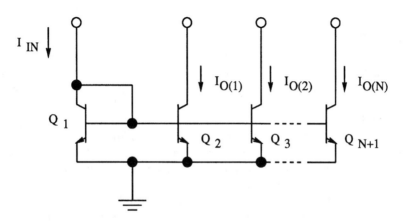

FIGURE 1.12   Multiple-output current mirror.

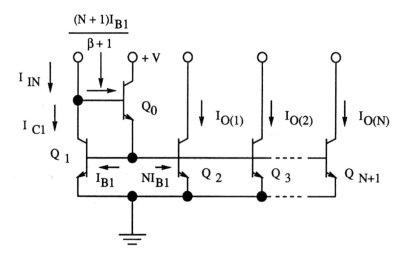

FIGURE 1.13   Improved multiple output current mirror.

The output current for each device is:

$$I_O = \frac{I_{IN}}{1 + \dfrac{N+1}{\beta(\beta+1)}} \tag{1.27}$$

The Wilson Current Mirror

In the simple current mirrors discussed, it was assumed that the collector voltage of the output stage was small compared to the Early voltage. When this is untrue, the output current will not remain constant, but will increase as output voltage (V_{CE}) increases. In other words, the output compliance range is limited with these circuits due to the finite output impedance of the BJT.

A modification of the improved output current mirror of Fig. 1.13 was proposed by Wilson[7] and is illustrated in Fig. 1.14.

The Wilson current mirror is connected such that $V_{CB2} = 0$ and $V_{BE1} = V_{BE0}$. Both Q_1 and Q_2 now operate with a near-zero collector–emitter bias although the collector of Q_0 might feed into a high-voltage point. It can be shown that the output impedance of the Wilson mirror is increased by a factor of $\beta/2$ over the simple mirror. This higher impedance translates into a higher output compliance. This circuit also reduces the difference between input and output current by means of the emitter follower stage.

1.10   The Basic BJT Switch

In digital circuits, the BJT is used as a switch to generate one of only two possible output voltage levels, depending on the input voltage level. Each voltage level is associated with one of the binary digits, 0 or 1. Typically, the high voltage level may fall between 2.8 V and 5 V while the low voltage level may fall between 0 V and 0.8 V.

Logic circuits are based on BJT stages that are either in cutoff with both junctions reverse-biased or in a conducting mode with the emitter–base junction forward-biased. When the BJT is "on" or conducting emitter current, it can be in the active region or the saturation region. If it is in the saturation region, the collector–base region is also forward-biased.

The three possible regions of operation are summarized in Table 1.1.

The BJT very closely approximates certain switch configurations. For example, when the switch of Fig. 1.15(a) is open, no current flows through the resistor and the output voltage is +12 V. Closing the switch

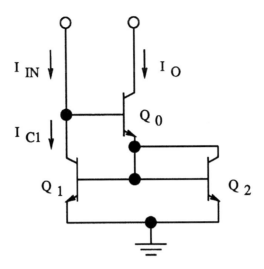

FIGURE 1.14 Wilson current mirror.

TABLE 1.1 Regions of Operation

Region	Cutoff	Active	Saturation
C–B bias	Reverse	Reverse	Forward
E–B bias	Reverse	Forward	Forward

causes the output voltage to drop to zero volts and a current of 12/R flows through the resistance. When the base voltage of the BJT of Fig. 1.15(b) is negative, the device is cut off and no collector current flows. The output voltage is +12 V, just as in the case of the open switch. If a large enough current is now driven into the base to saturate the BJT, the output voltage becomes very small, ranging from 20 mV to 500 mV, depending on the BJT used. The saturated state corresponds closely to the closed switch. During the time that the BJT switches from cutoff to saturation, the active region equivalent circuit applies. For high-speed switching of this circuit, appropriate reactive effects must be considered. For low-speed switching, these reactive effects can be neglected.

Saturation occurs in the basic switching circuit of Fig. 1.15(b) when the entire power supply voltage drops across the load resistance. No voltage, or perhaps a few tenths of volts, then appears from collector to emitter. This occurs when the base current exceeds the value

$$I_{B(sat)} = \frac{V_{CC} - V_{CE(sat)}}{\beta R_L} \tag{1.28}$$

When a transistor switch is driven into saturation, the collector–base junction becomes forward-biased. This situation results in the electron distribution across the base region shown in Fig. 1.16. The forward-bias of the collector–base junction leads to a non zero concentration of electrons in the base that is unnecessary to support the gradient of carriers across this region. When the input signal to the base switches to a lower level to either turn the device off or decrease the current flow, the excess charge must be removed from the base region before the current can begin to decrease.

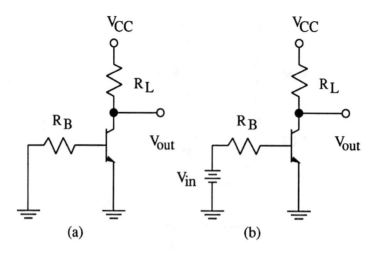

FIGURE 1.15 The BJT as a switch: (a) open switch, (b) closed switch.

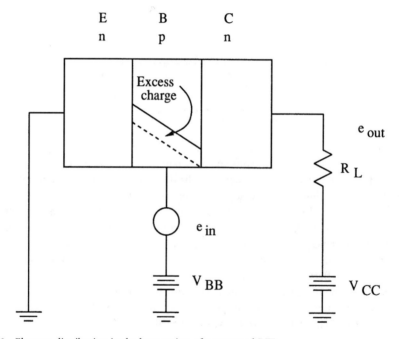

FIGURE 1.16 Electron distribution in the base region of a saturated BJT.

1.11 High-Speed BJT Switching

There are three major effects that extend switching times in a BJT:

1. The depletion-region or junction capacitances are responsible for delay time when the BJT is in the cutoff region.
2. The diffusion capacitance and the Miller-effect capacitance are responsible for the rise and fall times of the BJT as it switches through the active region.
3. The storage time constant accounts for the time taken to remove the excess charge from the base region before the BJT can switch from the saturation region to the active region.

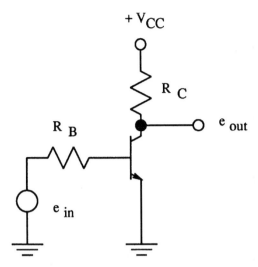

FIGURE 1.17   A simple switching circuit.

There are other second-order effects that are generally negligible compared to the previously listed time lags.

Since the transistor is generally operating as a large-signal device, the parameters such as junction capacitance or diffusion capacitance will vary as the BJT switches. One approach to the evaluation of time constants is to calculate an average value of capacitance over the voltage swing that takes place. Not only is this method used in hand calculations, but most computer simulation programs use average values to speed calculations.

Overall Transient Response

Before discussing the individual BJT switching times, it is helpful to consider the response of a common-emitter switch to a rectangular waveform. Figure 1.17 shows a typical circuit using an npn transistor.

A rectangular input pulse and the corresponding output are shown in Fig. 1.18. In many switching circuits, the BJT must switch from its "off" state to saturation and later return to the "off" state. In this case, the delay time, rise time, saturation storage time, and fall time must be considered in that order to find the overall switching time.

The total waveform is made up of five sections: delay time, rise time, on time, storage time, and fall time. The following list summarizes these points and serves as a guide for future reference:

t_d' = Passive delay time; time interval between application of forward base drive and start of collector-current response.

t_d = Total delay time; time interval between application of forward base drive and the point at which I_C has reached 10% of the final value.

t_r = Rise time; 10- to 90-% rise time of I_C waveform.

t_s' = Saturation storage time; time interval between removal of forward base drive and start of I_C decrease.

t_s = Total storage time; time interval between removal of forward base drive and point at which I_C = $0.9I_{C(sat)}$.

t_f = Fall time; 90- to 10-% fall time of I_C waveform

T_{on} = Total turn-on time; time interval between application of base drive and point at which I_C has reached 90% of its final value.

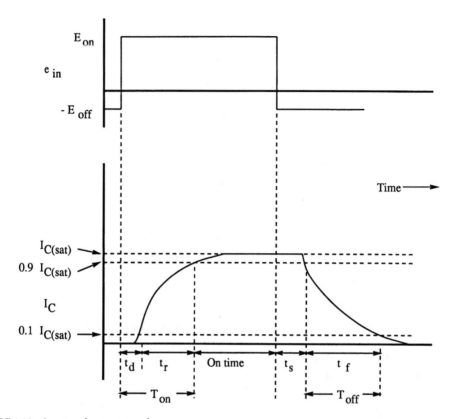

FIGURE 1.18   Input and output waveforms.

T_{off} = Total turn-off time; time interval between removal of forward base drive and point at which I_C has dropped to 10% of its value during on time.

Not all applications will require evaluation of each of these switching times. For instance, if the base drive is insufficient to saturate the transistor, t_s will be zero. If the transistor never leaves the active region, the delay time will also be zero.

The factors involved in calculating the switching times are summarized in the following paragraphs.[8] The passive delay time is found from:

$$t'_d = \tau_d \ln\left(\frac{E_{on} + E_{off}}{E_{on} - V_{BE(on)}}\right) \tag{1.29}$$

where τ_d is the product of the charging resistance and the average value of the two junction capacitances.

The active region time constant is a function of the diffusion capacitance, the collector–base junction capacitance, the transconductance, and the charging resistance. This time constant will be denoted by τ. If the transistor never enters saturation, the rise time is calculated from the well-known formula:

$$t_r = 2.2\tau \tag{1.30}$$

If the BJT is driven into saturation, the rise time is found from:[8]

$$t_r = \tau \ln\left(\frac{K - 0.1}{K - 0.9}\right) \tag{1.31}$$

where K is the overdrive factor or the ratio of forward base current drive to the value needed for saturation. The rise time for the case where K is large can be much smaller than the rise time for the nonsaturating case ($K < 1$). Unfortunately, the saturation storage time increases for large values of K.

The saturation storage time is given by:

$$t'_s = \tau_s \ln\left(\frac{I_{B1} - I_{B2}}{I_{B(sat)} - I_{B2}}\right) \tag{1.32}$$

where τ_s is the storage time constant, I_{B1} is the forward base current before switching, and I_{B2} is the current after switching and must be less than $I_{B(sat)}$. The saturation storage time can slow the overall switching time significantly. The higher speed logic gates utilize circuits that avoid the saturation region for the BJTs that make up the gate.

1.12   Simple Logic Gates

Although the resistor-transistor-logic (RTL) family has not been used since the late 1960s, it demonstrates the concept of a simple logic gate. Figure 1.19 shows a four-input RTL NOR gate.

If all four inputs are at the lower voltage level (e.g., 0 V), there is no conducting path from output to ground. No voltage will drop across R_L, and the output voltage will equal V_{CC}. If any or all of the inputs move to the higher voltage level (e.g., 4 V), any BJT with base connected to the higher voltage level will saturate, pulling the output voltage down to a few tenths of a volt. If positive logic is used, with the high voltage level corresponding to binary "1" and the low voltage level to binary "0," the gate performs the NOR function. Other logic functions can easily be constructed in the RTL family.

Over the years, the performance of logic gates has been improved by different basic configurations. RTL logic was improved by diode-transistor-logic (DTL). Then, transistor-transistor-logic (TTL) became very prominent. This family is still popular in the small-scale integration (SSI) and medium-scale integration (MSI) areas, but CMOS circuits have essentially replaced TTL in large-scale integration (LSI) and very-large-scale integration (VLSI) applications.

One popular family that is still prominent in very high-speed computer work is the emitter-coupled logic (ECL) family. While CMOS packs many more circuits into a given area than ECL, the frequency performance of ECL leads to its popularity in supercomputer applications.

1.13   Emitter-Coupled Logic

Emitter-coupled logic (ECL) was developed in the mid-1960s and remains the fastest silicon logic circuit available. Present ECL families offer propagation delays in the range of 0.2 ns.[9] The two major disadvantages of ECL are: (1) resistors which require a great deal of IC chip area, must be used in each gate, and. (2) the power dissipation of an ECL gate is rather high. These two shortcomings limit the usage of ECL in VLSI systems. Instead, this family has been used for years in larger supercomputers that can afford space and power to achieve higher speeds.

The high speeds obtained with ECL are primarily based on two factors. No device in an ECL gate is ever driven into the saturation region and, thus, saturation storage time is never involved as devices switch from one state to another. The second factor is that required voltage swings are not large. Voltage excursions necessary to change an input from the low logic level to the high logic level are minimal. Although noise margins are lower than other logic families, switching times are reduced in this way.

Figure 1.20 shows an older ECL gate with two separate outputs. For positive logic, X is the OR output while Y is the NOR output.

Often, the positive supply voltage is taken as 0 V and V_{EE} as −5 V due to noise considerations. The diodes and emitter follower Q_5 establish a temperature-compensated base reference for Q_4. When inputs A, B, and C are less than the voltage V_B, Q_4 conducts while Q_1, Q_2, and Q_3 are cut off. If any one of the

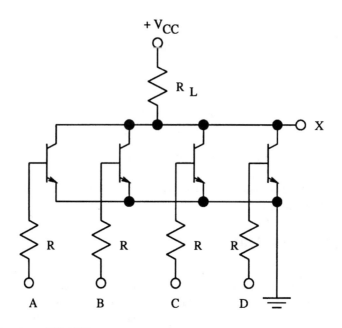

FIGURE 1.19   A four-input RTL NOR gate.

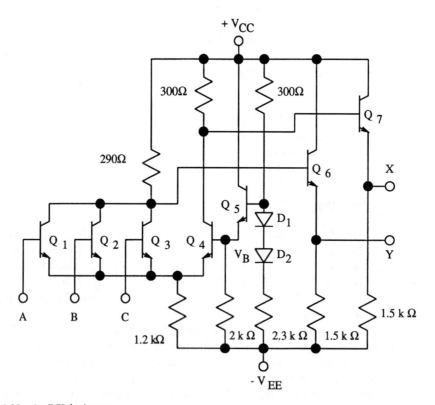

FIGURE 1.20   An ECL logic gate.

inputs is switched to the 1 level, which exceeds V_B, the transistor turns on and pulls the emitter of Q_4 positive enough to cut this transistor off. Under this condition, output Y goes negative while X goes positive. The relatively large resistor common to the emitters of Q_1, Q_2, Q_3, and Q_4 prevents these transistors from saturating. In fact, with nominal logic levels of −1.9 V and −1.1 V, the current through the emitter resistance is approximately equal before and after switching takes place. Thus, only the current path changes as the circuit switches. This type of operation is sometimes called *current mode switching*. Although the output stages are emitter followers, they conduct reasonable currents for both logic level outputs and, therefore, minimize the asymmetrical output impedance problem.

In an actual ECL gate, the emitter follower load resistors are not fabricated on the chip. The newer version of the gate replaces the emitter resistance of the differential stage with a current source, and replaces the bias voltage circuit with a regulated voltage circuit.

A Closer Look at the Differential Stage

Figure 1.21 shows a simple differential stage similar to the input stage of an ECL gate.[2] Both transistors are biased by a current source, I_T, called the *tail current*. The two input signals e_1 and e_2 make up a differential input signal defined as:

$$e_d = e_1 - e_2 \tag{1.33}$$

This differential voltage can be expressed as the difference between the base–emitter junction voltages as:

$$e_d = V_{BE1} - V_{BE2} \tag{1.34}$$

The collector currents can be written in terms of the base–emitter voltages as:

$$I_{C1} = \alpha I_{EO} e^{V_{BE1}/V_T} \approx I_{EO} e^{V_{BE1}/V_T} \tag{1.35}$$

$$I_{C2} = \alpha I_{EO} e^{V_{BE2}/V_T} \approx I_{EO} e^{V_{BE2}/V_T} \tag{1.36}$$

where matched devices are assumed.

A differential output current can be defined as the difference of the collector currents, or

$$I_d = I_{C1} - I_{C2} \tag{1.37}$$

Since the tail current is $I_T = I_{C1} + I_{C2}$, taking the ratio of I_d to I_T gives:

$$\frac{I_d}{I_T} = \frac{I_{C1} - I_{C2}}{I_{C1} + I_{C2}} \tag{1.38}$$

Since $V_{BE1} = e_d + V_{BE2}$, we can substitute this value for V_{BE1} into Eq. 1.35 to write:

$$I_{C1} = I_{EO} e^{(e_d + V_{BE2})/V_T} = I_{EO} e^{e_d/V_T} e^{V_{BE2}/V_T} \tag{1.39}$$

Substituting Eqs. 1.36 and 1.39 into Eq. 1.38 results in:

$$\frac{I_d}{I_T} = \frac{e^{e_d/V_T} - 1}{e^{e_d/V_T} + 1} = \tanh \frac{e_d}{2V_T} \tag{1.40}$$

or

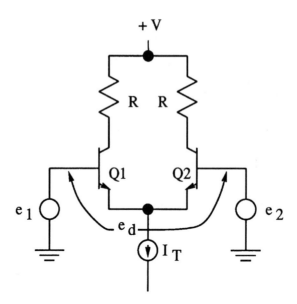

FIGURE 1.21   A simple differential stage similar to an ECL input stage.

$$I_d = I_T \tanh \frac{e_d}{2V_T} \tag{1.41}$$

This differential current is graphed in Fig. 1.22.

When e_d is zero, the differential current is also zero, implying equal values of collector currents in the two devices. As e_d increases, so also does I_d until e_d exceeds $4V_T$, at which time I_d has reached a constant value of I_T. From the definition of differential current, this means that I_{C1} equals I_T while I_{C2} is zero. As the differential input voltage goes negative, the differential current approaches $-I_T$ as the voltage reaches $-4V_T$. In this case, $I_{C2} = I_T$ while I_{C1} goes to zero.

The implication here is that the differential stage can move from a balanced condition with $I_{C1} = I_{C2}$ to a condition of one device fully off and the other fully on with an input voltage change of around 100 mV or $4V_T$. This demonstrates that a total voltage change of about 200 mV at the input can cause an ECL gate to change states. This small voltage change contributes to smaller switching times for ECL logic.

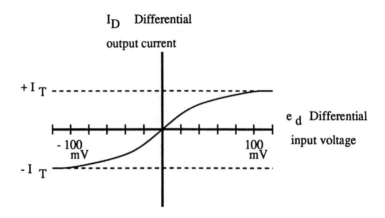

FIGURE 1.22   Differential output current as a function of differential input voltage.

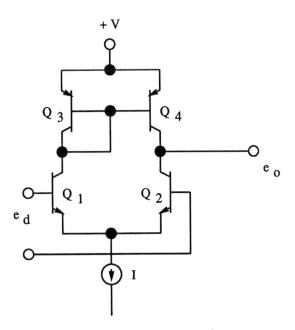

FIGURE 1.23   Differential input stage with current mirror load.

The ability of a differential pair to convert a small change in differential base voltage to a large change in collector voltage also makes it a useful building block for analog amplifiers. In fact, a differential pair with a pnp transistor current mirror load, as illustrated in Fig. 1.23, is widely used as an input stage for integrated circuit op-amps.

References

1. Brittain, J. E. (Ed.), *Turning Points in American Electrical History,* IEEE Press, New York, 1977, Sec. II-D.
2. Comer, D. T., *Introduction to Mixed Signal VLSI,* Array Publishing, New York, 1994, Ch. 7.
3. Sedra, A. S. and Smith, K. C., *Microelectronic Circuits, 4th ed.,* Oxford University Press, New York, 1998, Ch. 4.
4. Gray, P. R. and Meyer, R. G., *Analysis and Design of Analog Integrated Circuits, 3rd ed.,* John Wiley & Sons, Inc., New York, 1993, Ch. 1.
5. Vladimirescu, A., *The Spice Book,* John Wiley & Sons, Inc., New York, 1994, Ch. 3.
6. Streetman, B. G., *Solid State Electronic Devices, 4th ed.,* Prentice-Hall, Englewood Cliffs, NJ, 1995, Ch. 7.
7. Wilson, G. R., "A monolithic junction FET - NPN operational amplifier," *IEEE J. Solid-State Circuits,* Vol. SC-3, pp. 341-348, Dec. 1968.
8. Comer, D. J., *Modern Electronic Circuit Design,* Addison-Wesley, Reading, MA, 1977, Ch. 8.
9. Motorola Technical Staff, *High Performance ECL Data,* Motorola, Inc., Phoenix, AZ, 1993, Ch. 3.

2

RF Passive IC Components

Thomas H. Lee
Maria del MarHershenson
Sunderarajan S. Mohan
Hirad Samavati
C. Patrick Yue
Stanford University

2.1 Introduction

Passive energy storage elements are widely used in radio-frequency (RF) circuits. Although their impedance behavior often can be mimicked by compact active circuitry, it remains true that passive elements offer the largest dynamic range and the lowest power consumption. Hence, the highest performance will always be obtained with passive inductors and capacitors. Unfortunately, standard integrated circuit technology has not evolved with a focus on providing good passive elements. This chapter describes the limited palette of options available, as well as means to make the most use out of what is available.

2.2 Fractal Capacitors

Of capacitors, the most commonly used are parallel-plate and MOS structures. Because of the thin gate oxides now in use, capacitors made out of MOSFETs have the highest capacitance density of any standard IC option, with a typical value of approximately 7 fF/μm^2 for a gate oxide thickness of 5 nm. A drawback, however, is that the capacitance is voltage dependent. The applied potential must be well in excess of a threshold voltage in order to remain substantially constant. The relatively low breakdown voltage (on the order of 0.5 V/nm of oxide) also imposes an unwelcome constraint on allowable signal amplitudes. An additional drawback is the effective series resistance of such structures, due to the MOS channel resistance. This resistance is particularly objectionable at radio frequencies, since the impedance of the combination may be dominated by this resistive portion.

Capacitors that are free of bias restrictions (and that have much lower series resistance) may be formed out of two (or more) layers of standard interconnect metal. Such parallel-plate capacitors are quite linear and possess high breakdown voltage, but generally offer capacitance density two orders of magnitude lower than the MOSFET structure. This inferior density is the consequence of a conscious and continuing effort by technologists to keep low the capacitance between interconnect layers. Indeed, the vertical spacing between such layers generally does not scale from generation to generation. As a result, the disparity between MOSFET capacitance density and that of the parallel-plate structure continues to grow as technology scales.

A secondary consequence of the low density is an objectionably high capacitance between the bottom plate of the capacitor and the substrate. This bottom-plate capacitance is often a large fraction of the main capacitance. Needless to say, this level of parasitic capacitance is highly undesirable.

In many circuits, capacitors can occupy considerable area, and an area-efficient capacitor is therefore highly desirable. Recently, a high-density capacitor structure using lateral fringing and fractal geometries has been introduced.[1] It requires no additional processing steps, and so it can be built in standard digital processes. The linearity of this structure is similar to that of the conventional parallel-plate capacitor. Furthermore, the bottom-plate parasitic capacitance of the structure is small, which makes it appealing for many circuit applications. In addition, unlike conventional metal-to-metal capacitors, the density of a fractal capacitor increases with scaling.

Lateral Flux Capacitors

Figure 2.1(a) shows a lateral flux capacitor. In this capacitor, the two terminals of the device are built using a single layer of metal, unlike a vertical flux capacitor, where two different metal layers must be used. As process technologies continue to scale, lateral fringing becomes more important. The lateral spacing of the metal layers, s, shrinks with scaling, yet the thickness of the metal layers, t, and the vertical spacing of the metal layers, t_{ox}, stay relatively constant. This means that structures utilizing lateral flux enjoy a significant improvement with process scaling, unlike conventional structures that depend on vertical flux. Figure 2.1(b) shows a scaled lateral flux capacitor. It is obvious that the capacitance of the structure of Fig. 2.1(b) is larger than that of Fig. 2.1(a).

Lateral flux can be used to increase the total capacitance obtained in a given area. Figure 2.2(a) is a standard parallel-plate capacitor. In Fig. 2.2(b), the plates are broken into cross-connected sections.[2] As can be seen, a higher capacitance density can be achieved by using lateral flux as well as vertical flux. To emphasize that the metal layers are cross connected, the two terminals of the capacitors in Fig. 2.2(b) are identified with two different shadings. The idea can be extended to multiple metal layers as well.

Figure 2.3 shows the ratio of metal thickness to minimum lateral spacing, t/s, vs. channel length for various technologies.[3–5] The trend suggests that lateral flux will have a crucial role in the design of capacitors in future technologies.

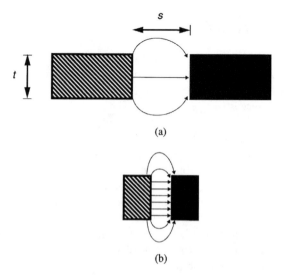

(a)

(b)

FIGURE 2.1   Effect of scaling on lateral flux capacitors: (a) before scaling and (b) after scaling.

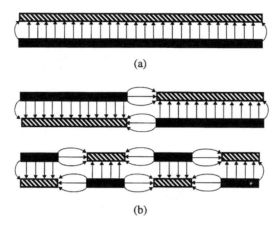

FIGURE 2.2   Vertical flux vs. lateral flux: (a) standard parallel-plate structure, and (b) cross-connected metal layers.

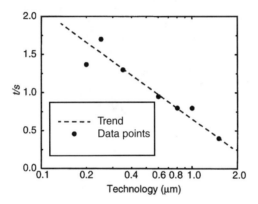

FIGURE 2.3   Ratio of metal thickness to horizontal metal spacing vs. technology (channel length).

The increase in capacitance due to fringing is proportional to the periphery of the structure; therefore, structures with large periphery per unit area are desirable. Methods for increasing this periphery are the subject of the following sections.

Fractals

A fractal is a mathematical abstract.[6] Some fractals are visualizations of mathematical formulas, while others are the result of the repeated application of an algorithm, or a *rule*, to a *seed*. Many natural phenomena can be described by fractals. Examples include the shapes of mountain ranges, clouds, coastlines, etc.

Some ideal fractals have finite area but infinite perimeter. The concept can be better understood with the help of an example. *Koch islands* are a family of fractals first introduced as a crude model for the shape of a coastline. The construction of a Koch curve begins with an *initiator*, as shown in the example of Fig. 2.4(a). A square is a simple initiator with $M = 4$ sides. The construction continues by replacing each segment of the initiator with a curve called a *generator*, an example of which is shown in Fig. 2.4(b) that has $N = 8$ segments. The size of each segment of the generator is $r = 1/4$ of the initiator. By recursively replacing each segment of the resulting curve with the generator, a fractal border is formed. The first step of this process is depicted in Fig. 2.4(c). The total area occupied remains constant throughout the succession of stages because of the particular shape of the generator. A more complicated Koch island can be seen in Fig. 2.5. The associated initiator of this fractal has four sides and its generator has 32 segments. It can be noted that the curve is self similar, that is, each section of it looks like the entire fractal. As we zoom in on Fig. 2.5, more detail becomes visible, and this is the essence of a fractal.

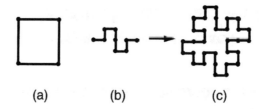

| (a) | (b) | (c) |

FIGURE 2.4   Construction of a Koch curve: (a) an initiator, (b) a generator, and (c) first step of the process.

FIGURE 2.5   A Koch island with $M = 4$, $N = 32$, and $r = 1/8$.

Fractal dimension, D, is a mathematical concept that is a measure of the complexity of a fractal. The dimension of a flat curve is a number between 1 and 2, which is given by

$$D = \frac{\log(N)}{\log\left(\frac{1}{r}\right)} \tag{2.1}$$

where N is the number of segments of the generator and r is the ratio of the generator segment size to the initiator segment size. The dimension of a fractal curve is not restricted to integer values, hence the term "fractal." In particular, it exceeds 1, which is the intuitive dimension of curves. A curve that has a high degree of complexity, or D, fills out a two-dimensional flat surface more efficiently. The fractal in Fig. 2.4(c) has a dimension of 1.5, whereas for the border line of Fig.2.5, $D = 1.667$.

For the general case where the initiator has M sides, the periphery of the initiator is proportional to the square root of the area:

$$P_0 = k \cdot \sqrt{A} \tag{2.2}$$

where k is a proportionality constant that depends on the geometry of the initiator. For example, for a square initiator, $k = 4$; and for an equilateral triangle, $k = 2 \cdot \sqrt[4]{27}$. After n successive applications of the generation rule, the total periphery is

$$P = k\sqrt{A} \cdot (Nr)^n \tag{2.3}$$

and the minimum feature size (the resolution) is

$$l = \frac{k\sqrt{A}}{M} \cdot r^n \tag{2.4}$$

Eliminating n from Eqs. 2.3 and 2.4 and combining the result with Eq. 2.1, we have

$$P = \frac{k^D}{M^{D-1}} \cdot \frac{(\sqrt{A})^D}{l^{D-1}} \tag{2.5}$$

Equation 2.5 demonstrates the dependence of the periphery on parameters such as the area and the resolution of the fractal border. It can be seen from Eq. 2.5 that as l tends toward zero, the periphery goes to infinity; therefore, it is possible to generate fractal structures with very large perimeters in any given area. However, the total periphery of a fractal curve is limited by the attainable resolution in practical realizations.

Fractal Capacitor Structures

The final shape of a fractal can be tailored to almost any form. The flexibility arises from the fact that a wide variety of geometries can be used as the initiator and generator. It is also possible to use different generators during each step. This is an advantage for integrated circuits where flexibility in the shape of the layout is desired.

Figure 2.6 is a three-dimensional representation of a fractal capacitor. This capacitor uses only one metal layer with a fractal border. For a better visualization of the overall picture, the terminals of this square-shaped capacitor have been identified using two different shadings. As was discussed before, multiple cross-connected metal layers may be used to improve capacitance density further.

One advantage of using lateral flux capacitors in general, and fractal capacitors in particular, is the reduction of the bottom-plate capacitance. This reduction is due to two reasons. First, the higher density of the fractal capacitor (compared to a standard parallel-plate structure) results in a smaller area. Second, some of the field lines originating from one of the bottom plates terminate on the adjacent plate, instead of the substrate, which further reduces the bottom-plate capacitance as shown in Fig. 2.7. Because of this

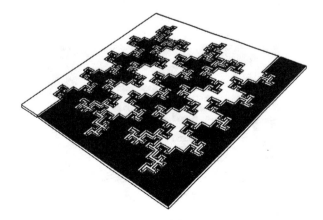

FIGURE 2.6 3-D representation of a fractal capacitor using a single metal layer.

property, some portion of the parasitic bottom-plate capacitor is converted into the more useful plate-to-plate capacitance.

The capacitance per unit area of a fractal structure depends on the dimension of the fractal. To improve the density of the layout, fractals with large dimensions should be used. The concept of fractal dimension is demonstrated in Fig. 2.8. The structure in Fig. 2.8(a) has a lower dimension compared to the one in Fig. 2.8(b), so the density (capacitance per unit area) of the latter is higher.

To demonstrate the dependence of capacitance density on dimension and lateral spacing of the metal layers, a first-order electromagnetic simulation was performed on two families of fractal structures. In Fig. 2.9, the boost factor is plotted vs. horizontal spacing of the metal layers. The *boost factor* is defined as the ratio of the total capacitance of the fractal structure to the capacitance of a standard parallel-plate structure with the same area. The solid line corresponds to a family of fractals with a moderate fractal

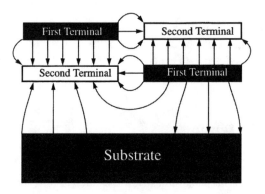

FIGURE 2.7   Reduction of the bottom-plate parasitic capacitance.

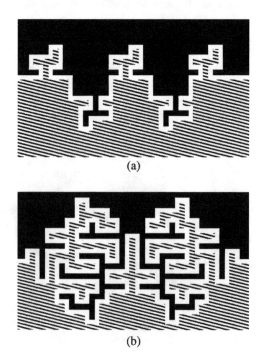

FIGURE 2.8   Fractal dimension of (a) is smaller than (b).

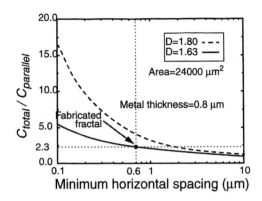

FIGURE 2.9   Boost factor vs. lateral spacing.

dimension of 1.63, while the dashed line represents another family of fractals with $D = 1.80$, which is a relatively large value for the dimension. In this first-order simulation, it is assumed that the vertical spacing and the thickness of the metal layers are kept constant at a 0.8-μm level. As can be seen in Fig. 2.9, the amount of boost is a strong function of the fractal dimension as well as scaling.

In addition to the capacitance density, the quality factor, Q, is important in RF applications. Here, the degradation in quality factor is minimal because the fractal structure automatically limits the length of the thin metal sections to a few microns, keeping the series resistance reasonably small. For applications that require low series resistance, lower dimension fractals may be used. Fractals thus add one more degree of freedom to the design of capacitors, allowing the capacitance density to be traded for a lower series resistance.

In current IC technologies, there is usually tighter control over the lateral spacing of metal layers compared to the vertical thickness of the oxide layers, from wafer to wafer and across the same wafer. Lateral flux capacitors shift the burden of matching away from oxide thickness to lithography. Therefore, by using lateral flux, matching characteristics can improve. Furthermore, the pseudo-random nature of the structure can also compensate, to some extent, the effects of non-uniformity of the etching process. To achieve accurate ratio matching, multiple copies of a unit cell should be used, as is standard practice in high-precision analog circuit design.

Another simple way of increasing capacitance density is to use an interdigitated capacitor depicted in Fig. 2.10.[2,7] One disadvantage of such a structure compared to fractals is its inherent parasitic inductance. Most of the fractal geometries randomize the direction of the current flow and thus reduce the effective series inductance; whereas for interdigitated capacitors, the current flow is in the same direction for all the parallel stubs. In addition, fractals usually have lots of rough edges that accumulate electrostatic energy more efficiently compared to interdigitated capacitors, causing a boost in capacitance (generally of the order of 15%). Furthermore, interdigitated structures are more vulnerable to non-uniformity of the etching process. However, the relative simplicity of the interdigitated capacitor does make it useful in some applications.

The woven structure shown in Fig. 2.11 may also be used to achieve high capacitance density. The vertical lines are in metal-2 and horizontal lines are in metal-1. The two terminals of the capacitor are identified using different shades. Compared to an interdigitated capacitor, a woven structure has much less inherent series inductance. The current flowing in different directions results in a higher self-resonant frequency. In addition, the series resistance contributed by vias is smaller than that of an interdigitated capacitor, because cross-connecting the metal layers can be done with greater ease. However, the capacitance density of a woven structure is smaller compared to an interdigitated capacitor with the same metal pitch, because the capacitance contributed by the vertical fields is smaller.

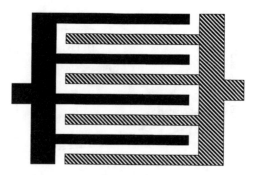

FIGURE 2.10   An interdigitated capacitor.

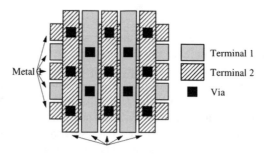

FIGURE 2.11   A woven structure.

2.3   Spiral Inductors

More than is so with capacitors, on-chip inductor options are particularly limited and unsatisfactory. Nevertheless, it is possible to build practical spiral inductors with values up to perhaps 20 nH and with Q values of approximately 10. For silicon-based RF ICs, Q degrades at high frequencies due to energy dissipation in the semiconducting substrate.[8] Additionally, noise coupling via the substrate at GHz frequencies has been reported.[9] As inductors occupy substantial chip area, they can potentially be the source and receptor of detrimental noise coupling. Furthermore, the physical phenomena underlying the substrate effects are complicated to characterize. Therefore, decoupling the inductor from the substrate can enhance the overall performance by increasing Q, improving isolation, and simplifying modeling.

Some approaches have been proposed to address the substrate issues; however, they are accompanied by drawbacks. Some[10] have suggested the use of high-resistivity (150 to 200 Ω-cm) silicon substrates to mimic the low-loss semi-insulating GaAs substrate, but this is rarely a practical option. Another approach selectively removes the substrate by etching a pit under the inductor.[11] However, the etch adds extra processing cost and is not readily available. Moreover, it raises reliability concerns such as packaging yield and long-term mechanical stability. For low-cost integration of inductors, the solution to substrate problems should avoid increasing process complexity.

In this section, we present the *patterned ground shield* (PGS),[23] which is compatible with standard silicon technologies, and which reduces the unwanted substrate effects. The great improvement provided by the PGS reduces the disparity in quality between spiral inductors made in silicon and GaAs IC technologies.

Understanding Substrate Effects

To understand why the PGS should be effective, consider first the physical model of an ordinary inductor on silicon, with one port and the substrate grounded, as shown in Fig. 2.12.[8] An on-chip inductor is

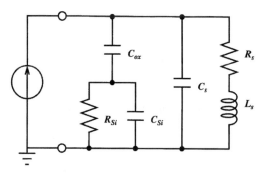

FIGURE 2.12 Lumped physical model of a spiral inductor on silicon.

physically a three-port element including the substrate. The one-port connection shown in Fig. 2.12 avoids unnecessary complexity in the following discussion and at the same time preserves the inductor characteristics. In the model, the series branch consists of L_s, R_s, and C_s. L_s represents the spiral inductance, which can be computed using the Greenhouse method[12] or well-approximated by simple analytical formulas to be presented later. R_s is the metal series resistance whose behavior at RF is governed by the eddy current effect. This resistance accounts for the energy loss due to the skin effect in the spiral interconnect structure as well as the induced eddy current in any conductive media close to the inductor. The series feedforward capacitance, C_s, accounts for the capacitance due to the overlaps between the spiral and the center-tap underpass.[13] The effect of the inter-turn fringing capacitance is usually small because the adjacent turns are almost at equal potentials, and therefore it is neglected in this model. The overlap capacitance is more significant because of the relatively large potential difference between the spiral and the center-tap underpass. The parasitics in the shunt branch are modeled by C_{ox}, C_{Si}, and R_{Si}. C_{ox} represents the oxide capacitance between the spiral and the substrate. The silicon substrate capacitance and resistance are modeled by C_{Si} and R_{Si}, respectively.[14,15] The element R_{Si} accounts for the energy dissipation in the silicon substrate.

Expressions for the model element values are as follows:

$$R_s = \frac{\rho l}{\delta w \left(1 - e^{-\frac{t}{\delta}}\right)} \tag{2.6}$$

$$C_s = nw^2 \cdot \frac{\varepsilon_{ox}}{t_{oxM1-M2}} \tag{2.7}$$

$$C_{ox} = \frac{\varepsilon_{ox}}{2t_{ox}} \cdot l \cdot w \tag{2.8}$$

$$C_{Si} = \frac{1}{2} \cdot l \cdot w \cdot C_{sub} \tag{2.9}$$

$$R_{Si} = \frac{2}{l \cdot w \cdot G_{sub}} \tag{2.10}$$

where ρ is the DC resistivity of the spiral; t is the overall length of the spiral windings; w is the line width; δ is the skin depth; n is the number of crossovers between the spiral and center-tap (and thus $n = N - 1$, where N is the number of turns); $t_{oxM1-M2}$ is the oxide thickness between the spiral and substrate; C_{sub} is the substrate capacitance per unit area; and G_{sub} is the substrate conductance per unit area. In general, one treats C_{sub} and G_{sub} as fitting parameters.

Exploration with the model reveals that the substrate loss stems primarily from the penetration of the electric field into the lossy silicon substrate. As the potential drop in the semiconductor (i.e., across R_{Si} in Fig. 2.12) increases with frequency, the energy dissipation in the substrate becomes more severe. It can be seen that increasing R_p to infinity reduces the substrate loss. It can be shown that R_p approaches infinity as R_{Si} goes either to zero or infinity. This observation implies that Q can be improved by making the silicon substrate *either* a perfect insulator or a perfect conductor. Using high-resistivity silicon (or etching it away) is equivalent to making the substrate an open circuit. In the absence of the freedom to do so, the next best option is to convert the substrate into a better conductor. The approach is to insert a ground plane to block the inductor electric field from entering the silicon. In effect, this ground plane becomes a pseudo-substrate with the desired characteristics.

The ground shield cannot be a solid conductor, however, because image currents would be induced in it. These image currents tend to cancel the magnetic field of the inductor proper, decreasing the inductance. To solve this problem, the ground shield is patterned with slots orthogonal to the spiral as illustrated in Fig. 2.13. The slots act as an open circuit to cut off the path of the induced loop current. The slots should be sufficiently narrow such that the vertical electric field cannot leak through the patterned ground shield into the underlying silicon substrate. With the slots etched away, the ground strips serve as the termination for the electric field. The ground strips are merged together around the four outer edges of the spiral. The separation between the merged area and the edges is not critical. However, it is crucial that the merged area not form a closed ring around the spiral since it can potentially support unwanted loop current. The shield should be strapped with the top layer metal to provide a low-impedance path to ground. The general rule is to prevent negative mutual coupling while minimizing the impedance to ground.

The shield resistance is another critical design parameter. The purpose of the patterned ground shield is to provide a good short to ground for the electric field. Since the finite shield resistance contributes to energy loss of the inductor, it must be kept small. Specifically, by keeping the shield resistance small compared to the reactance of the oxide capacitance, the voltage drop that can develop across the shield resistance is very small. As a result, the energy loss due to the shield resistance is insignificant compared

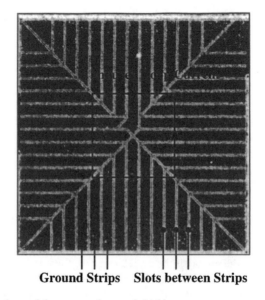

Ground Strips   Slots between Strips

FIGURE 2.13   A close-up photo of the patterned ground shield.

to other losses. A typical on-chip spiral inductor has parasitic oxide capacitance between 0.25 and 1 pF, depending on the size and the oxide thickness. The corresponding reactance due to the oxide capacitance at 1 to 2 GHz is of the order of 100 Ω, and hence a shield resistance of a few ohms is sufficiently small not to cause any noticeable loss.

With the PGS, one can expect typical improvements in Q ranging from 10 to 33%, in the frequency range of 1 to 2 GHz. Note that the inclusion of the ground shields increases C_p, which causes a fast roll-off in Q above the peak-Q frequency and a reduction in the self-resonant frequency. This modest improvement in inductor Q is certainly welcome, but is hardly spectacular by itself. However, a more dramatic improvement is evident when evaluating inductor-capacitor resonant circuits. Such LC tank circuits can absorb the parasitic capacitance of the ground shield. Since the energy stored in such parasitic elements is now part of the circuit, the overall circuit Q is greatly increased. Improvements of factors of approximately two are not unusual, so that tank circuits realized with PGS inductors possess roughly the same Q as those built in GaAs technologies.

As stated earlier, substrate noise coupling can be an issue of great concern owing to the relatively large size of typical inductors. Shielding by the PGS improves isolation by 25 dB or more at GHz frequencies. It should be noted that, as with any other isolation structure (such as a guard ring), the efficacy of the PGS is highly dependent on the integrity of the ground connection. One must often make a tradeoff between the desired isolation level and the chip area that is required to provide a low-impedance ground connection.

Simple, Accurate Expressions for Planar Spiral Inductances

In the previous section, a physically based model for planar spiral inductors was offered, and reference was made to the Greenhouse method as a means for computing the inductance value. This method uses as computational atoms the self- and mutual inductances of parallel current strips. It is relatively straightforward to apply, and yields accurate results. Nevertheless, simpler analytic formulas are generally preferred for design since important insights are usually more readily obtained.

As a specific example, square spirals are popular mainly because of their ease of layout. Other polygonal spirals have also been used to improve performance by more closely approximating a circular spiral. However, a quantitative evaluation of possible improvements is cumbersome without analytical formulas for inductance.

Among alternative shapes, hexagonal and octagonal inductors are used widely. Figures 2.14 through 2.16 show the layout for square, hexagonal, and octagonal inductors, respectively. For a given shape, an inductor is completely specified by the number of turns n, the turn width w, the turn spacing s, and any one of the following: the outer diameter d_{out}, the inner diameter d_{in}, the average diameter $d_{avg} = 0.5(d_{out} + d_{in})$, or the fill ratio, defined as $\rho = (d_{out} - d_{in})/(d_{out} + d_{in})$. The thickness of the inductor has only a very small effect on inductance and will therefore be ignored here.

We now present three approximate expressions for the inductance of square, hexagonal, and octagonal planar inductors. The first approximation is based on a modification of an expression developed by Wheeler[16]; the second is derived from electromagnetic principles by approximating the sides of the spirals as current sheets; and the third is a monomial expression derived from fitting to a large database of inductors (whose exact inductance values are obtained from a 3-D electromagnetic field solver). All three expressions are accurate, with typical errors of 2 to 3%, and very simple, and are therefore excellent candidates for use in design and optimization.

Modified Wheeler Formula

Wheeler[16] presented several formulas for planar spiral inductors, which were intended for discrete inductors. A simple modification of the original Wheeler formula allows us to obtain an expression that is valid for planar spiral integrated inductors:

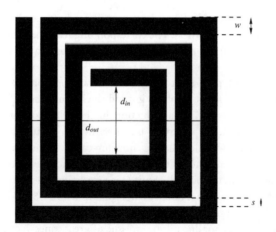

FIGURE 2.14 Square inductor.

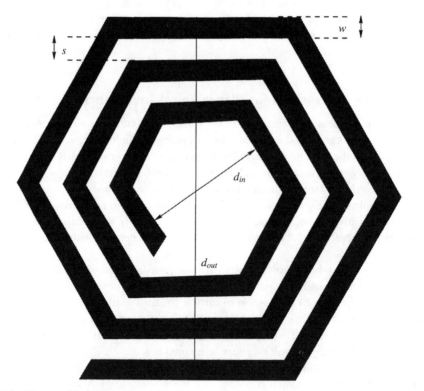

FIGURE 2.15 Hexagonal inductor.

$$L_{mw} = K_1 \mu_0 \frac{n^2 d_{avg}}{1 + K_2 \rho} \tag{2.11}$$

where ρ is the fill ratio defined previously. The coefficients K_1 and K_2 are layout dependent and are shown in Table 2.1.

The fill factor ρ represents how hollow the inductor is: for small ρ, we have a hollow inductor ($d_{out} \cong d_{in}$), and for a large ρ we have a filled inductor ($d_{out} >> d_{in}$). Two inductors with the same average diameter but different fill ratios will, of course, have different inductance values; the filled one has a smaller inductance because its inner turns are closer to the center of the spiral, and so contribute less positive

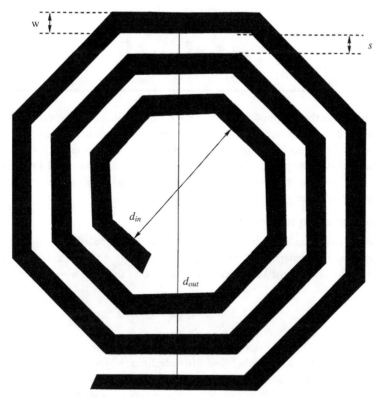

FIGURE 2.16 Octagonal inductor.

TABLE 2.1 Coefficients for Modified Wheeler Formula

Layout	K_1	K_2
Square	2.34	2.75
Hexagonal	2.33	3.82
Octagonal	2.25	3.55

mutual inductance and more negative mutual inductance. Some degree of hollowness is generally desired since the innermost turns contribute little overall inductance, but significant resistance.

Expression Based on Current Sheet Approximation

Another simple and accurate expression for the inductance of a planar spiral can be obtained by approximating the sides of the spirals by symmetrical current sheets of equivalent current densities.[17] For example, in the case of the square, we obtain four identical current sheets: the current sheets on opposite sides are parallel to one another, whereas the adjacent ones are orthogonal. Using symmetry and the fact that sheets with orthogonal current sheets have zero mutual inductance, the computation of the inductance is now reduced to evaluating the self-inductance of one sheet and the mutual inductance between opposite current sheets. These self- and mutual inductances are evaluated using the concepts of geometric mean distance (GMD) and arithmetic mean distance (AMD).[17,18] The resulting expression is:

$$L_{gmd} = \frac{\mu n^2 d_{avg}}{\pi}(c_1(\log c_2/\rho) + c_3\rho) \tag{2.12}$$

where the coefficients c_i are layout dependent and are shown in Table 2.2.

TABLE 2.2   Coefficients for Current-Sheet Inductance Formula

Layout	c_1	c_2	c_3
Square	2.00	2.00	0.54
Hexagonal	1.83	1.71	0.45
Octagonal	1.87	1.68	0.60

A detailed derivation of these formulas can be found in Ref. 19. Since this formula is based on a current sheet approximation, its accuracy worsens as the ratio *s/w* becomes large. In practice, this is not a problem since practical integrated spiral inductors are built with *s* < *w*. The reason is that a smaller spacing improves the inter-winding magnetic coupling and reduces the area consumed by the spiral. A large spacing is only desired to reduce the inter-winding capacitance. This is rarely a concern as this capacitance is always dwarfed by the under-pass capacitance.[8]

Data-Fitted Monomial Expression

Our final expression is based on a data-fitting technique, in which a population of thousands of inductors are simulated with an electromagnetic field solver. The inductors span the entire range of values of relevance to RF circuits. A monomial expression is then fitted to the data, which ultimately yields:

$$L_{mon} = \beta d_{avg}^{\alpha 1} w^{\alpha 2} d_{avg}^{\alpha 3} n^{\alpha 4} s^{\alpha 5} \tag{2.13}$$

where the coefficients β and α_i are layout dependent, and given in Table 2.3.

Of course, it is also possible to use other data-fitting techniques; for example, one which minimizes the maximum error of the fit, or one in which the coefficients must satisfy given inequalities or bounds. The monomial expression is useful since, like the other expressions, it is very accurate and very simple. Its real value, however, is that it can be used for the optimal design of inductors and circuits containing inductors, using geometric programming, which is a type of optimization method that requires monomial models.[20,21]

Figure 2.17 shows the absolute error distributions of these expressions. The plots show that typical errors are in the 1 to 2% range, and most of the errors are below 3%. These expressions for inductance, while quite simple, are thus sufficiently accurate that field solvers are rarely necessary.

These expressions can be included in a physical, scalable lumped-circuit model for spiral inductors where, in addition to providing design insight, they allow efficient optimization schemes to be employed.

2.4   On-Chip Transformers

Transformers are important elements in RF circuits for impedance conversion, impedance matching, and bandwidth enhancement. Here, we present an analytical model for monolithic transformers that is suitable for circuit simulation and design optimization. We also provide simple expressions for calculating the mutual coupling coefficient (*k*).

We first discuss different on-chip transformers and their advantages and disadvantages. We then present an analytical model along with expressions for the elements in it and the mutual coupling coefficient.

TABLE 2.3   Coefficients for Monomial Inductance Formula

Layout	b	α_1	α_2	α_3	α_4	α_5
Square	1.66×10^{-3}	-1.33	-0.13	2.50	1.83	-0.022
Hexagonal	1.33×10^{-3}	-1.46	-0.16	2.67	1.80	-0.030
Octagonal	1.34×10^{-3}	-1.35	-0.15	2.56	1.77	-0.032

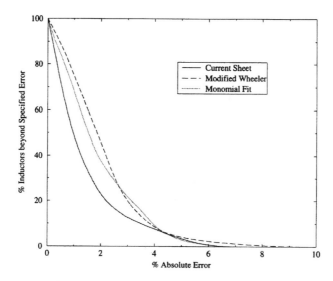

FIGURE 2.17 Error distribution for three formulas, compared to field solver simulations.

Monolithic Transformer Realizations

Figures 2.18 through 2.23 illustrate common configurations of monolithic transformers. The different realizations offer varying tradeoffs among the self-inductance and series resistance of each port, the mutual coupling coefficient, the port-to-port and port-to-substrate capacitances, resonant frequencies, symmetry, and area. The models and coupling expressions allow these tradeoffs to be systematically explored, thereby permitting transformers to be customized for a variety of circuit design requirements.

The characteristics desired of a transformer are application dependent. Transformers can be configured as three- or four-terminal devices. They may be used for narrowband or broadband applications. For example, in single-sided to differential conversion, the transformer might be used as a four-terminal narrowband device. In this case, a high mutual coupling coefficient and high self-inductance are desired, along with low series resistance. On the other hand, for bandwidth extension applications, the transformer is used as a broadband three-terminal device. In this case, a small mutual coupling coefficient and high series resistance are acceptable, while all capacitances need to be minimized.[22]

The tapped transformer (Fig. 2.18) is best suited for three-port applications. It permits a variety of tapping ratios to be realized. This transformer relies only on lateral magnetic coupling. All windings can be implemented with the top metal layer, thereby minimizing port-to-substrate capacitances. Since the two inductors occupy separate regions, the self-inductance is maximized while the port-to-port capacitance is minimized. Unfortunately, this spatial separation also leads to low mutual coupling ($k = 0.3$–0.5).

The interleaved transformer (Fig. 2.19) is best suited for four-port applications that demand symmetry. Once again, capacitances can be minimized by implementing the spirals with top level metal so that high resonant frequencies may be realized. The interleaving of the two inductances permit moderate coupling ($k = 0.7$) to be achieved at the cost of reduced self-inductance. This coupling may be increased at the cost of higher series resistance by reducing the turn width (w) and spacing (s).

The stacked transformer (Fig. 2.20) uses multiple metal layers and exploits both vertical and lateral magnetic coupling to provide the best area efficiency, the highest self-inductance, and highest coupling ($k = 0.9$). This configuration is suitable for both three- and four-terminal configurations. The main drawback is the high port-to-port capacitance, or equivalently a low self-resonant frequency. In some cases, such as narrowband impedance transformers, this capacitance may be incorporated as part of the resonant circuit. Also, in multi-level processes, the capacitance can be reduced by increasing the oxide thickness between spirals. For example, in a five-metal process, 50 to 70% reductions in port-to-port capacitance can be achieved by implementing the spirals on layers five and three instead of five and four.

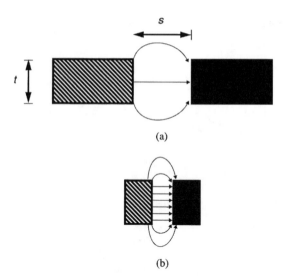

FIGURE 2.18   Tapped transformer.

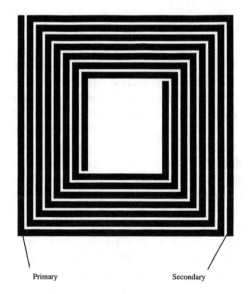

Primary   Secondary

FIGURE 2.19   Interleaved transformer.

The increased vertical separation will reduce k by less than 5%. One can also trade off reduced coupling for reduced capacitance by displacing the centers of the stacked inductors (Figs. 2.21 and 2.22).

Analytical Transformer Models

Figures 2.23 and 2.24 present the circuit models for tapped and stacked transformers, respectively. The corresponding element values for the tapped transformer model are given by the following equations (subscript o refers to the outer spiral, i to the inner spiral, and T to the whole spiral):

$$L_T = \frac{9.375\mu_0 n_T^2 A D_T^2}{11 O D_T - 7 A D_T} \qquad (2.14)$$

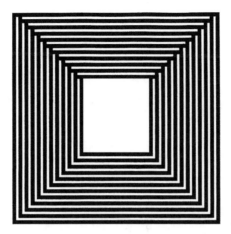

FIGURE 2.20 Stacked transformer with top spiral overlapping the bottom one.

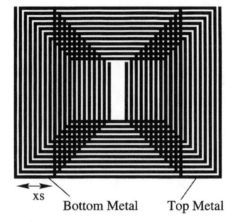

FIGURE 2.21 Stacked transformer with top and bottom spirals laterally shifted.

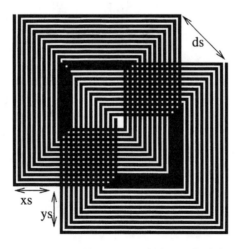

FIGURE 2.22 Stacked transformer with top and bottom spirals diagonally shifted.

$$L_o = \frac{9.375\mu_0 n_o^2 AD_o^2}{11OD_o - 7AD_o} \tag{2.15}$$

$$L_i = \frac{9.375\mu_0 n_i^2 AD_i^2}{11OD_i - 7AD_i} \tag{2.16}$$

$$M = \frac{L_T - L_o - L_i}{2\sqrt{L_o L_i}} \tag{2.17}$$

$$R_{so} = \frac{\rho l_o}{\delta w\left(1 - e^{-\frac{t}{\delta}}\right)} \tag{2.18}$$

$$R_{si} = \frac{\rho l_i}{\delta w\left(1 - e^{-\frac{t}{\delta}}\right)} \tag{2.19}$$

$$C_{ovo} = \frac{\varepsilon_{ox}}{t_{ox,t-b}} \cdot (n_o - 1)w^2 \tag{2.20}$$

$$C_{oxo} = \frac{\varepsilon_{ox}}{2t_{ox}} \cdot l_o w \tag{2.21}$$

$$C_{oxi} = \frac{\varepsilon_{ox}}{2t_{ox}} \cdot (l_o + l_i)w \tag{2.22}$$

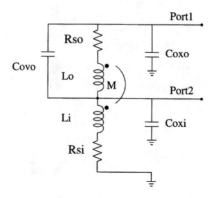

FIGURE 2.23   Tapped transformer model.

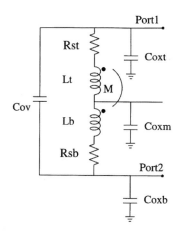

FIGURE 2.24   Stacked transformer model.

where ρ is the DC metal resistivity; δ is the skin depth; $t_{ox,t-b}$ is the oxide thickness from top level metal to bottom metal; n is the number of turns; OD, AD, and ID are the outer, average, and inner diameters, respectively; l is the length of the spiral; w is the turn width; t is the metal thickness; and A is the area.

Expressions for the stacked transformer model are as follows (subscript t refers to the top spiral and b to the bottom spiral):

$$L_t = \frac{9.375\mu_0 n^2 AD^2}{11OD_T - 7AD_T} \tag{2.23}$$

$$L_b = L_t \tag{2.24}$$

$$k = 0.9 - \frac{d_s}{AD} \tag{2.25}$$

$$M = k\sqrt{L_t L_b} \tag{2.26}$$

$$R_{st} = \frac{\rho_t l}{\delta_t w \left(1 - e^{-\frac{t_t}{\delta_t}}\right)} \tag{2.27}$$

$$R_{sb} = \frac{\rho_b l}{\delta_b w \left(1 - e^{-\frac{t_b}{\delta_b}}\right)} \tag{2.28}$$

$$C_{ov} = \frac{\varepsilon_{ox}}{2t_{ox, t-b}} \cdot l \cdot w \cdot \frac{A_{ov}}{A} \tag{2.29}$$

$$C_{oxt} = \frac{\varepsilon_{ox}}{2t_{oxt}} \cdot l \cdot w \cdot \frac{A - A_{ov}}{A} \tag{2.30}$$

$$C_{oxb} = \frac{\varepsilon_{ox}}{2t_{ox}} \cdot l \cdot w \tag{2.31}$$

$$C_{oxm} = C_{oxt} + C_{oxb} \tag{2.32}$$

where t_{oxt} is the oxide thickness from top metal to the substrate; t_{oxb} is the oxide thickness from bottom metal to substrate; k is the coupling coefficient; A_{ov} is the overlap area of the two spirals; and d_s is the center-to-center spiral distance.

The expressions for the series resistances (R_{so}, R_{si}, R_{sp}, and R_{sb}), the port-substrate capacitances (C_{oxo}, C_{oxi}, C_{oxt}, C_{oxb}, and C_{oxm}) and the crossover capacitances (C_{ovo}, C_{ovi}, and C_{ov}) are taken from Ref. 8. Note that the model accounts for the increase in series resistance with frequency due to skin effect. Patterned ground shields (PGS) are placed beneath the transformers to isolate them from resistive and capacitive coupling to the substrate.[23] As a result, the substrate parasitics can be neglected.

The inductance expressions in the foregoing are based on the modified Wheeler formula discussed earlier.[24] This formula does not take into account the variation in inductance due to conductor thickness and frequency. However, in practical inductor and transformer realizations, the thickness is small compared to the lateral dimensions of the coil and has only a small impact on the inductance. For typical conductor thickness variations (0.5 to 2.0 μm), the change in inductance is within a few percent for practical inductor geometries. The inductance also changes with frequency due to changes in current distribution within the conductor. However, over the useful frequency range of a spiral, this variation is negligible.[23] When compared to field solver simulations, the inductance expression exhibits a maximum error of 8% over a broad design space (outer diameter OD varying from 100 to 480 μm, L varying from 0.5 to 100 nH, w varying from 2 μm to 0.3OD, s varying from 2 μm to w, and inner diameter ID varying from 0.2 to 0.8OD).

For the tapped transformer, the mutual inductance is determined by first calculating the inductance of the whole spiral (L_T), the inductance of the outer spiral (L_o), the inductance of the inner spiral (L_i), and then using the expression $M = (L_T - L_o - L_i)/2$. For the stacked transformer, the spirals have identical lateral geometries and therefore identical inductances. In this case, the mutual inductance is determined by first calculating the inductance of one spiral (L_T), the coupling coefficient (k) and then using the expression $M = kL_T$. In this last case the coupling coefficient is given by $k = 0.9 - d_s/(AD)$ for $d_s < 0.7AD$, where d_s is the center-to-center spiral distance and AD is the average diameter of the spirals. As d_s increases beyond 0.7AD, the mutual coupling coefficient becomes harder to model. Eventually, k crosses zero and reaches a minimum value of approximately –0.1 at $d_s = AD$. As d_s increases further, k asymptotically approaches zero. At $d_s = 2AD$, $k = -0.02$, indicating that the magnetic coupling between closely spaced spirals is negligible.

The self-inductances, series resistances, and mutual inductances are independent of whether a transformer is used as a three- or four-terminal device. The only elements that require recomputation are the port-to-port and port-to-substrate capacitances. This situation is analogous to that of a spiral inductor being used as a single- or dual-terminal device.

As with the inductance formulas, the transformer models obviate the need for full field solutions in all but very rare instances, allowing rapid design and optimization.

References

1. Samavati, H. et al., "Fractal capacitors," *1998 IEEE ISSCC Dig. of Tech. Papers*, Feb. 1998.
2. Akcasu, O. E., "High capacitance structures in a semiconductor device," U.S. Patent 5 208 725, May 1993.
3. Bohr, M., "Interconnect scaling — The real limiter to high performance VLSI," *Intl. Electron Devices Meeting Tech. Digest*, pp. 241-244, 1995.
4. Bohr, M. et al., "A high performance 0.25 μm logic technology optimized for 1.8V operation," *Intl. Electron Devices Meeting Tech. Digest*, pp. 847-850, 1996.
5. Venkatesan, S. et al., "A high performance 1.8V, 0.20 μm CMOS technology with copper metallization," *Intl. Electron Devices Meeting Tech. Digest*, pp. 769-772, 1997.
6. Mandelbrot, B. B., *The Fractal Geometry of Nature*, W. H. Freeman, New York, 1983.
7. Pettenpaul, E. et al., "Models of lumped elements on GaAs up to 18 GHz," *IEEE Trans. Microwave Theory and Techniques*, vol. 36, no. 2, pp. 294-304, Feb. 1988.
8. Yue, C. P., Ryu, C., Lau, J., Lee, T. H., and Wong, S. S., "A physical model for planar spiral inductors on silicon," *Intl. Electron Devices Meeting Tech. Digest*, pp. 155-158, Dec. 1996.
9. Pfost, M., Rein, H.-M., and Holzwarth, T., "Modeling substrate effects in the design of high speed Si-bipolar IC's," *IEEE J. Solid-State Circuits*, vol. 31, no. 10, pp. 1493-1501, Oct. 1996.
10. Ashby, K. B., Koullias, I. A., Finley, W. C., Bastek, J. J., and Moinian, S., "High Q inductors for wireless applications in a complementary silicon bipolar process," *IEEE J. Solid-State Circuits*, vol. 31, no. 1, pp. 4-9, Jan. 1996.
11. Chang, J. Y.-C., Abidi, A. A., and Gaitan, M., "Large suspended inductors on silicon and their use in a 2-μm CMOS RF amplifier," *IEEE Electron Device Letters*, vol. 14, no. 5, pp. 246-248, May 1993.
12. Greenhouse, H. M., "Design of planar rectangular microelectronic inductors," *IEEE Trans. Parts, Hybrids, and Packing*, vol. PHP-10, no. 2, pp. 101-109, June 1974.
13. Wiemer, L. and Jansen, R. H., "Determination of coupling capacitance of underpasses, air bridges and crossings in MICs and MMICs," *Electronics Letters*, vol. 23, no. 7, pp. 344-346, Mar. 1987.
14. Ho, I. T. and Mullick, S. K., "Analysis of transmission lines on integrated-circuit chips," *IEEE J. Solid-State Circuits*, vol. SC-2, no. 4, pp. 201-208, Dec. 1967.
15. Hasegawa, H., Furukawa, M., and Yanai, H., "Properties of microstrip line on Si-SiO$_2$ system," *IEEE Trans. Microwave Theory and Techniques*, vol. MTT-19, no. 11, pp. 869-881, Nov. 1971.
16. Wheeler, H. A., "Simple inductance formulas for radio coils," *Proc. IRE*, vol. 16, no. 10, pp. 1398-1400, October 1928.
17. Rosa, E. B., "Calculation of the self-inductances of single-layer coils," *Bull. Bureau of Standards*, vol. 2, no. 2, pp. 161-187, 1906.
18. Maxwell, J. C., *A Treatise on Electricity and Magnetism*, Dover, 3rd ed., 1967.
19. Mohan, S. S., "Formulas for planar spiral inductances," *Tech. Rep., IC Laboratory*, Stanford University, Aug. 1998, http://www-smirc.stanford.edu.
20. Boyd, S. and Vandenberghe, L., "Introduction to convex optimization with engineering applications," Course Notes, 1997, http://www-leland.stanford.edu/class/ee364/.
21. Hershenson, M., Boyd, S. P., and Lee, T. H., "GPCAD: A tool for CMOS op-amp synthesis," in *Digest of Technical Papers, IEEE International Conference on Computer-Aided Design*, Nov. 1998.
22. Lee, T. H., *The Design of CMOS Radio-Frequency Integrated Circuits*, Cambridge University Press, Cambridge, 1998.
23. Yue, C. P. et al., "On-chip spiral inductors with patterned ground shields for Si-based RF ICs," *IEEE J. Solid-State Circuits*, vol. 33, pp. 743-752, May 1998.
24. Wheeler, H. A., "Simple inductance formulas for radio coils," *Proc. IRE*, vol. 16, no. 10, pp. 1398-1400, Oct. 1928.

3

CMOS Amplifier Design

Harry W. Li
University of Idaho at Moscow

R. Jacob Baker
University of Idaho at Boise

Donald C. Thelen
Analog Interfaces

3.1   Introduction

This chapter discusses the design, operation, and layout of CMOS analog amplifiers and subcircuits (current mirrors, biasing circuits, etc.). To make this discussion meaningful and clear, we need to define some important variables related to the DC operation of MOSFETs (Fig. 3.1). Figure 3.1(a) shows the simplified schematic representations of n- and p-channel MOSFETs. We say simplified because, when these symbols are used, it is *assumed* that the fourth terminal of the MOSFET (i.e., the body connection) is connected to either the lowest potential on the chip (V_{SS} or ground for the NMOS) or the highest potential (V_{DD} for the PMOS). Figure 3.1(b) shows the more general schematic representation of n- and p-channel MOSFETs. We are assuming that, although the drain and source of the MOSFETs are inter-changeable, drain current flows from the top of the device to the bottom. Because of the assumed direction of current flow, the drain terminal of the n-channel is on the top of the symbol, while the drain terminal of the p-channel is on the bottom of the schematic symbol. The following are short descriptions of some important characteristics of MOSFETs that will be useful in the following discussion.

The Threshold Voltage

Loosely defined, the threshold voltage, V_{THN} or V_{THP}, is the minimum gate-to-source voltage (V_{GS} for the n-channel or V_{SG} for the p-channel) that causes a current to flow when a voltage is applied between the drain and source of the MOSFET. As shown in Fig. 3.1(c) the threshold voltage is estimated by plotting the square root of the drain current against the gate-source voltage of the MOSFET and looking at the intersection of the line tangent with this plot with the x-axis (V_{GS} for the n-channel). As seen in the figure, a current does flow below the threshold voltage of the device. This current is termed, for obvious reasons, the *subthreshold current*. The subthreshold current is characterized by plotting the log of the

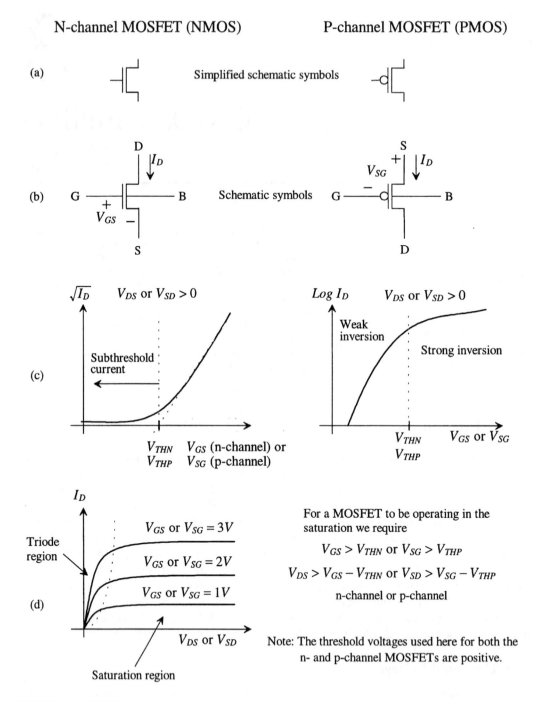

FIGURE 3.1   MOSFET device characteristics.

drain current against the gate-source voltage. The slope of the curve in the subthreshold region (sometimes also called the *weak inversion region*) is used to specify how the drain current changes with V_{GS}. A typical value for the reciprocal of the slope of this curve is 100 mV/dec. An equation relating the drain current of an n-channel MOSFET operating in the subthreshold region to V_{GS} is (assuming $V_{DS} > 100$ mV):

$$I_D = I_{D0} \cdot \frac{W}{L} \cdot \exp\left[\frac{q(V_{GS} - V_{THN})}{kT \cdot N}\right] \qquad (3.1)$$

where W and L are the width and length of the MOSFET, I_{D0} is a measured constant, k is Boltzmann's constant (1.38×10^{-23} J/K), T is temperature in Kelvin, q is the electronic charge (1.609×10^{-23} C), and N is the slope parameter. Note that the slope of the Log I_D vs. V_{GS} curve in the subthreshold region is

$$\frac{\Delta \log I_D}{V_{GS}} = \frac{q \cdot \log e}{N \cdot kT} \qquad (3.2)$$

The Body Effect

The threshold voltage of a MOSFET is dependent on the potential between the source of the MOSFET and its body. Consider Fig. 3.2, showing the situation when the body of an n-channel MOSFET is connected to ground and the source of the MOSFET is held V_{SB} above ground. As V_{SB} is increased (i.e., the potential on the source of the MOSFET increases relative to ground), the minimum V_{GS} needed to cause appreciable current to flow increases (V_{THN} increases as V_{SB} increases). We can relate V_{THN} to V_{SB} using the body-effect coefficient, γ, by

$$V_{THN} = V_{THN0} + \gamma\sqrt{|2\phi_F| + V_{SB}} - \sqrt{|2\phi_F|} \qquad (3.3)$$

where V_{THN0} is the zero-bias threshold voltage when $V_{SB} = 0$ and ϕ_F is the surface electrostatic potential[1] with a typical value of 300 mV. An important thing to notice here is that the threshold voltage tends to change less with increasing source-to-substrate (body) potential (increasing V_{SB}).

The Drain Current

In the following discussion, we will assume that the gate-source voltage of a MOSFET is greater than the threshold voltage so that a reasonably sized drain current can flow ($V_{GS} > V_{THN}$ or $V_{SG} > V_{THP}$). If this is the case, the MOSFET operates in either the triode region or the saturation region [Fig. 3.1(d)]. The drain current of a *long L* n-channel MOSFET operating in the *triode region*, is given by

$$I_D = \beta\left[(V_{GS} - V_{THN})V_{DS} - \frac{V_{DS}^2}{2}\right] \qquad (3.4)$$

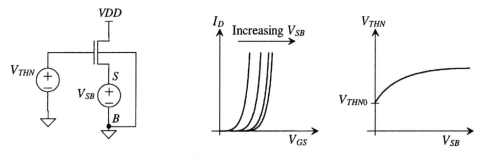

FIGURE 3.2 Illustration of the threshold voltage dependence on body-effect.

assuming long L with $V_{DS} \leq V_{GS} - V_{THN}$. Note that the MOSFET behaves like a voltage-controlled resistor when operating in the deep triode region with a channel resistance (the resistance is measured between the drain and source of the MOSFET) approximated by

$$R_{CH} = \frac{1}{\beta(V_{GS} - V_{THN}) - \beta V_{DS}} \text{ or } \frac{1}{\beta(V_{GS} - V_{THN})} \text{ if } V_{DS} \ll V_{GS} - V_{THN} \tag{3.5}$$

When doing analog design, it is often useful to implement a resistor whose value is dependent on a controlling voltage (a voltage-controlled resistor). When the NMOS device is operating in the saturation region, the drain current is given by

$$I_D = \beta(V_{GS} - V_{THN})^2[1 + \lambda(V_{DS} - V_{DS,sat})] \tag{3.6}$$

assuming long L with $V_{DS} \geq V_{GS} - V_{THN}$, where $V_{DS,sat} = V_{GS} V_{THN}$.

The *transconductance parameter*, β, is given by

$$\beta = KP \cdot \frac{W}{L} = \mu_{n,p} C_{ox} \cdot \frac{W}{L} \tag{3.7}$$

where $\mu_{n,p}$ is the mobility of either the electron or hole and C_{ox} is the oxide capacitance per unit area [ε_{ox}/t_{ox}, the dielectric constant of the gate oxide (35.4 aF/µm) divided by the gate oxide thickness]. Typical values for KP_n, KP_p, and C_{ox} for a 0.5-µm process are 150 µA/V², 50 µA/V², and 4 fF/µm², respectively. Also, an important thing to note in these equations for an n-channel MOSFET, is that V_{GS}, V_{DS}, and V_{THN} can be directly replaced with V_{SG}, V_{SD}, and V_{THP}, respectively, to obtain the operating equations for the p-channel MOSFET (keeping in mind that all quantities under normal conditions for operation are positive.) Also note that the saturation slope parameter λ (also known as the channel length/mobility modulation parameter) determines how changes in the drain-to-source voltage affect the MOSFET drain current and thus the MOSFET output resistance.

Short-Channel MOSFETs

As the channel length of a MOSFET is reduced, the electron and hole mobilities, μ_n and μ_p, start to get smaller. The mobility is simply a ratio of the electron or hole velocity to the applied electric field. Reducing the channel length increases the applied electric field while at the same time causing the velocity of the electron or hole to saturate (this velocity is labeled v_{sat}). This effect is called *mobility reduction* or *hot-carrier effects* (because the mobility also decreases with increasing temperature). The result, for a MOSFET with a short channel length L, is a reduction in drain current and a labeling of *short-channel MOSFET*. A short-channel MOSFET's current is, in general, linearly dependent on the MOSFET V_{GS} or

$$I_D = W \cdot v_{sat} \cdot C_{ox}(V_{GS} - V_{THN} - V_{DS,sat}) \tag{3.8}$$

To avoid short-channel effects (and, as we shall see, increase the output resistance of the MOSFET when doing analog design), the channel length of the MOSFET is made, generally, 2 to 5 times larger than the minimum allowable L. For a 0.5-µm CMOS process, this means we make the channel length of the MOSFETs 1.0 to 2.5 µm.

MOSFET Output Resistance

An important parameter of a MOSFET in analog applications is its output resistance. Consider the portion of a MOSFET's I-V characteristics shown in Fig. 3.3(a). When the MOSFET is operating in the saturation region, the slope of I_D, because of changes in the drain current with changes in the

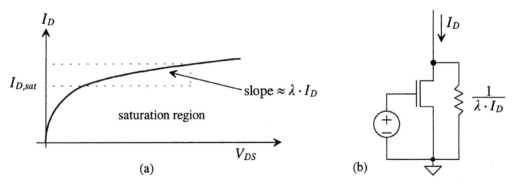

FIGURE 3.3 Output resistance of a MOSFET.

drain-source voltage, is relatively small. If this change were zero, the drain current would be a fixed value independent of the voltage between the drain and source of the MOSFET (in other words, the MOSFET would act like an ideal current source.) Even with the small change in current with V_{DS}, we can think of the MOSFET as a current source. To model the changes in current with V_{DS}, we can simply place a resistor across the drain and source of the MOSFET [Fig. 3.3(b)]. The value of this resistor is

$$r_o \approx \frac{1}{\lambda I_D} \tag{3.9}$$

where λ is in the range of 0.1 to 0.01 V^{-1}.

At this point, several practical comments should be made: (1) in general, to increase λ, the channel length is made 2 to 5 times the minimum allowable channel L (this was also necessary to reduce the short-channel effects discussed above); (2) the value of λ is normally determined empirically; trying to determine λ from an equation is, in general, not too useful; (3) the output resistance of a MOSFET is a function of the MOSFET's drain current. The exact value of this current is not important when estimating the output resistance. Whether $I_{D,sat}$ (the drain current at $V_{D,sat}$) or the actual operating point current, I_D, is used is not practically important when determining r_o.

MOSFET Transconductance

It is useful to determine how a change in the gate-source voltage changes the drain current of a MOSFET operating in the saturation region. We can relate the change in drain current, i_d, to the change in gate-source voltage, v_{gs}, using the MOSFET transconductance, g_m, or

$$g_m = \frac{\Delta i_D}{\Delta v_{GS}} \Rightarrow i_d = g_m v_{gs} \tag{3.10}$$

Neglecting the output resistance of a MOSFET, we can write the sum of the DC and AC (or changing) components of the drain current and gate-source voltage using

$$i_D = i_d(AC) + I_D(DC) = \frac{\beta}{2}(V_{GS}(DC) + v_{gs}(AC) - V_{THN})^2 \tag{3.11}$$

If we hold the DC values constant and assume they are large compared to the AC components, then by simply taking the derivative of i_D with respect to v_{gs}, we can determine g_m. Doing this results in

$$g_m = \beta(V_{GS} - V_{THN}) = \sqrt{2\beta I_D} \tag{3.12}$$

Following this same procedure for the MOSFET operating in the subthreshold region results in

$$g_m = \frac{I_D \cdot q}{kT} \tag{3.13}$$

Notice how the change in transconductance is linear with drain current when the device is operating in the subthreshold region. The larger incremental increase in g_m is due to the exponential relationship between I_D and V_{GS} when the device is operating in the weak inversion region (as compared to the square law relationship when the device is operating in the strong inversion region).

MOSFET Open-Circuit Voltage Gain

At this point, we can ask, "What's the largest possible voltage gain I can get from a MOSFET under ideal biasing conditions?" Consider the schematic diagram shown in Fig. 3.4(a) without biasing circuitry shown and with the effects of finite MOSFET output resistance modeled by an external resistor. The open-circuit voltage gain can be written as

$$|A| = \frac{v_{out}}{v_{gs}} = g_m r_o = \frac{\sqrt{2\beta I_D}}{\lambda I_D} = \frac{\sqrt{2\beta}}{\lambda\sqrt{I_D}} \text{ (for normal operation)} \tag{3.14}$$

which increases as the DC drain biasing current is reduced (and the MOSFET intrinsic speed decreases) until the MOSFET enters the subthreshold region. Once in the subthreshold region, the voltage gain flattens out and becomes

$$A_v = g_m r_o = \frac{I_D \cdot q}{kT} \cdot \frac{1}{\lambda I_D} = \frac{q}{\lambda \cdot kT} \text{ (in the subthreshold region)} \tag{3.15}$$

Layout of the MOSFET

Figure 3.5 shows the layout of both n-channel and p-channel devices. In this layout, we are assuming that a p-type substrate is used for the body of the NMOS (the body connection for the n-channel MOSFET is made through the p$^+$ diffusion on the left in the layout). An n-well is used for the body of the PMOS devices (the connection to the n-well is made through the n$^+$ diffusion on the right in the layout). An important thing to notice from this layout is that the intersection of polysilicon (poly for short) and n$^+$ (in the p-substrate) or p$^+$ (in the n-well) forms a MOSFET. The length and width of the MOSFET, as seen in Fig. 3.5, is determined by the size of this overlap. Also note that the four metal connections to the terminals of each MOSFET, in this layout, are floating; that is, not connected to anything but the

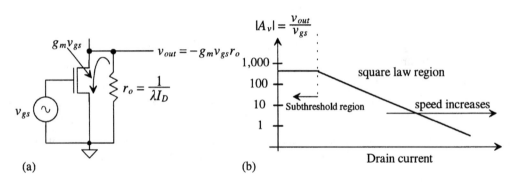

FIGURE 3.4   Open-circuit voltage gain (DC biasing not shown).

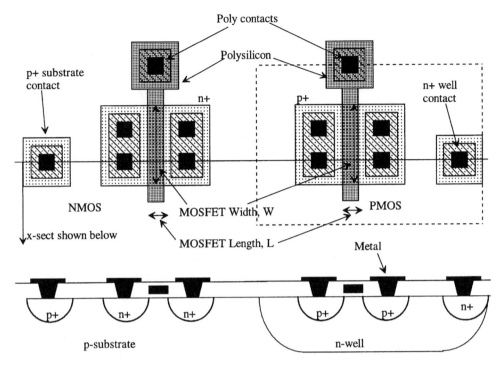

FIGURE 3.5   Layout and cross-sectional views of NMOS and PMOS devices.

MOSFETs themselves. It is important to understand, since the p-substrate is common to all MOSFETs on the chip, that this would require the body of the n-channel MOSFET (again the p$^+$ diffusion on the left in the layout) be connected to *VSS* (ground for most digital applications).

3.2   Biasing Circuits

A fundamental component of any CMOS amplifier is a biasing circuit. This section presents important design topologies and considerations used in the design of CMOS biasing circuits. We begin this section with a discussion of current mirrors.

The Current Mirror

The basic CMOS current mirror is shown in Fig. 3.6. For the moment, we will not concern ourselves with the implementation of I_{REF}. By tying M1's gate to its drain, we set V_{GS} at a value given by

$$V_{GS} = V_{THN} + \sqrt{\frac{2I_{REF}}{\beta_1}} \tag{3.16}$$

For most design applications, the term under the square root is set to a few hundred millivolts or less. Because M2's gate-source voltage, and thus its drain current (I_{OUT}), is set by I_{REF}, we say that M2 mirrors the current in M1. If M2's β is different from M1's, we can relate the currents by

$$\frac{I_{OUT}}{I_{REF}} = \frac{2\beta_2(V_{GS} - V_{THN})^2}{2\beta_1(V_{GS} - V_{THN})^2} = \frac{\beta_2}{\beta_1} = \frac{W_2}{W_1} \text{ if } L_1 = L_2 \tag{3.17}$$

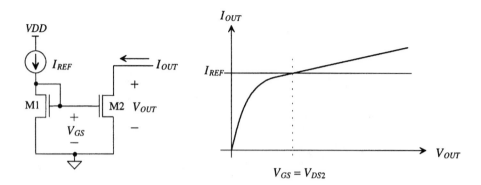

FIGURE 3.6   Basic CMOS current mirror.

Notice that we have neglected the finite output resistance of the MOSFETs in this equation. If we include the effects of finite output resistance, and assume M1 and M2 are sized the same, we see from Fig. 3.6 that the only time I_{REF} and I_{OUT} are equal is when $V_{GS} = V_{OUT} = V_{DS2}$.

Design Example

Design a 100-µA current sink and a 200-µA current source assuming that you are designing with a 0.5-µm CMOS process with $KP_n = 150$ µA/V², $KP_p = 50$ µA/V², $V_{THN} = 0.8$ V, and $V_{THP} = 0.9$ V. Assume λ was empirically determined (or determined from simulations) to be 0.05 V⁻¹ with $L = 2.0$ µm. Assume that an I_{REF} of 50 µA is available for the design. Determine the output resistance of the current source/sink and the minimum voltage across the current source/sink.

The schematic of the design is shown in Fig. 3.7. If we design so that the term $\sqrt{(2I_{REF})/\beta_1}$ is 300 mV, the width of M1, W_1, is 15 µm ($KP_n = 150$ µA/V², $L = 2$ µm). The MOSFET, M1, has a gate-source voltage 300 mV in *excess* of the threshold voltage, or, in other words, the MOSFET V_{GS} is 1.1 V. For M2 and M3 to sink 100 µA (twice the current in M1), we simply increase their widths to 30 µm for the same V_{GS} bias supplied by M1. Note the minimum voltage required across M3, V_{DS3}, in order to keep M3 out of the triode region ($V_{DS3}\ V_{GS} - V_{THN}$), is simply the *excess gate voltage*, $\sqrt{(2I_{REF})/\beta_1}$, or, for this example, 300 mV. (Note: simply put, 300 mV is the minimum voltage on the drain of M3 required to keep it in the saturation region.) Increasing the widths of the MOSFETs lowers the minimum voltage required across the MOSFETs (lowers the excess gate voltage) so they remain in saturation at the price of increased layout area. Also, differences in MOSFET threshold voltage become more significant, affecting the matching between devices.

The purpose of M2 should be obvious at this point; it provides a 100-µA bias for the p-channel current mirror M4/M5. Again, if we set M4's excess gate voltage to 300 mV (so that the V_{SG} of the p-channel MOSFET is 1.2 V), the width of M4 can be calculated (assuming that L is 2 µm and $KP_p = 50$ µA/V²) to be 45 µm (or a factor of 3 times the n-channel width due to the differences in the transconductance parameters of the MOSFETs). Since the design required a current source of 200 µA, we increase the width of the p-channel, M5, to 90 µm (so that it mirrors twice the current that flows in M4). The output resistance of the current source, M5, is 100 kΩ, while the maximum voltage on the drain of M5 is 3 V (in order to keep M5 in saturation).

Layout of Current Mirrors

In order to get the best matching between devices, we need to lay the MOSFETs out so that differences in the mirrored MOSFETs' widths and lengths are minimized. Figure 3.8 shows the layout of the M1 (15/2) and M2 (30/2) MOSFETs of Fig. 3.7. Notice how, instead of laying M2 out in a fashion similar to M1 (i.e., a single poly over active strip), we split M2 into two separate MOSFETs that have the same shape as M1.

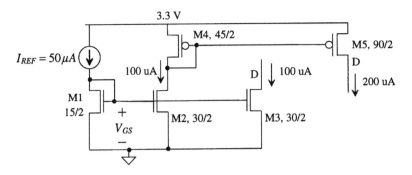

FIGURE 3.7 Design example.

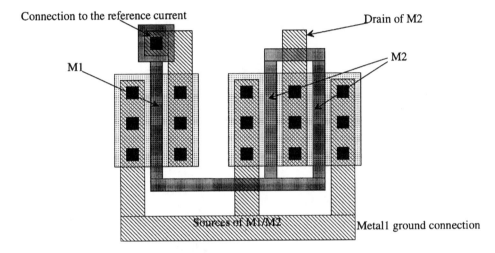

FIGURE 3.8 Layout of MOSFET mirror M1/M2 in Fig. 3.7.

The matching between current mirrors can also be improved using a *common-centroid* layout (Fig. 3.9). Parameters such as the threshold voltage and *KP* in practice can have a somewhat linear change with position on the wafer. This may be the result of varying temperature on the top of the wafer during processing, or fluctuations in implant dose with position. If we lay MOSFETs M2 and M1 out as shown in Fig. 3.9(a) (which is the same way they were laid out in Fig. 3.8), M1 has a "weight" of 1 while M2 has a weight of 5. These numbers may correspond to the threshold voltages of three individual MOSFETs with numerical values 0.81, 0.82, and 0.83 V. By using the layout shown in Fig. 3.9(b), M1's or M2's average weight is 2. In other words, using the threshold voltages as an example, M1's threshold voltage is 0.82 V while the average of M2's threshold voltage is also 0.82 V. Similar discussions can be made if the transconductance parameters vary with position. Figure 3.9(c) shows how three devices can be matched using a common-centroid layout (M2 and M1 are the same size while M3 is 4 times their size.) A good exercise at this point is to modify the layout of Fig. 3.9(c) so that M2 is twice the size of M1 and one half the size of M3.

The Cascode Current Mirror

We saw that in order to improve the matching between the two currents in the basic current mirror of Fig. 3.6, we needed to force the drain-source voltage of M2 to be the same as the drain-source voltage of M1 (which is also V_{GS} in Fig. 3.6). This can be accomplished using the cascode connection of MOSFETs shown in Fig. 3.10. The term "cascode" comes from the days of vacuum tube design where a cascode of a common-cathode amplifier driving a common-grid amplifier was used to increase the speed and gain of an overall amplifier design.

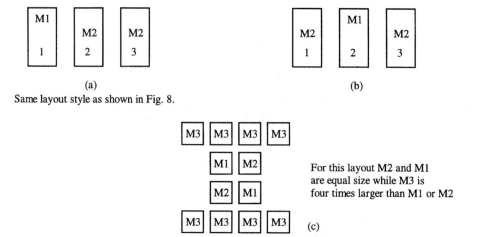

FIGURE 3.9 Common-centroid layout used to improve matching.

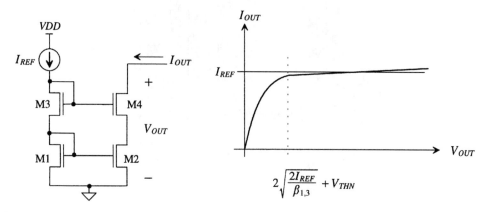

FIGURE 3.10 Basic cascode CMOS current mirror.

Using the cascode configuration results in higher output resistance and thus better matching. For the following discussion, we will assume that M1 and M3 are the same size, as are M2 and M4. Again, remember that M1 and M2 form a current mirror and operate in the same way as previously discussed. The addition of M3 and M4 helps force the drain-source voltages of M1/M2 to the same value. The minimum V_{OUT} allowable, in order to keep M4 out of the triode region, across the current mirror increases to

$$V_{OUT, min} = 2\sqrt{\frac{2I_{REF}}{\beta_{1,3}}} + V_{THN}$$ (3.18)

which is basically an increase of V_{GS} over the basic current mirror of Fig. 3.6.

The output resistance of the cascode configuration can be derived with the circuit model of Fig. 3.11. Here, we assume that the gates of M4 and M2 are at fixed DC potentials (which are set by I_{REF} flowing through M1/M3). Since the source of M2 is held at ground, we know that the AC component of v_{gs2} is 0. Therefore, we can replace M2 with a small-signal resistance $r_o = (1/\lambda I_{OUT})$. To determine the output resistance of the cascode current mirror, we apply an AC test voltage, v_{test}, and measure the AC test current that flows into the drain of M4. Ideally, only the DC component will flow through v_{test}. We can write the AC gate-source voltage of M4 as $v_{gs4} = -i_{test} \cdot r_o$. The drain current of M4 is then $g_{m4} v_{gs4} = -i_{test} \cdot g_{m4} r_o$ while

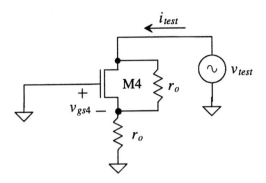

FIGURE 3.11   Determining the small-signal output resistance of the cascode current mirror.

the current through the small-signal output resistance of M4, r_o, is $(v_{test} - (-i_{test}r_o))/r_o$. Combining these equations yields the cascode output resistance of

$$R_{out,\,cascode} = \frac{v_{test}}{i_{test}} = r_o(1 + g_m r_o) + r_o \approx g_m r_o^2. \qquad (3.19)$$

The cascode current source output resistance is $g_m r_o$ (the open-circuit voltage gain of a MOSFET) times larger than r_o (the simple current mirror output resistance.) The main drawback of the cascode configuration is the increase in the minimum required voltage across the current sink in order for all MOSFETs to remain in the saturation region of operation.

Low-Voltage Cascode Current Mirror

If we look at the cascode current mirror of Fig. 3.10, we see that the drain of M2 is held at the same potential as the drain of M1, that is, VGS or $\sqrt{(2I_{REF})/\beta} + V_{THN}$. We know that the voltage on the drain of M2 can be as low as $\sqrt{(2I_{REF})/\beta}$ before it starts to enter the triode region. Knowing this, consider the *wide-swing current mirror* shown in Fig. 3.12. Here, "wide-swing" means the minimum voltage across the current mirror is $2\sqrt{(2I_{REF})/\beta}$, the sum of the excess gate voltages of M2 and M4. To understand the operation of this circuit, assume that M1 through M4 have the same W/L ratio (their βs are all equal). We know that the V_{GS} of M1 and M2 is $\sqrt{(2I_{REF})/\beta} + V_{THN}$. It is desirable to keep M2's drain at $\sqrt{(2I_{REF})/\beta}$ for wide-swing operation. This means, since M3/M4 are the same size as M1/M2, the gate voltage of M3/M4 must be $V_{GS} + \sqrt{(2I_{REF})/\beta}$ or $2\sqrt{(2I_{REF})/\beta} + V_{THN}$. By sizing M5's channel width so

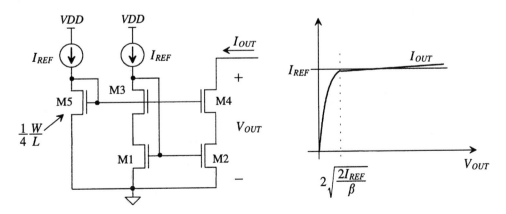

FIGURE 3.12   Wide-swing CMOS cascode current mirror.

that it is one fourth of the size of the other transistor widths and forcing I_{REF} through the diode connected M5, we can generate this voltage. We should point out that the size (its W/L ratio) of M5 can be further decreased, say to 1/5, in order to keep M1/M2 from entering the triode region (and the output resistance from decreasing). The cost is an increase in the minimum allowable voltage across M2/M4, that is, V_{OUT}.

Simple Current Mirror Biasing Circuits

Figure 3.13 shows two simple circuits useful for generating the reference current, I_{REF}, used in the current mirrors discussed in the previous section. The circuit shown in Fig. 3.13(a) uses a simple resistor with a gate-drain connected MOSFET to generate a reference current. Note how, by adding MOSFETs mirroring the current in M1 or M2, we can generate any multiple of I_{REF} needed. The reference current, of Fig. 3.13(a), can be determined by solving

$$I_{REF} = \frac{VDD - V_{GS}}{R} = \frac{\beta}{2}(V_{GS} - V_{THN})^2 \tag{3.20}$$

Figure 3.13(b) shows a MOSFET-only bias circuit. Since the same current flows in M1 and M2, we can mirror off of either MOSFET to generate our bias currents. The current flowing in M1/M2 is designed using

$$I_{REF} = \frac{\beta_1}{2}(V_{GS} - V_{THN})^2 = \frac{\beta_2}{2}(VDD - V_{GS} - V_{THN})^2 \tag{3.21}$$

Notice in both equations above that the reference current is a function of the power supply voltage. Fluctuations, as a result of power supply noise, in *VDD* directly affect the bias currents. In the next section, we will present a method for generating currents that reduces the currents' sensitivity to changes in *VDD*.

Temperature Dependence of Resistors and MOSFETS

Figure 3.14 shows how a resistor changes with temperature, assuming a linear dependence. The temperature coefficient is used to relate the value of a resistor at room temperature, or some known temperature T_0, to the value at a different temperature. This relationship can be written as

$$R(T) = R(T_0) \cdot [1 + TCR(T - T_0)] \tag{3.22}$$

where *TCR* is the temperature coefficient of the resistor *ppm/°C* (parts per million, a multiplier of 10^{-6}, per degree C). Typical values for TCRs for n-well, n⁺, p⁺, and poly resistors are 2000, 500, 750, and 100 ppm/°C, respectively.

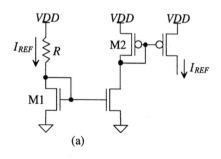

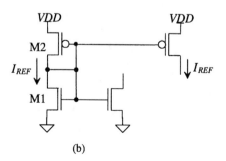

(a)   (b)

FIGURE 3.13 Simple biasing circuits.

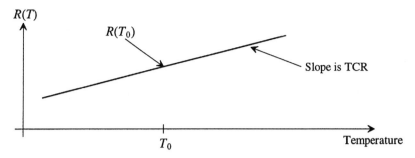

FIGURE 3.14 Variation of a resistor with temperature.

Figure 3.15 shows how the drain current of a MOSFET changes with temperature. At low gate-source voltages, the drain current increases with increasing temperature. This is a result of the threshold voltage decreasing with increasing temperature which dominates the I-V characteristics of the MOSFET. The temperature coefficient of the threshold voltage (NMOS or PMOS), TCV_{TH}, is generally around −3000 ppm/°C. We can relate the threshold voltage to temperature using

$$V_{TH}(T) = V_{TH}(T_0)[1 + TCV_{TH}(T - T_0)] \tag{3.23}$$

At larger gate-source voltages, the drain current decreases with increasing temperature as a result of the electron or hole mobility decreasing with increasing temperature. In other words, at low gate-source voltages, the temperature changing the threshold voltage dominates the I-V characteristics of the MOSFET; while at larger gate-source voltages, the mobility changing with temperature dominates. Note that at around 1.8 V, for a typical CMOS process, the drain current does not change with temperature. The mobility can be related to temperature by

$$\mu(T) = \mu(T_0)\left(\frac{T}{T_0}\right)^{-1.5} \tag{3.24}$$

The Self-Biased Beta Multiplier Current Reference

Figure 3.16 shows the self-biased beta multiplier current reference. This circuit employs positive feedback, with a gain less than one, to reduce the sensitivity of the reference current to power supply changes. MOSFET M2 is made K times wider than MOSFET M1 (in other words, $\beta_2 = K\beta_1$; hence, the name beta multiplier). We know from this figure that $V_{GS1} = V_{GS2} + I_{REF}R$, where $V_{GS1} = \sqrt{2I_{REF}/\beta_1} + V_{THN}$ and $V_{GS2} = \sqrt{(2I_{REF})/K\beta_1} + V_{THN}$; therefore, we can write the current in the circuit as

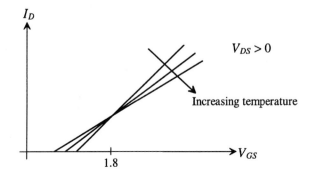

FIGURE 3.15 Temperature characteristics of a MOSFET.

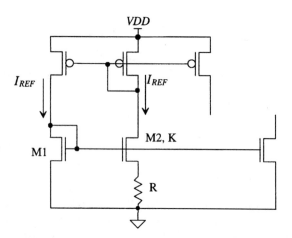

FIGURE 3.16 Beta multiplier current reference.

$$I_{-REF} = \frac{2}{R^2 \beta_1}\left[1 - \sqrt{\frac{1}{K}}\right]^2 \tag{3.25}$$

which shows no first-order dependence on the power supply voltage.

Start-up Circuit

One of the drawbacks of using the reference circuit of Fig. 3.16 is that it has two stable operating points [Fig. 3.17(a)]. The desirable operating point, point A, occurs when the current flowing in the circuit is I_{REF}. The undesirable operating point occurs when zero current flows in the circuit, point B. Because of the possibility of zero current flowing in the reference, a start-up circuit should always be used when using the beta multiplier. The purpose of the start-up circuit is to ensure that point B is avoided. When designed properly, the start-up circuit [Fig. 3.17(b)] does not affect the beta multiplier operation when I_{REF} is non-zero (M3 is off when operating at point A).

A Comment about Stability

Since the beta multiplier employs positive feedback, it is possible that the circuit can become unstable and oscillate. However, with the inclusion of the resistor in series with the source of M2, the gain around the loop, from the gate of M2 to the drain/gate of M1, with the loop broken between the gates of M1 and M2, is less than one, keeping the reference stable. Adding a large capacitance across R, however, can increase the loop gain to the point of instability. This situation could easily occur if R is bonded out to externally set the current.

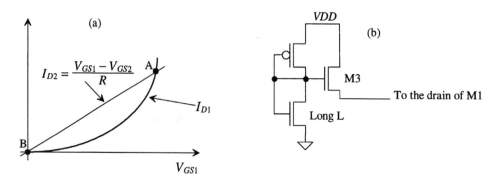

FIGURE 3.17 Start-up circuit for the beta multiplier circuit shown in Fig. 3.16.

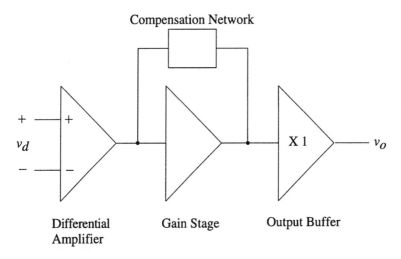

FIGURE 3.18   Block diagram for a generic op-amp.

3.3   Amplifiers

Now that we have introduced biasing circuits and MOS device characteristics, we will immediately dive into the design of operational amplifiers. Since space is very limited, we will employ a top-down approach in which a series of increasingly complex circuits are dissected stage by stage and individual blocks analyzed.

Operational amplifiers typically are composed of either two or three stages consisting of a differential amplifier, a gain stage and an output stage as seen in Fig. 3.18. In some applications, the gain stage and the output stage are one and the same if the load is purely capacitive. However, if the output is to drive a resistive load or a large resistive load, then a high current gain buffer amplifier is used at the output. Each stage plays an important role in the performance of the amplifier.

The differential amplifier offers a variety of advantages and is always used as the input to the overall amplifier. Since it provides common-mode rejection, it eliminates noise common on both inputs, while at the same time amplifying any differences between the inputs. The limit for which this common mode rejection occurs is called *common-mode range* and signifies the upper and lower common mode signal values for which the devices in the diff-amp are saturated. The differential amplifier also provides gain. The gain stage is typically a common-source or cascode type amplifier. So that the amplifier is stable, a compensation network is used to intentionally lower the gain at higher frequencies. The output stage provides high current driving capability for driving either large capacitive or resistive loads. The output stage typically will have a low output impedance and high signal swing characteristics. In some cases, it may be advantageous to add bipolar devices to improve the performance of the circuitry. These will be presented as the block level circuits are analyzed.

The Simple Unbuffered Op-Amp

Examine the simple operational amplifier shown in Fig. 3.19. Here, the amplifier can be segregated into a biasing block, a differential input stage, an output stage, and a compensation capacitor.

The biasing circuit is a simple current mirror driver, consisting of the resistor R_{bias} and the transistor M8. The current through M8 is mirrored through both M5 and M7. Thus, the current through the entire circuit is set by the value of R_{bias} and the relative W/Ls of M8, M5, and M7. The actual values of this current will be discussed a little bit later on. When designing with R_{bias}, one must be careful not to ignore the effect of temperature on R_{bias}, and thus the values of the currents through the circuit. We will see later on how the bias current greatly affects the performance of the amplifier. For the commercial

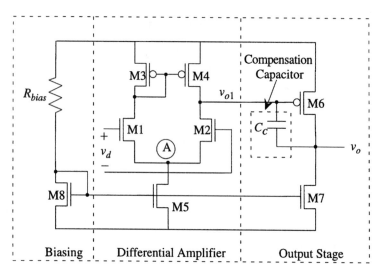

FIGURE 3.19   Basic two-stage op-amp.

temperature range of 0°C to 125°C, the current through M8 should be simulated with R_{bias} at values of ±30% of its nominal value. Other, more sophisticated voltage references (as discussed earlier) can be used in place of the resistor reference, and will be presented as we progress.

The Differential Amplifier

The differential amplifier is composed of M1, M2, M3, M4, and M5, with M1 matching M2 and M3 matching M4. The transistor M5 can be replaced by a current source in the ideal case to enhance one's understanding of the circuit's functionality. The node labeled as node A can be thought of as a virtual ground, since the current through M5 is constant and thus the voltage at node A is also constant.

Now assume that the gate of M2 is tied to ground as seen in Fig. 3.20. Any small signal on the gate of M1 will result in a small signal current i_{d1}, which will flow from drain to source of M1 and will also be mirrored from M3 to M4. Note that since M5 can be thought of as an ideal current source, it can be replaced with its ideal impedance of an open circuit. Therefore, i_{d1} will also flow from source to drain of M2. Remember that the small signal current is different from the DC current flowing through M2. The small signal current can be thought of as the current generated from the instantaneous (AC + DC) voltage at the output. If the instantaneous voltage at the output node swings up, then the small signal current through M2 will flow from drain to source. However, if the instantaneous voltage swings down,

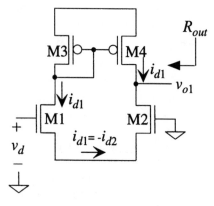

FIGURE 3.20   Pseudo-AC circuit showing small signal currents.

the change in voltage will be considered negative and thus the small signal current will assume an identical polarity, thus flowing from source to drain of M2. The small signal voltage produced at the output node will then be $2i_{d1}$ times the output impedance of the amplifier. In this case, R_{out} is simply $r_{o2}||r_{o4}$. The value of the small signal current, i_{d1}, is simply

$$i_{d1} = g_{m1}v_{gs1} \tag{3.26}$$

and the differential voltage, $v_d = v_{gs1} + v_{sg2}$. Therefore, since $v_{gs1} = v_{gs2}$, then,

$$\frac{i_{d1}}{v_d} = \frac{g_{m1}}{2} \tag{3.27}$$

the small signal output voltage is simply

$$v_{o1} = 2 \cdot i_{d1} \cdot r_{o2}||r_{o4} \tag{3.28}$$

and the small signal voltage gain can then be easily derived as

$$\frac{v_{o1}}{v_d} = g_{m1,2}(r_{o2}||r_{o4}) \tag{3.29}$$

Now let us examine this equation more carefully. Substituting the expressions for g_m and r_o, the previous equation becomes

$$\frac{v_{o1}}{v_d} \approx \sqrt{2\beta_{1,2}I_{D1,2}} \cdot \frac{1}{2\lambda I_{D1,2}} = K' \cdot \sqrt{\frac{W_{1,2}}{L_{1,2}I_{D1,2}}} \cdot \frac{1}{\lambda} \tag{3.30}$$

where K' is a constant, which is uncontrollable by the designer. When designing analog circuits, it is just as important to understand the effects of the controllable variables on the specification as it is to know the absolute value of the gain using hand calculations. This is because the hand analysis and computer simulations will vary a great deal because of the complex modeling used in today's CAD tools. Examining Eq. (3.30), and knowing that the effect of λ on the gain diminishes as L increases such that $1/\lambda$ is directly proportional to channel length. Then, a proportionality can be established between $W/L_{1,2}$ and the drain current vs. the small signal gain such that

$$\frac{v_{o1}}{v_d} \propto \sqrt{\frac{W_{1,2} \cdot L_{1,2}}{I_{D1,2}}} \tag{3.31}$$

Notice that the constant was not included since the value is not dependent on anything the designer can adjust. The importance of this equation tells us that the gain is proportional to the square root of the product of W and L and inversely proportional to the square root of the drain current through M1 and M2, which is also $1/2$ I_{D6}. So to increase the gain, one must increase W or L or decrease the value I_{D6}, which is dependent on the value of R_{bias}.

The Gain Stage

Now examine the output stage consisting of M6 and M7. Here, the driving transistor, M6, is a simple inverting amplifier with a current source load as seen in equivalent circuit shown in Fig. 3.21. The value of the small signal current is defined by the AC signal v_{o1}, which is equal to v_{sg6}. Therefore,

$$\frac{i_{d6}}{v_{o1}} = -g_{m6} \tag{3.32}$$

and the gain of the stage is simply

$$\frac{i_{d6}}{v_{o1}} \cdot \frac{v_o}{i_{d6}} = \frac{v_o}{v_{o1}} = -g_{m6} \cdot r_{o6}||r_{o7} \tag{3.33}$$

Again, we can write the gain in terms of the designer's variables, so that the gain of the amplifier can be expressed as a proportion of

$$\frac{v_o}{v_{o1}} \propto \sqrt{\frac{W_6 \cdot L_{6,7}}{I_{D6,7}}} \tag{3.34}$$

Therefore, overall, the gain of the entire amplifier is

$$\frac{v_o}{v_d} = g_{m1,2}(r_{o2}||r_{o4}) \cdot -g_{m6}(r_{o6}||r_{o7}) \tag{3.35}$$

or as the proportionalities

$$\frac{v_o}{v_d} \propto \sqrt{\frac{W_{1,2} \cdot L_{1,2}}{I_{D1,2}}} \cdot \sqrt{\frac{W_6 \cdot L_{6,7}}{I_{D6,7}}} \tag{3.36}$$

So, the key variables for adjusting gain are the drain currents (the smaller the better) and the W and L ratios of M1, M2, and M6 (the larger the better). Of course, there are lower and upper limits to the drain currents and the W/Ls, respectively, that we will examine as we analyze other specifications that will ultimately determine the bounds of adjustability.

Frequency Response

Now examine the amplifier circuit shown in Fig. 3.22. In this particular circuit, the compensation capacitor, C_C, is removed and will be re-added shortly. The capacitors C_1 and C_2 represent the total lumped capacitance from each ground. Since the output nodes associated with each output is a high

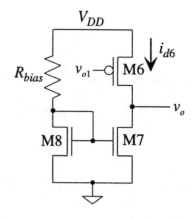

FIGURE 3.21   Output stage circuit.

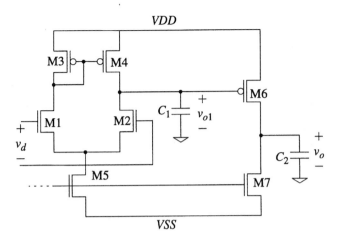

FIGURE 3.22 Two-stage op-amp with lumped parasitic capacitors.

impedance, these nodes will be the dominant frequency-dependent nodes in the circuit. Since each node in a circuit contributes a high-frequency pole, the frequency response will be dominated by the high-impedance nodes.

An additional capacitor could have been included at the gate of M4 to ground. However, the equivalent impedance from the gate of M4 to ground is approximately equal to the impedance of a gate-drain connected device or $1/g_{m3}$ as seen in Fig. 3.23 (a)–(c). If a controlled source has the controlling voltage directly across its terminals [Fig. 3.23(b)], then the effective resistance is simply the controlling voltage (in this case, v_{gs}) divided by the controlled current ($g_m v_{gs}$), which is $1/g_m \| r_o$ or approximately $1/g_m$ if r_o is much greater than $1/g_m$ [Fig. 3.23(c)]. Therefore, the impedance seen from the gate of M4 to ground is low, and the pole associated with the node will be at a much higher frequency than the high-impedance nodes. The same holds true for the node associated with the drain of M5. That node is considered an AC ground, so it has no effect on the frequency response of the small signal circuit. The only remaining node is the node which defines the current mirrors (the gate of M8). Since this node is not in the small signal path, and is a DC bias voltage, it can also be considered an AC ground.

One can approximate the frequency response of the amplifier by examining both the effective impedance and parasitic capacitance at the output of each stage. The parasitic capacitances can be seen in the small signal model of a MOSFET in Fig. 3.24. The capacitors C_{gb}, C_{sb}, and C_{db} represent the bulk depletion capacitors of the transistors, while C_{gd} and C_{gs} represent the overlap capacitances from gate to drain and

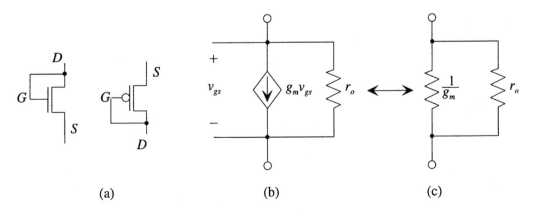

(a) (b) (c)

FIGURE 3.23 Equivalent resistance for a gate-drain connected MOSFET device.

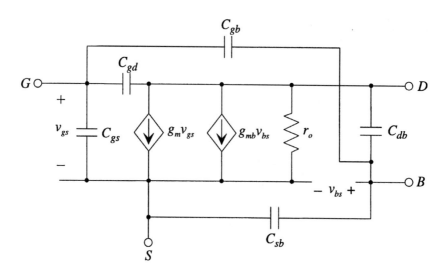

FIGURE 3.24   High-frequency, small signal model with parasitic capacitors.

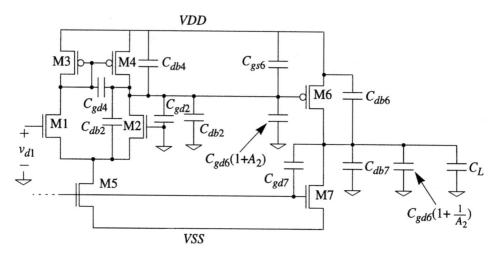

FIGURE 3.25   Two-stage op-amp with parasitics shown explicitly.

gate to source, respectively. Referring to Fig. 3.25, which shows the parasitic capacitors explicitly, C_1 can now be written as

$$C_1 = C_{db4} + C_{gd4} + C_{db2} + C_{gd2} + C_{gs6} + C_{gd6}(1 + A_2) \qquad (3.37)$$

Note that C_{gd4} is included as capacitor to ground since the low impedance caused by the gate-drain device of M3 can be considered equivalent to an AC ground. The capacitor C_{db2} is also connected to AC ground at the source-coupled node consisting of M1 and M2.

Miller's theorem was used to determine the effect of the bridging capacitor C_{db6}, connected from the gate to the drain of M6. Miller's theorem approximates the effects of the gate-drain capacitor by replacing the bridging capacitor with an equivalent input capacitor of value $C_{db6} \cdot (1 - A_2)$ and an equivalent output capacitor with a value of $C_{db6} \cdot (1 - 1/A_2)$. The term A_2 is the gain across the original bridging capacitor and is $-g_{m6} \cdot r_{o6} \| r_{o7}$. The reader should consult Ref. 2 for a proof of Miller's theorem.

The capacitor C_2 can also be determined by examining Fig. 3.25,

$$C_2 = C_{db6} + C_{db7} + C_{gd7} + C_{gd6} \cdot \left(1 + \frac{1}{A_2}\right) + C_L \tag{3.38}$$

Now assume that C_1 is greater than C_2. This means that the pole associated with the diff-amp output will be lower in frequency than the pole associated with the output of the output stage. Thus, a good model for the op-amp can be seen in Fig. 3.26. The transfer function, ignoring C_C, will then become

$$\frac{v_o(s)}{v_{d1}(s)} = -g_{m1}(r_{o2}||r_{o4}) \cdot g_{m6}(r_{o6}||r_{o7}) \cdot \frac{1}{\left(j\frac{f}{f_{p1}} + 1\right)\left(\frac{f}{f_{p2}} + 1\right)} \tag{3.39}$$

where

$$f_{p1} = \frac{1}{2\pi R_{out1} C_1}, f_{p2} = \frac{1}{2\pi R_{out} C_2} \tag{3.40}$$

The plot of the transfer function can be seen in Fig. 3.27. Note that in examining the frequency response, the phase margin is virtually zero. Remember that phase margin is the difference between phase at the frequency at which the magnitude plot reaches 0 dB (also known as the gain-bandwidth product) and the phase at the frequency at which the phase has shifted −180°. It is recommended for stability reasons that the phase margin of any amplifier be at least 45° (60° is recommended). A phase margin below 45° will result in long settling times and increased propagation delays. The system can also be thought of as a simple second-order linear controls system with the phase margin directly affecting the transient response of the system.

Compensation

Now we will include C_C in the circuit. If C_C is much greater than C_{gd6}, then the C_C will dominate the value of C_1 [especially since it is multiplied by $(1 − A_2)$] and will cause the pole, f_{p1}, to roll off much earlier than without C_C to a new location, f_{p1}'. One could solve, using circuit analysis, the circuit shown in Fig. 3.26 with C_C included to also prove that the second pole, f_{p2}, moves further out[3] to a higher frequency, f_{p2}'. Ideally, the addition of C_C will cause an equivalent single pole roll off of the overall frequency response. The second pole should not begin to affect the frequency response until after the magnitude response is below 0 dB. The new values of the poles are

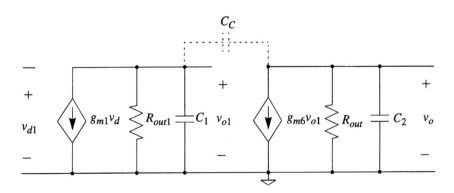

FIGURE 3.26 Model used to determine the frequency response of the two-stage op-amp.

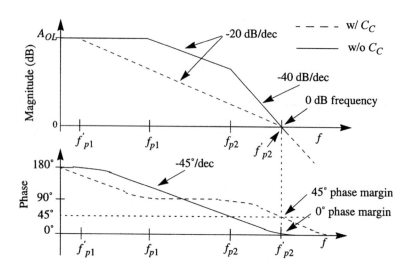

FIGURE 3.27   Magnitude and phase of the two-stage op-amp with and without compensation.

$$f'_{p1} \approx \frac{1}{2\pi(g_{m6}R_{out})C_C R_{out1}}, f'_{p2} \approx \frac{g_{m6}C_C}{2\pi \cdot (C_2 C_1 + C_2 C_C + C_C C_1)} \approx \frac{g_{m6}}{2\pi \cdot C_2} \qquad (3.41)$$

If the previously discussed analysis was performed, one would also see that by using Miller's theorem, we are neglecting a right-hand plane (RHP) zero that could have negative consequences on our phase margin, since the phase of an RHP zero is similar to the phase of a left-hand plane (LHP) zero. An RHP zero behaves similarly to an LHP zero when examining the magnitude response; however, the phase response will cause the phase plot to shift $-180°$ more quickly. The RHP zero is at a value of

$$f_{z1} = \frac{g_{m6}}{C_C} \qquad (3.42)$$

To avoid effects of the RHP zero, one must try to move the zero out well beyond the point at which the magnitude plot reaches 0 dB (suggested rule of thumb: factor of 10 greater). The comparison of the frequency response of the two-stage op-amp with and without C_C can be seen in Fig. 3.27.

One remedy to the zero problem is to add a resistor, R_z, in series with compensation capacitor as seen in Fig. 3.28. The expression of the zero after adding the resistor becomes

$$f_{z1} = \frac{1}{C_C\left(\dfrac{1}{g_{m6}} - R_Z\right)} \qquad (3.43)$$

and the zero can be pushed into the LHP where it adds phase shift and increases phase margin if $R_z > 1/g_{m6}$. Fig. 3.29[4] shows the root locus plot for the zero as R_z is introduced. With $R_z = 0$, the zero location is on the RHP real axis. As R_z increases in value, the zero gets pushed to infinity at the point at which $R_z = 1/g_{m6}$. Once $R_z > 1/g_{m6}$, the zero appears in the LHP where its phase shift adds to the overall phase response, thus improving phase margin. This type of compensation is commonly referred to as *lead compensation,* and is commonly used as a simple method for improving the phase margin. One should be careful about using R_z, since the absolute values of the resistors are not well predicted. The value of the resistor should be simulated over its maximum and minimum values to ensure that no matter if the

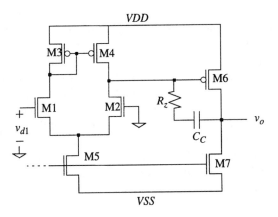

FIGURE 3.28   Compensation including a nulling resistor.

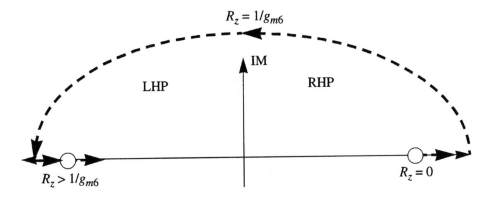

FIGURE 3.29   Root locus plot of how the zero shifts from the RHP to the LHP as R_z increases in value.

zero is pushed into the LHP or the RHP, the value of the zero is always 10 times greater than the gain-bandwidth product.

Other AC Specifications

Referring back to the equations for gain and the first pole location, we can now see adjustable parameters for affecting frequency response by plugging in the proportionalities for their corresponding factors. The values of the resistors are directly proportional to the length of L (the longer the L, the higher the resistance, since longer channel lengths diminish the effects of channel length modulation, λ).

The *gain-bandwidth product* for the compensated op-amp is the open-loop gain multiplied by the bandwidth of the amplifier (as set by f_{p1}). Therefore, the gain-bandwidth product is

$$GBW = (g_{m1} \cdot r_{o2}||r_{o4})(g_{m6} \cdot r_{o6}||r_{o7})\left(\frac{1}{2\pi(g_{m6}R_{out})C_C R_{out1}}\right) = \frac{g_{m1}}{2\pi C_C} \qquad (3.44)$$

Or we can write the expression as

$$GBW \propto \frac{\sqrt{\dfrac{W}{L}_{1,2} I_{D1,2}}}{C_C} \qquad (3.45)$$

which shows that the most efficient way to increase *GBW* is to decrease C_C. One could increase the $W/L_{1,2}$ ratio, but its increase is only by the square root of $W/L_{1,2}$. Increasing $I_{D1,2}$ will also yield an increase (also to the square root) in *GBW*, but one must remember that it simultaneously decreases the open-loop gain. The value of C_C must be large enough to affect the initial roll-off frequency as a larger C_C improves phase margin. One could easily determine the value of C_C by first knowing the gain-bandwidth specification, then iteratively choosing values for $W/L_{1,2}$ and $I_{D1,2}$ and then solving for C_C.

One other word of warning: the designer must not neglect the effects of loading on the circuit. Practically speaking, the load capacitor usually dominates the value of the capacitor, C_2 in Fig. 3.22. The value of the load capacitor directly affects the phase shift through the amplifier by changing the pole location associated with the output node. The second pole has a value of approximately $g_{m6}/2\pi C_L$ as was seen earlier. Since the second pole needs to be greater than the gain-bandwidth product, the following relationship can be deduced:

$$\frac{g_{m6}}{C_L} > \frac{g_{m1,2}}{C_C} \tag{3.46}$$

or

$$C_C > \frac{g_{m1,2}}{g_{m6}} \cdot C_L \tag{3.47}$$

Thus, one can see the effect of C_L on phase margin in that the minimum size of the compensation capacitor is directly dependent on the size of the load capacitor. For a phase margin of 60°, it can be shown[3] that the second pole must be 2.2 times greater than the *GBW*.

The *common-mode rejection ratio* (CMRR) measures how well the amplifier can reject signals common to both inputs and is written by

$$\text{CMRR} = 20\log\left|\frac{A_d}{A_{cm}}\right| = 20\log\left|\frac{v_o/v_d}{v_o/v_{cm}}\right| \tag{3.48}$$

where V_{cm} is a common mode input signal, which in this case is composed of an AC and a DC component (the higher the value of CMRR, the better the rejection). The circuit for calculating common mode gain can be seen in Fig. 3.30. The gain through the output stage will be the same for both the differential gain, A_d, and the common-mode gain, A_{cm}. The last stage will cancel itself out in the expression for CMRR; thus, the differential amplifier will determine how well the entire amplifier rejects common mode signals. This rejection is one of the most advantageous reasons for using a diff-amp as an input stage. If the inputs are subjected to the same noise source, the diff-amp has the ability to reject the noise signal and only amplify the difference in the inputs. For the diff-amp used in this example, the common-mode gain is

$$\frac{v_{o1}}{v_{cm}} = -\frac{1}{2g_{m4}r_{o5}} \tag{3.49}$$

This equation bears some explanation. Since the inputs are tied together, the source-coupled node can no longer be considered an AC ground. Therefore, the resistance of the tail current device, M5, must be considered in the analysis. The AC currents flowing through both M1 and M2 are equal and $v_{gs3} = v_{gs4}$. The output impedance of the diff-amp will drop considerably when using a common-mode signal due to the feedback loop consisting of M1, M3, M4, and M2. As a result, the common-mode signal appearing on the drains of M3 and M4 will be identical.

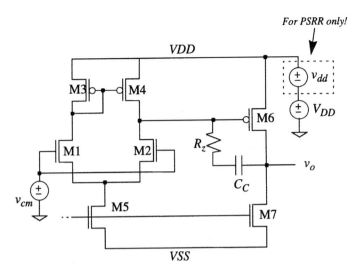

FIGURE 3.30 Circuit used to determine CMRR and PSRR.

One can determine this gain by using half-circuit analysis (Fig. 3.31). This is equivalent to a common source amplifier with a large source resistance of value $2r_{o5}$. When using half-circuit analysis, the current source resistance doubles because the current through the driving device M1 is one half of the original tail current. Therefore, the gain of this circuit can be approximately determined as the negative ratio between the resistance attached to the drain of M1 divided by the resistance attached to the source of M1, or

$$\frac{v_{o1}}{v_{cm}} = -\frac{1/g_{m3,4}}{2r_{o5}} \tag{3.50}$$

Adding the expression for the differential gain of the diff-amp, v_{o1}/v_d, we can write the expression for the CMRR as

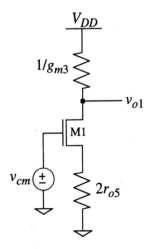

FIGURE 3.31 Half-circuit used to determine the common-mode gain.

$$\text{CMRR} = 20\log(2g_{m1,2}g_{m3,4}(r_{o2}||r_{o4})r_{o5}) \tag{3.51}$$

or as a proportion,

$$\text{CMRR} \propto 20\log\left(\sqrt{\frac{W_{1,2} \cdot L_{1,2}}{I_{D1,2}}} \cdot \sqrt{\frac{(W/L)_{3,4}}{2I_{D5}}} \cdot L_5\right) \tag{3.52}$$

So, it can be seen that the most efficient manner in which to increase the *CMRR* of this amplifier is to increase the channel length of M5 (the tail current device). This, too, has a large signal implication that we will discuss later in this section.

The *power supply rejection ratio* (PSRR) measures how well the amplifier can reject changes in the power supply. This is also a critical specification because it would be desirable to reject noise on the power supply outright. PSRR from the positive supply is defined as:

$$\text{PSRR}^{v_{dd}} = \frac{v_o/v_d}{v_o/v_{dd}} \tag{3.53}$$

where the gain v_o/v_{dd} is the small signal gain from v_{dd} (refer back to Fig. 3.30) to the output of the amplifier with the input signal, v_d, equal to zero. PSRR can also be measured from V_{SS} by inserting a small signal source in series with ground. One should be careful, however, when simulating this specification to make sure that the inputs are properly biased so as to ensure that they are in saturation before simulating. It is best to use a fully differential (differential input and differential output) to most effectively reject power supply noise (to be discussed later).

Large Signal Considerations

Other considerations that must be discussed are the large signal tradeoffs. One cannot ignore the effects of adjusting the small signal specifications on the large signal characteristics. The large signal characteristics that are important include the common-mode range, slew rate, and output signal swing.

Slew rate is defined as the maximum rate of change of the output voltage due to a change in the input voltage. For this particular amplifier, the maximum output voltage is ultimately limited by how fast the tail current device (M5) can charge and discharge the compensation capacitor. The slew rate can then be approximated as

$$SR = \frac{dVo}{dt} \approx \frac{I_{D5}}{C_C} \tag{3.54}$$

Typically, the diff-amp is the major limitation when considering slew rate. However, the tradeoff issues again come into play. If I_{D5} is increased too much, the gain of the diff-amp may decrease below a satisfactory amount. If C_C is made too small, then the phase margin may decrease below an acceptable amount.

The *common-mode range* is defined as the range between the maximum and minimum common-mode voltages for which the amplifier behaves linearly. Referring to Fig. 3.32, suppose that the common-mode voltage is DC value and that the differential signal is also as shown. If the common-mode voltage is swept from ground to V_{DD}, there will be a range for which the amplifier will behave normally and where the gain of the amplifier is relatively constant. Above or below that range, the gain drops considerably because the common-mode voltage forces one or more devices into the triode region.

The maximum common-mode voltage is limited by both M1 and M2 going into triode. This point can be defined by a borderline equation in which $V_{DS1,2} = V_{GS1,2} - V_{THN}$ or, in this case, $V_{D1,2} = V_{G1,2} -$

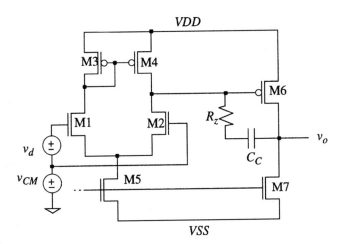

FIGURE 3.32 Determining the CMR for the two-stage op-amp.

V_{THN}. Substituting $V_{DD} - V_{SG3}$ for $V_{D1,2}$, and solving for $V_{G1,2}$, which now represents the maximum common-mode voltage, the expression becomes

$$V_{G1,2(max)} = V_{DD} - \left[\sqrt{\frac{I_{D5}}{\beta_3}} + V_{THP} \right] + V_{THN} \qquad (3.55)$$

where the value of V_{SG3} is written in terms of its drain current using the saturation equation. If the threshold voltages are assumed to be approximately the same value, then the equation can be written as

$$V_{G1,2(max)} = V_{DD} - \sqrt{\frac{I_{D5}}{\beta_3}} = V_{DD} - \sqrt{\frac{L_3 \cdot I_{D5}}{W_3 \cdot K_3}} \qquad (3.56)$$

The minimum voltage is limited by M5 being driven into nonsaturation by the common-mode voltage source. The borderline equation ($V_{D5} = V_{G5} - V_{THN}$) for this transistor can then be used with $V_{D5} = V_{G1,2} - V_{GS1,2}$ and writing both V_{G5} and $V_{GS1,2}$ in terms of its drain current yields

$$V_{G1,2(min)} = V_{SS} + \sqrt{\frac{2I_{D5}}{\beta_5}} + \sqrt{\frac{I_{D5}}{\beta_{1,2}}} = V_{SS} + \sqrt{\frac{2L_5 \cdot I_{D5}}{W_5 \cdot K_5}} + \sqrt{\frac{L_{1,2} \cdot I_{D5}}{W_{1,2} \cdot K_{1,2}}} \qquad (3.57)$$

Now notice the influencing factors for improving the common-mode range ($V_{G1,2(max)}$ is increased and $V_{G1,2(min)}$ is decreased). Assume that V_{DD} and V_{SS} are defined by the circuit application and are not adjustable. To make $V_{G1,2(max)}$ as large as possible, I_{D5} and L_3 should be made as small as possible while W_3 is made as large as possible. And to make $V_{G1,2(min)}$ as small as possible, L_5, I_{D5}, and $L_{1,2}$ should be made as small as possible while increasing W_5 and $W_{1,2}$ as large as possible. Making the drain current as small as possible is in direct conflict with the slew rate. Decreasing L_5 will also degrade the common-mode rejection ratio, and increasing W_3 will affect the pole location of the output node associated with the diff-amp, thus altering the phase margin. All these tradeoffs must be considered as the designer chooses a circuit topology and begins the process of iterating to a final design.

The output swing of the amplifier is defined as the maximum and minimum values that can appear on the output of the amplifier. In analog applications, we are concerned with producing the largest swing possible while keeping the output driver, M6, in saturation. This can be determined by inspecting the output stage. If the output voltage exceeds the gate voltage of M6 by more than a threshold voltage, then M6 will go into triode. Thus, since the gate voltage of M6 is defined as $V_{DD} - V_{SG3,4}$, the maximum output voltage will be determined by the size of M3 and M4. The larger the channel width of M3 and M4, the smaller the value of $V_{SG3,4}$ and the higher the output can swing. However, again the tradeoff of making M3 and M4 too large is the reduction in bandwidth due to the increased parasitic capacitance associated with the output of the diff-amp.

The minimum value of the output swing is limited by the gate voltage of M7, which is defined by biasing circuitry. Again using the borderline equation, the drain of M7 may not go below the gate of M7 by more than a V_{THN}. Thus, to improve the swing, the value of V_{GS7} must be made small, which implies that the value of V_{GS8} and V_{GS5} also be made small, resulting in large values of M5, M8, and M7. This is not a very wise option because increasing M5 causes all the PMOS devices to increase by the same factor, resulting in large devices for the entire circuit. By carefully designing the bias device M8, one can design V_{GS8} to be around 0.3 V above VTHN. Thus, the output can swing to within 0.3 V of V_{SS}.

Tradeoff Example

When designing amplifiers, the tradeoff issues that occur are many. For example, there are many effects that occur just by increasing the drain current through the diff-amp: the open-loop gain goes down (by the square root of I_D) while the bandwidth increases (by I_D) due to the fact that the resistors r_{o2} and r_{o4} decrease by $1/I_D$. The overall effect is an increase (by the square root of I_D) in GBW, as predicted earlier. A table summarizing the various tradeoffs that occur from attempting to increase the DC gain can be seen in Table 3.1. It is assumed that if a designer only takes the one action listed that the following secondary effects will occur. In fact, the designer should understand the secondary effects well enough to take a counteraction to offset the secondary effects.

The key for the entire circuit design is the size of M5; the remaining transistors can be written as factors of W_5. The minimum amount of current flowing through M5 is determined by the slew rate. Since M3 and M4 carry half the current of M5, then the widths of M3 and M4 can be determined by assuming that $V_{SG3} = V_{SG4} = V_{GS5}$.

$$\frac{2I_{D3,4}}{I_{D5}} \approx \frac{2K_P \cdot (W/L)_{3,4} \cdot (V_{SG3,4} - |V_{THP}|)^2}{K_n \cdot (W/L)_5 \cdot (V_{GS5} - V_{THN})^2} \approx \frac{2K_P \cdot (W/L)_{3,4}}{K_n \cdot (W/L)_5} \tag{3.58}$$

and since $L_3 = L_5$, and $K_n = 3K_p$, then that leads to the conclusion that $W_{3,4} = 1.5 \cdot W_5$. If the nulling resistor is used in the compensation network, the values for M6 and M7 are determined by the amount of load capacitance attached to the output. If a large capacitance is present, the widths of M6 and M7 will need to be large so as to provide enough sinking and sourcing current to and from the load capacitor. Suppose

TABLE 3.1　Tradeoff Issues for Increasing the Gain of the Two-Stage Op-Amp

Desire	Action(s)	Secondary effects
Increase DC gain	Increase W/L1,2	Decreases phase margin
		Increases GBW
		Increases CMRR
		Decreases CMR
	Decrease ID5	Decreases SR
		Increases CMR
		Increases CMRR
		Increases phase margin
	Increase W/L6	Increases phase margin
		Increases output swing
	Decrease ID6	Decreases output current drive
		Decreases phase margin

it was decided that the amount of current needed for M6 and M7 was twice that of M5. Then, W_7 would be twice as large as W_5, and W_6 would be six times larger than W_5 to account for the differing K values. Alternatively, if everything is saturated, then $I_{D3} = I_{D4}$, and the drain voltage at the output of the diff-amp is identical to the drain voltage of M3. This implies that under saturation conditions, the gate of M6 is at the same potential as the gate of M4; thus, again it must be emphasized that we are talking about quiescent conditions here, and the current through M6 will be defined by the ratio of M6 to M4. Therefore, since M6 is carrying four times as much current as M3, then $W_6 = 4W_{3,4}$, which is six times the value of W_5. Thus, every device except M1 and M2 is written in terms of M5.

The sizes of M1 and M2 are the most critical of the amplifier. If $W_{1,2}$ are made too large, then C_C will be large due to its relationship with C_L and the ratio of g_{m1} and g_{m6} [Eq. (3.48)]. However, if $W_{1,2}$ is made too small, the gain may not meet the requirements needed.

One word about high-impedance nodes. If two current sources are in a series as shown in Fig. 3.33(a), then the value of the voltage between them is difficult to predict. In the ideal case, both current sources have infinite impedances, so any slight mismatch between I_1 and I_2 will result in large swings in v_A. The same holds true for Fig. 3.33(b). Since the two devices form a high impedance at the output, and each device can be considered a current source, any mismatches in the currents defined by $\beta_2(V_{SG2} - V_{THP})^2$ and $\beta_1(V_{GS1} - V_{THN})^2$ will result in large voltage offsets at the output, with the device with the larger defined current being driven into triode. Thus, the smaller of the two defined currents will be the one flowing through both devices. Another way to visualize this condition is to place a large resistor representing the output impedance from v_o to ground. Any difference between the two transistor currents will flow into or out of the resistor, creating a large voltage offset. Feedback is typically used around the op-amp to stabilize the DC output value.

A Word about Circuit Simulation

Circuit simulators have become powerful design tools for analysis of complicated analog circuits. However, the designer must be very careful about the role of the simulator in the design. When simulating high-gain amplifier circuits, it is important to understand the trends and inner working of the circuit before simulations begin. One should always *interpret* rather than blindly trust the simulation results (the latter is a guaranteed recipe for disaster!). For example, the previously mentioned high-impedance nodes should always be given careful consideration when simulating the circuit. Because these nodes are highly dependent on λ, predicting the actual DC value of the nodes either throughsimulation or hand analysis is a near impossibility. Before any AC analysis is performed, check the values of the DC points in the circuit to ensure that every device is in saturation. Failure to do so will result in very wrong answers.

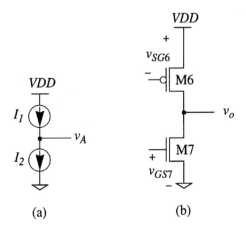

FIGURE 3.33 Illustration of a high-impedance node.

Other Output Stages

With the preceding design, the output stage was a high-impedance driver, capable of handling only capacitive loads. If a resistive load is present, an additional stage should be added that has a low output impedance and high current drive capability. An example of the output stage can be seen in Fig. 3.34. Here, the output impedance is simply $1/g_{m9}||1/g_{m10}$. Since we do not wish to have a large output impedance, the values for L_9 and L_{10} should be made as small as possible. The transistors M11 and M12 are used to help bias the output devices such that M9 and M10 are just barely on under quiescent conditions. This kind of amplifier is known as a class AB output stage and has limitations in CMOS due to the body effect.

In some cases, it is advantageous to use a bipolar output driver as seen in Fig. 3.35. Since most BiCMOS processes provide only one flavor of BJT (an npn), the transistor Q1 can be used for extra current drive. This results in a dual-sloped transfer curve characteristic as the output stage goes from sourcing to sinking. It should be noted that one could use this output stage with the complementary version of the two-stage amplifier previously discussed.

Another BiCMOS output stage can be seen in Fig. 3.36.[5] This is known as a "pseudo-push-pull" output stage. M6 and M7 can be output of the previously discussed two-stage op-amp. With the new pseudo-push-pull output attached, the amplifier is able to achieve high output swing with large current drive

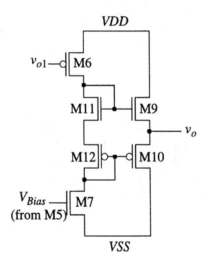

FIGURE 3.34   A low-impedance output stage.

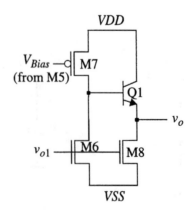

FIGURE 3.35   Using an npn BJT as an output driver.

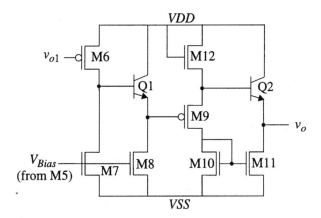

FIGURE 3.36 A "pseudo" push-pull npn only output driver.

capability using very little silicon area. The transistor MQ1 is for level shifting purposes only. When the output signal needs to swing in the positive direction, Q2 behaves just like the BJT output driver shown in Fig. 3.35. When the output swings in the negative direction, Q1 drives the gate of M9 down, thus increasing I_{D9}. The increase in current is mirrored via M10 to M11, which is able to sink a large amount of current. The output voltage at its lowest is approximately the same as the emitter voltage of Q2. The transistor Q2 provides the needed low output impedance.

Another advantage of using BJT devices in the design is to provide large g_ms as compared to the MOS counterpart. BiCMOS circuits can be constructed, which offer high input impedance and gain bandwidth products. If both npn and pnp devices are available in the BiCMOS process, then the output stage seen in Fig. 3.37 can be used. This circuit functions as a buffer circuit with very low output impedance.

High-Performance Operational-Amplifier Considerations

In the commercial world, it seems there are not many applications that require simple, easy-to-design op-amps. High bit rate communication systems and over-sampled data converters push the bandwidth and slew rate capabilities of CMOS op-amps, while battery-powered systems are required to squeeze just enough performance out of micro-amps of current, and a volt or two of power supply. Sensor interfaces and audio systems demand low noise and distortion. To further complicate things, CMOS analog circuits are often integrated with large digital circuits, making isolation from switching noise a major concern.

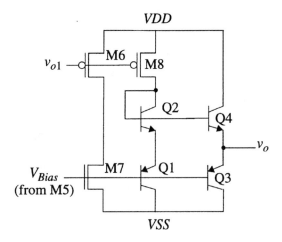

FIGURE 3.37 A low-impedance output stage using both npn and pnp devices.

This section will present solutions to some of the problems faced by op-amp designers who, because of budget constraints or digital compatibility, do not have the option to use bipolar junction transistors in their design. We will then give some hints on where the designer might use bipolar transistors if they are available.

Power Supply Rejection

Fully differential circuits like the OTA shown in Fig. 3.38 are used in mixed signal circuits because they provide good rejection of substrate noise and power supply noise. As long as the noise coupled from the substrate or power supply is equal for both outputs, the difference between the two signals is noise-free (differential component of the noise is zero). This is illustrated in Fig. 3.39. The top two traces in Fig. 3.39 are a differential signal corrupted with common-mode noise. The bottom trace is the difference between these two noisy signals. If the next circuit in the path has good common-mode rejection, the substrate and power supply noise will be ignored. In practical circuits, mismatches between the transistors of symmetrical halves of the differential circuit will lead to imperfect matching of noise on the outputs, and therefore reduced rejection of power supply noise. Common centroid layouts and large geometry transistors are necessary to minimize mismatches. Differential circuits are capable of twice the signal swing of single-ended circuits, making them especially welcome in low-voltage and low-noise applications.

Single-stage or multiple-stage op-amps can be made differential, but each stage requires a common-mode feedback circuit to give the differential output a common-mode reference. Consider the folded cascode OTA shown in Fig. 3.38. If the threshold voltage of M5A is slightly larger than the threshold voltage of M5B and M5C, the pull-down currents will be larger than the pull-up currents. This small current difference, in combination with the very high output impedance of the cascode current mirrors, will cause the output voltages to be pegged at the negative power supply. This common-mode error cannot be corrected by applying feedback to the differential pair. A common-mode feedback circuit is needed to find the average of the output voltages, and control the pull-up or pull-down current in the outputs to maintain this average at the desired reference. A center-tapped resistor between the outputs could be used to sense the common-mode voltage if the outputs were buffered to drive such a load. Since a folded cascode OTA cannot drive resistors, a switched capacitor would be a better choice to sense the common-mode voltage as shown in Fig. 3.40. The PH1 and PH2 clock signals must be non-overlapping.

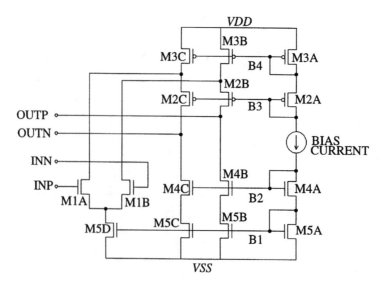

FIGURE 3.38   Folded cascode OTA.

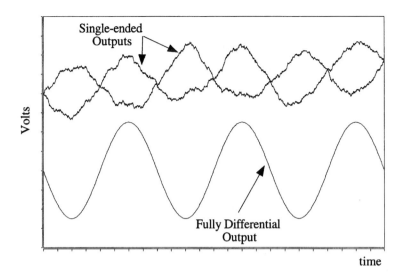

FIGURE 3.39 Simulation output illustrating the difference between single-ended and fully differential signals.

When the PH1 switches are closed, C_{1A} and C_{1B} are discharged to zero, while C_{2A} and C_{2B} provide feedback to the common-mode amplifier. The PH1 switches are then opened, and a moment later, the PH2 switches closed. The charge transfer that takes place moves the center tap between C_{2A} and C_{2B} toward the average of the two output voltages. After many clock cycles, the input to the common-mode feedback amplifier will be the average of the two voltages. C_{1A} and C_{1B} can be precharged to a bias voltage to provide a level shift. This allows direct feedback from the common-mode circuit to the pull-up bias as shown in Fig. 3.41.

A BiCMOS version of the folded cascode amplifier can be seen in Fig. 3.42.[6] Here, it is again assumed that only npn devices are available. Note that to best utilize the npn devices, the folded cascode uses a diff-amp with P-channel input devices. The amplifier also uses the high swing current mirror presented in Section 3.2.

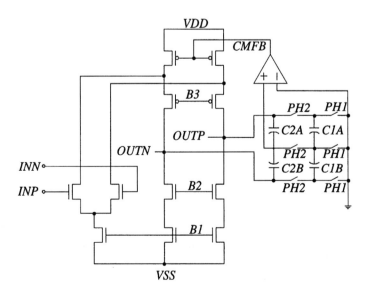

FIGURE 3.40 Folded cascode OTA using switched capacitor common mode feedback.

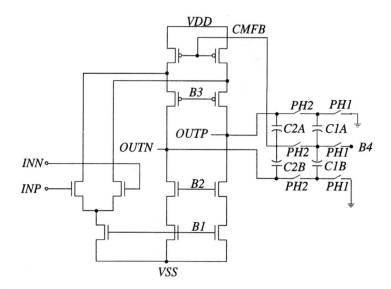

FIGURE 3.41 Using a bias voltage to precharge the switched capacitor common mode feedback capacitors.

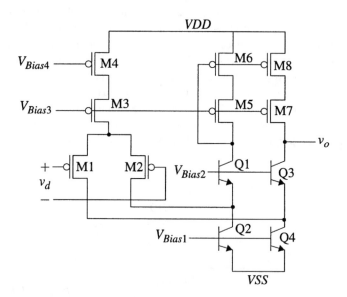

FIGURE 3.42 BiCMOS folded cascode amplifier.

Slew Rate

The slew rate of a single-stage class A OTA is the maximum current the output can source or sink, divided by the capacitive load. The slew rate is therefore proportional to the steady-state power consumed. Class AB amplifiers give a better tradeoff between power and slew rate. The amplifier's maximum output current is not the same as the steady-state current. An example of a class AB single-stage op-amp is shown in Fig. 3.43. The differential pair is replaced by M1A, M1B, M2A, and M2B. A level shifter consisting of M3A, M3B, M4A, and M4B couples the input signal to M2A and M2B, and sets up the zero input bias current. If the width of M2A and M2B are three times the width of M1A and M1B, the small signal voltage on the nodes between M1 and M2 will be approximately zero. The current available from one of the input transistors is approximately

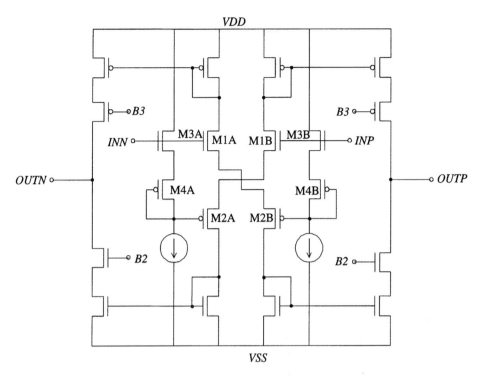

FIGURE 3.43 A class AB single stage op-amp with high slew rate capability.

$$I_{\_out} = KP\frac{W}{L}(V_{in} + V_{bias} + V_{THN})^2 \tag{3.59}$$

The differential current from the input stage is

$$I_{OUTP} - I_{OUTN} = 2 \cdot \beta(V_{in} + V_{BIAS} + V_{THN}) \tag{3.60}$$

It is interesting to note that the non-linearities cancel. The output current becomes non-linear again as soon as one of the input transistors turns off. The maximum current available from the input transistors is not limited by a current source as is a differential pair. It should be noted that it is impossible to keep all transistors saturated for low power supplies and large input common-mode swings. A similar BiCMOS circuit with high slew rate can be seen in Fig. 3.44.[7]

Adaptive bias is another method to reduce the ratio of supply current to slew rate. Adaptive bias senses the current in each side of a conventional differential pair, and increases the bias current when the current on one side falls below a preset value. This guarantees that neither side of the differential pair will turn off. The side that is using most of the current must therefore be supplied with much more than the zero input amount. The differential pair current can be sensed by measuring the gate-to-source voltages as shown in Fig. 3.45, or by measuring the current in the load as shown in Fig. 3.46. Both of these adaptive bias schemes depend on a feedback loop to control the current in the differential pair. These circuits improve settling time for low power, low bandwidth circuits, but the delay in the feedback path is a problem for high-speed circuits.

A second form of adaptive bias can be used when it is known that increased output current is needed just after clock edges, such as in switched capacitor filters. This type of adaptive bias can be realized using the switched capacitor bias boost shown in Fig. 3.47. In this circuit, $I_{D1}{\cdot}L_1/W_1 > I_{D2}{\cdot}L_2/W_2$, which makes $V_1 > V_2$. When the PH1 switches are closed, the steady-state voltage across the capacitor is $V_1 - V_2$, and the current in M3 is set by I_{D2}. When the PH2 switches close, V_2 is momentarily increased, while V_1 is

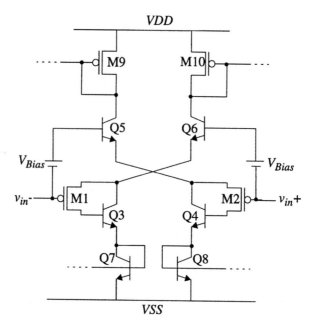

FIGURE 3.44   A BiCMOS class AB input stage.

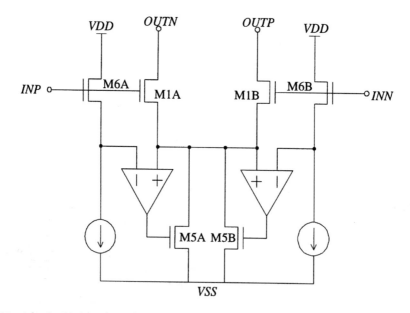

FIGURE 3.45   Adaptive biasing scheme for improved slew rate performance.

decreased. The transient is repeated when the PH1 switches close. The net effect of the switched capacitor bias boost is that the current in M3 increases after both clock edges. Notice that the current is increased, whether it is needed or not. This circuit is fast because there is no feedback loop, but it is not the most efficient because it does not actually sense the current in the differential pair.

In all three approaches, the quiescent current is less than the current when the output is required to slew. Therefore, output swing and gain are not degraded when the bias current returns to its quiescent value. All three adaptive bias circuits will cause larger current spikes to be put on the supplies by the op-amps. The width of the power supply lines should be increased to compensate for the increased *IR* drop

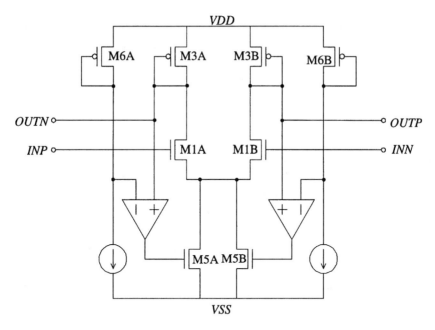

FIGURE 3.46 Another adaptive biasing circuit.

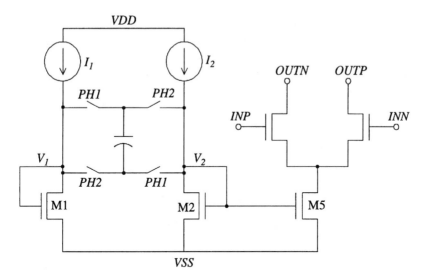

FIGURE 3.47 Adaptive biasing using a switched capacitor circuit.

and crosstalk. Enhanced slew rate circuits are not linear time invariant systems because the transconductance and output impedance of the transistors are bias current dependent, and the bias is time varying. A transient analysis is the most dependable way to evaluate settling time in this case.

Output Swing

Decreasing power supply voltages put an uncomfortable squeeze on the design engineer. To maintain the desired signal-to-noise ratio with a smaller signal swing, circuit impedances must decrease, which often cancels any power savings that may be gained by a lower supply voltage. To get the best signal swing, differential circuits are used. If the output stages use cascode current mirrors, bias voltages must be generated, which keep both the mirror and cascode transistors in saturation with the minimum voltage on the output. Another example of a high swing bias circuit (refer also to Section 3.2) and a current mirror is shown in Fig. 3.48. First, let the W/L ratio of M2, M4, and M6 be equal, and the W/L ratio of M3 and M5 be equal. Now recall that to keep a MOSFET in saturation

$$V_{DS} \geq V_{GS} - V_{THN} \qquad (3.61)$$

The minimum output voltage that will keep both M5 and M6 in saturation with proper biasing is

$$V_{out} \geq V_{GS6} - 2V_{THN} + V_{GS5} \qquad (3.62)$$

ignoring the bulk effects. For a given current, the minimum drain voltage can be rewritten as

$$V_{DS} \geq \sqrt{\frac{I_D \cdot L}{K \cdot W}} \qquad (3.63)$$

The equation for the minimum V_{OUT} can be rewritten as

$$V_{OUT} \geq \sqrt{\frac{I_{OUT} \cdot L_6}{K_6 \cdot W_6}} + \sqrt{\frac{I_{OUT} \cdot L_5}{K_5 \cdot W_5}} \qquad (3.64)$$

The trick to making this bias generator work is setting V_{DS} of M1 equal to the minimum V_{DS} required by M3 and M5. M1 is biased in the linear region, while we wish to keep M3 and M5 saturated. It is a good idea to set $L_1 = L_3 = L_5$ to match etching tolerances. A second trick is to make sure M3 and M4 stay saturated. M4 is inserted between the gate and drain connections of M3 to make $V_{DS3} = V_{DS5}$. If the W/L ratio of M3 is too small, M3 will be forced out of saturation by the source of M4. If the W/L_3 is too large, the gate voltage of M3 will not be large enough to keep M4 in saturation.

DC Gain

Start by calculating the DC gain of the folded cascode OTA in Fig. 3.38. If we assume the output impedance of the n-channel cascode current mirror is much greater than the output impedance of the individual transistors, then the DC gain is approximately

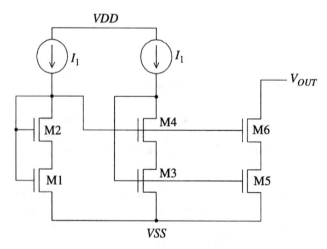

FIGURE 3.48   A high swing biasing circuit for low power supply applications.

$$\frac{v_o}{v_{in}} = -g_{m1} \cdot r_{o1} || r_{o3} \cdot g_{m2} \cdot r_{o2} \tag{3.65}$$

If we assume that the current from the M3 splits equally to M1 and M2, the gain can be written as

$$\frac{v_o}{v_{in}} \propto \frac{\sqrt{W_1 L_1 W_2 L_2}}{I_{D1}} \cdot \frac{L_3}{2L_1 + L_3} \tag{3.66}$$

We can see that the gate area of the differential pair and cascode transistors must both double each time current is doubled to maintain the same gain. We also note that it is desirable to make $L_3 > L_1$. If the current in the amplifier were raised to increase gain-bandwidth, or slew rate, it would be desirable to increase the widths of the transistors by the same factor to maintain output swing.

Regulated gate cascode outputs increase the gain of the OTA by effectively multiplying the g_m of M4 by the gain of the RGC amplifier. The stability of the RGC amplifier loop must be considered. An example of a gain boosted output stage is shown in Fig. 3.49.

Gain Bandwidth and Phase Margin

Again, start with the transfer function for the folded cascode OTA of Fig. 3.38. If we assume the output impedances of the n-channel cascode current mirrors are very large, and the gain of M2 is much greater than one, we have

$$\frac{v_o}{v_{in}} = \frac{g_{m1} r_{o1} g_{m2} r_{o2}}{(C_1 \cdot C_{out} \cdot r_{o1} \cdot r_{o2}) \left(s^2 + \left(\frac{g_{m2}}{C_1} + \frac{1}{C_1 \cdot r_{o1}} + \frac{1}{C_1 \cdot r_{o2}} + \frac{1}{C_{out} \cdot r_{o2}}\right)s + \frac{1}{C_1 \cdot C_{out} \cdot r_{o1} \cdot r_{o2}}\right)} \tag{3.67}$$

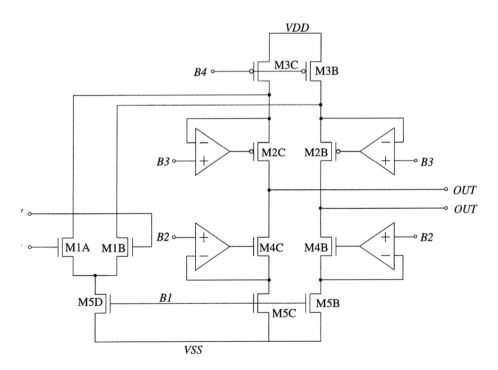

FIGURE 3.49 OTA with regulated gate cascode output.

where ro_1 is now the parallel combination of r_{o1} and r_{o3}. If we further assume that the poles are spaced far apart, and that g_{m2} is much larger than $1/r_{o1}$ and $1/r_{o2}$, then the gain-bandwidth product is g_{m1}/C_{load}. The second pole, which will determine phase margin, is approximately

$$\omega = \frac{g_{m2}}{C_1} + \frac{1}{C_{out} \cdot r_{o2}}$$

The depletion capacitance of the drains of M1 and M3 will also add to this capacitance. As a first cut, let $C_1 = K_c \cdot W_2 \cdot L_2$. Now the equation for the second pole boils down to

$$\omega \approx \frac{\sqrt{K \cdot I_{D2}}}{K_C \cdot L_2 \cdot \sqrt{W_2 \cdot L_2}} + \frac{I_{D2}}{L_2 \cdot C_{out}}$$

To get maximum phase margin, we clearly want to use as short a channel length as the gain specification will allow. The folded cascode OTA and two-stage OTA both have n-channel and p-channel transistors in the signal path. Since holes have lower mobility than electrons, it is necessary to make a silicon p-channel transistor about three times wider than an n-channel transistor of the same length to get the same transconductance. The added parasitic capacitance of the wider p-channel transistor is a hindrance for high-speed design. The telescopic OTA shown in Fig. 3.50 has only n-channel transistors in the signal path, and can therefore achieve very high bandwidths with acceptable phase margin. Its main drawback is that the output common-mode voltage must be more positive than the input common-mode voltage. This amplifier can achieve even wider bandwidth with acceptable phase margin if M2 is replaced by an npn bipolar transistor.

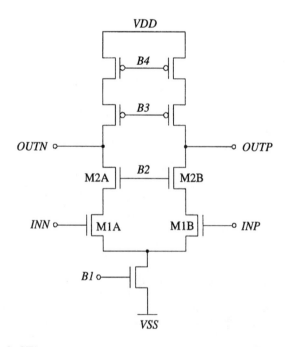

FIGURE 3.50   A telescopic OTA.

References

1. R. J. Baker, H. W. Li, and D. E. Boyce, *CMOS: Circuit Design, Layout, and Simulation*, IEEE Press, 1998.
2. A. S. Sedra and K. C. Smith, *Microelectronic Circuits, fourth edition*, Oxford University Press, London, 1998.
3. P. E. Allen and D. R. Holberg, *CMOS Analog Circuit Design*, Saunders College Publishing, Philadelphia, 1987.
4. P. R. Gray, *Basic MOS Operational Amplifier Design — An Overview, Analog MOS Integrated Circuits*, IEEE Press, 1980.
5. H. Qiuting, A CMOS power amplifier with a novel output structure, *IEEE J. Solid-State Circuits*, vol. 27, no. 2, pp. 203-207, Feb. 1992.
6. M. Ismail, and T. Fiez, *Analog VLSI: Signal and Information Processing*, McGraw-Hill, Inc., New York, 1994.
7. S. Sen and B. Leung, A class-AB high-speed low-power operational amplifier in BiCMOS technology, *IEEE J. Solid-State Circuits*, vol. 31, no. 9, pp. 1325-1330, Sept. 1996.

4

Bipolar Amplifier Design

Geert A. De Veirman
Silicon Systems, Inc.

4.1   Introduction

This chapter gives an overview of amplifier design techniques using bipolar transistors. An elementary understanding of the operation of the bipolar junction transistor is assumed. Section 4.2 reviews the basic principles of amplification and details the proper selection of an operating point. This section also introduces the three fundamental single-transistor amplifier stages: the common-emitter, the common-collector, and the common-base configurations. Section 4.3 deals with the problem of amplification of dc and difference signals. The emitter-coupled differential pair is discussed in great depth. Issues specific to output stages are presented in Section 4.4. Section 4.5 briefly touches on supply-independent biasing techniques. Next, Section 4.6 combines all the acquired building block knowledge in a condensed overview of operational amplifiers. A short conclusion is presented in Section 4.7, followed by a list of references.

4.2   Single-Transistor Amplifiers

Basic Principles

Bipolar Transistor Operation

Although prior exposure to bipolar transistor fundamentals is expected, a few elementary concepts are briefly reviewed here. The *bipolar junction transistor* (BJT) contains two back-to-back pn junctions,

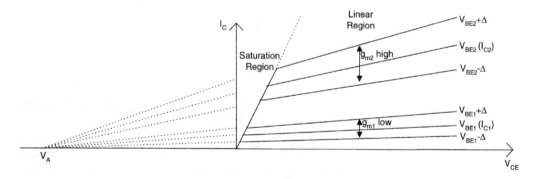

FIGURE 4.1 $I_C(I_{through})$ vs. $V_{CE}(V_{across})$ characteristics of the npn bipolar junction transistor.

known as the base–emitter and base–collector junctions. In this chapter, it is assumed that the BJT is operated in the so-called normal mode with a forward-bias applied at the base–emitter junction and a reverse voltage across the base–collector interface. Then, the collector current I_C, which is β times the base current I_B, is exponentially related to the base–emitter voltage V_{BE}. The mathematical expression of this dependency is known as the *Ebers–Moll equation*. For small signal variations, the relationship between the current *through* the transistor (I_C) and the *control input* voltage (V_{BE}) is expressed by the *transconductance* g_m.

When designing bipolar transistor amplifiers, it is also essential to understand the relationship between I_C and the voltage *across* the transistor (the collector–emitter voltage V_{CE}). For each input voltage V_{BE}, a different characteristic exists in the I_C–V_{CE} diagram, as shown in Fig. 4.1. As long as V_{CE} is high enough to keep the transistor in the linear region, the I_C vs. V_{CE} characteristics are fairly flat. When V_{BE} is held constant, increasing V_{CE} corresponds to applying more reverse voltage across the base–collector junction. As a result, the base width is modulated and this causes the I_C curves to increase approximately linearly with V_{CE}. When the I_C curves are extrapolated to negative V_{CE} values, they eventually all go through a single pivot point, which is referred to as the transistor's *Early voltage* V_A. For very small values of V_{CE}, the base–collector junction becomes forward-biased. As a result, the transistor saturates and the collector current drops off rapidly.

In Fig. 4.1, V_{BE1} is less than V_{BE2}. In the linear operating region, the exponential nature of the BJT renders $g_{m1} < g_{m2}$, which means that for an identical incremental input voltage Δ, I_{C2} changes by a larger amount than I_{C1}. Also, since both curves merge at V_A, I_{C2} is a stronger function of V_{CE} than I_{C1}.

Mathematical expressions for the relationships discussed above will be introduced as needed in the following sections.

Basic Bipolar Amplifier Concepts

In order to amplify a voltage with a bipolar transistor, a load is placed in series with the device. In its simplest form, this load consists of a resistor R_L. Taking into account the polarity of electrons and the direction of current flow, one ends up with a transistor and supply arrangement as shown in Fig. 4.2, where the voltage to be amplified (V_i) drives the base. Hence, the only available output node (V_o) in Fig. 4.2 is the voltage between the transistor and the load resistor. This voltage can be determined in the I_C–V_{CE} diagram of Fig. 4.1 by subtracting the voltage drop across R_L from the supply voltage V_{CC}. The result is a straight line, which is known as the *load line*. The load line, shown in Fig. 4.3, graphically represents the possible operating points of the bipolar transistor. The line is straight simply because the resistor R_L is a linear element. Every point on the load line represents a state, which is determined by the voltage at the transistor's base. This state in turn determines how much current flows through the transistor and how much voltage there exists across this device.

To determine the circuit's voltage amplification, all one has to do is to plot the output voltage as a function of the input voltage. For different values of R_L, this process results in the *transfer characteristics* depicted in Fig. 4.4. The transfer characteristics allow us to visually determine two important items: first,

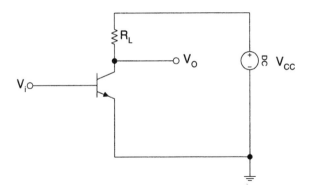

FIGURE 4.2 Conceptual schematic of single-transistor amplifier.

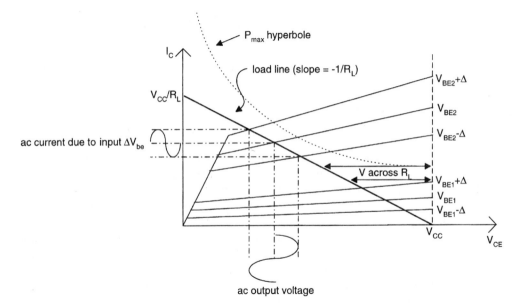

FIGURE 4.3 Load line characteristic.

the range in which the input voltage V_i should lie in order to obtain the maximum possible voltage variation at the output; and second, the optimum value of R_L. Since the transfer characteristics in Fig. 4.4 are far from linear, the best one can do is to identify a region where the curves can be linearized, provided the signal swings are not too large. Also, note that the transfer curves do not pass through the (0,0) coordinate. This means that the desired proportionality between input and output is not present. Consequently, one would want to move the V_o and V_i axes, so that their origin coincides with a desired point on the curves, as is the case for the V_o'/V_i' coordinate system illustrated in Fig. 4.4. The next subsection will discuss how such a suitable *operating point* (or *quiescent point*) can be established.

Figure 4.4 furthermore graphically clarifies how the bipolar transistor can be used in three distinct ways: first, as a switch; second, as a linear amplifier for small signals; and third, as a linear amplifier for large signals. When operating the BJT as a switch, the (logic) designer is primarily concerned about the speed of the transitions between the OFF and ON states. Such obviously non-linear modus operandi is a topic outside the scope of this chapter. Large signals pose particular difficulties, which will be covered a bit later in the sections on differential pairs and output stages. In the next several sections, we will deal mainly with linear small-signal operation.

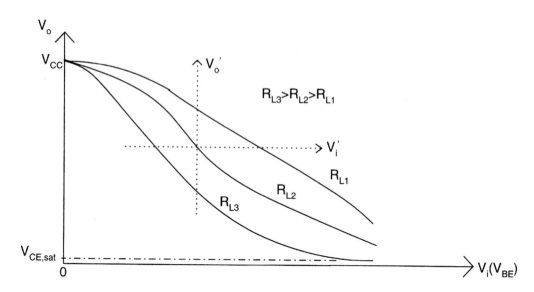

FIGURE 4.4   Transfer characteristics.

Setting a Suitable Operating Point

The choice of a proper operating point depends on several factors. Three key decision criteria are:

1. Maximum allowable power dissipation of the active element. Each transistor has a maximum power limit. If this limit is exceeded, permanent damage (e.g., degradation of β) occurs or, in the worst case, the device may be destroyed. (Actually, maximum power ratings can be temporarily exceeded, as long as such transgressions are very fast. But such operation goes beyond this discussion.)   In the I_C–V_{CE} diagram, constant power curves are represented by hyperboles. One such hyperbole, representing the maximum allowable power dissipation P_{max}, is included in Fig. 4.3. The designer must guarantee that excursions from the operating point along the amplifier's load line do not cross into the forbidden zone above the P_{max} curve.
2. Proper location on the transfer curve. As mentioned above during the discussion of Fig. 4.4, this choice directly affects the achievable linearity and signal swing.
3. Bias current. Figure 4.5 illustrates how the vertical axis in the V_o/V_i diagram can be moved. All one has to do is apply a proper dc bias voltage and superpose a small signal input. Making use of a separate bias voltage, as shown in Fig. 4.5(a), is one possibility. Alternatively, the dc voltage at the base can be set by means of a resistive divider from V_{CC}, and the ac signal can be capacitively coupled through C_{ci}, as depicted in Fig. 4.5(b). Similarly, a coupling capacitor C_{co} can be employed to separate the dc bias at V_o from the ac signal swing. As such, C_{co} moves the horizontal axis in the V_o/V_i diagram.

In Fig. 4.6(a), a second load resistor R_L' is added. Since the capacitor C_{co} acts as an open circuit for dc signals, the operating point remains solely determined by the intersection of the chosen transistor characteristic and the *static* load line, which results from the combination of V_{CC} and R_L. Assuming C_{co} is large, this coupling capacitor behaves as a short-circuit for ac signals. Hence, R_L' is effectively in parallel with R_L. Therefore, signal excursions occur along a new *dynamic* load line, as illustrated in Fig. 4.6(b).

Common-Emitter Amplifier

The circuits in Figs. 4.5 or 4.6(a) are known as *common-emitter amplifiers*. This name is derived from the fact that the emitter terminal is common with the ground node.

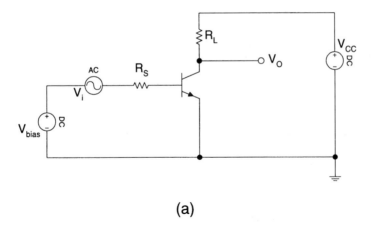

(a)

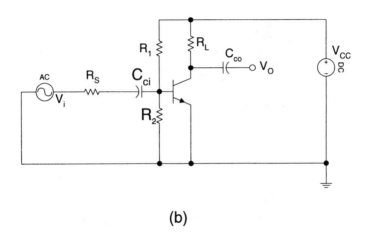

(b)

FIGURE 4.5   Common-emitter amplifier: (a) V_{bias} in series with ac signal source, (b) arrangement with coupling capacitors.

Small-Signal Gain

The small-signal equivalent circuit of the common-emitter amplifier is shown in Fig. 4.7. This circuit is derived based on the observation that for ac signals, fixed dc bias sources are identical to ac ground. We have also assumed (for now) that the coupling capacitors C_{ci} and C_{co} are so large that for any frequencies of interest, they effectively behave as short-circuits. Moreover, parasitic capacitances internal to the transistor are ignored. R_L' has been eliminated for simplicity (or one could assume that R_L represents the effective parallel resistance). R_S is the series output resistance of the ac source V_i and r_b stands for the physical resistance of the base. $R_{.\pi}$ models the linearized input voltage–current characteristic of the base–emitter junction. In other words, r_π is the ratio between v_{be} and i_b for small-signal excursions from the operating point set by V_{BE}. Although the parallel combination of R_1 and R_2 shunts r_π, we assume here that both resistors have such high ohmic values that their effect on r_π can be regarded as immaterial. From the Ebers–Moll equation (with $V_{BE} \gg V_T$),

$$I_C = \beta I_B = I_S \exp\left(\frac{V_{BE}}{V_T}\right)\left(1 + \frac{V_{CE}}{V_A}\right) \qquad (4.1)$$

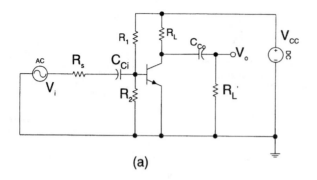

(a)

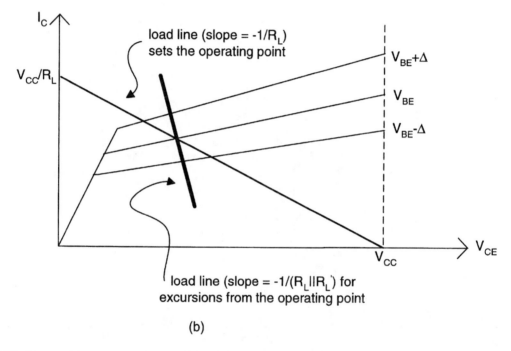

(b)

FIGURE 4.6   (a) Common-emitter amplifier with ac coupled load R'_L, (b) I_C vs. V_{CE} diagram with static and dynamic load lines.

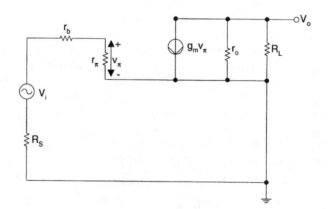

FIGURE 4.7   Small-signal equivalent circuit for the common-emitter amplifier.

where I_S is the transistor's saturation current and $V_T = kT/q$ the thermal voltage, one readily derives the following expressions for the transconductance g_m, the input resistance r_π, and the output resistance r_o.

$$g_m = \frac{dI_C}{dV_{BE}} = \frac{I_C}{V_T} \tag{4.2}$$

$$r_\pi = \left(\frac{dI_B}{dV_{BE}}\right)^{-1} = \left(\frac{I_C}{\beta V_T}\right)^{-1} = \frac{\beta}{g_m} \tag{4.3}$$

$$r_o = \left(\frac{dI_C}{dV_{CE}}\right)^{-1} = \left(\frac{I_S}{V_A}\exp\left(\frac{V_{BE}}{V_T}\right)\right)^{-1} \tag{4.4}$$

In general, $V_A \gg V_{CE}$. Therefore,

$$r_o \approx \frac{V_A}{I_C} \tag{4.5}$$

The following two equations can be written for the circuit in Fig. 4.7.

$$v_\pi = V_i \frac{r_\pi}{r_b + R_S + r_\pi} \tag{4.6}$$

$$V_o = -g_m v_\pi (r_o \| R_L) \tag{4.7}$$

where $\|$ denotes the parallel combination of two resistors. Combining Eqs. (4.6) and (4.7) leads to the expression for the small-signal gain

$$A = \frac{V_o}{V_i} = -g_m \frac{r_\pi}{r_b + R_S + r_\pi} \frac{r_o R_L}{r_o + R_L} \tag{4.8}$$

Assuming $r_b + R_S \ll r_\pi$ and $R_L \ll r_o$,

$$A \approx -g_m R_L \tag{4.9}$$

Equation (4.9) implies that selecting a higher-valued load resistor R_L results in higher gain. However, this conclusion is only valid as long as $R_L \ll r_o$. Conversely, for a given R_L, the gain can be increased by choosing a higher g_m. According to Eq. (4.2), this corresponds to a higher I_C. However, r_o goes down with increasing I_C (see Eq. (4.5) and Fig. 4.1), causing some reduction in gain. Furthermore, assuming β is constant over the range of operating points, Eq. (4.3) specifies that r_π also decreases with increasing g_m, augmenting the reduction in gain. To compound the problem, β actually starts rolling off at high current levels.

The best choice of operating point is most often determined by ensuring the largest possible output signal swing. In that case,

$$I_C R_L = \frac{V_{CC} - V_{CE, sat}}{2} \approx \frac{V_{CC}}{2} \tag{4.10}$$

Under this condition, the gain expression reduces to

$$A \approx -\frac{V_{CC}}{2V_T} \qquad (4.11)$$

Since V_T is about 25 mV at room temperature, a 5-V supply yields a gain A approximately equal to 100.

The apparent paradox (a high R_L limits g_m and vice versa) could be circumvented if a scheme were conceived that combined a static low-R_L load line (allowing a high bias current and hence g_m) with a dynamic high-R_L load line. One potential solution is to steer some dc current away from R_L by means of a current source, as illustrated in Fig. 4.8(a). Figure 4.8(b) depicts the corresponding vertical movement of the load line. The current source can be made with a pnp transistor and three resistors, as shown in Fig. 4.8(c). Why this particular configuration makes a good current source will become clear later from the discussion of the reciprocal npn circuit.

The logical next step consists of the complete removal of R_L and replacing this *passive* component with an *active* pnp load. A conceptual drawing, including a current generator and mirror, is given in Fig. 4.9(a). Figure 4.9(b) shows that the operating point is now determined by the intersection of the npn and pnp I_C–V_{CE} curves. For the amplifier in Fig. 4.9(a),

$$A \approx -g_m(r_{o,n} \| r_{o,p}) \approx -\frac{I_C}{V_T}\left(\frac{V_{A,n}}{I_C} \| \frac{V_{A,p}}{I_C}\right) \approx -\frac{V_{A,n}V_{A,p}}{V_T(V_{A,n}+V_{A,p})} \qquad (4.12)$$

Similar to Eq. (4.11), the gain expressed by Eq. (4.12) is independent of the bias current I_C. Furthermore, A no longer depends on the supply voltage V_{CC}. The circuit in Fig. 4.9(a), however, is only conceptual in nature. In practice, it is dc unstable. Indeed, due to the small slopes of the intersecting transistor characteristics (or, equivalently, the high output resistances of both transistors), minor mismatches between the two currents, small temperature differences, or simply basic variations in processing will cause a sizable shift of the operating point. Intuitively, some kind of feedback network is needed to guarantee quiescent stability. The discussion of such circuitry is deferred to Section 4.3 and Fig. 4.32.

Stabilizing the Common-Emitter's Operating Point

Up to this point, we have (implicitly) assumed that the operating point is set by means of a fixed dc voltage V_B at the base of the transistor. However, due to the exponential nature of the BJT, such arrangement leaves both I_C and the operating point very sensitive to even small changes in this base voltage. Indeed,

$$\frac{dI_C}{dV_{BE}} = \frac{d\left(I_S \exp\left(\frac{V_{BE}}{V_T}\right)\right)}{dV_{BE}} = \frac{I_S}{V_T}\exp\left(\frac{V_{BE}}{V_T}\right) = \frac{I_C}{V_T} \qquad (4.13)$$

Therefore, $\Delta I_C/I_C = \Delta V_{BE}/V_T = \Delta V_B/V_T$. At room temperature, a 1-mV change in V_B translates into a 4% collector current change. Furthermore, the strong temperature dependence of V_T as well as I_S requires V_B to decrease by about 2 mV per degree rise in temperature if a constant current I_C is to be maintained. Rather than fixing the voltage V_B at the base, one could try to force a fixed base current I_B and rely on the $I_C = \beta I_B$ relationship to set the collector current. However, β is strongly temperature dependent as well as subject to wide process variations. Moreover, β varies according to the transistor's current bias.

Consequently, the only way a stable operating point can be secured is by fixing the emitter current I_E. Since

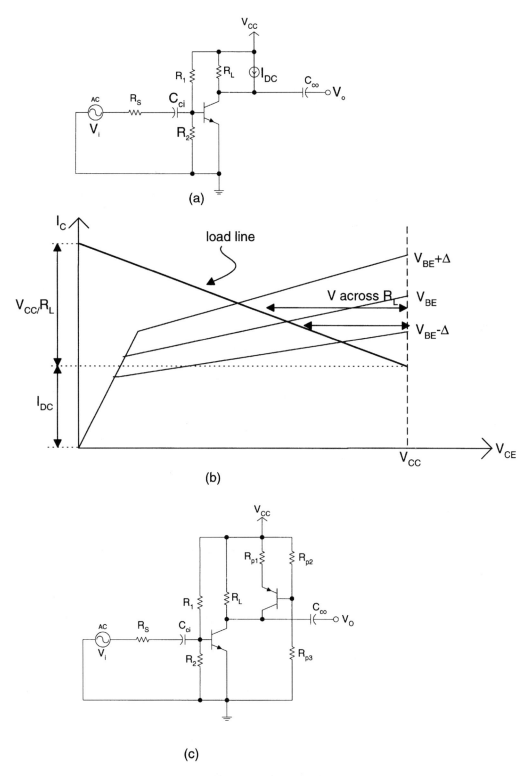

FIGURE 4.8 (a) Common-emitter amplifier with dc current bypass of R_L; (b) I_C vs. V_{CE} diagram with load line; (c) current source implemented by means of pnp and three resistors.

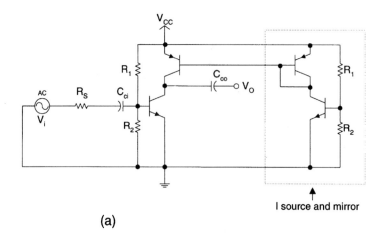

(a)

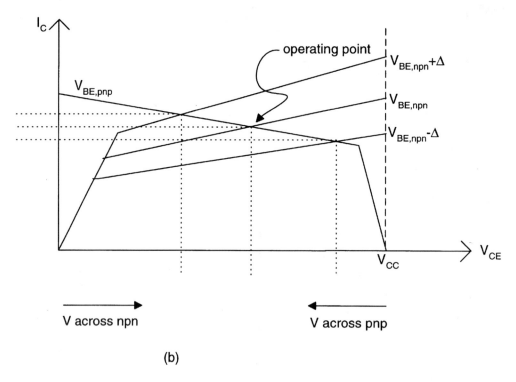

(b)

FIGURE 4.9   (a) Common-emitter amplifier with active load; (b) I_C vs. V_{CE} diagram.

$$I_C = \alpha I_E = \frac{\beta}{\beta + 1} I_E \approx I_E \qquad (4.14)$$

fixing I_E yields a predictable collector current I_C. In discrete realizations, the value of R_L is well-determined (e.g., better than 1%). In integrated circuits (ICs), on the other hand, resistors are subject to wide absolute value tolerances. However, resistor ratios are usually tightly controlled. If one is able to set $I_C = V_{ref}/R_C$,

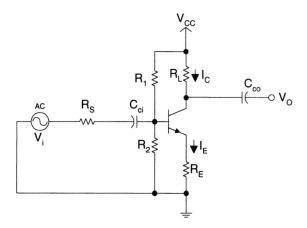

FIGURE 4.10 Common-emitter amplifier with degeneration resistor.

where V_{ref} is a well-defined voltage, such as a bandgap reference (see Section 4.5), the product $I_C R_L$ only depends on V_{ref} and a resistor ratio. Thus, the operating point can be firmly established.

A simple way to set I_E consists of inserting a resistor R_E in series with the emitter, as shown in Fig. 4.10. The insertion of R_E introduces negative feedback, which makes the biasing circuit self-correcting. By inspection,

$$V_B = V_{BE} + R_E I_E \approx V_{BE} + R_E I_C \qquad (4.15)$$

If the base current I_B is assumed to be much smaller than the current through the resistive divider network consisting of R_1 and R_2, then

$$V_B \approx \frac{R_2}{R_1 + R_2} V_{CC} \qquad (4.16)$$

Combined with the Ebers–Moll identity

$$V_{BE} = V_T \ln\left(\frac{I_C}{I_S}\right) \qquad (4.17)$$

Eqs. (4.15) and (4.16) can be solved for I_C. However, no closed-form solution is possible and numerical iterations must be performed. Intuitively, when I_C increases, the voltage drop across R_E also increases. Since V_B is constant, V_{BE} must go down, forcing I_C to decrease instead. Conversely, when I_C begins to decrease, the voltage across R_E goes down, V_{BE} is increased, which in turn leads to a reversal in the direction of the change in I_C. Eventually, an equilibrium point is reached. The dc voltage at the output can be expressed as

$$V_{o,dc} = V_{CC} - I_C R_C = V_{CC} - (V_B - V_{BE})\frac{R_L}{R_E} \qquad (4.18)$$

Apparently, the arrangement of R_1, R_2, R_E and the npn transistor approximates a current source quite well. The approximation more closely approaches an ideal current source as the voltage drop across R_E is increased. In practice, however, the available supply voltage V_{CC} is fixed (or at least limited) and the

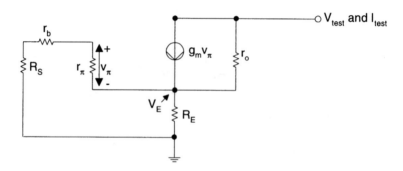

FIGURE 4.11 Small-signal equivalent circuit for calculation of the output resistance of the common-emitter configuration with degeneration.

emitter degeneration voltage subtracts from the achievable signal swing. Consequently, an acceptable compromise must be made.

The fact that the transistor's collector current is more or less constant, independent of the voltage at the output, implies that the configuration possesses a high output impedance. A mathematical expression for the output resistance of the degenerated common-emitter configuration can be derived from the small-signal equivalent circuit depicted in Fig. 4.11. If a test voltage V_{test} and a test current I_{test} are applied, the following equations can be written

$$I_{test} = g_m v_\pi + \frac{V_{test} - V_E}{r_o} \tag{4.19}$$

$$I_{test} = \frac{V_E}{R_E} + \frac{V_E}{r_b + R_S + r_\pi} \tag{4.20}$$

$$v_\pi = -\frac{r_\pi}{r_b + R_S + r_\pi} V_E \tag{4.21}$$

The output resistance R_o is simply the ratio of V_{test} and I_{test}. Thus,

$$R_o = \frac{V_{test}}{I_{test}} = r_o \left(1 + \frac{g_m r_\pi R_E}{r_b + R_S + r_\pi + R_E} + \frac{R_E(r_b + R_S + r_\pi)}{r_o(r_b + R_S + r_\pi + R_E)} \right) \tag{4.22}$$

Assuming typical values for each of the resistances, Eq. (4.22) can be simplified to

$$R_o \approx r_o \left(1 + g_m R_E + \frac{R_E}{r_o} \right) \approx r_o (1 + g_m R_E) \tag{4.23}$$

In conclusion, the applied degeneration significantly boosts the transistor's output resistance r_o, more specifically by a multiplication factor $(1 + g_m R_E)$. The reader should take note that Fig. 4.11 and Eq. (4.23) only represent the degenerated transistor. The output resistance of the complete degenerated common-emitter amplifier of Fig. 4.10 is the shunt combination of R_o and R_L.

The effect of R_E on the input impedance can similarly be derived from the small-signal equivalent circuit in Fig. 4.12.

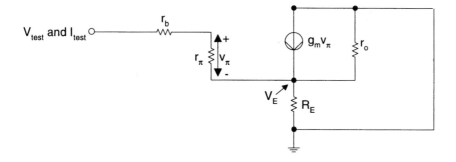

FIGURE 4.12 Small-signal equivalent circuit for calculation of the input resistance of the common-emitter configuration with degeneration.

$$I_{test} = \frac{V_E}{R_E} + \frac{V_E}{r_o} - g_m v_\pi \tag{4.24}$$

$$V_{test} = (r_b + r_\pi)I_{test} + V_E \tag{4.25}$$

$$v_\pi = r_\pi I_{test} \tag{4.26}$$

Combining Eqs. (4.24) through (4.26) yields

$$R_i = \frac{V_{test}}{I_{test}} = r_b + r_\pi + (1 + g_m r_\pi)\frac{r_o R_E}{r_o + R_E} \approx r_b + r_\pi + (\beta + 1)R_E \tag{4.27}$$
$$\approx r_\pi + \beta R_E \approx r_\pi(1 + g_m R_E)$$

The desirable multiplication factor found in Eq. (4.23) is likewise recognized in Eq. (4.27).

Lastly, the effect of R_E on the small-signal gain is calculated from the circuit in Fig. 4.13.

$$\frac{V_E}{R_E} + \frac{V_E - V_i}{r_b + R_S + r_\pi} + \frac{V_E - V_o}{r_o} - g_m v_\pi = 0 \tag{4.28}$$

$$\frac{V_o}{R_L} + \frac{V_o - V_E}{r_o} + g_m v_\pi = 0 \tag{4.29}$$

$$v_\pi = (V_i - V_E)\frac{r_\pi}{r_b + R_S + r_\pi} \tag{4.30}$$

After some manipulation, and assuming r_o is quite high, as well as r_π being much larger than either r_b, R_S, or R_L, one finds

$$A = \frac{V_o}{V_i} \approx -\frac{g_m R_L}{1 + g_m R_E} \approx -\frac{R_L}{R_E} \tag{4.31}$$

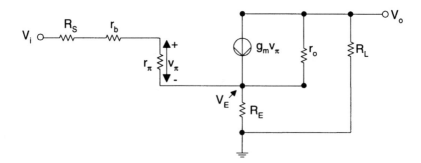

FIGURE 4.13 Small-signal equivalent circuit for calculation of the gain of the common-emitter amplifier with degeneration.

Again, the factor $(1 + g_m R_E)$ shows up. However, this time it appears in the denominator, which translates into a large unwanted reduction in gain. A possible solution lies in the use of a decoupling capacitor C_E placed in parallel with R_E. Then, Eq. (4.31) becomes

$$A \approx -\frac{g_m R_L (1 + sC_E R_E)}{1 + g_m R_E + sC_E R_E} \tag{4.32}$$

Eq. (4.32) has a zero at

$$s_z = -\frac{1}{C_E R_E} \tag{4.33}$$

and a pole at a higher frequency given by

$$s_p = -\frac{1 + g_m R_E}{C_E R_E} \approx -\frac{g_m}{C_E} \tag{4.34}$$

If C_E is chosen sufficiently large, the desirable dc degeneration resistance properties can be combined with a small-signal gain, which for the frequencies of interest approaches the $-g_m R_L$ value realized by the undegenerated common-emitter amplifier.

    The stabilization method shown in Fig. 4.10 is very commonly used in discrete amplifier realizations. As mentioned before, integrated circuit implementations have the advantage of near-perfect matching between like components. Rather than stabilizing the current in each transistor by the method described above, usually a single accurate bias current generator is incorporated on a chip and the resulting reference current is mirrored throughout. The discussion of suitable voltage and current reference generators is postponed until Section 4.5. Figure 4.14(a) shows how a reference current I_{REF} is mirrored to the common-emitter amplifier by means of a basic npn current mirror. Following the previous derivations

$$R_{E,\,eff} = r_o(1 + g_m R_{E2}) = r_o\left(1 + \frac{I_{REF}}{V_T}R_{E2}\right) = \frac{V_A}{I_{REF}}\left(1 + \frac{I_{REF}}{V_T}R_{E2}\right) \approx \frac{V_A}{V_T}R_{E2} \tag{4.35}$$

$R_{E,eff}$ is much higher than could be obtained by a resistor only, under the restriction of an equal voltage drop. Thus, we have a higher performance current source. Also, for a given C_E, the pole and zero, which obey Eqs. (4.33) and (4.34), move to lower frequencies. Conversely, a lower valued C_E could be used. This property is quite useful since integrated circuit capacitors consume a rather large amount of silicon real estate, and die size directly relates to cost.

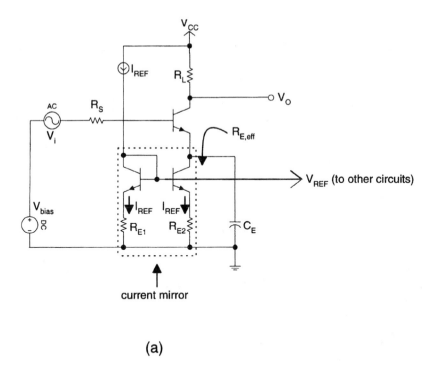

(a)

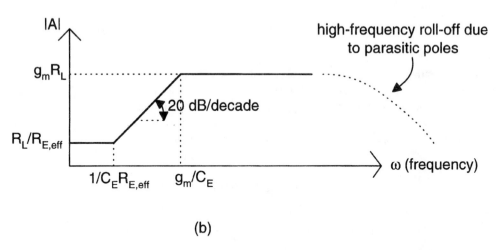

(b)

FIGURE 4.14 (a) Common-emitter amplifier with bias current generator and mirror, and (b) magnitude response vs. frequency.

Frequency Response of the Common-Emitter Amplifier

With the exception of C_E, we have thus far ignored the effect of the capacitors. At low frequencies, the ac coupling capacitances C_{ci} and C_{co} result in a gain reduction. In that sense, they behave similarly to C_E. At high frequencies, the internal transistor parasitics lead to a decrease in gain. Three parasitic capacitors must be accounted for: the emitter–base capacitance C_π, the collector–base capacitance C_μ, and the collector–substrate capacitance C_{cs}. Figure 4.15 shows the high-frequency small-signal equivalent circuit

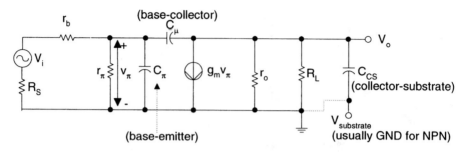

FIGURE 4.15 High-frequency small-signal equivalent circuit for the common-emitter amplifier.

of the common-emitter amplifier. With R_i representing the parallel combination of $(R_S + r_b)$ with r_π, and R_C similarly designating the parallel combination of R_L with r_o, Eq. (4.9) must be rewritten as

$$A = -\frac{g_m R_i R_C N(s)}{R_S + r_b D(s)} \tag{4.36}$$

where

$$
\begin{aligned}
\frac{N(s)}{D(s)} =& \\
&\left(1 - \frac{sC_\mu}{g_m}\right)\Big/\left(1 + s(C_\pi R_i + C_\mu R_i + C_\mu R_C + C_\mu R_i R_C g_m + C_{cs} R_C)\right. \\
&\left. + s^2(C_\pi C_{cs} R_i R_C + C_\pi C_\mu R_i R_C + C_\mu C_{cs} R_i R_C)\right)
\end{aligned}
\tag{4.37}
$$

One observes that Eq. (4.37) contains a right half-plane zero located at $s_z = g_m/C_\mu$, resulting from the capacitive feedthrough from input to output. However, this right half-plane zero is usually at such a high frequency that it can be ignored in most applications. Furthermore, in most cases, one can assume that $D(s)$ contains a dominant pole p_1 and a second pole p_2 at a substantially higher frequency. If the dominant pole assumption is valid, $D(s)$ can be factored in the following manner

$$D(s) = \left(1 - \frac{s}{p_1}\right)\left(1 - \frac{s}{p_2}\right) \approx 1 - \frac{s}{p_1} + \frac{s^2}{p_1 p_2} \tag{4.38}$$

Equating Eqs. (4.37) and (4.38) yields

$$p_1 = -\frac{1}{R_i}\frac{1}{C_\pi + C_{cs}\dfrac{R_C}{R_i} + C_\mu\left(1 + g_m R_C + \dfrac{R_C}{R_i}\right)} \tag{4.39}$$

$$p_2 = -\frac{1}{R_C(C_\mu + C_{cs})}\frac{C_\pi + C_{cs}\dfrac{R_C}{R_i} + C_\mu\left(1 + g_m R_C + \dfrac{R_C}{R_i}\right)}{C_\pi + \dfrac{C_\mu C_{cs}}{C_\mu + C_{cs}}} \tag{4.40}$$

In Eqs. (4.39) and (4.40), the collector–base capacitance C_μ appears with a multiplication factor. One readily recognizes that this factor is dominated by the term $g_m R_C$, which is equal to the amplifier's gain A. A very important conclusion is that, looking into the common-emitter amplifier's input, the base–collector capacitance C_μ appears approximately A times larger than its actual physical value. This phenomenon is widely referred to as the *Miller effect*. As a result, C_μ often constitutes the predominant frequency limiter in high-gain common-emitter amplifiers.

If the dominant pole assumption is a valid model for the amplifier's high-frequency response, the -3 dB frequency can be more easily calculated by the method of *time constants*, rather than by deriving the complete transfer function. Indeed, it can be shown that

$$\omega_{3dB,HF} = \frac{1}{\sum_{i=1}^{n} C_i R_i} \tag{4.41}$$

where C_i are the capacitors in the high-frequency equivalent scheme (Fig. 4.15) and R_i are the resistances measured across each C_i when all other capacitors are replaced by open circuits. The reader can quickly verify that

$$R_\pi = r_\pi \,\|\, (r_b + R_S) = R_i \tag{4.42}$$

$$R_\mu = (r_\pi \,\|\, (r_b + R_S)) + (R_L \,\|\, r_o)(1 + g_m(r_\pi \,\|\, (r_b + R_S))) = R_i + R_C(1 + g_m R_i) \tag{4.43}$$

$$R_{Ccs} = R_L \,\|\, r_o = R_C \tag{4.44}$$

Then, according to Eq. (4.41),

$$\omega_{3dB,HF} = \frac{1}{R_\pi C_\pi + R_\mu C_\mu + R_{Ccs} C_{cs}} \tag{4.45}$$

which is identical to Eq. (4.39).

Likewise, one can show that if the low-frequency behavior can be described by a single-pole model,

$$\omega_{3dB,LF} = \sum_{j=1}^{m} \frac{1}{C_j R_j} \tag{4.46}$$

where C_j are the decoupling and coupling capacitors C_E, C_{ci}, C_{co}, and R_j are the resistances individually measured across their terminals when all other capacitors are simultaneously shorted.

This concludes the analysis of the common-emitter amplifier. The bipolar transistor, however, is a three-terminal device (actually there are four terminals if the substrate node is included), so there are obviously different ways to hook the BJT up into a circuit. Based on an understanding of the transistor's operation, it clearly does not make much sense to consider the base as an output, nor to use the collector as an input. This leaves us with two other meaningful possibilities, which will be studied in further detail.

Common-Collector Amplifier (Emitter Follower)

In the amplifier of Fig. 4.16, the input signal is provided to the base of the transistor, while the output is taken at the emitter. Similar to the common-emitter amplifier, this circuit configuration derives its

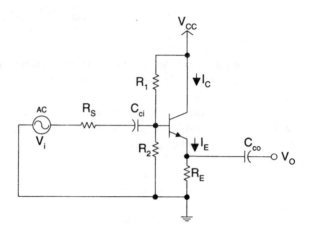

FIGURE 4.16 Common-collector amplifier (emitter follower).

name based on the terminal tied to a common ac ground node. In Fig. 4.16, the collector is connected directly to the supply voltage V_{CC}, hence the name *common-collector* amplifier.

Small-Signal Gain

The common-collector amplifier's mid-frequency small-signal equivalent circuit is shown in Fig. 4.17. By inspection,

$$\frac{V_o}{R_E} + \frac{V_o - V_i}{r_b + R_S + r_\pi} + \frac{V_o}{r_o} - g_m v_\pi = 0 \tag{4.47}$$

$$v_\pi = (V_i - V_o)\frac{r_\pi}{r_b + R_S + r_\pi} \tag{4.48}$$

$$A = \frac{V_o}{V_i} = \frac{1}{1 + \dfrac{r_b + R_S + r_\pi}{(\beta + 1)(R_E \parallel r_o)}} \tag{4.49}$$

$$A \approx \frac{1}{1 + \dfrac{r_\pi}{\beta R_E}} \approx \frac{g_m R_E}{1 + g_m R_E} \tag{4.50}$$

In most cases, $g_m R_E \gg 1$ and, thus,

$$A \approx 1 \tag{4.51}$$

This means that the output voltage at the emitter nearly follows the input at the base. Eq. (4.51) is the reason why the common-collector amplifier is most often referred to as the *emitter follower* configuration.

The emitter follower's input resistance is identical to that of the degenerated common-emitter's. Thus, R_i is high and given by Eq. (4.27). Its output resistance R_o can be calculated from Fig. 4.17 by applying V_{test} and I_{test}, as demonstrated previously for the common-emitter circuit.

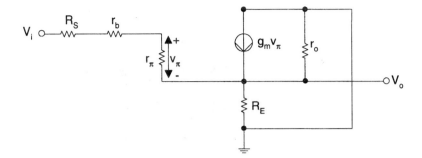

FIGURE 4.17 Small-signal equivalent circuit for calculation of the gain of the common-collector amplifier (emitter follower).

$$R_o = R_E \parallel r_o \parallel \frac{r_b + R_S + r_\pi}{\beta + 1} \approx \frac{1}{g_m} + \frac{R_S}{\beta} \tag{4.52}$$

Generally, Eq. (4.52) means that R_o is low. When the emitter follower is driven from a low-impedance (near-ideal) voltage source, R_o approaches its lower limit $1/g_m$. At room temperature, for example, a 1-mA bias current yields $R_o \approx 25\ \Omega$.

Since there is no gain, the emitter follower does not experience the aforementioned Miller effect and, correspondingly, has a wide bandwidth.

At this time, the reader may interject, "Why not use a (non-degenerated) common-emitter amplifier with a $1/g_m$ load, rather than the emitter follower?" Indeed, both circuits have identical input and output resistances as well as a gain of 1. The answer, however, does not lie in these small-signal parameters, but depends on the signal level. For example, for the common-emitter, a 1-V input is large (actually it would steer the transistor from the OFF state all the way into saturation). For the emitter follower, on the other hand, such a signal is considered small. Thus, the emitter follower is particularly suited as an output stage for large signals. As such, the emitter follower circuit will be revisited in Section 4.4.

In Fig. 4.18, R_E is replaced by a current source. As derived above, the effective degeneration resistor is now equal to $r_o(1 + g_m R_E)$, and thus much higher. According to Eq. (4.50), the circuit's gain is therefore even closer to 1. The drawback of using a current source is the introduction of the parasitic capacitor C_E, which adversely affects the frequency response.

Common-Base Amplifier

Figure 4.19 introduces an amplifier, where the input is applied at the emitter and the output taken at the collector. Since the base is connected to ac ground, this circuit is known as a *common-base* configuration.

Subsequent to the analysis of the common-emitter and emitter follower configurations, it is straight-forward to derive an expression for the input resistance R_i.

$$R_i \approx \frac{r_b + r_\pi}{\beta + 1} \approx \frac{1}{g_m} \tag{4.53}$$

According to Eq. (4.53), R_i is very low. Thus, it seems logical that the common-base circuit should be driven by a source with an equally low impedance. As we have seen above, such a source exists in the form of an emitter follower. Figure 4.20 shows the resulting combined circuit. Especially noteworthy in Fig. 4.20 is the symmetry in the arrangement of both transistors and the resistor R_E. This particular building block is called an *emitter-coupled pair*, as the reader might have expected based on the device connections. The emitter-coupled pair is by far the most frequently used configuration in integrated

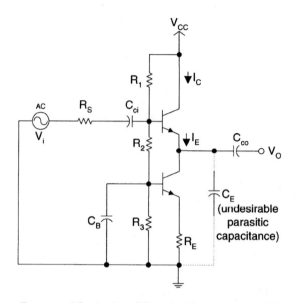

FIGURE 4.18   Common-collector amplifier (emitter follower) with current source bias.

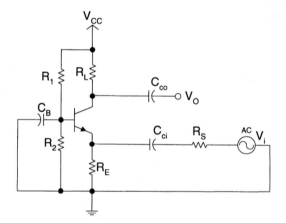

FIGURE 4.19   Common-base amplifier.

circuit bipolar amplifiers. It forms the cornerstone in the amplification of dc and difference signals and, as such, is the subject of an in-depth study in Section 4.3.

The common-base amplifier, however, can also be driven from a high-impedance source, as illustrated in Fig. 4.21. There, the outlined two-transistor arrangement is known as a *cascode* configuration. In Fig. 4.21, the common-emitter circuit at the bottom is loaded by the low-input impedance ($1/g_m$) of the cascode common-base stage. Therefore, it provides no voltage amplification, but only current gain. The common-base stage, in turn, has a current gain approximately equal to 1 (to be exact, $\alpha = \beta/(\beta + 1)$), but contributes voltage gain by means of the load resistor R_L. The overall voltage transfer function is

$$A = \frac{V_o}{V_i} \approx -g_m R_L$$

(4.54)

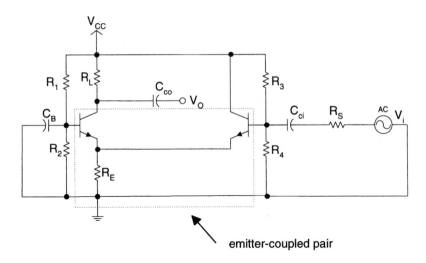

FIGURE 4.20 Common-base amplifier driven by an emitter follower.

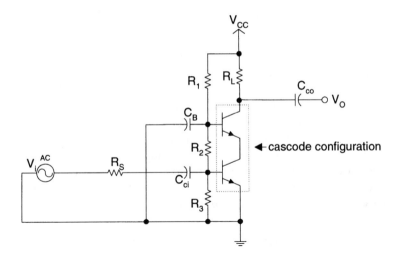

FIGURE 4.21 Common-base amplifier driven from common-emitter stage.

The gain expressed in Eq. (4.54) is identical to the gain realized by the single-transistor common-emitter circuit. The reader may wonder, "What is the benefit of adding a second transistor in the common-base configuration?" The answer and advantage are related to the circuit's high-frequency response. Indeed, the common-base cascode stage effectively eliminates the Miller effect and, thereby, broadbands the circuit.

Darlington and Pseudo-Darlington Pairs

Whereas this chapter section was entitled single-transistor amplifiers, two two-transistor configurations, namely the emitter-coupled pair and the cascode arrangement, were briefly presented. Figure 4.22 introduces two other important two-transistor combinations: the *Darlington* and *pseudo-Darlington pairs*. The Darlington pair consists of two npn transistors connected as shown in Fig. 4.22(a). Combined, the two transistors effectively behave as a single npn with a (high) current gain $\beta = \beta_1\beta_2$. This improved current gain is traded off against a higher voltage requirement: the base–emitter voltage is twice that of

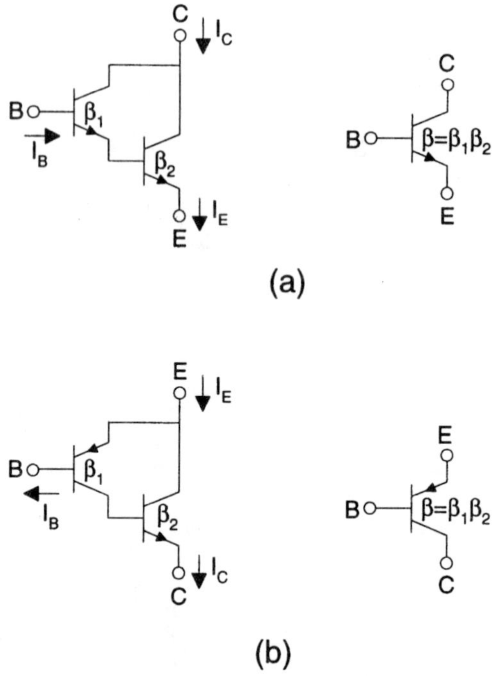

(a)

(b)

FIGURE 4.22   (a) Darlington pair, and (b) pseudo-Darlington pair.

a single transistor and the minimum base–collector voltage equals ($V_{CE,sat} + V_{BE}$). The pseudo-Darlington pair, shown in Fig. 4.22(b), is made out of the combination of a *npn* and a *pnp*. Both transistors jointly act as a single *pnp* with $\beta = \beta_1\beta_2$. The pseudo-Darlington's improved current gain is an important positive feature, as pnp's are generally far inferior in this respect compared to npn's. Similar to the Darlington pair, the required collector–emitter voltage of the pseudo configuration is equal to ($V_{CE,sat} + V_{BE}$). Its base–emitter voltage requirement equals ($V_{BE} + V_{CE,sat}$). While higher than for a single transistor, this is not quite as high as needed by the cascade of two transistors.

4.3   Differential Amplifiers[*]

Introduction: Amplification of dc and Difference Signals

Differential amplifiers represent a class of amplifiers that amplify the difference between two signals, including dc. We will show that to obtain their desired characteristics, differential amplifiers critically depend on component matching. Therefore, discrete implementations necessitate careful component pairing, which is not only painstaking to the design engineer, but also significantly raises the amplifier's cost. In contrast, integrated circuit technology with its inherent small relative component tolerances is particularly suited for this application.

Section 4.2 emphasized that the active elements used for amplification are far from linear devices. To circumvent the problems associated with the bipolar transistor's non-linear input–output relationship, the amplifier circuits were linearized through the selection of a suitable operating point. Recalling Fig. 4.10, the input bias and dc level at the output were separated from the desired input–output signals by means of coupling capacitors C_{ci} and C_{co}. Further, an emitter degeneration resistor R_E reduced the drift of the operating point and a decoupling capacitor C_E was added to counteract the associated undesired gain reduction.

---

[*] This section largely parallels the bipolar technology part of Ref. 12 by the same author.

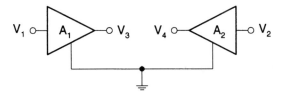

FIGURE 4.23 Initial concept of difference amplifier.

Obviously, the presence of coupling and decoupling capacitors makes the circuit in Fig. 4.10 unsuitable for dc amplification. But, even at low frequencies, this amplifier scheme is not usable due to the large capacitor values required, which in turn give rise to large RC time constants and, consequently, slow recovery times from any transient disturbances.

The requirement to avoid capacitors in low-frequency or dc amplifiers unavoidably leads to a mixing of the concepts of bias and signal. A second characteristic of such amplifiers, as will become evident, is the application of symmetry to compensate for the drift of the active components.

An intuitive solution appears to lie in the use of two amplifiers connected in a difference configuration, as illustrated in Fig. 4.23. For this circuit, one can write the following equations

$$V_3 = A_1 V_1 \tag{4.55}$$

$$V_4 = A_2 V_2 \tag{4.56}$$

Assuming A_1 approximately equal to A_2 (i.e., $A_1 = A + \Delta/2$ and $A_2 = A - \Delta/2$) yields

$$V_3 - V_4 = A(V_1 - V_2) + \frac{\Delta}{2}(V_1 + V_2) \tag{4.57}$$

$$V_3 + V_4 = A(V_1 + V_2) + \frac{\Delta}{2}(V_1 - V_2) \tag{4.58}$$

Clearly, when both amplifiers are perfectly matched or $\Delta = 0$, the difference mode is completely separated from the sum mode.

Upon further reflection, however, the difference amplifier of Fig. 4.23 is not really the solution to the amplification problem under consideration. Indeed, in many instances, small signals, which sit on top of large pedestal voltages, need to be amplified. For example, assume $A = 100$, a difference signal of 10 mV, and a bias voltage equal to 10 V. The amplifier scheme of Fig. 4.23 would result in a 1-V difference signal in addition to a 1000-V output voltage common to both amplifiers. It is clearly unrealistic to assume that both amplifiers will remain linear and matched over such an extended voltage range. Instead, the real solution is obtained by coupling the amplifiers. The resulting arrangement is known as the *differential pair* amplifier.

In the mathematical description of the differential pair, four amplification factors are generally defined to express the relationship between the differential or difference (subscript D) and common or sum mode (subscript C) input and output signals. Applied to the circuit in Fig. 4.23, one can write

$$V_3 - V_4 = V_{oD} = A_{DD} V_{iD} + A_{DC} V_{iC} \tag{4.59}$$

$$V_3 + V_4 = 2V_{oC} = 2A_{CC} V_{iC} + 2A_{CD} V_{iD} \tag{4.60}$$

where $V_{iD} = V_1 - V_2$ and $V_{iC} = (V_1 + V_2)/2$.

The ratio A_{DD}/A_{CC} is commonly referred to as the amplifier's *common-mode rejection ratio* or CMRR. While an amplifier's CMRR is an important characteristic, maximizing its value is not a designer's goal in itself. Rather, the real purpose is to suppress large sum signals so that the two amplifiers exhibit a small output swing and, thereby, indeed operate as a matched pair. Furthermore, the ultimate goal is to avoid the application of a common-mode input signal resulting in a differential signal at the output. This objective can only be accomplished through a combination of careful device matching, precise selection of the amplifier's bias and operating point, as well as by a high common-mode rejection. While we have only considered differential output signals up to this point, in some instances a single-ended output is desired. Eqs. (4.59) and (4.60) can be rearranged as

$$V_3 = V_{oC} + \frac{1}{2}V_{oD} = \left(A_{CD} + \frac{1}{2}A_{DD}\right)V_{iD} + \left(A_{CC} + \frac{1}{2}A_{DC}\right)V_{iC} \tag{4.61}$$

$$V_4 = V_{oC} - \frac{1}{2}V_{oD} = \left(A_{CD} - \frac{1}{2}A_{DD}\right)V_{iD} + \left(A_{CC} - \frac{1}{2}A_{DC}\right)V_{iC} \tag{4.62}$$

One concludes that in the single-ended case, all three ratios A_{DD}/A_{CC}, A_{DD}/A_{CD}, and A_{DD}/A_{DC} must be high to yield the desired result.

Bipolar Differential Pairs (Emitter-Coupled Pairs)

Emitter-Coupled Pairs

Figure 4.24 depicts the basic circuit diagram of a bipolar differential pair. A differential signal V_{iD} is applied between the bases of two transistors, which, unless otherwise noted, are assumed to be identical. The dc bias voltage V_{bias} and a common-mode signal V_{iC} are also present. The transistors' common-emitter node (hence the name *emitter-coupled pair*) is connected to ground through a biasing network, which for simplicity is represented by a single resistor R_O. The amplifier output is taken differentially across the two collectors, which are tied to the power supply V_{CC} by means of a matched pair of load resistors R_L.

Low-Frequency Large-Signal Analysis

Applying the bipolar transistor's Ebers–Moll relationship with $V_{BE} \gg V_T$ (where $V_T = kT/q$ is the thermal voltage) and assuming that both transistors are matched (i.e., the saturation currents $I_{S1} = I_{S2}$), the difference voltage V_{iD} can be expressed as follows

$$V_{iD} = V_{BE1} - V_{BE2} = V_T \ln\left(\frac{I_{C1}}{I_{C2}}\right) \tag{4.63}$$

After some manipulation and substituting $I_{C1} + I_{C2} = \alpha I_{EE}$ (the total current flowing through R_O), one gets

$$I_{C1} = \frac{\alpha I_{EE}}{1 + \exp\left(\dfrac{-V_{iD}}{V_T}\right)} \tag{4.64}$$

$$I_{C2} = \frac{\alpha I_{EE}}{1 + \exp\left(\dfrac{V_{iD}}{V_T}\right)} \tag{4.65}$$

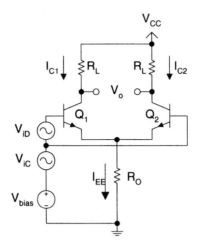

FIGURE 4.24 Emitter-coupled pair.

where α is defined as $\beta/(\beta + 1)$.

Since $V_{oD} = -R_L (I_{C1} - I_{C2})$, the expression for the differential output voltage V_{oD} becomes

$$V_{oD} = -\alpha R_L I_{EE} \frac{\exp\left(\dfrac{V_{iD}}{2V_T}\right) - \exp\left(\dfrac{-V_{iD}}{2V_T}\right)}{\exp\left(\dfrac{V_{iD}}{2V_T}\right) + \exp\left(\dfrac{-V_{iD}}{2V_T}\right)} = -\alpha R_L I_{EE} \tanh\left(\frac{V_{iD}}{2V_T}\right) = -\alpha R_L I_{EE} \tanh x \quad (4.66)$$

The transfer function expressed in Eq. (4.66) is quite non-linear. A graphical representation is given in Fig. 4.25. When $V_{iD} > 2V_T$, the current through one of the two transistors is almost completely cut off and for further increases in V_{iD} the differential output signal eventually clips at $-\alpha R_L I_{EE}$. On the other hand, for small values of x, $\tanh x \approx x$. Under this small-signal assumption,

$$A_{DD} = \frac{V_{oD}}{V_{iD}} = -\frac{\alpha I_{EE}}{2V_T} R_L = -g_m R_L \quad (4.67)$$

While the next subsection contains a more rigorous small-signal analysis, a noteworthy observation here is that, under conditions of equal power dissipation, the differential amplifier of Fig. 4.24 has only one half the transconductance value and hence only one half the gain of a single-transistor common-emitter amplifier. From Eq. (4.67), one furthermore concludes that when the tail current I_{EE} is derived from a voltage source, which is proportional to absolute temperature (PTAT), and a resistor of the same type as R_L, the transistor pair's differential gain is determined solely by a resistor ratio. As such, the gain is well-controlled and insensitive to absolute process variations. A similar observation was made in Section 4.2 during the discussion of the common-emitter amplifier's operating point stability. In Section 4.5, a suitable PTAT reference will be presented.

An intuitive analysis of the common-mode gain can be carried out under the assumption that R_O is large (e.g., assume R_O represents the output resistance of a current source). Then, a common-mode input signal V_{iC} results only in a small current change i_C through R_O and therefore V_{BE} remains approximately constant. With $i_C \approx V_{iC}/R_O$ and $V_{oC} = -R_L i_C/2$, the common-mode gain can be expressed as

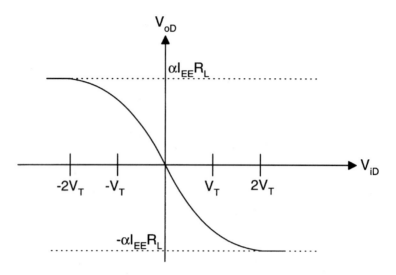

FIGURE 4.25   Emitter-coupled pair's dc transfer characteristic.

$$A_{CC} = \frac{V_{oC}}{V_{iC}} \approx -\frac{R_L}{2R_o} \tag{4.68}$$

Combining Eqs. (4.67) and (4.68) yields

$$\text{CMRR} = \frac{A_{DD}}{A_{CC}} \approx 2g_m R_O \tag{4.69}$$

Half-Circuit Equivalents

Figure 4.26 illustrates the derivation of the emitter-coupled pair's differential mode half-circuit equivalent representation. For a small differential signal, the sum of the currents through both transistors remains constant and the current through R_O is unchanged. Therefore, the voltage at the emitters remains constant. The transistors operate as if no degeneration resistor were present, resulting in a high gain. In sum mode, on the other hand, the common resistor R_O provides negative feedback, which significantly lowers the common-mode gain. In fact, with identical signals at both inputs, the symmetrical circuit can be split into two halves, each with a degeneration resistor $2R_O$ as depicted in Fig. 4.27.

Low-Frequency Small-Signal Analysis

Figure 4.28 represents the low-frequency small-signal differential mode equivalent circuit wherein R_{SD} models the corresponding source impedance. Under the presumption of matched devices,

$$A_{DD} = \frac{V_{oD}}{V_{iD}} = -g_m R_L \frac{r_\pi}{r_\pi + r_b + \frac{1}{2}R_{SD}} \tag{4.70}$$

With $r_b \ll r_\pi$ and assuming a low-impedance voltage source, Eq. (4.70) can be simplified to $-g_m R_L$ as was previously derived in Eqs. (4.67) or (4.9).

The low-frequency common-mode equivalent circuit is shown in Fig. 4.29. Under similar assumptions as in Eq. (4.70) and with R_{SC} representing the common-mode source impedance, one finds

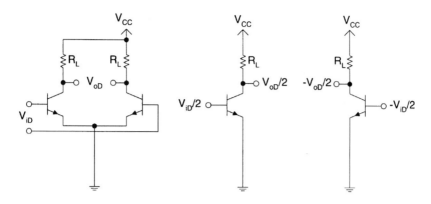

FIGURE 4.26 Difference mode.

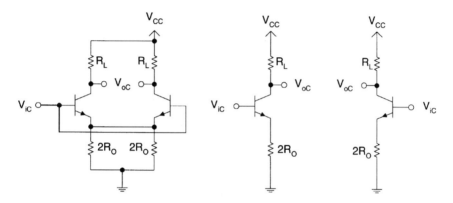

FIGURE 4.27 Sum mode.

$$A_{CC} = \frac{V_{oC}}{V_{iC}} = -g_m R_L \frac{r_\pi}{r_\pi + r_b + 2R_{SC} + 2(\beta + 1)R_O} \tag{4.71}$$

Upon substitution of $\beta = g_m r_\pi \gg 1$, Eq. (4.71) reduces to $-R_L/2R_O$, the intuitive result obtained earlier in Eq. (4.68). For $R_E = 2R_O$, this result is also identical to Eq. (4.31).

The combination of Eqs. (4.70) and (4.71) leads to

$$CMRR = \frac{A_{DD}}{A_{CC}} = \frac{r_\pi + r_b + 2R_{SC} + 2(\beta + 1)R_O}{r_\pi + r_b + \frac{1}{2}R_{SD}} \approx 2g_m R_O \tag{4.72}$$

Let us consider the special case where R_O models the output resistance of a current source, implemented by a single bipolar transistor. Then, $R_O = V_A/I_{EE}$, where V_A is the transistor's Early voltage. With $g_m = \alpha I_{EE}/2V_T$,

$$CMRR = \frac{\alpha V_A}{V_T} \tag{4.73}$$

which is independent of the amplifier's bias conditions, but only depends on the process technology and temperature. At room temperature, with $\alpha \approx 1$ and $V_A \approx 25$ V, the amplifier's CMRR would be

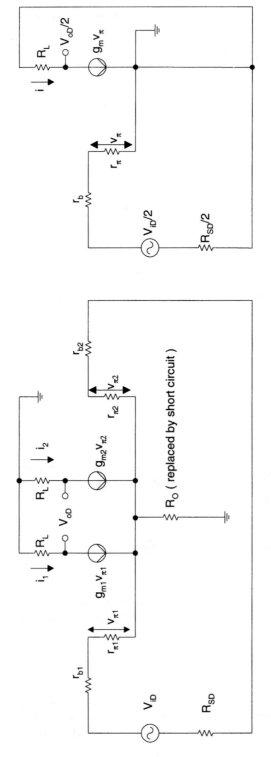

FIGURE 4.28   Small-signal equivalent circuit for difference mode.

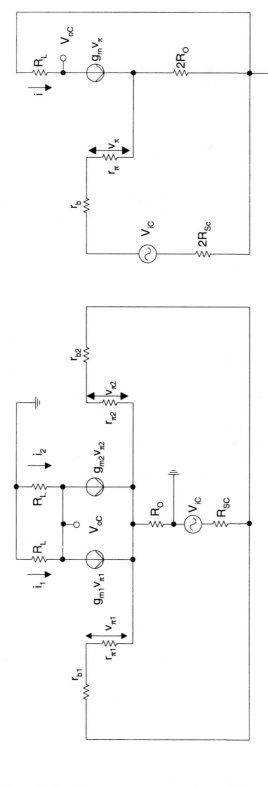

FIGURE 4.29   Small-signal equivalent circuit for sum mode.

approximately 60 dB. The use of an improved current source, for example a bipolar transistor in series with an emitter degeneration resistor R_D, can significantly increase the CMRR. More specifically,

$$\text{CMRR} = \frac{\alpha V_A}{V_T}\left(1 + \frac{I_{EE}R_D}{V_T}\right) \tag{4.74}$$

For $I_{EE}R_D = 250$ mV, the CMRR in Eq. (4.74) is eleven times higher than in Eq. (4.73).

In addition to expressions for the gain, the emitter-coupled pair's differential and common-mode input resistances can readily be derived from the small-signal circuits in Figs. 4.28 and 4.29.

$$R_{inD} = 2r_\pi \tag{4.75}$$

$$R_{inC} = \frac{1}{2}r_\pi + R_O(\beta + 1) \tag{4.76}$$

Taking into account the thermal noise of the transistors' base resistances and the load resistors R_L, as well as the shot noise caused by the collector currents, the emitter-coupled pair's total input referred squared noise voltage per Hertz is given by

$$\frac{V_{iN}^2}{\Delta f} = 8kT\left(r_b + \frac{1}{2g_m} + \frac{1}{g_m^2 R_L}\right) \tag{4.77}$$

Due to the presence of base currents, there is also a small input noise current, which however will be ignored here and in further discussions.

Small-Signal Frequency Response

When the emitter-base capacitance C_π, the collector-base capacitance C_μ, the collector-substrate capacitance C_{cs}, and the transistor's output resistance r_o are added to the transistor's hybrid-π small-signal model in Fig. 4.28, the differential gain transfer function becomes frequency dependent. Although the differential-mode small-signal equivalent circuit is identical to that of the non-degenerated common-emitter amplifier analyzed in Section 4.2, the high-frequency analysis is repeated here for the sake of completeness. With R_i representing the parallel combination of $(R_{SD}/2 + r_b)$ with r_π, and R_C similarly designating the parallel combination of R_L with r_o, Eq. (4.70) must be rewritten as

$$A_{DD} = -\frac{g_m R_i R_C}{\frac{1}{2}R_{SD} + r_b}\frac{N(s)}{D(s)} \tag{4.78}$$

where

$$\frac{N(s)}{D(s)} =$$

$$\left(1 - \frac{sC_\mu}{g_m}\right) \Big/ (1 + s(C_\pi R_i + C_\mu R_i + C_\mu R_C + C_\mu R_i R_C g_m + C_{cs}R_C) \tag{4.79}$$

$$+ s^2(C_\pi C_{cs}R_i R_C + C_\pi C_\mu R_i R_C + C_\mu C_{cs}R_i R_C))$$

As mentioned before, the right half-plane zero located at $s_z = g_m/C_\mu$ results from the capacitive feedthrough from input to output. This right half-plane zero is usually at such a high frequency that it can be ignored in most applications. Unlike in a single-ended amplifier, in a differential pair this zero can easily be canceled. One only has to add two capacitors $C_C = C_\mu$ between the bases of the transistors and the opposite collectors, as illustrated in Fig. 4.30(a). Rather than using physical capacitors, perfect tracking can be achieved by making use of the base–collector capacitances of transistors, whose emitters are either floating or shorted to the bases, as illustrated in Figs. 4.30(b) and (c). Note, however, that the compensating transistors contribute additional collector-substrate parasitics, which to some extent counteract the intended broadbanding effect.

If the dominant pole assumption is valid, $D(s)$ can be factored in the following manner

$$D(s) = \left(1 - \frac{s}{p_1}\right)\left(1 - \frac{s}{p_2}\right) \approx 1 - \frac{s}{p_1} + \frac{s^2}{p_1 p_2} \tag{4.80}$$

Equating Eqs. (4.79) and (4.80) yields

$$p_1 = -\frac{1}{R_i} \frac{1}{C_\pi + C_{cs}\dfrac{R_C}{R_i} + C_\mu\left(1 + g_m R_C + \dfrac{R_C}{R_i}\right)} \tag{4.81}$$

$$p_2 = -\frac{1}{R_C(C_\mu + C_{cs})} \frac{C_\pi + C_{cs}\dfrac{R_C}{R_i} + C_\mu\left(1 + g_m R_C + \dfrac{R_C}{R_i}\right)}{C_\pi + \dfrac{C_\mu C_{cs}}{C_\mu + C_{cs}}} \tag{4.82}$$

Rather than getting into a likewise detailed analysis, the discussion of the emitter-coupled pair's common-mode frequency response is limited here to the effect of the unavoidable capacitor C_O (representing, for instance, the collector–base and collector–substrate parasitic capacitances of the BJT), which shunts R_O. The parallel combination of R_O and C_O yields a zero in the common-mode transfer function. Correspondingly, a pole appears in the expression for the amplifier's CMRR. Specifically,

$$\text{CMRR} = 2g_m \frac{R_O}{1 + sC_O R_O} \tag{4.83}$$

The important conclusion from Eq. (4.83) is that at higher frequencies, the amplifier's CMRR rolls off by 20 dB per decade.

dc Offset

Input Offset Voltage

Until now, perfect matching between like components has been assumed. While ratio tolerances in integrated circuit technology can be very tightly controlled, minor random variations between "equal" components are unavoidable. These minor mismatches result in a differential output voltage, even if no differential input signal is applied. When the two bases in Fig. 4.24 are tied together, but the transistors and load resistors are slightly mismatched, the resulting differential output offset voltage can be expressed as

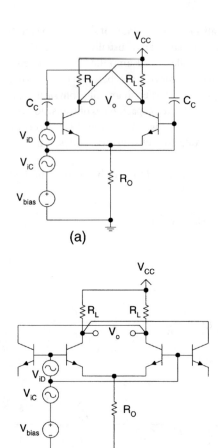

(a)

(b)

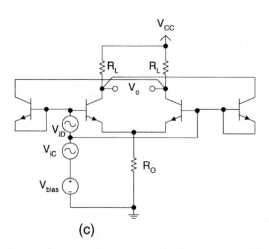

(c)

FIGURE 4.30 C_μ cancellation: (a) capacitor implementation, (b) transistor with floating emitter, and (c) transistor with shorted base–emitter junction.

$$V_{oO} = -(R_L + \Delta R_L)(I_C + \Delta I_C) + R_L I_C =$$

$$- (R_L + \Delta R_L)(I_S + \Delta I_S) \exp\left(\frac{V_{BE}}{V_T}\right) + R_L I_S \exp\left(\frac{V_{BE}}{V_T}\right) \qquad (4.84)$$

or

$$V_{oO} \approx -\left(\frac{\Delta R_L}{R_L} + \frac{\Delta I_S}{I_S}\right) R_L I_S \exp\left(\frac{V_{BE}}{V_T}\right) = -\left(\frac{\Delta R_L}{R_L} + \frac{\Delta I_S}{I_S}\right) R_L I_C \qquad (4.85)$$

Conversely, the output offset can be referred back to the input through a division by the amplifier's differential gain.

$$V_{iO} = \frac{V_{oO}}{-g_m R_L} = V_T\left(\frac{\Delta R_L}{R_L} + \frac{\Delta I_S}{I_S}\right) \qquad (4.86)$$

The input referred offset voltage V_{iO} represents the voltage that must be applied between the input terminals in order to nullify the differential output voltage. In many instances, the absolute value of the offset voltage is not important because it can easily be measured and canceled, either by an auto-zero technique or by trimming. Rather, when offset compensation is applied, the offset stability under varying environmental conditions becomes the primary concern. The drift in offset voltage over temperature can be calculated by differentiating Eq. (4.86):

$$\frac{dV_{iO}}{dT} = \frac{V_{iO}}{T} \qquad (4.87)$$

From Eq. (4.87), one concludes that the drift is proportional to the magnitude of the offset voltage and inversely related to the change in temperature.

Input Offset Current

Since in most applications the differential pair is driven by a low-impedance voltage source, its input offset voltage is an important parameter. Alternatively, the amplifier can be controlled by high-impedance current sources. Under this condition, the input offset current I_{iO}, which originates from a mismatch in the base currents, is the offset parameter of primary concern.

Parallel to the definition of V_{iO}, I_{iO} is the value of the current source that must be placed between the amplifier's open-circuited input terminals to reduce the differential output voltage to zero.

$$I_{iO} = \frac{I_C + \Delta I_C}{\beta + \Delta \beta} - \frac{I_C}{\beta} \approx \frac{I_C}{\beta}\left(\frac{\Delta I_C}{I_C} - \frac{\Delta \beta}{\beta}\right) \qquad (4.88)$$

The requirement of zero voltage difference across the output terminals can be expressed as

$$(R_L + \Delta R_L)(I_C + \Delta I_C) = (R_L I_C) \qquad (4.89)$$

Eq. (4.89) can be rearranged as

$$\frac{\Delta I_C}{I_C} \approx -\frac{\Delta R_L}{R_L} \qquad (4.90)$$

Substituting Eq. (4.90) into Eq. (4.88) yields

$$I_{iO} = -\frac{I_{EE}}{2\beta}\left(\frac{\Delta R_L}{R_L} + \frac{\Delta\beta}{\beta}\right) \tag{4.91}$$

I_{iO}'s linear dependence on the bias current and its inverse relationship to the transistors' current gain β as expressed by Eq. (4.91) intuitively make sense.

Gain Enhancement Techniques

From Eq. (4.67), one concludes that there are two ways to increase the emitter-coupled pair's gain: namely, an increase in the bias current or the use of a larger valued load resistor. Similar to the earlier discussion of the common-emitter amplifier, practical limitations of the available supply voltage and the corresponding limit on the allowable I-R voltage drop across the load resistors (in order to avoid saturating either of the two transistors), however, limit the maximum gain that can be achieved by a single stage. This section introduces two methods that generally allow the realization of higher gain while avoiding the dc bias limitations.

Negative Resistance Load

In the circuit of Fig. 4.31, a gain boosting positive feedback circuit is connected between the output terminals. The output dc bias voltage is simply determined by V_{CC}, together with the product of R_L and the current flowing through it, which is now equal to $(I_E + I_R)/2$. However, for ac signals, the added circuit — consisting of two transistors with cross-coupled base–collector connections and the resistors R_C between the emitters — represents a negative resistance of value $-2(R_C + 1/g_{mc})$, where $g_{mc} = \alpha I_R/2V_T$. The amplifier's differential gain can now be expressed as

$$A_{DD} \approx -g_m R_L \frac{1}{1 - \dfrac{g_{mc}R_L}{1 + g_{mc}R_C}} \tag{4.92}$$

Active Load

Another approach to increase the gain consists of replacing the load resistors by active elements, such as pnp transistors. Figure 4.32 shows a fully differential realization of an emitter-coupled pair with active loads. The differential gain is determined by the product of the transconductance of the input devices and the parallel combination of the output resistances of the npn and pnp transistors. Since $g_m = I_C/V_T$, $r_{on} = V_{An}/I_C$, and $r_{op} = V_{Ap}/I_C$, the gain becomes

$$A_{DD} = -g_m \frac{V_{An}V_{Ap}}{(V_{An} + V_{Ap})I_C} = -\frac{1}{V_T}\frac{V_{An}V_{Ap}}{V_{An} + V_{Ap}} \tag{4.93}$$

Consequently, the gain is relatively high and independent of the bias conditions. The disadvantage of the fully differential realization with active loads is that the output common-mode voltage is not well-defined. This problem also exists for the corresponding single-ended implementation (see Fig. 4.9(a)), as mentioned in Section 4.2. If one were to use a fixed biasing scheme for both types of transistors in Fig. 4.32, minor, but unavoidable mismatches between the currents in the npn and pnp transistors will result in a significant shift of the operating point. The solution lies in a common-mode feedback (CMFB) circuit that controls the bases of the active loads and forces a predetermined voltage at the output nodes. The CMFB circuit has high gain for common-mode signals, but does not respond to differential signals present at its inputs. A possible realization of such a CMFB circuit is seen in the right portion of Fig.

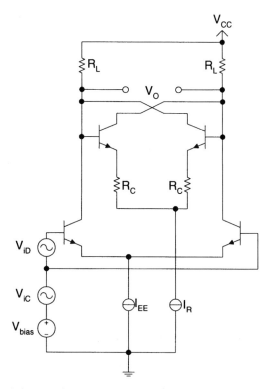

FIGURE 4.31 Emitter-coupled pair with negative resistor load.

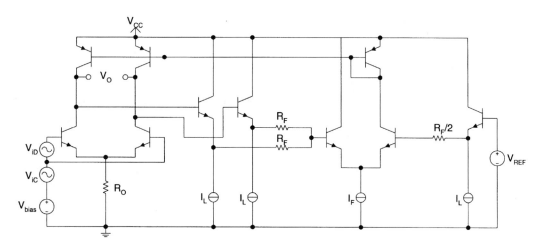

FIGURE 4.32 Emitter-coupled pair with active pnp load.

4.32. Via emitter followers and resistors R_F, the output nodes are connected to one input of a differential pair, whose other input terminal is similarly tied to a reference voltage V_{REF}. The negative feedback provided to the pnp load transistors forces an equilibrium state where the dc voltages at the output terminals of the differential pair gain stage are equal to V_{REF}. An alternative active load implementation with a single-ended output is shown in Fig. 4.33. Contrary to the low CMRR of a single-ended realization with resistive loads, the circuit in Fig. 4.33 inherently possesses the same CMRR as a differential realization since the output voltage depends on a current differencing as a result of the pnp mirror configuration.

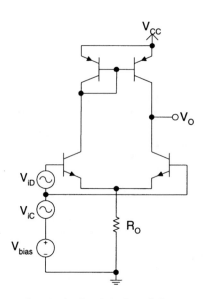

FIGURE 4.33   Emitter-coupled pair with active load and single-ended output.

The drawback of the single-ended circuit is a lower frequency response, particularly when low-bandwidth lateral pnp transistors are used.

Linearization Techniques

As derived previously, the linear range of operation of the emitter-coupled pair is limited to approximately $V_{iD} \approx 2V_T$. This section describes two techniques that can be used to extend the linear range of operation.

Emitter Degeneration

The most common technique to increase the linear range of the emitter-coupled pair relies on the inclusion of emitter degeneration resistors, as shown in Fig. 4.34. The analysis of the differential gain transfer function proceeds as before; however, no closed-form expression can be derived. Intuitively, the inclusion of R_E introduces negative feedback, which lowers the gain and extends the amplifier's linear operating region to a voltage range approximately equal to the product of $R_E I_E$. The small-signal differential gain can be expressed as

$$A_{DD} \approx -G_M R_L \tag{4.94}$$

where G_M is the effective transconductance of the degenerated input stage. Therefore,

$$G_M = \frac{g_m}{1 + g_m R_E} \approx \frac{1}{R_E} \tag{4.95}$$

Consequently,

$$A_{DD} \approx -\frac{g_m R_L}{1 + g_m R_E} \tag{4.96}$$

In case $g_m R_E \gg 1$,

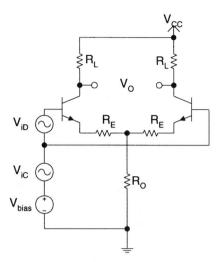

FIGURE 4.34 Differential pair with degeneration resistors.

$$A_{\text{DD}} \approx -\frac{R_L}{R_E} \tag{4.97}$$

In comparison to the undegenerated differential pair, the gain is reduced by an amount $(1 + g_m R_E) \approx g_m R_E$, which is proportional to the increase in linear input range. The common-mode gain transfer function for the circuit in Fig. 4.34 is

$$A_{\text{CC}} \approx -\frac{R_L}{2R_O + R_E} \tag{4.98}$$

For practical values of R_E, A_{CC} remains relatively unchanged compared to the undegenerated prototype. As a result, the amplifier's CMRR is reduced approximately by the amount $g_m R_E$. Also, the input referred squared noise voltage per Hertz can be derived as

$$\frac{V_{\text{iN}}^2}{\Delta f} = 8kT\left[r_b + \frac{1}{2g_m}(g_m^2 R_E^2) + \frac{1}{g_m^2 R_L}(g_m^2 R_E^2) + R_E\right] \tag{4.99}$$

This means that, to a first order, the noise also increases by the factor $g_m R_E$. Consequently, although the amplifier's linear input range is increased, its signal-to-noise ratio (SNR) remains unchanged. To complete the discussion of the emitter degenerated differential pair, the positive effect emitter degeneration has on the differential input resistance R_{inD}, and, to a lesser extent, on R_{inC} should be mentioned. For the circuit in Fig. 4.34,

$$R_{\text{inD}} = 2[r_\pi + (\beta + 1)R_E] \tag{4.100}$$

$$R_{\text{inC}} = \frac{1}{2}r_\pi + \frac{(\beta + 1)R_E}{2} + R_O(\beta + 1) \tag{4.101}$$

Parallel Combination of Asymmetrical Differential Pairs

A second linearization technique consists of adding the output currents of two parallel asymmetrical differential pairs with respective transistor ratios 1:*r* and *r*:1 as shown in Fig. 4.35. The reader will observe

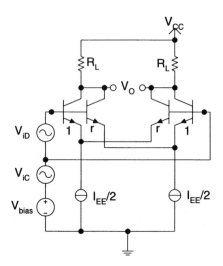

FIGURE 4.35   Parallel asymmetrical pairs.

that each differential pair in Fig. 4.35 is biased by a current source of magnitude $I_{EE}/2$ so that the power dissipation as well as the output common-mode voltage remain the same as for the prototype circuit in Fig. 4.24. Assuming, as before, an ideal exponential input voltage–output current relationship for the bipolar transistors, the following voltage transfer function can be derived:

$$V_{oD} = -\frac{\alpha}{2}I_{EE}R_L\left[\tanh\left(\frac{V_{iD}}{2V_T} - \frac{\ln r}{2}\right) + \tanh\left(\frac{V_{iD}}{2V_T} + \frac{\ln r}{2}\right)\right] \qquad (4.102)$$

After Taylor series expansion and some manipulation, Eq. (4.102) can be rewritten as

$$V_{oD} = -\alpha I_{EE}R_L(1-d)\left[\frac{V_{iD}}{2V_T} + \left(d - \frac{1}{3}\right)\left(\frac{V_{iD}}{2V_T}\right)^3 + \ldots\right] \qquad (4.103)$$

where

$$d = \left(\frac{r-1}{r+1}\right)^2 \qquad (4.104)$$

Equation (4.103) indicates that the dominant third-harmonic distortion component can be canceled by setting $d = 1/3$ or $r = 2 + \sqrt{3} = 3.732$. The presence of parasitic resistances within the transistors tends to require a somewhat higher ratio r for optimum linearization. In practice, the more easily realizable ratio $r = 4$ (or $d = 9/25$) is frequently used. When the linear input ranges at a 1% total harmonic distortion (THD) level of the single symmetrical emitter-coupled pair in Fig. 4.24 and the dual circuit with $r = 4$ in Fig. 4.35 are compared, a nearly threefold increase is noted. For $r = 4$ and neglecting higher-order terms, Eq. (4.103) becomes

$$A_{DD} \approx -0.64 g_m R_L \qquad (4.105)$$

where $g_m = \alpha I_{EE}/2V_T$ as before. Equation (4.105) means that the tradeoff for the linearization is a reduction in the differential gain to 64% of the value obtained by a single symmetrical emitter-coupled pair with equal power dissipation. The squared input referred noise voltage per Hertz for the two parallel asymmetrical pairs can be expressed as

$$\frac{V_{iN}^2}{\Delta f} = \frac{8kT}{(0.64)^2}\left(\frac{r_b}{5} + \frac{1}{2g_m} + \frac{1}{g_m^2 R_L}\right) \qquad (4.106)$$

The factor $r_b/5$ appears because of an effective reduction in the base resistance by a factor $(r + 1)$ due to the presence of five transistors vs. one in the derivation of Eq. (4.77). If the unit transistor size in Fig. 4.35 is scaled down accordingly, a subsequent comparison of Eqs. (4.77) and (4.106) reveals that the input referred noise for the linearized circuit of Fig. 4.35 is 1/0.64, or 1.56 times higher than for the circuit in Fig. 4.24. Combined with the nearly threefold increase in linear input range, this means that the SNR nearly doubles. The approximately 6-dB increase in SNR is a distinct advantage over the emitter degeneration linearization technique. Moreover, the linearization approach introduced in this section can be extended to a summation of the output currents of three, four, or more parallel asymmetrical pairs. However, there is a diminished return in the improvement. Also, for more than two pairs, the required device ratios become quite large and the sensitivity of the linear input range to small mismatches in the actual ratios versus their theoretical values increases as well.

Rail-to-Rail Common-Mode Inputs and Minimum Supply Voltage Requirement

With the consistent trend toward lower power supplies, the usable input common-mode range as a percentage of the supply voltage is an important characteristic of differential amplifiers. Full rail-to-rail input compliance is a highly desirable property. Particularly for low-power applications, the ability to operate from a minimal supply voltage is equally important. For the basic emitter-coupled pair in Fig. 4.24, the input is limited on the positive side when the npn transistors saturate. Therefore,

$$V_{iC, pos} = VCC - \frac{1}{2}R_L I_{EE} + V_{bc, forward} \qquad (4.107)$$

If one limits $R_L I_{EE}/2 < V_{bc,forward}$, $V_{iC,pos}$ can be as high as V_{CC} or even slightly higher. On the negative side, the common-mode input voltage is limited to that level, where the tail current source starts saturating. Assuming a single bipolar transistor is used as the current source,

$$V_{iC, neg} > V_{BE} + V_{CE, sat} \approx 1 \text{ V} \qquad (4.108)$$

The opposite relationships hold for the equivalent pnp transistor-based circuit. As a result, the rail-to-rail common-mode input requirement can be resolved by putting two complementary stages in parallel. In general, as the input common-mode traverses between V_{CC} and ground, three distinct operating conditions can occur: (1) at high voltage levels, only the npn stage is active; (2) at intermediate voltage levels, both the npn and pnp differential pairs are enabled; and finally, (3) for very low input voltages, only the pnp stage is operating. If care is not taken, three distinct gain ranges can occur: based on g_{mn} only; resulting from $g_{mn} + g_{mp}$; and, contributed by g_{mp} only. Non-constant g_m and gain that depends on the input common-mode is usually not desirable for several reasons, not the least of which is phase compensation if the differential pair is used as the first stage in an operational amplifier. Fortunately, the solution to this problem is straightforward if one recognizes that the transconductance of the bipolar transistor is proportional to its bias current. Therefore, the only requirement for a bipolar constant-g_m complementary circuit with full rail-to-rail input compliance is that under all operating conditions the sum of the bias currents of the npn and pnp subcircuits remains constant. A possible implementation is shown in Fig. 4.36. If $V_{iC} < V_{REF}$, the pnp input stage is enabled and the npn input transistors are off. When $V_{iC} > V_{REF}$, the bias current is switched to the npn pair and the pnp input devices

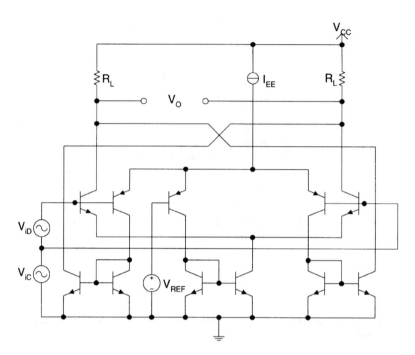

FIGURE 4.36　Low-voltage rail-to-rail input circuit.

turn off. For $R_L I_{EE}/2 < V_{cb,forward,n}$, the minimum required power supply voltage is $V_{BE,n} + V_{BE,p} + V_{CE,sat,n} + V_{CE,sat,p}$, which is lower than 2 V.

4.4　Output Stages

Output stages are specially designed to deliver large signals (as close to rail-to-rail as possible) and a significant amount of power to a specified load. The load to be driven is often very low-ohmic in nature; for example, 4 to 8 ohms in the case of audio loudspeakers. Therefore, output stages must possess a low output impedance and be able to supply high amounts of current — without distorting the signal. The output stage also needs to have a relatively wide bandwidth, so that it does not contribute major frequency limitations to the overall amplifier. Equally desirable is a high efficiency in the power transfer. Preferably, the output stage consumes no (or very little) power in the quiescent state, when no input signal is present. In this section, two major classes of amplifiers, distinguished by their quiescent power needs, are discussed. *Class A amplifiers* consume the same amount of power regardless of the presence of an ac signal. *Class B amplifiers*, on the other hand, only consume power while activated by an input signal and dissipate absolutely no power in stand-by mode. A hybrid between these two distinct cases is *Class AB* operation, which consumes only a small amount of quiescent power. Because output stages deliver high power levels, care must be taken to guarantee that the transistors do not exceed their maximum power ratings, even under unintended operating conditions, such as when the output is shorted to ground. To avoid permanent damage or total destruction, output stages often include some sort of overload protection circuitry.

Class A Operation

The emitter follower, analyzed in Section 4.2, immediately comes to mind as a potential output stage configuration. The circuit is revisited here with an emphasis on its signal-handling capability and power efficiency. Figure 4.37(a) shows an emitter follower transistor Q_1 biased by a current source Q_2 and loaded by a resistor R_L. For generality, separate positive (V_{CC}) and negative ($-V_{EE}$) supplies, as well as ground are used in Fig. 4.37(a) and the analysis below.

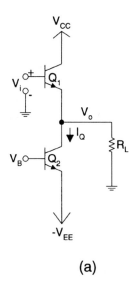

(a)

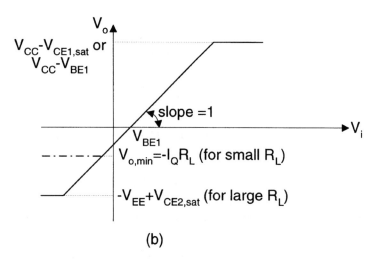

(b)

FIGURE 4.37 (a) Emitter-follower output stage (Class A), and (b) voltage transfer diagram.

The following identities can readily be derived

$$V_i = V_{BE,1} + V_o \tag{4.109}$$

$$V_i = V_T \ln \frac{I_1}{I_S} + V_o = V_T \ln \frac{I_Q + \dfrac{V_o}{R_L}}{I_S} + V_o \tag{4.110}$$

where I_Q is the quiescent current supplied by the current source Q_2.

If one assumes that R_L is quite large, the output current $I_o = V_o/R_L$ is relatively small and $V_{BE,1}$ is approximately constant. Then,

$$V_o \approx V_i - V_T \ln\frac{I_Q}{I_S} \tag{4.111}$$

or there is approximately a fixed voltage drop between input and output.

The positive output excursion is limited by the eventual saturation of Q_1. This means that V_o cannot exceed $(V_{CC} - V_{CEsat,1})$. $V_{Cesat,1}$ is less than $V_{BE,1}$. However, since V_i is most often provided by a previous gain stage, its voltage level can generally not be raised above the supply. In practice, this means that the maximum output voltage is limited to

$$V_{o,\,max} = V_{CC} - V_{BE,\,1} \approx V_{CC} - V_T \ln\frac{I_Q}{I_S} \tag{4.112}$$

Similarly, the negative excursion is limited by the saturation of Q_2. Therefore,

$$V_{o,\,min} = -V_{EE} + V_{CEsat,\,2} \tag{4.113}$$

If the assumption about the load resistor is not valid and R_L is small, the slope of the transfer characteristic is not exactly 1 and some curvature occurs for larger signal excursions. Also, negative output clipping can occur sooner. Indeed, the maximum current flow through R_L during the negative excursion is bounded by the bias I_Q. Consequently,

$$V_{o,\,min} = -I_Q R_L \tag{4.114}$$

Figure 4.37(b) graphically represents the emitter follower's transfer characteristic for both R_L assumptions. If V_{CC} and V_{EE} are much larger than V_{BE} and V_{CEsat}, the maximum symmetrical swing one can obtain is

$$V_{o,\,peak} \approx \frac{V_{CC} + V_{EE}}{2} \tag{4.115}$$

provided V_i has the proper dc bias and the quiescent current is equal to or greater than the optimum value

$$I_{Q,\,opt} = I_{o,\,peak} = \frac{V_{o,\,peak}}{R_L} \tag{4.116}$$

Under the conditions of Eqs. (4.115) and (4.116), the emitter follower's power dissipation is

$$P_{supply} = (V_{CC} + V_{EE})I_{Q,\,opt} \approx 2V_{o,\,peak}I_{o,\,peak} \tag{4.117}$$

For sinusoidal input conditions, the average power delivered to the load is expressed as

$$P_{load} = \frac{1}{2}V_{o,\,peak}I_{o,\,peak} \tag{4.118}$$

Therefore, the highest achievable power efficiency is limited to

$$\eta = \frac{P_{load}}{P_{supply}} \approx 25\% \tag{4.119}$$

In conclusion, while the emitter follower can be used as an output stage, it suffers from two major limitations. First, the output swing is asymmetrical and not rail-to-rail. Second, the operation is Class

A and thus consumes dc power. The emitter follower's maximum power efficiency, which is only reached at full signal swing, is very poor.

Class B and Class AB Operation

Figure 4.38(a) shows a symmetrical configuration with the emitters of an npn and a pnp transistor tied together at the output node, while the input is provided to their joint bases. This dual emitter follower combination is frequently referred to as a *push-pull* arrangement. When no input is applied, clearly no current flow is possible and, thus, the operation is Class B. The push-pull configuration, however, does not solve all the problems of the single-ended emitter follower. Indeed, the output swing is still not rail-to-rail, but is limited to one V_{BE} drop from either supply rail (assuming V_i cannot exceed V_{CC} or V_{EE}). The circuit's transfer characteristic is shown in Fig. 4.38(b). Note that both transistors are off, not just for zero input as desired, but they also do not turn on when small inputs are applied. In effect, for $V_{BE,p} < V_i < V_{BE,n}$, the output has a dead zone. Such *hard non-linearity* leads to undesirable *cross-over distortion*. A method to overcome this problem will be discussed shortly. For larger inputs, only one of the transistors conducts.

The push-pull configuration's power efficiency under sinusoidal input conditions can be derived as follows. The dissipated power is equal to

$$P_{supply} = (V_{CC} + V_{EE})I_{supply} = (V_{CC} + V_{EE})\frac{V_{o,\,peak}}{\pi R_L} \tag{4.120}$$

while the power delivered to the load is expressed as

$$P_{load} = \frac{V_{o,\,peak}^2}{2R_L} \tag{4.121}$$

Thus,

$$\eta = \frac{P_{load}}{P_{supply}} = \frac{\pi}{2}\frac{V_{o,\,peak}}{V_{CC} + V_{EE}} \tag{4.122}$$

Equation (4.122) says that the efficiency is directly proportional to the peak output amplitude. In case the base–emitter voltage can be neglected relative to the supply voltages,

$$V_{o,\,peak,\,max} = \frac{V_{CC} + V_{EE} - V_{BE,\,n} - V_{BE,\,p}}{2} \approx \frac{V_{CC} + V_{EE}}{2} \tag{4.123}$$

The power efficiency's upper bound is therefore given by

$$\eta_{max} \approx \frac{\pi}{4} \approx 78.6\% \tag{4.124}$$

The dead zone and resulting cross-over distortion can be eliminated by inserting two conducting diodes between the bases of the npn and pnp output devices, as shown in Fig. 4.39(a). Strictly speaking, the push-pull circuit is then no longer Class A, but rather becomes Class AB as a small stand-by current constantly flows through both output devices. The resulting linear transfer characteristic is shown in Fig. 4.39(b). One should, however, note that the actual quiescent current is not well-defined, as it depends on matching between the base–emitter voltage drops of the diodes and the output transistors. For discrete

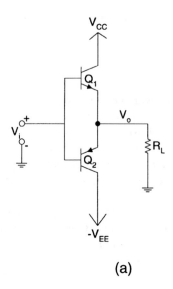

(a)

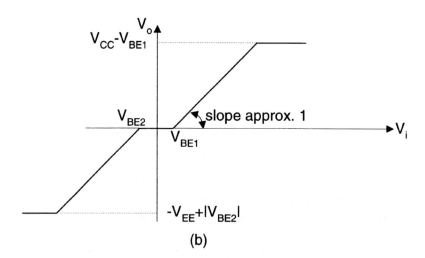

(b)

FIGURE 4.38 (a) Emitter-follower push-pull output stage (Class B), and (b) voltage transfer diagram.

implementations, this configuration would clearly be too sensitive. Even in integrated circuits, it is often desirable to stabilize the operating point through the inclusion of degeneration resistors, as illustrated in Fig. 4.40(a). In addition, these series resistors act as passive current limiters. Indeed, a potential problem occurs in the circuit of Fig. 4.39(a) when the output node is shorted to ground ($R_L = 0$). If the input voltage is large and positive, Q_2 and the diode string are off, forcing all the current I_B to flow into the base of Q_1, where it gets multiplied by the (high) current gain β_1. The resulting collector current may be so large as to cause Q_1 to self-destruct. Obviously, the voltage drop, which builds across either degeneration resistor, limits the maximum current and, as such, can prevent damage. Unfortunately, the inclusion of series resistors is a far from perfect solution as the values needed for quiescent current stabilization and overdrive protection are often unacceptably large. Thus, the degeneration resistors reduce the power efficiency, limit the output swing, and raise the circuit's output resistance. Fortunately, this issue can be circumvented if a diode is added in parallel with the degeneration resistor, as shown in Fig. 4.40(a). For low current values, the diode is off and the circuit is characterized by the high resistance of R_1 (R_2). At higher current levels, the diode turns on and provides a low dynamic resistance. For small input signals, the degeneration resistor is in series with the load and voltage division occurs. At higher input levels when the diode is on, the output again follows the input. The transfer characteristic, illustrated in Fig. 4.40(b), shows that rather than being completely eliminated as in the circuit of Fig. 4.39, the dead zone is replaced by a "slow zone." In other words, the hard non-linearity of the simple push-pull circuit in Fig. 4.38 is transformed into a more acceptable *soft non-linearity*. The transfer characteristics can be further linearized by applying negative feedback around the complete amplifier. The reader should observe, however, that the diode voltage drop further limits the maximum signal swing. While the diodes may reduce the passive current limiting provided by the resistors, superior active limiting is achieved when they are replaced by transistors, as depicted in Fig. 4.40(c). When the voltage drop across R_1 becomes high enough to forward-bias the base–emitter junction of Q_{D1}, the transistor starts to pull current away from the base of Q_1, delivering it harmlessly to the load, without additional multiplication. One should note that the limiting effect for large negative currents is not nearly as effective in this scheme, since this is largely determined by the current sinking capability of the driving transistor (not shown in Fig. 4.40), which pulls current out of the base of Q_2. On the other hand, pnp transistors are usually lateral or substrate devices with relatively low current gain, which furthermore rolls off very quickly at higher current levels. Consequently, the potential problem of negative current overload is inherently less severe. Whereas these lateral or substrate pnp's are generally adequate for low to moderate power applications, if high power must be delivered, a complementary bipolar process with isolated vertical pnp's is required. When such a more complicated and expensive process is not available, alternatively quasi-complementary structures can be used. In Fig. 4.41, for example, the pnp transistor is replaced by a pseudo-Darlington pair (see Section 4.2).

In summary, the Class AB push-pull configuration meets the requirements of an output stage with relatively high efficiency in its power transfer and low stand-by dissipation. For very low voltage applications, however, the voltage drop across the base–emitter junction (and the series diodes) constitutes a serious limitation. A true rail-to-rail output swing (apart from an unavoidable, but low V_{CEsat}) can only be achieved by a complementary common-emitter output stage, as drawn in Fig. 4.42. Following the discussion in Section 4.2, the reader will likely interject that this arrangement suffers from a high output impedance and potential frequency limitations. An in-depth treatment of low-voltage common-emitter output configurations is beyond the scope of this chapter. The interested reader is referred to Ref. 5.

4.5   Bias Reference

It is definitely not the author's intention to present an in-depth discussion of voltage and current reference design. However, on several occasions, the terms "bandgap voltage" and "PTAT voltage" were mentioned. It was also noted that a current, which is proportional to absolute temperature and inversely related to a resistor, is quite often desired in order to stabilize an amplifier's gain. A circuit that provides such supply-independent voltages and currents is shown in Fig. 4.43. By inspection,

$$V_{BG} = V_{BE1} + R_2 I = V_{BE1} + V_{PTAT} \tag{4.125}$$

The current mirror consisting of Q_3 and Q_4 forces the current I to split evenly between Q_1 and Q_2. As a result, the following identity is valid:

$$V_{BE1} = V_{BE2} + R_1 \frac{I}{2} \tag{4.126}$$

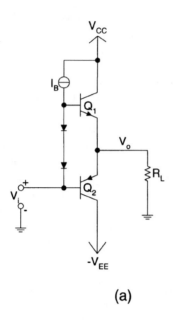

(a)

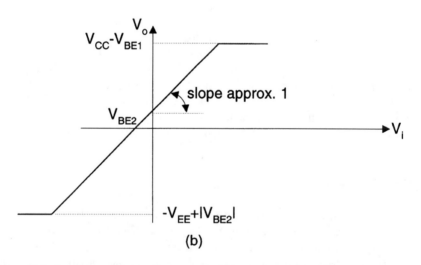

(b)

FIGURE 4.39 (a) Emitter-follower push-pull output stage (Class AB), and (b) voltage transfer diagram.

Substituting the Ebers–Moll identity into Eq. (4.126), where I_{S2} is N times I_{S1}, yields

$$I = \frac{2}{R_1}\left(V_T \ln\frac{I}{2I_{S1}} - V_T \ln\frac{I}{2I_{S2}}\right) = \frac{2}{R_1} V_T \ln N \qquad (4.127)$$

By combining Eqs. (4.125) and (4.127), one gets

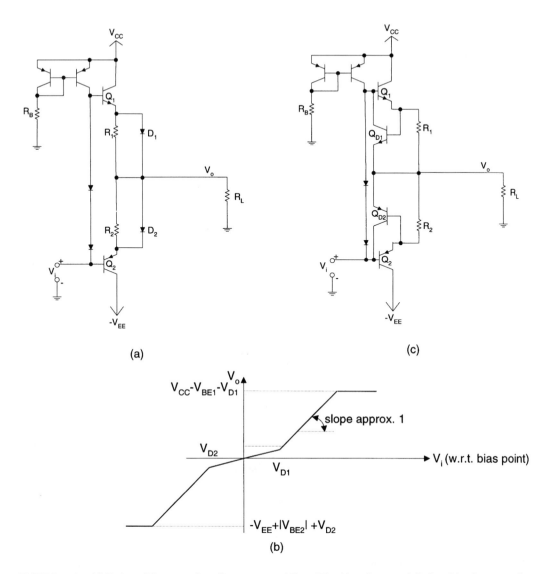

FIGURE 4.40 (a) Emitter-follower push-pull output stage (Class AB) with resistors and diodes; (b) voltage transfer diagram; and (c) emitter-follower push-pull output stage (Class AB) with resistors and transistors.

$$V_{BG} = V_{BE1} + 2\frac{R_2}{R_1}V_T \ln N \qquad (4.128)$$

Also,

$$V_{PTAT} = 2\frac{R_2}{R_1}V_T \ln N = 2\frac{R_2}{R_1}\frac{kT}{q}\ln N \qquad (4.129)$$

From Eq. (4.129), one concludes that V_{PTAT} is indeed proportional to absolute temperature. Apart from the absolute temperature T, V_{PTAT} only depends on physical constants (k and q), an area multiple N, and the ratio of two resistors. In addition to V_{PTAT}, the expression for V_{BG} (Eq. (4.128)) contains the term V_{BE1}, which decreases by 2 mV per degree increase in temperature. Through an appropriate selection of the resistors R_1 and R_2, the voltage V_{BG} can be made temperature independent. This occurs for $V_{BG} \approx$

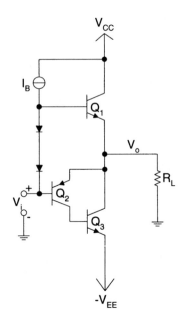

FIGURE 4.41 Quasi-complementary push-pull output stage.

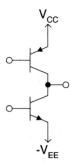

FIGURE 4.42   Rail-to-rail complementary common-emitter output stage.

1.25 V, known as the BJT bandgap voltage. The currents I_{SOURCE} and I_{SINK} in Fig. 4.43 are simply mirrored copies of I, and thus exhibit the desired PTAT and resistor dependence.

Further observation of the circuit in Fig. 4.43 reveals that it possesses a second (although unstable) operating point. Indeed, the circuit is also in equilibrium when $V_{BG} = V_{PTAT} = 0$ V and there is no current flow. To prevent the bias reference from being stuck in this undesired state, an initial start-up circuit is added as shown. When power is first applied, the start-up circuit injects a small current into the mirror Q_3-Q_4, forcing the circuit to wake up and drift away from the zero state. As V_{BG} increases toward 1.25 V, the differential pair eventually switches and the start-up current is simply thrown away. The reader should realize that the circuit in Fig. 4.43 is conceptual in nature and specifically drawn to show that a supply voltage $V_{CC} < 2$ V suffices. However, it suffers from non-idealities due to base currents and poor supply rejection resulting from the simple current mirrors with relatively low output impedances. At the expense of added circuit complexity and the need for a higher supply voltage, significant improvements can be made. Such specialized bias reference discussion, however, goes beyond the scope of this chapter.

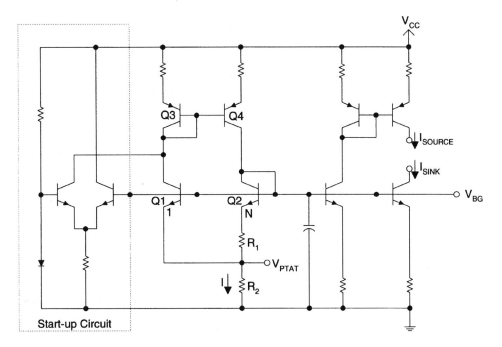

FIGURE 4.43   Bias reference.

4.6   Operational Amplifiers

Introduction

Operational amplifiers (or op-amps) are key analog circuit building blocks that find widespread use in a variety of applications, such as precision amplification circuits (e.g., instrumentation amplifiers), continuous-time and switched-capacitor filters, etc. The traditional symbolic representation of the operational amplifier is shown in Fig. 4.44. The op-amp is a five-terminal device, with inverting and non-inverting input terminals (hence, accommodating a differential input signal), a single-ended output terminal, as well as positive and negative supply terminals. Most commercially available op-amps require a dual supply system of equal, but opposite value (e.g. ±15 V or ±5 V); however, asymmetrical or single-supply circuits are also available (e.g., +5 V and ground). Special-purpose operational amplifiers with differential or fully balanced outputs also exist. The internal circuitry of op-amps combines the different building blocks, which were previously discussed. Op-amps typically consist of two or three stages. The input stage, based on a differential pair, provides the initial amplification. A second or intermediate stage may be included to boost the amplifier's gain. Differential to single-ended conversion is also accomplished in the first stage, or, if applicable, in the second stage. The output stage, typically an emitter follower push-pull configuration, provides a low impedance, large swing, and high current drive. An elementary op-amp schematic can be found in Fig. 4.47(a).

Ideal Op-Amps

The op-amp's input–output relationship can be expressed as

$$V_o = A(V_i^+ - V_i^-) \tag{4.130}$$

In the case of an ideal op-amp, A is assumed to have infinite magnitude as well as bandwidth. As such, there is no phase shift over frequency, as illustrated in Fig. 4.45(a). Since the output V_o must remain

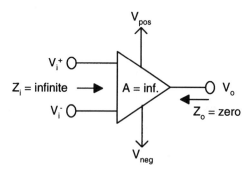

FIGURE 4.44 Op-amp symbol.

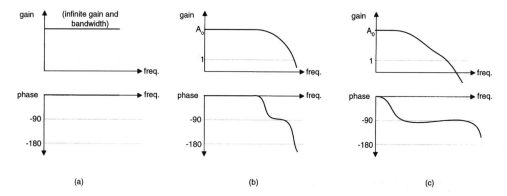

FIGURE 4.45 Magnitude and phase responses: (a) ideal op-amp; (b) real op-amp (potentially unstable in unity-gain feedback loop); (c) real op-amp with internal compensation (guaranteed stable, phase margin approx. 90 degrees).

bounded, the assumption of infinite gain leads to the fundamental principle of *virtual ground* (virtual short). In other words, if the op-amp is ideal, there cannot exist a voltage difference between the input terminals, and V_i^+ must equal V_i^-. The assumption of ideality also calls for an infinite input impedance (i.e., the op-amp does not load the driving source) and a zero output impedance (i.e., the op-amp can accommodate arbitrarily small loads).

Op-Amp Non-idealities

Real operational amplifiers generally approximate the ideal op-amp model reasonably well; however, they naturally have finite gain, finite bandwidth, and finite input as well as output impedances. Specific non-idealities are temized below.

Finite Gain

The op-amp's (low-frequency) gain can be made quite high, particularly when multiple gain stages are cascaded. The absolute gain value, however, is not very well-defined as it depends on widely varying process parameters, such as the transistor's current gain β. If a precise gain is required, the op-amp must be configured into a feedback network; for example, the non-inverting and inverting gain stages shown in Figs. 4.46(a) and (b), respectively. Assuming finite op-amp gain but infinite input impedance, the gain of the non-inverting amplifier in Fig. 4.46(a) can be expressed as

$$\frac{V_o}{V_i} = A\frac{R_i + R_f}{R_i + R_f + AR_i} \tag{4.131}$$

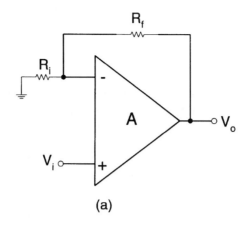

(a)

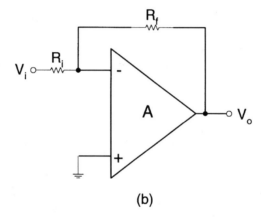

(b)

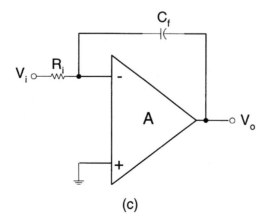

(c)

FIGURE 4.46   (a) Noninverting amplifier; (b) inverting amplifier; and (c) inverting integrator.

If A is high, Eq. (4.131) can be approximated by

$$\frac{V_o}{V_i} \approx \frac{A}{A+1} \frac{R_i + R_f}{R_i} \approx 1 + \frac{R_f}{R_i} \tag{4.132}$$

For practical resistor values, the *closed-loop* gain in Eq. (4.132) is not very high (at least compared to the op-amp's open-loop gain A), but it can be accurately set, even over process corners, by the ratio of two like components. Similarly, the inverting amplifier of Fig. 4.46(b) provides a closed-loop gain

$$\frac{V_o}{V_i} = -A\frac{R_f}{R_i + R_f + AR_i} \approx -\frac{A}{A+1}\frac{R_f}{R_i} \approx -\frac{R_f}{R_i} \tag{4.133}$$

An important building block in active filter design is the inverting integrator circuit displayed in Fig. 4.46(c). By inspection,

$$\frac{V_o}{V_i} = \frac{-A}{(A+1)sR_iC_f + 1} \approx \frac{-1}{sR_iC_f} \tag{4.134}$$

Unlike for the previous two amplifiers, the gain in Eq. (4.134) depends on the product of absolute resistor and capacitor values. When implemented as a monolithic circuit, the integrator's time constant is therefore subject to large tolerances, which need to be compensated for by trimming or an automatic tuning loop.

Finite Bandwidth

The op-amp's internal electronics are characterized by parasitic poles and/or zeros. As previously discussed, these unavoidable parasitics cause the gain to roll off at high frequencies and also introduce phase shifts. A typical op-amp's magnitude and phase responses are shown in Fig. 4.45(b). Assuming the op-amp has a dominant pole, the gain initially decreases by 20 dB/decade and the phase shift approaches −90°. At higher frequencies, the gain starts to decrease faster due to the combined effect of additional non-dominant parasitics and the phase increases as well. Since op-amps are used in gain stabilizing negative feedback circuits, as discussed in the previous sub-section, this high-frequency behavior can lead to instability. The worst possible situation occurs for unity feedback configurations. To avoid potential oscillation problems even under these conditions, commercial op-amps are nearly always internally compensated. Typically, a low-frequency dominant pole is introduced, which causes the gain to drop below unity at the −180° phase cross-over frequency, as shown in Fig. 4.45(c). The difference between the actual phase shift at the op-amp's unity-gain frequency and −180° is referred to as the *phase margin*. Whereas an op-amp is unconditionally stable when its phase margin is positive, values well above 45° are highly desirable to obtain quick settling to input transients. Conversely, the ratio between unity and the actual gain at the frequency corresponding to −180° phase shift is called the *gain margin*. If the dominant pole assumption is valid, the op-amp's gain A should be rewritten as

$$A(s) = A(j\omega) = \frac{A_o\omega_o}{s + \omega_o} \approx \frac{A_o\omega_o}{s} \tag{4.135}$$

where A_o is the dc gain and ω_o is the radial −3 dB frequency. $A_o\omega_o$ is referred to as the op-amp's *gain–bandwidth product*. When Eq. (4.135) is substituted into Eqs. (4.133) and (4.134), the latter gain expressions become frequency dependent. If Eq. (4.135) is likewise combined with the integrator gain in Eq. (4.134), the latter's denominator turns into a second-order polynomial.

Finite Input Impedance

An op-amp's input stage is usually a differential pair. Expressions for its differential and common-mode input resistances have been previously derived. If a high differential input resistance is desired, several design options exist. First, npn input pairs are better than pnp's, thanks to the higher β. Also, the input impedance increases as the input transconductance is lowered, either by reducing the bias current or by adding degeneration resistors. A Darlington pair can be used when an even higher input resistance is required. At high frequencies, the input impedance becomes capacitive (and thus decreases) due to the BJT's C_π and C_μ.

Input Bias Current

The bipolar transistors in the input differential pair require base current. For npn transistors, the current flows into the base, whereas current must be pulled out of the base in the case of pnp transistors. As a result of the higher β, for a given transconductance, the required input bias current is lower in absolute value in the case of an npn input stage compared to a pnp. Darlington or pseudo-Darlington input pairs can further reduce the input bias current requirement. Alternatively, input bias or base current cancellation techniques are sometimes applied.

Input Offset Voltage

Unavoidable device mismatches require the application of a small difference voltage between the input terminals in order to get zero volts at the output terminal. The reader is referred back to Section 4.3 (Input Offset Voltage).

Input Offset Current

Similar to the input offset voltage, when the inputs are currents rather than voltages, small component mismatches require the application of a small input difference current in order to get zero volts at the output node. See Section 4.3 (Input Offset Current) for more details.

Finite Output Impedance

A typical output stage consists of an emitter follower push-pull configuration. Hence, the output resistance depends on $1/g_m$, plus the resistance of the driving stage divided by β. Since β rolls off at high frequencies, R_o increases accordingly. As such, the output impedance appears inductive. This phenomenon can lead to stability problems when driving capacitive loads.

Finite Common-Mode Rejection Ratio

In Section 4.3, the common-mode rejection ratio was defined as the ratio between the differential and common-mode gains. However, an op-amp's CMRR can be more meaningfully explained in terms of the input offset voltage. In this way, the CMRR can be defined as the change in input offset voltage due to a unit change in common-mode input voltage. The CMRR of commercial op-amps typically measures 100 to 120 dB.

Finite Power Supply Rejection Ratio

Similar to the CMRR, the *power supply rejection ratio* (PSRR) is defined as the change in input offset voltage due to a unit change in the supply voltage. An op-amp's PSRR is nearly always different with respect to its positive and negative supplies. Hence, two individual performance numbers are specified. Typical values are in the range of the CMRR.

Slew Rate, Full-Power Bandwidth, and Unity-Gain Frequency

When a step input is applied, the op-amp's output cannot change instantaneously. Rather, a finite transition time is needed. The maximum rate of change in output voltage is referred to as the op-amp's *slew rate* (SR). To determine an expression for the slew rate, consider the elementary op-amp in Fig. 4.47(a). This basic circuit consists of an input differential pair gain stage, which also converts the input to a single-ended signal. An intermediate stage provides additional gain. In addition, a dominant pole

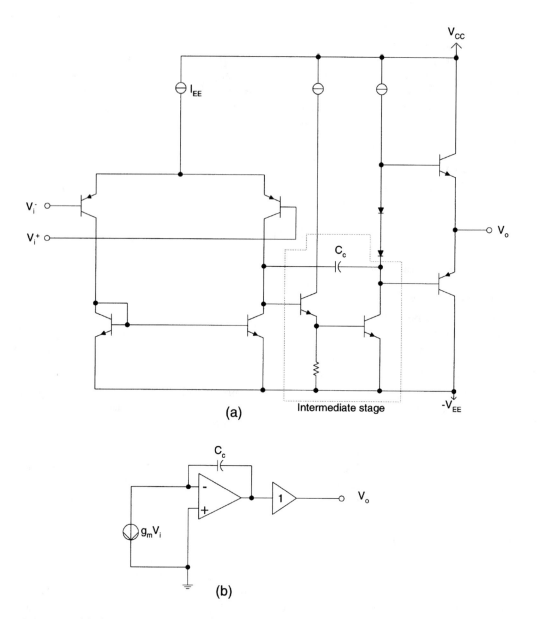

FIGURE 4.47   (a) Elementary op-amp, and (b) equivalent circuit for slew rate calculation.

is introduced by means of the Miller capacitor C_c. The output stage is a Class AB emitter follower push-pull configuration. For the purpose of slew rate calculation, the elementary op-amp can be replaced by the equivalent circuit in Fig. 4.47(b). Then,

$$SR = \left. \frac{dV_o}{dt} \right|_{max} = \left. \frac{g_m V_i}{C_c} \right|_{max} \tag{4.136}$$

The maximum value of $g_m V_i$ is equal to α times the tail current source I_{EE}. Thus,

$$SR = \frac{\alpha I_{EE}}{C_c} \tag{4.137}$$

Based on the equivalent circuit in Fig. 4.47(b), the op-amp's radial unity-gain frequency is readily determined as

$$\omega_u = \frac{g_m}{C_c} \tag{4.138}$$

By combining Eqs. (4.137) and (4.138), the following relationship can be identified between the op-amp's slew rate and its unity-gain frequency

$$SR = \frac{\alpha I_{EE}}{g_m} \omega_u \tag{4.139}$$

From Eq. (4.137), one could conclude that to improve the slew rate, I_{EE} must be increased and/or C_c lowered. However, for the elementary op-amp in Fig. 4.47(a), I_{EE} is also directly proportional to g_m. Thus, both methods of increasing the slew rate at the same time increase the unity-gain bandwidth ω_u. Obviously, a higher ω_u would also be desirable. However, ω_u has an upper limit, which is determined by the op-amp's non-dominant poles. Indeed, less than $-180°$ phase shift must be maintained at ω_u in order to guarantee stability when external feedback is applied. This requirement for ω_u severely limits our flexibility to enhance the op-amp's slew rate. For a given ω_u, Eq. (4.139) suggests that the slew rate can be improved by increasing the I_{EE}/g_m ratio. Two approaches that fall into this category have been previously described in Section 4.3 on differential pair linearization. First, instead of a simple emitter-coupled pair, a differential pair with degeneration resistors can be used. Unfortunately, as pointed out before, the degeneration resistors negatively impact the circuit's noise and also degrade its offset performance. A better approach for improved slew rate is the use of parallel asymmetrical differential pairs, such as two pairs with a 1:r emitter ratio. Although the emphasis in Section 4.3 was on linearization, and a ratio $r = 4$ was chosen specifically to eliminate the third harmonic distortion, the reader may recall that the tradeoff was a reduction in current efficiency (g_m/I_{EE}) to 64% of that of a simple differential pair. In case of an op-amp, a wide linear input range is of lesser or no concern (when feedback is applied, there is virtually no differential signal across the input terminals) and therefore different emitter ratios can be chosen. g_m/I_{EE} further decreases with larger r values. As such, the slew rate can be improved, while keeping ω_u constant. The reader is referred to Ref. 7 for further details.

Another parameter that can directly be correlated to the slew rate is the full-power bandwidth ω_{max}. The full-power bandwidth is defined as the maximum radian frequency for which the op-amp achieves a full output swing under the assumption of a sinusoidal signal. Let

$$V_o = \hat{V} \sin \omega t \tag{4.140}$$

Then,

$$\frac{dV_o}{dt} = \hat{V} \omega \cos \omega t \tag{4.141}$$

and

$$\frac{dV_o}{dt}\bigg|_{max} = \hat{V} \omega_{max} \tag{4.142}$$

The left-hand side in Eq. (4.142) is, per definition, the slew rate. Hence, the slew rate and full-power bandwidth are simply related by the amplitude of the sinusoidal output signal.

4.7   Conclusion

In this chapter, the basic concepts behind signal amplification using bipolar transistors were introduced. Different circuit configurations, which fulfill distinct roles as building blocks in multi-stage designs or operational amplifiers, were presented. During their analysis, no parallel was drawn nor was a comparison made with respect to competing CMOS designs. This chapter would, however, not be complete if the main pros and cons of bipolar amplifiers were not very briefly mentioned. Bipolar amplifiers typically enjoy an advantage by offering higher gain, wider bandwidth, lower noise, and smaller offsets compared to their CMOS counterparts. The transconductance of an MOS transistor, while proportional to the device's size, is generally much lower than for bipolar and, rather than linearly, only increases with the square root of the bias current.

One disadvantage of bipolar amplifiers is the need for input bias currents, coupled with unavoidable input offset currents. Second, MOS transistors in saturation approximate a square-law I–V relationship and are therefore inherently more linear than bipolar devices, which are exponential in nature. Additionally, the linearity of MOS designs can be improved simply by proper device sizing. But, perhaps the most significant drawback of bipolar technology is a more complicated and expensive process (especially in the case of a complementary process with true isolated pnp transistors). Furthermore, bipolar's incompatibility with mainstream VLSI processes prevents higher levels of system integration.

References

Textbooks

1. R. D. Middlebrook, *Differential Amplifiers*, Wiley, New York, 1963.
2. L. J. Giacoletto, *Differential Amplifiers*, Wiley, New York, 1970.
3. P. R. Gray and R. G. Meyer, *Analysis and Design of Analog Integrated Circuits*, 2nd ed., Wiley, New York, 1984.
4. A. B. Grebene, *Bipolar and MOS Analog Integrated Circuit Design*, Wiley, New York, 1984.
5. J. Fonderie, *Design of Low-Voltage Bipolar Operational Amplifiers*, Delft University Press, Delft, 1991.

Articles

6. J. E. Solomon, The monolithic opamp: a tutorial study, *IEEE J. Solid-State Circuits*, vol. SC-9, pp. 314-332, Dec. 1974.
7. J. Schmoock, An input stage transconductance reduction technique for high-slew rate operational amplifiers, *IEEE J. Solid-State Circuits*, vol. SC-10, pp. 407-411, Dec. 1975.
8. J. O. Voorman, W. H. A. Bruls, and P. J. Barth, Bipolar integration of analog gyrator and laguerre type filters, *Proc. ECCTD*, 1983, Stuttgart, pp. 108-110.
9. J. H. Huijsing and D. Linebarger, Low-voltage operational amplifier with rail-to-rail input and output ranges, *IEEE J. Solid-State Circuits*, vol. SC-20, pp. 1144-1150, Dec. 1985.
10. R. J. Widlar and M. Yamatake, A fast settling opamp with low supply currents, *IEEE J. Solid-State Circuits*, vol. SC-24, pp. 796-802, June 1989.
11. G. A. De Veirman, S. Ueda, J. Cheng, S. Tam, K. Fukahori, M. Kurisu, and E. Shinozaki, A 3.0 V 40 Mbit/s hard disk drive read channel IC, *IEEE J. Solid-State Circuits*, vol. SC-30, pp. 788-199, July 1995.
12. G. A. De Veirman, Differential circuits, in *Encyclopedia of Electrical and Electronics Engineering*, J. G. Webster, Editor, Wiley, New York, 1999.

5

High-Frequency Amplifiers

Chris Toumazou
Alison Payne
Imperial College,
University of London

5.1   Introduction

As the operating frequency of communication channels for both video and wireless increases, there is an ever-increasing demand for high-frequency amplifiers. Furthermore, the quest for single-chip integration has led to a whole new generation of amplifiers predominantly geared toward CMOS VLSI. In this chapter, we will focus on the design of high-frequency amplifiers for potential applications in the front-end of video, optical, and RF systems. Figure 5.1 shows, for example, the architecture of a typical mobile phone transceiver front-end. With channel frequencies approaching the 2-GHz range, coupled with demands for reduced chip size and power consumption, there is an increasing quest for VLSI at microwave frequencies. The shrinking feature size of CMOS has facilitated the design of complex analog circuits and systems in the 1- to 2-GHz range, where more traditional low-frequency lumped circuit techniques are now becoming feasible. Since the amplifier is the core component in such systems, there has been an abundance of circuit design methodologies for high-speed, low-voltage, low-noise, and low distortion operation.

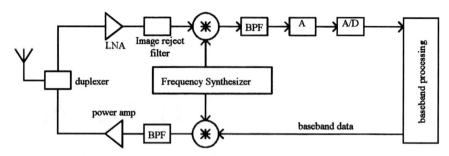

FIGURE 5.1   Generic wireless transceiver architecture.

This chapter will present various amplifier designs that aim to satisfy these demanding requirements. In particular, we will review, and in some cases present new ideas for power amps, LNAs, and transconductance cells, which form core building blocks for systems such as Fig. 5.1. Section 5.2 begins by reviewing the concept of current-feedback, and shows how this concept can be employed in the development of low-voltage, high-speed, constant-bandwidth CMOS amplifiers. The next two sections of the chapter focus on amplifiers for wireless receiver applications, investigating performance requirements and design strategies for optical receiver amplifiers (Section 5.3) and high-frequency low-noise amplifiers (Section 5.4). Section 5.5 considers the design of amplifiers for the transmitter side, and in particular the design and feasibility of Class E power amps are discussed. Finally, Section 5.6 reviews a very recent low-distortion amplifier design strategy termed "log-domain," which has shown enormous potential for high-frequency, low-distortion tunable filters.

5.2   The Current Feedback Op-Amp

Current Feedback Op-Amp Basics

The operational amplifier (op-amp) is one of the fundamental building blocks of analog circuit design.[1,2] High-performance signal processing functions such as amplifiers, filters, oscillators, etc., can be readily implemented with the availability of high-speed, low-distortion op-amps. In the last decade, the development of complementary bipolar technology has enabled the implementation of single-chip video op-amps.[3–7] The emergence of op-amps with non-traditional topologies, such as the current feedback op-amp, has improved the speed of these devices even further.[8–11] Current feedback op-amp structures are well known for their ability to overcome (to a first-order approximation) the gain-bandwidth tradeoff and slew rate limitation that characterizes traditional voltage feedback op-amps.[12]

Figure 5.2 shows a simple macromodel of a current feedback op-amp (CFOA), along with a simplified circuit diagram of the basic architecture. The topology of the current feedback op-amp differs from the conventional voltage feedback op-amp (VOA) in two respects. First, the input stage of a CFOA is a unity-gain voltage buffer connected between the inputs of the op-amp. Its function is to force V_n to follow V_p, very much like a conventional VOA does via negative feedback. In the case of the CFOA, because of the low output impedance of the buffer, current can flow in or out of the inverting input, although in normal operation (with negative feedback) this current is extremely small. Secondly, a CFOA provides a high open-loop transimpedance gain $Z(j\omega)$, rather than open-loop voltage gain as with a VOA. This is shown in Fig. 5.2, where a current-controlled current source senses the current I_{INV} delivered by the buffer to the external feedback network, and copies this current to a high impedance $Z(j\omega)$. The voltage conveyed to the output is given by Eq. 5.1:

$$V_{OUT} = Z(j\omega) \cdot I_{INV} \Rightarrow \frac{V_{OUT}}{I_{INV}}(j\omega) = Z(j\omega) \tag{5.1}$$

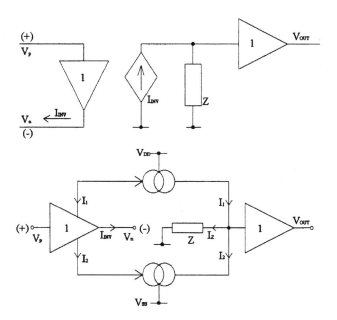

FIGURE 5.2 Current feedback op-amp macromodel.

When the negative feedback loop is closed, any voltage imbalance between the two inputs due to some external agent, will cause the input voltage buffer to deliver an error current I_{INV} to the external network. This error current $I_{INV} = I_1 - I_2 = I_Z$ is then conveyed by the current mirrors to the impedance Z, resulting in an output voltage as given by Eq. 5.1. The application of negative feedback ensures that V_{OUT} will move in the direction that reduces the error current I_{INV} and equalizes the input voltages.

We can approximate the open-loop dynamics of the current feedback op-amp as a single pole response. Assuming that the total impedance $Z(j\omega)$ at the gain node is the combination of the output resistance of the current mirrors R_o in parallel with a compensation capacitor C, we can write:

$$Z(j\omega) = \frac{R_o}{1 + j\omega R_o C} = \frac{R_o}{1 + j\dfrac{\omega}{\omega_o}} \tag{5.2}$$

where $\omega_o = 1/R_o \cdot C$ represents the frequency where the open-loop transimpedance gain is 3 dB down from its low frequency value R_o. In general, R_o is designed to be very high in value.

Referring to the non-inverting amplifier configuration shown in Fig. 5.3:

$$I_{INV} = \frac{V_{IN}}{R_G} - \frac{V_{OUT} - V_{IN}}{R_F} = \frac{V_{IN}}{R_G // R_F} - \frac{V_{OUT}}{R_F} \tag{5.3}$$

Substituting Eq. 5.1 into Eq. 5.3 yields the following expression for the closed-loop gain:

$$A_{CL}(j\omega) = \left(1 + \frac{R_F}{R_G}\right) \cdot \frac{Z(j\omega)}{R_F + Z(j\omega)} = \left(1 + \frac{R_F}{R_G}\right) \cdot \frac{1}{1 + \dfrac{R_F}{Z(j\omega)}} \tag{5.4}$$

Combining Eqs. 5.2 and 5.4, and assuming that the low frequency value of the open-loop transimpedance is much higher than the feedback resistor ($R_o \gg R_F$) gives:

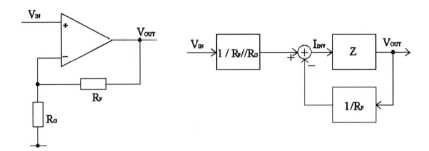

FIGURE 5.3   CFOA non-inverting amplifier configuration.

$$A_{CL}(j\omega) = \left(1 + \frac{R_F}{R_G}\right) \cdot \frac{1}{1 + j\dfrac{R_F \cdot \omega}{R_o \cdot \omega_o}} = \frac{A_{Vo}}{1 + j\dfrac{\omega}{\omega_\alpha}} \tag{5.5}$$

Referring to Eq. 5.5, the closed-loop gain $A_{Vo} = 1 + R_F/R_G$, while the closed-loop –3 dB frequency ω_α is given by:

$$\omega_\alpha = \frac{R_o}{R_F} \cdot \omega_o \tag{5.6}$$

Eq. 5.6 indicates that the closed-loop bandwidth does not depend on the closed-loop gain as in the case of a conventional VOA, but is determined by the feedback resistor R_F. Explaining this intuitively, the current available to charge the compensation capacitor at the gain node is determined by the value of the feedback resistor R_F and not R_o, provided that $R_o \gg R_F$. So, once the bandwidth of the amplifier is set via R_F, the gain can be independently varied by changing R_G. The ability to control the gain independently of bandwidth constitutes a major advantage of current feedback op-amps over conventional voltage feedback op-amps.

The other major advantage of the CFOA compared to the VFOA is the inherent absence of slew rate limiting. For the circuit of Fig. 5.3, assume that the input buffer is very fast and thus a change in voltage at the non-inverting input is instantaneously converted to the inverting input. When a step ΔV_{IN} is applied to the non-inverting input, the buffer output current can be derived as:

$$I_{INV} = \frac{V_{IN} - V_{OUT}}{R_F} + \frac{V_{IN}}{R_G} \tag{5.7}$$

Eq. 5.7 indicates that the current available to charge/discharge the compensation capacitor is proportional to the input step regardless of its size, that is, there is no upper limit. The rate of change of the output voltage is thus:

$$\frac{dV_{OUT}}{dt} = \frac{I_{INV}}{C} \Rightarrow V_{OUT}(t) = \Delta V_{IN} \cdot \left(1 + \frac{R_F}{R_G}\right) \cdot (1 - e^{-t/R_f \cdot C}) \tag{5.8}$$

Eq. 5.8 indicates an exponential output transition with time constant $\tau = R_F \cdot C$. Similar to the small-signal frequency response, the large-signal transient response is governed by R_F alone, regardless of the magnitude of the closed-loop gain. The absence of slew rate limiting allows for faster settling times and eliminates slew rate-related non-linearities.

In most practical bipolar realizations, Darlington-pair transistors are used in the input stage to reduce input bias currents, which makes the op-amp somewhat noisier and increases the input offset voltage.

This is not necessary in CMOS realizations due to the inherently high MOSFET input impedance. However, in a closed-loop CFOA, R_G should be much larger than the output impedance of the buffer. In bipolar realizations, it is fairly simple to obtain a buffer with low output resistance, but this becomes more of a problem in CMOS due to the inherently lower gain of MOSFET devices. As a result, R_G typically needs to be higher in a CMOS CFOA than in a bipolar realization, and consequently, R_F needs to be increased above the value required for optimum high-frequency performance. Additionally, the fact that the input buffer is not in the feedback loop imposes linearity limitations on the structure, especially if the impedance at the gain node is not very high. Regardless of these problems, current feedback op-amps exhibit excellent high-frequency characteristics and are increasingly popular in video and communications applications.[13]

The following sections outline the development of a novel low-output impedance CMOS buffer, which is then employed in a CMOS CFOA to reduce the minimum allowable value of R_G.

CMOS Compound Device

A simple PMOS source follower is shown in Fig. 5.4. The output impedance seen looking into the source of M1 is approximately $Z_{out} = 1/g_m$, where g_m is the small signal transconductance of M1. To increase g_m, the drain current of M1 could be increased, which leads to an increased power dissipation. Alternatively, the dimensions of M1 can be increased, resulting in additional parasitic capacitance and hence an inferior frequency response. Figure 5.5 shows a configuration that achieves a higher transconductance than the simple follower of Fig. 5.3 for the same bias current.[11] The current of M2 is fed back to M1 through the a:1 current mirror. This configuration can be viewed as a compound transistor whose gate is the gate of M1 and whose source is the source of M2. The impedance looking into the compound source can be approximated as $Z_{out} = (g_{m1} - a \cdot g_{m2})/(g_{m1} \cdot g_{m2})$, where g_{m1} and g_{m2} represent the small signal transconductance of M1 and M2, respectively. The output impedance can be made small by setting the current mirror ransfer ratio $a = g_{m1}/g_{m2}$.

The p-compound device is practically implemented as in Fig. 5.6. In order to obtain a linear voltage transfer function from node 1 to 2, the gate-source voltages of M1 and M3 must cancel. The current mirror (M4-M2) acts as an NMOS-PMOS gate-source voltage matching circuit[14] and compensates for the difference in the gate-source voltages of M1 and M3, which would normally appear as an output offset. DC analysis, assuming a square law model for the MOSFETs, shows that the output voltage exactly follows the input voltage. However, in practice, channel length modulation and body effects preclude exact cancellation.[15]

Buffer and CFOA Implementation

The current feedback op-amp shown in Fig. 5.7 has been implemented in a single-well 0.6-μm digital tCMOS process[11]; the corresponding layout plot is shown in Fig. 5.8. The chip has an area of 280 μm by

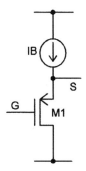

FIGURE 5.4 Simple PMOS source follower.

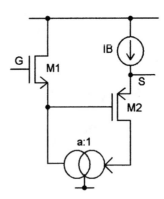

FIGURE 5.5   Compound MOS device.

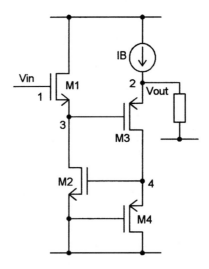

FIGURE 5.6   Actual p-compound device implementation.

330 μm and a power dissipation of 12 mW. The amplifier comprises two voltage followers (input and output) connected by cascoded current mirrors to enhance the gain node impedance. A compensation capacitor (Cc = 0.5 pF) at the gain node ensures adequate phase margin and thus closed-loop stability. The voltage followers have been implemented with two compound transistors, p-type and n-type, in a push-pull arrangement. Two such compound transistors in the output stage are shown shaded in Fig. 5.7. The input voltage follower of the current feedback op-amp was initially tested open-loop, and measured results are summarized in Table 5.1. The load is set to 10 kΩ/10 pF, except where mentioned otherwise, 10 kΩ being a limit imposed by overall power dissipation of the chip. Intermodulation distortion was measured with two tones separated by 200 kHz. The measured output impedance of the buffer is given in Fig. 5.9. It remains below 80 Ω up to a frequency of about 60 MHz, when it enters an inductive region. A maximum impedance of 140 Ω is reached around 160 MHz. Beyond this frequency, the output impedance is dominated by parasitic capacitances. The inductive behavior is characteristic of the use of feedback to reduce output impedance, and can cause stability problems when driving capacitive loads. Small-signal analysis (summarized in Table 5.2) predicts a double zero in the output impedance.[15]

Making factor G in Table 5.2 small will reduce the output impedance, but also moves the double zero to lower frequencies and intensifies the inductive behavior. The principal tradeoff in this configuration is between output impedance magnitude and inductive behavior. In practice, the output impedance can be reduced by a factor of 3 while still maintaining good stability when driving capacitive loads. Figure

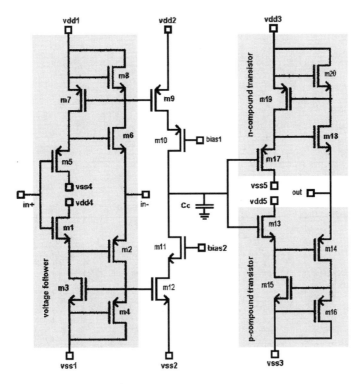

FIGURE 5.7 Current feedback op-amp schematic.

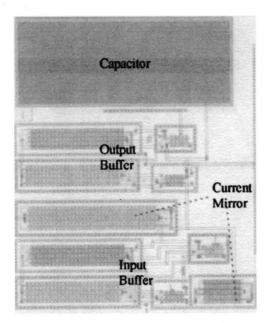

FIGURE 5.8 Current feedback op-amp layout plot.

5.10 shows the measured frequency response of the buffer. Given the low power dissipation, excellent slew rates have been achieved (Table 5.2).

After the characterization of the input buffer stage, the entire CFOA was tested to confirm the suitability of the compound transistors for the implementation of more complex building blocks. Open-loop

TABLE 5.1 Voltage Buffer Performance

Power Supply	5 V	Dissipation	5 mW
DC gain (no load)	−3.3dB	Bandwidth	140 MHz
Output impedance	75Ω	Min. load resistance	10 KΩ
HD2 (V_{in} = 200 mV$_{rms}$)	1 MHz	−50 dB	
	10 MHz	−49 dB	
	20 MHz	−45 dB	
IM3 (V_{in}=200 mV$_{rms}$)	20 MHz, Δf = 200 KHz	−53 dB	
Slew rate	(Load = 10 pF)	+ 130 V/μs	−72 V/μs
Input referred noise	10 $nV\sqrt{Hz}$		

Note: Load = 10 kΩ/10 pF, except for slew rate measurement.

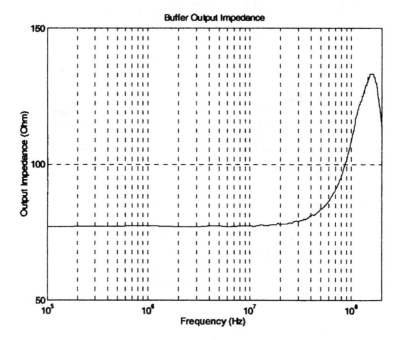

FIGURE 5.9 Measured buffer output impedance characteristics.

TABLE 5.2 Voltage Transfer Function and Output Impedance of Compound Device

$$Z_{out} = \frac{G}{(g_{m1} + g_{ds1} + g_{ds2}) \cdot (g_{m3} + g_{ds3}) \cdot (g_{m4} + g_{ds4})}$$

$$\frac{V_{out}}{V_{in}} = \frac{g_{m1} \cdot g_{m3} \cdot (g_{m4} + g_{ds4})}{(g_{m1} + g_{ds1} + g_{ds2}) \cdot (g_{m3} + g_{ds3}) \cdot (g_{m4} + g_{ds4}) + g_L \cdot G}$$

$$G = (g_{m1} + g_{ds1} + g_{ds2}) \cdot (g_{m4} + g_{ds4} + g_{ds3}) - g_{m2} \cdot g_{m3}$$

transimpedance measurements are shown in Fig. 5.11. The bandwidth of the amplifier was measured at gain settings of 1, 2, 5, and 10 in a non-inverting configuration, and the feedback resistor was trimmed to achieve maximum bandwidth at each gain setting separately. CFOA measurements are summarized in Table 5.3, loading conditions are again 10 kΩ/10 pF.

Fig. 5.12 shows the measured frequency response for various gain settings. The bandwidth remains constant at 110 MHz for gains of 1, 2, and 5, consistent with the expected behavior of a CFOA. The

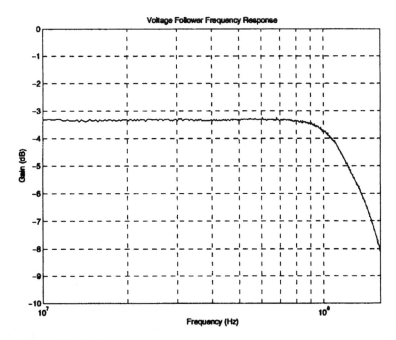

FIGURE 5.10   Measured buffer frequency response.

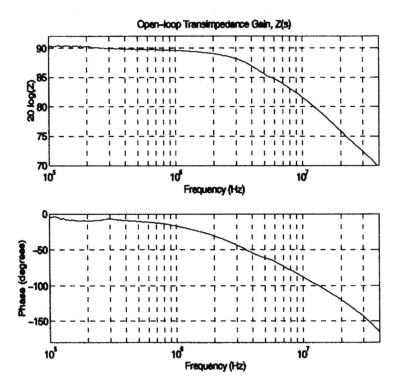

FIGURE 5.11   Measured CFOA open-loop transimpedance gain.

TABLE 5.3   Current Feedback Op-Amp Measurement Summary

Power Supply	5 V	Power Dissipation	12 mW
Gain	Bandwidth (MHz)		
1	117		
2	118		
5	113		
10	42		

Frequency	Input (mV rms)	Gain	HD2 (dB)
1 MHz	140	2	−51
	40	5	−50
	10	10	−49
10 MHz	80	2	−42
	40	5	−42
	13	10	−43

bandwidth falls to 42 MHz for a gain of 10 due to the finite output impedance of the input buffer stage which series as the CFOA inverting input. Figure 5.13 illustrates the step response of the CFOA driving a 10 kΩ/10 pF load at a voltage gain of 2. It can be seen that the inductive behavior of the buffers has little effect on the step response. Finally, distortion measurements were carried out for the entire CFOA for gain settings 2, 5, and 10 and are summarized in Table 5.3. HD2 levels can be further improved by employing a double-balanced topology. A distortion spectrum is shown in Fig. 5.14; the onset of HD3 is due to clipping at the test conditions.

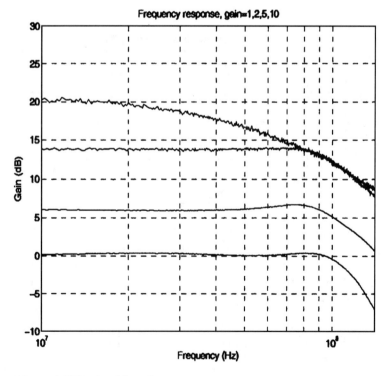

FIGURE 5.12   Measured CFOA closed-loop frequency response.

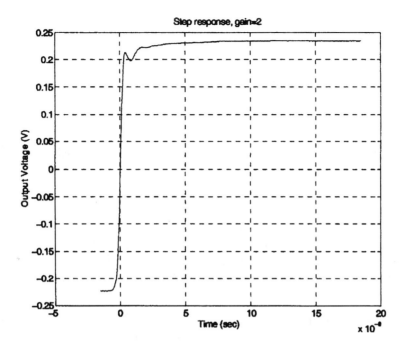

FIGURE 5.13   Measured CFOA step response.

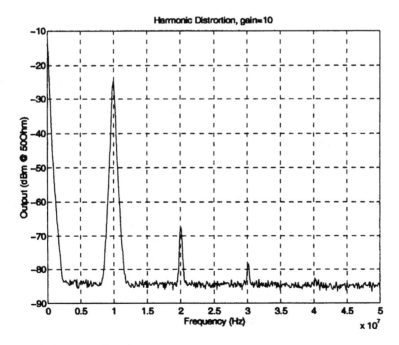

FIGURE 5.14   CFOA harmonic distortion measurements.

5.3 RF Low-Noise Amplifiers

This section reviews the important performance criteria demanded of the front-end amplifier in a wireless communication receiver. The design of CMOS LNAs for front-end wireless communication receiver applications is then addressed. Section 5.4 considers the related topic of low-noise amplifiers for optical receiver front-ends.

Specifications

The front-end amplifier in a wireless receiver must satisfy demanding requirements in terms of noise, gain, impedance matching, and linearity.

Noise

Since the incoming signal is usually weak, the front-end circuits of the receiver must possess very low noise characteristics so that the original signal can be recovered. Provided that the gain of the front-end amplifier is sufficient so as to suppress noise from the subsequent stages, the receiver noise performance is determined predominantly by the front-end amplifier. Hence, the front-end amplifier should be a low-noise amplifier (LNA).

Gain

The voltage gain of the LNA must be high enough to ensure that noise contributions from the following stages can be safely neglected. As an example, Fig. 5.15 shows the first three stages in a generic front-end receiver, where the gain and output-referred noise of each stage are represented by G_i and N_i ($i = 1, 2, 3$), respectively. The total noise at the third stage output is given by:

$$N_{out} = N_{in} G_1 G_2 G_3 + N_1 G_2 G_3 + N_2 G_3 + N_3 \tag{5.9}$$

This output noise (N_{out}) can be referred to the input to derive an equivalent input noise (N_{eq}):

$$N_{eq} = \frac{N_{out}}{Gain} = \frac{N_{out}}{G_1 G_2 G_3} = N_{in} + \frac{N_1}{G_1} + \frac{N_2}{G_1 G_2} + \frac{N_3}{G_1 G_2 G_3} \tag{5.10}$$

According to Eq. 22.10, the gain of the first stage should be high in order to reduce noise contributions from subsequent stages. However, if the gain is too high, a large input signal may saturate the subsequent stages, yielding intermodulation products which corrupt the desired signal. Thus, optimization is inevitable.

Input Impedance Matching

The input impedance of the LNA must be matched to the antenna impedance over the frequency range of interest, in order to transfer the maximum available power to the receiver.

Linearity

Unwanted signals at frequencies fairly near the frequency band of interest may reach the LNA with signal strengths many times higher than that of the wanted signal. The LNA must be sufficiently linear to

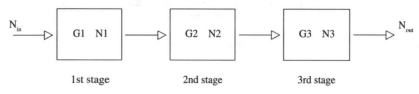

FIGURE 5.15   Three-stage building block with gain G_i and noise N_i per stage.

prevent these out-of-band signals from generating intermodulation products within the wanted frequency band, and thus degrading the reception of the desired signal. Since third-order mixing products are usually dominant, the linearity of the LNA is related to the "third-order intercept point" (IP3), which is defined as the input power level that results in equal power levels for the output fundamental frequency component and the third-order intermodulation components. The dynamic range of a wireless receiver is limited at the lower bound by noise and at the upper band by non-linearity.

CMOS Common-Source LNA: Simplified Analysis

Input Impedance Matching by Source Degeneration

For maximum power transfer, the input impedance of the LNA must be matched to the source resistance, which is normally 50 Ω. Impedance-matching circuits consist of reactive components and therefore are (ideally) lossless and noiseless. Figure 5.16 shows the small signal equivalent circuit of a CS LNA input stage with impedance-matching circuit, where the gate-drain capacitance C_{gd} is assumed to have negligible effect and is thus neglected.[16,17] The input impedance of this CS input stage is given by:

$$Z_{in} = j\omega(L_g + L_s) + \frac{1}{j\omega C_{gs}} + \frac{g_m}{C_{gs}}L_s \qquad (5.11)$$

Thus, for matching, the two conditions below must be satisfied:

$$(i) \qquad \omega_o^2 = \frac{1}{(L_g + L_s)C_{gs}} \qquad \text{and} \qquad (ii) \qquad \frac{g_m}{C_{gs}}L_s = Rs \qquad (5.12)$$

Noise Figure of CS Input Stage

Two main noise sources exist in a CS input stage as shown in Fig. 5.17; thermal noise from the source resistor R_s (denoted $\overline{v_{Rs}^2}$) and channel thermal noise from the input transistor (denoted $\overline{i_d^2}$). The output noise current due to $\overline{v_{Rs}^2}$ can be determined from Fig. 5.17 as:

$$\overline{i_{nout1}^2} = \frac{g_m^2 \overline{v_{Rs}^2}}{\omega^2(g_m L_s + R_s C_{gs})^2} = \frac{g_m^2}{4\omega^2 R_s^2 C_{gs}^2}\overline{v_{Rs}^2} \qquad (5.13)$$

while the output noise current due to $\overline{i_d^2}$ can be evaluated as:

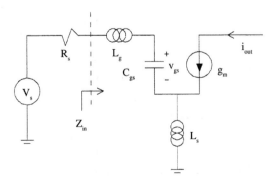

FIGURE 5.16 Simplified small-signal equivalent circuit of the CS stage.

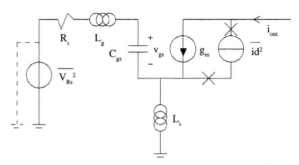

FIGURE 5.17 Simplified noise equivalent circuit of the CS stage. $\overline{V_{Rs}}^2 = 4kT_{Rs}$; $\overline{id}^2 = KT\Gamma g_{dc}$.

$$\overline{i_{nout2}} = \frac{\overline{i_d}}{\left(1 + \dfrac{g_m L_s}{R_s C_{gs}}\right)} = \frac{1}{2}\overline{i_d} \qquad \therefore \overline{i_{nout2}}^2 = \frac{1}{4}\overline{i_d}^2 \tag{5.14}$$

From Eqs. 5.13 and 5.14, the noise figure of the CS input stage is determined as:

$$NF = 1 + \frac{\overline{i_{nout2}}^2}{\overline{i_{nout1}}^2} = 1 + \Gamma\left(\frac{\omega_o^2 R_s C_{gs}^2}{g_m}\right) = 1 + \Gamma\left(\frac{L_s}{L_s + L_g}\right) \tag{5.15}$$

In practice, any inductor (especially a fully integrated inductor) has an associated resistance that will contribute thermal noise, degrading the noise figure in Eq. 5.15.

Voltage Amplifier with Inductive Load

Referring to Fig. 5.15, the small signal current output is given by:

$$i_{out} = \frac{g_m v_s}{[1 - \omega^2 C_{gs}(L_g + L_s)] + j\omega(g_m L_s + R_s C_{gs})} \tag{5.16}$$

For an inductive load (L_1) with a series internal resistance r_{L1}, the output voltage is thus:

$$v_{out} = -i_{out}(r_{L1} + j\omega L_1) = \frac{-(r_{L1} + j\omega L_1)g_m v_s}{[1 - \omega^2 C_{gs}(L_g + L_s)] + j\omega(g_m L_s + R_s C_{gs})} \tag{5.17}$$

Assuming that the input is impedance matched, the voltage gain at the output is given by:

$$\left|\frac{v_{out}}{v_s}\right| = \frac{\sqrt{(r_{L1})^2 + (\omega_o)^2(L_1)^2}}{2\omega_o L_s} = \frac{r_{L1}}{2\omega_o L_s}\sqrt{1 + \left(\frac{\omega_o L_1}{r_{L1}}\right)^2} \cong \frac{1}{2}\omega_o\left(\frac{L_1}{L_s}\right)\left(\frac{L_1}{r_{L1}}\right) \tag{5.18}$$

CMOS Common-Source LNA: Effect of C_{gd}

In the analysis so far, the gate-drain capacitance (C_{gd}) has been assumed to be negligible. However, at very high frequencies, this component cannot be neglected. Figure 5.18 shows the modified input stage of a CS LNA including C_{gd} and an input ac-coupling capacitance C_{in}. Small signal analysis shows that the input impedance is now given by:

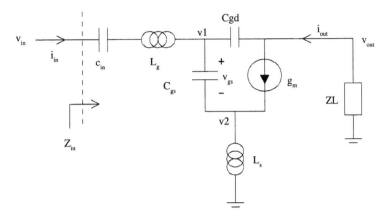

FIGURE 5.18 Noise equivalent circuit of the CS stage, including effects of C_{gd}.

$$Z_{in} = \frac{g_m L_s}{C_{gs} + C_{gd}\left[j\omega L_s g_m + \frac{g_m(1-\omega^2 L_s C_{gd})}{\frac{1}{Z_L} + j\omega C_{gd}}\right]} \qquad (5.19)$$

Equation 5.19 exhibits resonance frequencies that occur when:

$$1 - \omega^2 L_s C_{gs} = 0 \qquad \text{and} \qquad 1 - \omega^2 L_g C_{in} = 0 \qquad (5.20)$$

Equation 5.19 indicates that the input impedance matching is degraded by the load Z_L when C_{gd} is included in the analysis.

Input Impedance with Capacitive Load

If the load Z_L is purely capacitive, that is,

$$Z_L = \frac{1}{j\omega C_L} \qquad (5.21)$$

then the input impedance can be easily matched to the source resistor R_s. Substituting Eq. 5.21 for Z_L, the bracketed term in the denominator of Eq. 5.19 becomes:

$$d_1 = j\omega L_s g_m + \frac{g_m(1-\omega^2 L_s C_{gd})}{j\omega(C_{gd} + C_L)} = 0 \qquad (5.22)$$

under the condition that

$$1 - \omega^2 L_s(2C_{gd} + C_L) = 0 \qquad (5.23)$$

The three conditions in Eqs. 5.20 and 5.23 should be met to ensure input impedance matching. However, in practice, we are unlikely to be in the situation of using a load capacitor.

Input Impedance with Inductive Load

If $Z_L = j\omega L_L$, the CS LNA input impedance is given by:

$$Z_{in} = \frac{g_m L_s}{C_{gs} + j\omega C_{gd} g_m \left[L_s + L_L \left(\frac{1 - \omega^2 L_s C_{gd}}{1 - \omega^2 L_L C_{gd}} \right) \right]} \tag{5.24}$$

In order to match to a purely resistive input, the value of the reactive term in Eq. 5.24 must be negligible, which is difficult to achieve.

Cascode CS LNA

Input Matching

As outlined in the paragraph above, the gate-drain capacitance (C_{gd}) degrades the input impedance matching and therefore reduces the power transfer efficiency. In order to reduce the effect of C_{gd}, a cascoded structure can be used.[18–20] Figure 5.19 shows a cascode CS LNA. Since the voltage gain from the gate to the drain of M1 is unity, the gate-drain capacitance (C_{gd1}) no longer sees the full input-output voltage swing which greatly improves the input-output isolation. The input impedance can be approximated by Eq. 5.11, thus allowing a simple matching circuit to be employed.[18]

Voltage Gain

Figure 5.20 shows the small-signal equivalent circuit of the cascode CS LNA. Assuming that input is fully matched to the source, the voltage gain of the amplifier is given by:

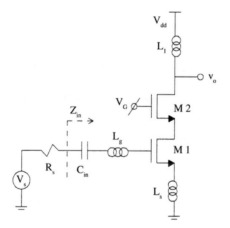

FIGURE 5.19   Cascode CS LNA.

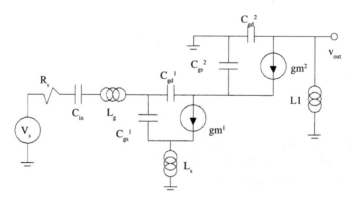

FIGURE 5.20   Equivalent circuit of cascode CS LNA.

$$\frac{v_{out}}{v_s} = -\frac{1}{2}\left(\frac{j\omega L_1}{1 - \omega^2 L_1 C_{gd2}}\right)\left(\frac{g_{m2}}{g_{m2} + j\omega C_{gs2}}\right)\left[\frac{g_{m1}}{(1 - \omega^2 L_s C_{gs1}) + j\omega L_s g_{m1}}\right] \quad (5.25)$$

At the resonant frequency, the voltage gain is given by:

$$\frac{v_{out}}{v_s}(\omega_o) = -\frac{1}{2}\left(\frac{L_1}{L_s}\right)\left(\frac{1}{1 - \omega_o^2 L_1 C_{gd2}}\right) \times \frac{1}{1 + j\omega_o\left(\frac{C_{gs1}}{g_{m2}}\right)} \approx -\frac{1}{2}\left(\frac{L_1}{L_s}\right) \times \frac{1}{1 + j\left(\frac{\omega_o}{\omega_T}\right)} \quad (5.26)$$

From Eq. 5.26, the voltage gain is dependent on the ratio of the load and source inductance values. Therefore, high gain accuracy can be achieved since this ratio is largely process independent.

Noise Figure

Figure 5.21 shows an equivalent circuit of the cascode CS LNA for noise calculations. Three main noise sources can be identified: the thermal noise voltage from R_s, and the channel thermal noise currents from M1 and M2. Assuming that the input impedance is matched to the sources, the output noise current due to $\overline{v_{RS}^2}$ can be derived as:

$$\overline{i_{out1}} = \frac{1}{2j\omega_o L_s(1 - \omega_o^2 L_1 C_{gd2})}\left(\frac{g_{m2}}{g_{m2} + j\omega_o C_{gs2}}\right)\overline{v_{RS}} \quad (5.27)$$

The output noise current contribution due to $\overline{i_{d1}^2}$ of M1 is given by:

$$\overline{i_{out2}} = \frac{1}{2(1 - \omega_o^2 L_1 C_{gd2})}\left(\frac{g_{m2}}{g_{m2} + j\omega_o C_{ds2}}\right)\overline{i_{d1}} \quad (5.28)$$

The output noise current due to $\overline{i_{d2}^2}$ of M2 is given by:

$$\overline{i_{out3}} = \frac{j\omega_o C_{gs2}}{(1 - \omega_o^2 L_1 C_{gd2})(g_{m2} + j\omega_o C_{gs2})}\overline{i_{d2}} \quad (5.29)$$

The noise figure of the cascode CS LNA can thus be derived as:

$$NF = 1 + \frac{\overline{i_{out2}^2}}{\overline{i_{out1}^2}} + \frac{\overline{i_{out3}^2}}{\overline{i_{out1}^2}} = 1 + \Gamma\left(1 + \frac{4\omega_o^2 C_{gs2}^2}{g_{m1}g_{m2}}\right) \quad (5.30)$$

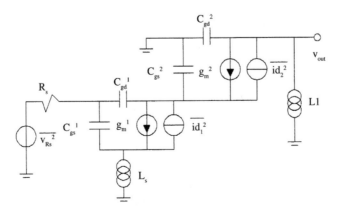

FIGURE 5.21   Noise equivalent circuit of cascode CS LNA.

In order to improve the noise figure, the transconductance values (g_m) of M1 and M2 should be increased. Since the gate-source capacitance (C_{gs2}) of M2 is directly proportional to the gate width, the gate width of M2 cannot be enlarged to increase the transconductance. Instead, this increase should be realized by increasing the gate bias voltage.

5.4   Optical Low-Noise Preamplifiers

Figure 5.22 shows a simple schematic diagram of an optical receiver, consisting of a photodetector, a preamplifier, a wide-band voltage amplifier, and a pre-detection filter. Since the front-end transimpedance preamplifier is critical in determining the overall receiver performance, it should possess a wide bandwidth so as not to distort the received signal, high gain to reject noise from subsequent stages, low noise to achieve high sensitivity, wide dynamic range, and low inter-symbol-interference (ISI).

Front-End Noise Sources

Receiver noise is dominated by two main noise sources: the detector (PIN photodiode) noise and the amplifier noise. Figure 5.23 illustrates the noise equivalent circuit of the optical receiver.

PIN Photodiode Noise

The noise generated by a PIN photodiode arises mainly from three shot noise contributions: quantum noise $S_q(f)$, thermally generated dark-current shot noise $S_D(f)$, and surface leakage-current shot noise $S_L(f)$. Other noise sources in a PIN photodiode, such as series resistor noise, are negligible in comparison. The quantum noise $S_q(f)$, also called signal-dependent shot noise, is produced by the light-generating nature of photonic detection and has a spectral density $S_q(f) = 2q I_{pd} \Delta f$, where I_{pd} is the mean signal current arising from the Poisson statistics. The dark-current shot noise $S_D(f)$ arises in the photodiode bulk material. Even when there is no incident optical power, a small reverse leakage current still flows, resulting in shot noise with a spectral density $S_D(f) = 2q I_{DB} \Delta f$, where I_{DB} is the mean thermally generated dark current. The leakage shot noise $S_L(f)$ occurs because of surface effects around the active region, and is described by $S_L(f) = 2q I_{SL} \Delta f$, where I_{SL} is the mean surface leakage current.

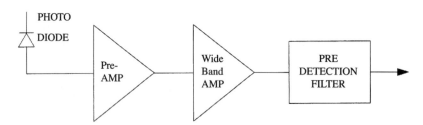

FIGURE 5.22   Front-end optical receiver.

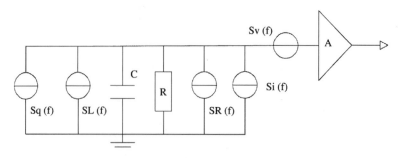

FIGURE 5.23   Noise equivalent circuit of the front-end optical receiver.

Amplifier Noise

For a simple noise analysis, the pre- and post-amplifiers in Fig. 5.22 are merged to a single amplifier with a transfer function of $A_v(\omega)$. The input impedance of the amplifier is modeled as a parallel combination of R_{in} and C_{in}.

If the photodiode noise is negligibly small, the amplifier noise will dominate the whole receiver noise performance, as can be inferred from Fig. 5.23. The equivalent noise current and voltage spectral densities of the amplifier are represented as $S_i(A^2/Hz)$ and $S_v(V^2/Hz)$, respectively.

Resistor Noise

The thermal noise generated by a resistor is directly proportional to the absolute temperature T and is represented by a series noise voltage generator or by a shunt noise current generator[21] of value:

$$\overline{v_R^2} = 4kTR\Delta f \qquad \text{or} \qquad \overline{i_R^2} = 4kT\frac{1}{R}\Delta f \tag{5.31}$$

where k is Boltzmann's constant and R is the resistance.

Receiver Performance Criteria

Equivalent Input Noise Current $\langle \overline{i_{eq}^2} \rangle$

The transfer function from the current input to the amplifier output voltage is given by:

$$Z_T(\omega) = \frac{V_{out}}{I_{pd}} = Z_{in}A_v(\omega) = \frac{R_{in}}{1\,j\omega R_{in}(C_{pd} + C_{in})}A_v(\omega) \tag{5.32}$$

where C_{pd} is the photodiode capacitance, and R_{in} and C_{in} are the input resistance and capacitance of the amplifier, respectively. Assuming that the photodiode noise contributions are negligible and that the amplifier noise sources are uncorrelated, the equivalent input noise current spectral density can be derived from Fig. 5.23 as:

$$S_{eq}(f) = S_i + \frac{S_v}{[Z_{in}]^2} = S_i + S_v\left[\frac{1}{R_{in}^2} + (2\pi f)^2(C_{pd} + C_{in})^2\right] \tag{5.33}$$

The total mean-square noise output voltage $\langle \overline{v_{no}^2} \rangle$ is calculated by combining Eqs. 5.32 and 5.33 as follows:

$$\langle \overline{v_{no}^2} \rangle = \int_0^\infty S_{eq}(f)|Z_T(f)|^2 df \tag{5.34}$$

This total noise voltage can be referred to the input of the amplifier by dividing it by the squared dc gain $|Z_T(0)|^2$ of the receiver, to give an equivalent input mean-square noise current:

$$\langle \overline{i_{eq}^2} \rangle = \frac{\langle \overline{v_{no}^2} \rangle}{|Z_T(0)|^2} = \left(S_i + \frac{S_v}{R_{in}^2}\right)\int_0^\infty \frac{|Z_T(f)|^2}{|Z_T(0)|^2}df + S_v[2\pi(C_{pd} + C_{in})]^2\int_0^\infty f^2\frac{|Z_T(f)|^2}{|Z_T(0)|^2}df \tag{5.35}$$

$$= \left(S_i + \frac{S_v}{R_{in}^2}\right)I_2B + [2\pi(C_{pd} + C_{in})]^2 I_3 B^3 S_v$$

where B is the operating bit-rate, and $I_2(= 0.56)$ and $I_3(= 0.083)$ are the Personick second and third integrals, respectively, as given in Ref. 22.

According to Morikoni et al.,[23] the Personick integral in Eq. 5.35 is correct only if a receiver produces a raised-cosine output response from a rectangular input signal at the cut-off bit rate above which the frequency response of the receiver is zero. However, the Personick integration method is generally preferred when comparing the noise (or sensitivity) performance of different amplifiers.

Optical Sensitivity

Optical sensitivity is defined as the minimum received optical power incident on a perfectly efficient photodiode connected to the amplifier, such that the presence of the amplifier noise corrupts on average only one bit per 10^9 bits of incoming data. Therefore, a detected power greater than the sensitivity level guarantees system operation at the desired performance. The optical sensitivity is predicted theoretically by calculating the equivalent input noise spectral density of the receiver, and is calculated[24] via Eq. 5.36:

$$S = 10\log_{10}\left(Q\frac{hc}{q\lambda}\sqrt{\langle \overline{i_{eq}^2}\rangle} \cdot \frac{1}{1mW}\right) \text{ (dBm)} \tag{5.36}$$

where h is Planck's constant, c is the speed of light, q is electronic charge, and λ (μm) is the wavelength of light in an optical fiber. $Q(= \sqrt{SNR})$, where SNR represents the required signal-to-noise ratio (SNR). The value of Q should be 6 for a bit error rate (BER) of 10^{-9}, and 7.04 for a BER of 10^{-12}. The relation between Q and BER is given by:

$$BER = \frac{\exp(-Q^2/2)}{\sqrt{2\pi}Q} \tag{5.37}$$

Since the number of photogenerated electrons in a single bit is very large (more than 10^4) for optoelectronic integrated receivers,[25] Gaussian statistics of the above BER equation can be used to describe the detection probability in PIN photodiodes.

SNR at the Photodiode Terminal[22]

Among the photodiode noise sources, quantum noise is generally dominant and can be estimated as:

$$\langle \overline{i_n^2}\rangle_q = 2qI_{pd}B_{eq} \tag{5.38}$$

where I_{pd} is the mean signal current and B_{eq} is the equivalent noise bandwidth. The signal-to-noise-ratio (SNR) referred to the photodiode terminal is thus given by:

$$SNR = \frac{I_{pd}^2}{\langle \overline{i_n^2}\rangle_{pd} + \dfrac{4kTB_{eq}}{R_B} + \langle \overline{i_{eq}^2}\rangle_{amp}} \tag{5.39}$$

where all noise contributions due to the amplifier are represented by the equivalent noise current $\langle \overline{i_{eq}^2}\rangle_{amp}$. It is often convenient to combine the noise contributions from the amplifier and the photodiode with the thermal noise from the bias resistor, by defining a noise figure NF:

$$\langle \overline{i_n^2}\rangle_{pd} + \frac{4kTB_{eq}}{R_B} + \langle \overline{i_{eq}^2}\rangle_{amp} = \frac{4kTB_{eq}NF}{R_B} \tag{5.40}$$

The SNR at the photodiode input is thus given by:

$$\text{SNR} \cong \frac{I_{pd}^2 R_B}{4kTB_{eq}\text{NF}} \tag{5.41}$$

Inter-Symbol Interference (ISI)

When a pulse passes through a band-limited channel, it gradually disperses. When the channel bandwidth is close to the signal bandwidth, the expanded rise and fall times of the pulse signal will cause successive pulses to overlap, deteriorating the system performance and giving higher error rates. This pulse overlapping is known as inter-symbol interference (ISI). Even with raised signal power levels, the error performance cannot be improved.[26]

In digital optical communication systems, sampling at the output must occur at the point of maximum signal in order to achieve the minimum error rate. The output pulse shape should therefore be chosen to maximize the pulse amplitude at the sampling instant and give a zero at other sampling points; that is, at multiples of $1/B$, where B is the data-rate. Although the best choice for this purpose is the sinc-function pulse, in practice a raised-cosine spectrum pulse is used instead. This is because the sinc-function pulse is very sensitive to changes in the input pulse shape and variations in component values, and because it is impossible to generate an ideal sinc-function.

Dynamic Range

The dynamic range of an optical receiver quantifies the range of detected power levels within which correct system operation is guaranteed. Dynamic range is conventionally defined as the difference between the minimum input power (which determines sensitivity) and the maximum input power (limited by overload level). Above the overload level, the BER rises due to the distortion of the received signal.

Transimpedance (TZ) Amplifiers

High-impedance (HZ) amplifiers are effectively open-loop architectures, and exhibit a high gain but a relatively low bandwidth. The frequency response is similar to that of an integrator, and thus HZ amplifiers require an output equalizer to extend their frequency capabilities. In contrast, the transimpedance (TZ) configuration exploits resistive negative feedback, providing an inherently wider bandwidth and eliminating the need for an output equalizer. In addition, the use of negative feedback provides a relatively low input resistance and thus the architecture is less sensitive to the photodiode parameters. In a TZ amplifier, the photodiode bias resistor R_B can be omitted, since bias current is now supplied through the feedback resistor.

In addition to wider bandwidth, TZ amplifiers offer a larger dynamic range because the transimpedance gain is determined by a linear feedback resistor, and not by a non-linear open-loop amplifier as is the case for HZ amplifiers. The dynamic range of TZ amplifiers is set by the maximum voltage swing available at the amplifier output, provided no integration of the received signal occurs at the front end. Since the TZ output stage is a voltage buffer, the voltage swing at the output can be increased with high current operation. The improvement in dynamic range in comparison to the HZ architecture is approximately equal to the ratio of open-loop to closed-loop gain.[27] Conclusively, the TZ configuration offers the better performance compromise compared to the HZ topology, and hence this architecture is preferred in optical receiver applications.

A schematic diagram of a TZ amplifier with PIN photodiode is shown in Fig. 5.24. With an open-loop, high-gain amplifier and a feedback resistor, the closed-loop transfer function of the TZ amplifier is given by:

$$Z_T(s) = \frac{-R_f}{\left(\dfrac{1+A}{A}\right) + sR_f\left[\dfrac{C_{in} + (1+A)C_f}{A}\right]} \cong \frac{R_f}{1 + sR_f\left(\dfrac{C_{in}}{A} + C_f\right)} \tag{5.42}$$

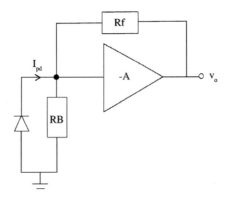

FIGURE 5.24   Schematic diagram of a transimpedance amplifier with photodiode.

where A is the open-loop mid-band gain of the amplifier which is assumed to be greater than unity, R_f is the feedback resistance, C_{in} is the total input capacitance of the amplifier including the photodiode and the parasitic capacitance, and C_f represents the stray feedback capacitance. The −3 dB bandwidth of the TZ amplifier is approximately given by:

$$f_{-3d} = \frac{(1+A)}{2\pi R_f C_T} \qquad (5.43)$$

where C_T is the total input capacitance including the photodiode capacitance. The TZ amplifier can thus have wider bandwidth by increasing the open-loop gain, although the open-loop gain cannot be increased indefinitely without stability problems.

However, a tradeoff between low noise and wide bandwidth exists, since the equivalent input noise current spectral density of TZ amplifier is given by:

$$S_{eq}(f) = \frac{4kT}{R_f} + \frac{4kT}{R_B} + S_i(f) + S_v(f)\left[\left(\frac{1}{R_f} + \frac{1}{R_B}\right)^2 + (2\pi f)^2 (C_{pd} + C_{in})^2\right] \qquad (5.44)$$

where C_{in} is the input capacitance of the input transistor. Increasing the value of R_f reduces the noise current in Eq. 5.44 but also shrinks the bandwidth in Eq. 5.43. This conflict can be mitigated by making A in Eq. 5.43 as large as the closed-loop stability allows.[28] However, the feedback resistance R_f cannot be increased indefinitely due to the dynamic range requirements of the amplifier, since too large a feedback resistance causes the amplifier to be overloaded at high signal levels. This overloading can be avoided by using automatic gain control (AGC) circuitry, which automatically reduces the transimpedance gain in discrete steps to keep the peak output signal constant.[27]

The upper limit of R_f is set by the peak amplitude of the input signal. Since the dc transimpedance gain is approximately equal to the feedback resistance R_f, the output voltage is given by $I_{pd} \times R_f$, where I_{pd} is the signal photocurrent. If this output voltage exceeds the maximum voltage swing at the output, the amplifier will be saturated and the output will be distorted, yielding bit errors. The minimum value of R_f is determined by the output signal level at which the performance of the receiver is degraded due to noise and offsets. For typical fiber-optic communication systems, the input signal power is unknown, and may vary from just above the noise floor to a value large enough to generate 0.5 mA at the detector diode.[29]

The TZ configuration has some disadvantages over HZ amplifiers. The power consumption is fairly high, partly due to the broadband operation provided by negative feedback. A propagation delay exists in the closed-loop of the feedback amplifier that may reduce the phase margin of the amplifier and cause peaking in the frequency response. Additionally, any stray feedback capacitance C_f will further deteriorate the ac performance.

Among three types of TZ configuration in CMOS technology (common-source, common-drain, and common-gate TZ amplifiers), the common-gate configuration has potentially the highest bandwidth due to its inherently lower input resistance. Using a common-gate input configuration, the resulting amplifier bandwidth can be made independent of the photodiode capacitance (which is usually the limiting factor in achieving GHz preamplifier designs). Recently, a novel common-gate TZ amplifier has been demonstrated, which shows superior performance compared to various other configurations.[30,31]

Layout for HF Operation

Wideband high-gain amplifiers have isolation problems irrespective of the choice of technology. Coupling from output to input, from the power supply rails, and from the substrate are all possible. Therefore, careful layout is necessary, and special attention must be given to stray capacitance, both on the integrated circuit and associated with the package.[32]

Input/Output Isolation

For stable operation, a high level of isolation between I/O is necessary. Three main factors degrade the I/O isolation[33,34]: (1) capacitive coupling between I/O signal paths through the air and through the substrate; (2) feedback through the dc power supply rails and ground-line inductance; and (3) the package cavity resonance since at the cavity resonant frequency, the coupling between I/O can become very large.

In order to reduce the unwanted coupling (or to provide good isolation, typically more than 60 dB) between I/O, the I/O pads should be laid out to be diagonally opposite each other on the chip with a thin 'left-to-right' geometry between I/O. The small input signal enters on the left-hand side of the chip, while the large output signal exits on the far right-hand side. This helps to isolate the sensitive input stages from the larger signal output stages.[35,36]

The use of fine line-widths and shielding are effective techniques to reduce coupling through the air. Substrate coupling can be reduced by shielding and by using a thin and low-dielectric substrate. Akazawa et al.[33] suggest a structure for effective isolation: a coaxial-like signal-line for high shielding, and a very thin dielectric dc feed-line structure for low characteristic impedance.

Reduction of Feedback through the Power Supply Rails

Careful attention should be given to layout of power supply rails for stable operation and gain flatness. Power lines are generally inductive; thus, on-chip capacitive decoupling is necessary to reduce the high-frequency power line impedance. However, a resonance between these inductive and capacitive components may occur at frequencies as low as several hundred MHz, causing a serious dip in the gain-frequency response and an upward peaking in the isolation-frequency characteristics. One way to reduce this resonance is to add a series damping resistor to the power supply line, making the Q factor of the LC resonance small. Additionally, the power supply line should be widened to reduce the characteristic impedance/inductance. In practice, if the characteristic impedance is as small as several ohms, the dip and peaking do not occur, even without resistive termination.[33]

Resonance also occurs between the IC pad capacitance (C_{pd}) and the bond-wire inductance (L_{bond}). This resonance frequency is typically above 2 GHz in miniature RF packages. Also in layout, the power supply rails of each IC chip stage should be split from the other stages in order to reduce the parasitic feedback (or coupling effect through wire-bonding inductance), which causes oscillation.[34] This helps to minimize crosstalk through power supply rail. The IC is powered through several pads and each pad is individually bonded to the power supply line.

I/O Pads

The bond pads on the critical signal path (e.g., input pad and output pads) should be made as small as possible to minimize the pad-to-substrate capacitance.[35] A floating n-well placed under the pad will further reduce the pad capacitance since the well capacitance will appear in series with the pad capacitance. This floating well also prevents the pad metal from spiking into the substrate.

High-Frequency (HF) Ground

The best possible HF grounds to the sources of the driver devices (and hence the minimization of inter-stage crosstalk) can be obtained by separate bonding of each source pad of the driver MOSFETs to the ground plane that is very close to the chip.[36] A typical bond-wire has a self-inductance of a few nH, which can cause serious peaking within the bandwidth of amplifiers or even instability. By using multiple bond-wires in parallel, the ground-line inductance can be reduced to less than 1 nH.

Flip-Chip Connection

In noisy environments, the noise-insensitive benefits of optical fibers may be lost at the receiver connection between the photodiode and the preamplifier. Therefore, proper shielding, or the integration of both components onto the same substrate, is necessary to prevent this problem. However, proper shielding is costly, while integration restricts the design to GaAs technologies.

As an alternative, the flip-chip interconnection technique using solder bumps has been used.[37,38] Small solder bumps minimize the parasitics due to the short interconnection lengths and avoid damages by mechanical stress. Also, it needs relatively low-temperature bonding and hence further reduces damage to the devices. Easy alignment and precise positioning of the bonding can be obtained by a self-alignment effect. Loose chip alignment is sufficient because the surface tension of the molten solder during re-flow produces precise self-alignment of the pads.[34] Solder bumps are fabricated onto the photodiode junction area to reduce parasitic inductance between the photodiode and the preamplifier.

5.5   Fundamentals of RF Power Amplifier Design

PA Requirements

An important functional block in wireless communication transceivers is the power amplifier (PA). The transceiver PA takes as input the modulated signal to be transmitted, and amplifies this to the power level required to drive the antenna. Because the levels of power required to transmit the signal reliably are often fairly high, the PA is one of the major sources of power consumption in the transceiver. In many systems, power consumption may not be a major concern, as long as the signal can be transmitted with adequate power. For battery-powered systems, however, the limited amount of available energy means that the power consumed by all devices must be minimized so as to extend the transmit time. Therefore, power efficiency is one of the most important factors when evaluating the performance of a wireless system.

The basic requirement for a power amplifier is the ability to work at low supply voltages as well as high operating frequencies, and the design becomes especially difficult due to the tradeoffs that can be made between supply voltage, output power, distortion, and power efficiency. Moreover, since the PA deals with large signals, small-signal analysis methods cannot be applied directly. As a result, both the analysis and the design of PAs are challenging tasks.

This section will first present a study of various configurations employed in the design of state-of-the-art non-linear RF power amplifiers. Practical considerations toward achieving full integration of PAs in CMOS technology will also be highlighted.

Power Amplifier Classification

Power amplifiers currently employed for wireless communication applications can be classified into two categories: linear power amplifiers and non-linear power amplifiers. For linear power amplifiers, the output signal is controlled by the amplitude, frequency, and phase of the input signal. Conversely, for non-linear power amplifiers, the output signal is only controlled by the frequency of input signal.

Conventionally, linear power amplifiers can be classified as Class A, Class B, or Class AB. These PAs produce a magnified replica of the input signal voltage or current waveform, and are typically used where

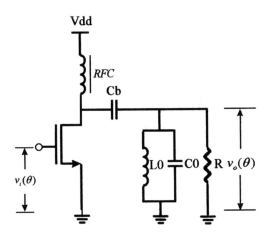

FIGURE 5.25   Single-ended Class A power amplifier.

accurate reproduction of both the envelope and the phase of the signal is required. However, either poor power efficiency or large distortion prevents them from being extensively employed in wireless communications.

Many applications do not require linear RF amplification. Gaussian Minimum Shift Keying (GMSK),[39] the modulation scheme used in the European standard for mobile communications (GSM), is an example of constant envelope modulation. In this case, the system can make use of the greater efficiency and simplicity offered by non-linear PAs. The increased efficiency of non-linear PAs, such as Class C, Class D, and Class E, results from techniques that reduce the average collector voltage–current product (i.e., power dissipation) in the switching device. Theoretically, these switching-mode PAs have 100% power efficiency since, ideally, there is no power loss in the switching device.

Linear Power Amplifiers

Class A
The basic structure of the Class A power amplifier is shown in Fig. 5.25.[40] For Class A amplification, the conduction angle of the device is 360°, that is, the transistor is in its active region for the entire input cycle. The serious shortcoming with Class A PAs is their inherently poor power efficiency, since the transistor is always dissipating power. The efficiency of a single-ended Class A PA is ideally limited to 50%. However, in practice, few designs can reach this ideal efficiency due to additional power loss in the passive components. In an inductorless configuration, the efficiency is only about 25%.[41]

Class B
A PA is defined as Class B when the conduction angle for each transistor of a push-pull pair is 180° during any one cycle. Figure 5.26 shows an inductorless Class B power amplifier. Since each transistor only conducts for half of the cycle, the output suffers crossover distortion due to the finite threshold voltage of each transistor. When no signal is applied, there is no current flowing; as a result, any current through either device flows directly to the load, thereby maximizing the efficiency. The ideal efficiency can reach 78%,[41] allowing this architecture to be of use in applications where linearity is not the main concern.

Class AB
The basic idea of Class AB amplification is to preserve the Class B push-pull configuration while improving the linearity by biasing each device slightly above threshold. The implementation of Class AB PAs is similar to Class B configurations. By allowing the two devices to conduct current for a short period, the output voltage waveform during the crossover period can be smoothed, which thus reduces the crossover distortion of the output signal.

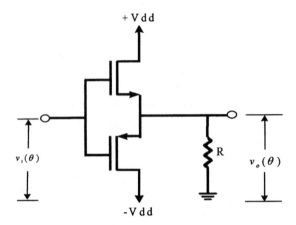

FIGURE 5.26   Inductorless Class B power amplifier.

Non-linear Power Amplifiers

Class C

A Class C power amplifier is the most popular non-linear power amplifier used in the RF band. The conduction angle is less than 180° since the switching transistor is biased on the verge of conduction. A portion of the input signal will make the transistor operate in the amplifying region, and thus the drain current of the transistor is a pulsed signal. Figures 5.27(a) and (b) show the basic configuration of a Class C power amplifier and its corresponding waveforms; clearly, the input and output voltages are not linearly related.

The efficiency of an ideal Class C amplifier is 100% since at any point in time, either the voltage or the current waveforms are zero. In practice, this ideal situation cannot be achieved, and the power efficiency should be maximized by reducing the power loss in the transistor. That is, minimize the current through the transistor when the voltage across the output is high, and minimize the voltage across the output when the current flows through the device.

Class D

A Class D amplifier employs a pair of transistors and a tuned output circuit, where the transistors are driven to act as a two-pole switch and the output circuit is tuned to the switching frequency. The theoretical power efficiency is 100%. Figure 5.28 shows the voltage-switching configuration of a Class D amplifier. The input signals of transistors Q_1 and Q_2 are out of phase, and consequently when Q_1 is on, Q_2 is off, and vice versa. Since the load network is a tuned circuit, we can assume that it provides little impedance to the operating frequency of the voltage v_d and high impedance to other harmonics. Since v_d is a square wave, its Fourier expansion is given by

$$v_d(\omega t) = V_{dc}\left[\frac{1}{2} + \frac{2}{\pi}\sin(\omega t) + \frac{2}{3\pi}\sin(3\omega t)...\right]\qquad(5.45)$$

The impedance of the RLC series load at resonance is equal to R_L, and thus the current is given by:

$$i_L(\omega t) = \frac{2V_{dc}}{\pi R_L}\sin(\omega t)\qquad(5.46)$$

Each of the devices carries the current during one half of the switching cycle. Therefore, the output power is given by:

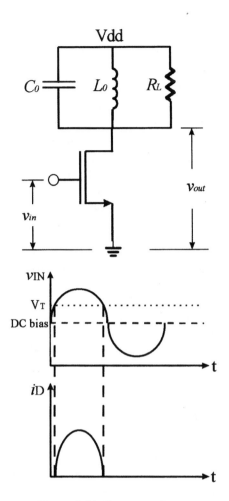

FIGURE 5.27 (a) Class C power amplifier, and (b) Class C waveforms.

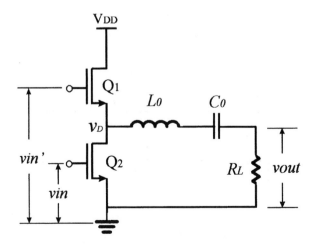

FIGURE 5.28 Class D power amplifier.

$$P_o = \frac{2}{\pi^2} \frac{V_{dc}^2}{R_L} \tag{5.47}$$

Design efforts should focus on reducing the switching loss of both transistors as well as generating the input driving signals.

Class E

The idea behind the Class E PA is to employ non-overlapping output voltage and output current waveforms. Several criteria for optimizing the performance can be found in Ref. 42. Following these guidelines, Class E PAs have high power efficiency, simplicity, and relatively high tolerance to circuit variations.[43] Since there is no power loss in the transistor as well as in the other passive components, the ideal power efficiency is 100%. Figure 5.29 shows a Class E PA, and the corresponding waveforms are given in Fig. 5.30.

The Class E waveforms indicate that the transistor should be completely off before the voltage across it changes, and that the device should be completely on before it starts to allow current to flow through it. Refs. 44 and 45 demonstrate practical Class E operation at RF frequencies using a GaAs process.

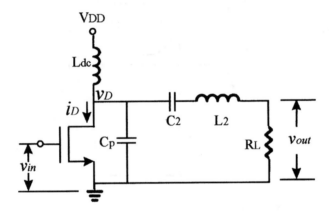

FIGURE 5.29   Class E power amplifier.

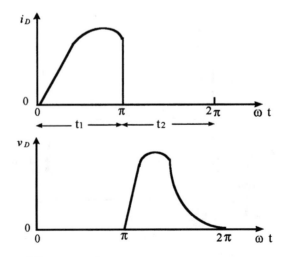

FIGURE 5.30   Waveforms of Class E operation.

Practical Considerations for RF Power Amplifiers

More recently, single-chip solutions for RF transceivers have become a goal for modern wireless communications due to potential savings in power, size, and cost. CMOS must clearly be the technology of choice for a single-chip transceiver due to the large amount of digital baseband processing required. However, the power amplifier design presents a bottleneck toward full integration, since CMOS power amplifiers are still not available. The requirements of low supply voltage, gigahertz-band operation, and high output power make the implementation of CMOS PAs very demanding. The proposal of "microcell" communications may lead to a relaxed demand for output power levels that can be met by designs such as that described in Ref. 46, where a CMOS Class C PA has demonstrated up to 50% power efficiency with 20 mW output power.

Non-linear power amplifiers seem to be popular for modern wireless communications due to their inherent high power efficiency. Since significant power losses occur in the passive inductors as well as the switching devices, the availability of on-chip, low-loss passive inductors is important. The implementation of CMOS on-chip spiral inductors has therefore become an active research topic.[47]

Due to the poor spectral efficiency of a constant envelope modulation scheme, the high power efficiency benefit of non-linear power amplifiers is eliminated. A recently proposed linear transmitter using a non-linear power amplifier may prove to be an alternative solution.[48] The development of high mobility devices such as SiGe HBTs has led to the design of PAs demonstrating output power levels up to 23 dBm at 1.9 GHz with power-added efficiency of 37%.[49] Practical power amplifier designs require that much attention be paid to issues of package and harmonic terminations. Power losses in the matching networks must be absolutely minimized, and tradeoffs between power-added efficiency and linearity are usually achieved through impedance matching. Although GaAs processes provide low-loss impedance matching structures on the semi-insulating substrate, good shielding techniques for CMOS may prove to be another alternative.

Conclusions

Although linear power amplifiers provide conventional "easy-design" characteristics and linearity for modulation schemes such as $\pi/4$-DQPSK, modern wireless transceivers are more likely to employ non-linear power amplifiers due to their much higher power efficiency. As the development of high-quality on-chip passive components makes progress, the trend toward full integration of the PA is becoming increasingly plausible.

The rapid development of CMOS technology seems to be the most promising choice for PA integration, and vast improvements in frequency performance have been gained through device scaling. These improvements are expected to continue as silicon CMOS technologies scale further, driven by the demand for high-performance microprocessors. The further development of high mobility devices such as SiGe HBTs may finally see GaAs MOSFETs being replaced by wireless communication applications, since SiGe technology is compatible with CMOS.

5.6 Applications of High-Q Resonators in IF-Sampling Receiver Architectures

Transconductance-C (gm-C) filters are currently the most popular design approach for realizing continuous-time filters in the intermediate frequency range in telecommunications systems. This section will consider the special application area of high-Q resonators for receiver architectures employing IF sampling.

IF Sampling

A design approach for contemporary receiver architectures that is currently gaining popularity is IF digitization, whereby low-frequency operations such as second mixing and filtering can be performed

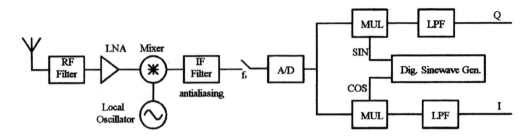

FIGURE 5.31   IF-sampling receiver.

more efficiently in the digital domain. A typical architecture is shown in Fig. 5.31. The IF signal is digitized, multiplied with the quadrature phases of a digital sinusoid, and lowpass filtered to yield the quadrature baseband signals. Since processing takes place in the digital domain, I/Q mismatch problems are eliminated. The principal issue in this approach, however, is the performance required from the A/D converter (ADC). Noise referred to the input of the ADC must be very low so that selectivity remains high. At the same time, the linearity of the ADC must be high to minimize corruption of the wanted signal through intermodulation effects. Both the above requirements should be achieved at an input bandwidth commensurate with the value of the IF frequency, and at an acceptable power budget.

Oversampling has become popular in recent years because it avoids many of the difficulties encountered with conventional methods for A/D and D/A conversion. Conventional converters are often difficult to implement in fine-line, very large-scale integration (VLSI) technology, because they require precise analog components and are very sensitive to noise and interference. In contrast, oversampling converters trade off resolution in time for resolution in amplitude, in such a way that the imprecise nature of the analog circuits can be tolerated. At the same time, they make extensive use of digital signal processing power, taking advantage of the fact that fine-line VLSI is better suited for providing fast digital circuits than for providing precise analog circuits. Therefore, IF-digitization techniques utilizing oversampling Sigma-Delta modulators are very well suited to modern sub-micron CMOS technologies, and their potential has made them the subject of active research.

Most Delta-Sigma modulators are implemented with discrete-time circuits, switched-capacitor (SC) implementations being by far the most common. This is mainly due to the ease with which monolithic SC filters can be designed, as well as the high linearity which they offer. The demand for high-speed ΣΔ oversampling ADCs, especially for converting bandpass signals, makes it necessary to look for a technique that is faster than switched-capacitor. This demand has stimulated researchers to develop a method for designing continuous-time ΔΣ ADCs. Although continuous-time modulators are not easy to integrate, they possess a key advantage over their discrete-time counterparts. The sampling operation takes place inside the modulator loop, making it possible to "noise-shape" the errors introduced by sampling, and provide a certain amount of anti-aliasing filtering at no cost. On the other hand, they are sensitive to memory effects in the DACs and are very sensitive to jitter. They must also process continuous-time signals with high linearity. In communications applications, meeting the latter requirement is complicated by the fact that the signals are located at very high frequencies.

As shown in Fig. 5.32, integrated bandpass implementations of continuous-time modulators require integrated continuous-time resonators to provide the noise shaping function. The gm-C approach of realizing continuous-time resonators offers advantages of complete system integration and total design freedom. However, the design of CMOS high-Q high-linearity resonators at the tens of MHz is very challenging. Since the linearity of the modulator is limited by the linearity of the resonators utilized, the continuous-time resonator is considered to be the most demanding analog sub-block of a bandpass continuous-time Sigma-Delta modulator. Typical specifications for a gm-C resonator used to provide the noise-shaping function in a ΣΔ modulator in a mobile receiver (see Fig. 5.32) are summarized in Table 5.4.

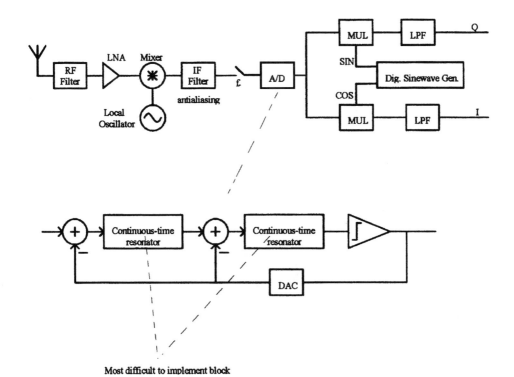

FIGURE 5.32 Continuous-time $\Sigma\Delta$ A/D in IF-sampling receiver.

TABLE 5.4 Fully Integrated Continuous-Time Resonator Specifications

Resonator Specifications	
Center frequency	50 MHz
Quality factor	50
Spurious free dynamic range	>30 dB
Power dissipation	Minimal

Linear Region Transconductor Implementation

The implementation of fully integrated, high-selectivity filters operating at tens to hundreds of MHz provides benefits for wireless transceiver design, including chip area economy and cost reduction. The main disadvantages of on-chip active filter implementations when compared to off-chip passives include increased power dissipation, deterioration in the available dynamic range with increasing Q, and Q and resonant frequency integrity (because of process variations, temperature drifts, and aging, automatic tuning is often unavoidable, especially in high-Q applications). The transconductor-capacitor (gm-C) technique is a popular technique for implementing high-speed continuous time filters and is widely used in many industrial applications.[52] Because gm-C filters are based on integrators built from an open-loop transconductance amplifier driving a capacitor, they are typically very fast but have limited linear dynamic range. Linearization techniques that reduce distortion levels can be used, but often lead to a compromise between speed, dynamic range, and power consumption.

As an example of the tradeoffs in design, consider the transconductor shown in Fig. 5.33. This design consists of a main transconductor cell (M1, M2, M3, M4, M10, M11, and M14) with a negative resistance load (M5, M6, M7, M8, M9, M12, and M13). Transistors M1 and M2 are biased in the triode region of operation using cascode devices M3 and M4 and determine the transconductance gain of the cell. In the triode region of operation, the drain current versus terminal voltage relation can be approximated (for

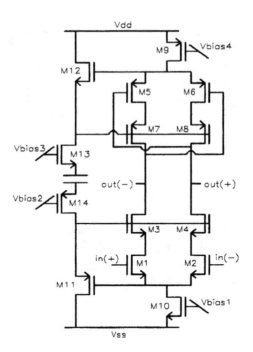

FIGURE 5.33   Triode region transconductor.

simple hand calculations) as $I_D = K[2(V_{GS} - V_T)V_{DS} - V_{DS}^2]$, where K and V_T are the transconductance parameter and the threshold voltage, respectively. Assuming that V_{DS} is constant for both M1 and M2, both the differential mode and the common mode transconductance gains can be derived as $G_{DM} = G_{CM} = 2KV_{DS}$, which can thus be tuned by varying V_{DS}.

The high value of common-mode transconductance is undesirable since it may result in regenerative feedback loops in high-order filters. To improve the CMRR transistor and avoid the formation of such loops, M10 is used to bias the transconductor, thus transforming it from a pseudo-differential to a fully differential transconductor.[53] Transistors M11 and M14 constitute a floating voltage source, thus maintaining a constant drain-source voltage for M1 and M2.

The non-linearities in the voltage-to-current transfer of this stage are mainly due to three effects. The first is the finite impedance levels at the sources of the cascode devices, which cause a signal-dependent variation of the corresponding drain-source voltages of M1 and M2. A fast floating voltage source and large cascode transistors therefore need to be used to minimize this non-linearity. The second cause of non-linearity is the variation of carrier mobility μ of the input devices M1 and M2 with $V_{GS} - V_T$, which becomes more apparent when short-channel devices are used ($K = \mu \cdot C_{ox} \cdot W/2 \cdot L$). A simple first-order model for transverse-field mobility degradation is given by $\mu = \mu_0/(1 + \theta \cdot (V_{GS} - V_T))$, where μ_0 and θ are the zero field mobility and the mobility reduction parameter, respectively. Using this model, the third-order distortion can be determined by a Maclaurin series expansion as $\theta^2/4(1 + \theta(V_{CM} - V_T))$.[54] This expression cannot be regarded as exact, although it is useful to obtain insight. Furthermore, it is valid only at low frequencies, where reactive effects can be ignored and the coefficients of the Maclaurin series expansion are frequency independent. At high frequencies or when very low values of distortion are predicted by the Maclaurin series method, a generalized power series method (Volterra series) must be employed.[55,56] Finally, a further cause of non-linearity is mismatch between M1 and M2, which can be minimized by good layout. A detailed linearity analysis of this transconductance stage is presented in Ref. 60.

To provide a load for the main transconductor cell, a similar cell implemented by p-devices is used. The gates of the linear devices M5 and M6 are now cross-coupled with the drains of the cascode devices M7 and M8. In this way, weak positive feedback is introduced. The differential-mode output resistance can now become negative and is tuned by the V_{DS} of M5 and M6 (M12 and M13 form a floating voltage source), while the common-mode output resistance attains a small value.

When connected to the output of the main transconductor cell as shown in Fig. 5.33, the cross-coupled p-cell forms a high-ohmic load for differential signals and a low-ohmic load for common-mode signals, resulting in a controlled common-mode voltage at the output.[54,57] CMRR can be increased even further using M10, as described previously. Transistor M9 is biased in the triode region of operation and is used to compensate the offset common-mode voltage at the output.

The key performance parameter of an integrator is the phase shift at its unity-gain frequency. Deviations from the ideal –90° phase include phase lead due to finite dc gain and phase lag due to high-frequency parasitic poles. In the transconductor design of Fig. 5.33, dc gain is traded for phase accuracy, thus compensating the phase lag introduced by the parasitic poles. The reduction in dc gain for increased phase accuracy is not a major problem for bandpass filter applications, since phase accuracy at the center frequency is extremely important while dc gain has to be adequate to ensure that attenuation specifications are met at frequencies below the passband.

From simulation results using parameters from a 0.8-μm CMOS process, with the transconductor unity gain frequency set at 50 MHz, third-order intermodulation components were observed at –78 dB with respect to the fundamental signals (two input signals at 49.9 MHz and 50.1 MHz were applied, each at 50 mVpp).

A gm-C Bandpass Biquad

Filter Implementation

The implementation of on-chip high-Q resonant circuits presents a difficult challenge. Integrated passive inductors have generally poor quality factors, which limits the Q of any resonant network in which they are employed. For applications in the hundreds of MHz to a few GHz, one approach is to implement the resonant circuit using low-Q passive on-chip inductors with additional Q-enhancing circuitry. However, for lower frequencies (tens of MHz), on-chip inductors occupy a huge area and this approach is not attractive.

As discussed above, an alternative method is to use active circuitry to eliminate the need for inductors. gm-C-based implementations are attractive due to their high-speed potential and good tunability. A bandpass biquadratic section based upon the transconductor of Fig. 5.33 is shown in Fig. 5.34. The transfer function of Fig. 5.34 is given by:

$$\frac{V_o}{V_i} = \frac{g_{mi} \cdot R_o}{(R_o \cdot C)^2} \cdot \frac{(1 + s \cdot R_o \cdot C)}{\left\{ s^2 + s\dfrac{2 \cdot R_o + g_m^{\,2} \cdot R_o^{\,2} \cdot R}{R_o^{\,2} \cdot C} + \dfrac{1 + g_m^{\,2} \cdot R_o^{\,2}}{R_o^{\,2} \cdot C^2} \right\}} \tag{5.48}$$

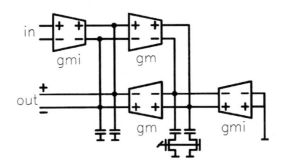

FIGURE 5.34 Biquad bandpass.

R_o represents the total resistance at the nodes due to the finite output resistance of the transconductors. R represents the effective resistance of the linear region transistors in the transconductor (see Fig. 5.33), and is used here to introduce damping and control the Q. From Eq. 5.48, it can be shown that $\omega_o \approx g_m/C$, $Q \approx g_m \cdot R_o/(2 + R_o \cdot R \cdot g_m^2)$, $Q_{max} = Q|_{r=0} = (g_m \cdot R_o)/2$ and $A_o = g_{mi} \cdot Q$. Thus, g_m is used to set the central frequency, R is used to control the Q, and g_{mi} controls the bandpass gain A_o. A dummy g_{mi} is used to provide symmetry and thus better stability due to process variations, temperature, and aging.

One of the main problems when implementing high-Q high-frequency resonators is maintaining the stability of the center frequency ω_o and the quality factor Q. This problem calls for very careful layout and the implementation of an automatic tuning system. Another fundamental limitation regarding available dynamic range occurs: namely, that the dynamic range (DR) of high-Q gm-C filters has been found to be inversely proportional to the filter Q.[57] The maximum dynamic range is given by:

$$DR = \frac{V_{max}^2}{V_{noise}^2} = \frac{V_{max}^2 \cdot C}{4 \cdot k \cdot T \cdot \xi \cdot Q} \tag{5.49}$$

where V_{max} is the maximum rms voltage across the filter capacitors, C is the total capacitance, k is Boltzmann's constant, T is the absolute temperature, and ξ is the noise factor of the active circuitry ($\xi=1$ corresponds to output noise equal to the thermal noise of a resistor of value $R = 1/g_m$, where g_m is the transconductor value used in the filter).

In practice, the dynamic range achieved will be less than this maximum value due to the amplification of both noise and intermodulation components around the resonant frequency. This is a fundamental limitation, and the only solution is to design the transconductors for low noise and high linearity. The linearity performance in narrowband systems is characterized by the spurious-free dynamic range (SFDR). SFDR is defined as the signal-to-noise ratio when the power of the third-order intermodulation products equals the noise power. As shown in Ref. 60, the SFDR of the resonator in Fig. 5.34 is given by:

$$SFDR = \frac{1}{2(k \cdot T)^{2/3}} \left(\frac{3 \cdot V_{o, peak}^2 \cdot C}{4 \cdot \xi \cdot IM_{3, int}} \right)^{2/3} \frac{1}{Q^2} \tag{5.50}$$

where $IM_{3, int}$ is the third-order intermodulation point of the integrator used to implement the resonator. The spurious free dynamic range of the resonator thus deteriorates by 6 dB if the quality factor is doubled, assuming that the output swing remains the same. In contrast, implementing a resonant circuit using low-Q passive on-chip inductors with additional Q-enhancing circuitry leads to a dynamic range amplified by a factor Q_o, where Q_o is the quality factor of the on-chip inductor itself.[59] However, as stated above, for frequencies in the tens of MHz, on-chip inductors occupy a huge area and thus the Q_o improvement in dynamic range is not high enough to justify the area increase.

Simulation Results

To confirm operation, the filter shown in Fig. 5.34 has been simulated in HSPICE using process parameters from a commercial 0.8-μm CMOS process. Figure 5.35 shows the simulated frequency and phase response of the filter for a center frequency of 50 MHz and a quality factor of 50. Figure 5.36 shows the simulated output of the filter when the input consists of two tones at 49.9 MHz and 50.1 MHz, respectively, each at 40 mVpp. At this level of input signal, the third-order intermodulation components were found to be at the same level as the noise. Thus, the predicted SFDR is about 34 dB with Q = 50. Table 5.5 summarizes the simulation results.

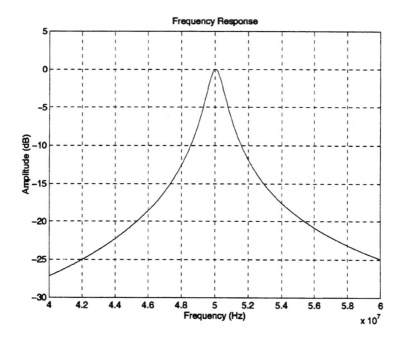

FIGURE 5.35   Simulated bandpass frequency response.

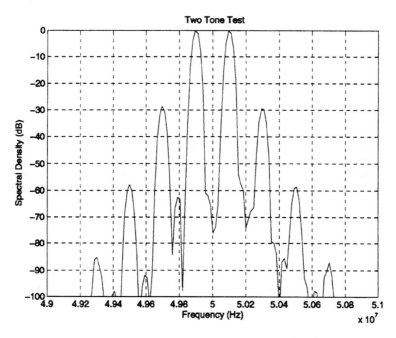

FIGURE 5.36   Simulated two-tone intermodulation test.

TABLE 5.5   Simulation Results

Power dissipation (Supply voltage = 5 V)	12.5 mW
Common-mode output offset	<1 mV
Center frequency	50 MHz
Quality factor	50
Output noise voltage (integrated over the band from 40 MHz to 60 MHz with Q = 50)	500 μV_{rms}
Output signal voltage (so that intermodulation components are at the same level as the noise, Q = 50)	25.2 mV_{rms}
Spurious free dynamic range (Q = 50)	34 dB

5.7   Log-Domain Processing

Instantaneous Companding

The concept of *instantaneous companding* is an emerging area of interest within the field of analog integrated circuit design. Currently, the main area of application for this technique is the implementation of continuous-time, fully integrated filters with wide dynamic range, high-frequency potential, and wide tunability.

With the drive toward lower supply voltages and higher operating frequencies, traditional analog integrated circuit design methodologies are proving inadequate. Conventional techniques to linearize inherently non-linear devices require an overhead in terms of increased power consumption or reduced operating speed. Recently, the use of companding, originally developed for audio transmission, has been proposed as an elegant solution to the problem of maintaining dynamic range and high-frequency operation under low supply voltage.[50] In this approach, non-linear amplifiers are used to compress the dynamic range of the input signal to ensure, for example, that signal levels always remain above a certain noise threshold but below levels that may cause overload. The overall system must thus adapt its operation to ensure that input-output linearity is maintained as the instantaneous signal level alters. Although the system is input-output linear, signals internal to the system are inherently non-linear. Companding has traditionally been realized in two ways — syllabic and instantaneous — depending on how the system adapts in response to the changing input signal level. In *syllabic* companding, the level of compression (and expansion) is a non-linear function of a slowly varying property of the signal (e.g., envelope or power). In contrast, *instantaneous* companding adapts the compression and expansion ratios instantaneously with the changing input signal amplitude.

Perhaps the best-known recent example of this approach is the log-domain technique, where the exponential I–V characteristics of bipolar junction transistors (BJTs) are directly exploited to implement input-output linear filters. Since the large signal device equations are utilized, there is no need for small signal operation or local linearization techniques, and the resulting circuits have the potential for wide dynamic range and high-frequency operation under low power supply voltages. Log-domain circuits are generally implemented using BJTs and capacitors only and thus are inherently suitable for integration. The filter parameters are easily tunable, which makes them robust. In addition, the design procedure is systematic, which suggests that these desirable features can be reproduced for different system implementations. The following section provides an introduction to the synthesis of log-domain filters and highlights various performance issues. Interested readers are advised to consult Refs. 62 through 77 for a more detailed treatment of the subject.

Log-Domain Filter Synthesis

The synthesis of log-domain filters using state-space transformations was originally proposed by Frey.[63] As an example of this methodology, consider a biquad filter with the following transfer function:

$$Y(s) = \frac{s\omega_o U_1(s) + \omega_o^2 U_2(s)}{s^2 + (\omega_o/Q)s + \omega_o^2} \tag{5.51}$$

Thus, using u_1 as input gives a bandpass response at the output y, while using u_2 as input gives a lowpass response. This system can also be described by the following state space equations:

$$\dot{x}_1 = -(\omega_o/Q)x_1 - \omega_o x_2 + \omega_o u_1$$
$$\dot{x}_2 = -\omega_o x_1 - \omega_o u_2 \tag{5.52}$$
$$y = x_1$$

where a "dot" denotes differentiation in time. To transform these linear state equations into non-linear nodal equations that can be directly implemented using bipolar transistors, the following exponential transformations are defined:

$$x_1 = I_S \exp\left(\frac{V_1}{V_t}\right) \qquad u_1 = \left(\frac{I_S^2}{I_o}\right)\exp\left(\frac{V_{in1}}{V_t}\right)$$
$$x_2 = I_o \exp\left(\frac{V_2}{V_t}\right) \qquad u_2 = I_S \exp\left(\frac{V_{in2}}{V_t}\right) \tag{5.53}$$

These transformations map the linear state variables to currents flowing through bipolar transistors (BJTs) biased in the active region. I_S represents the BJT reverse saturation current, while V_t is the thermal voltage. Substituting these transformations into the state (Eq. 5.52) gives:

$$C\dot{V}_1 = -\frac{I_o}{Q} - \frac{I_o^2}{I_S}\exp\left(\frac{V_2 - V_1}{V_t}\right) + I_S\exp\left(\frac{V_{o1} - V_1}{V_t}\right)$$
$$C\dot{V}_2 = I_S\exp\left(\frac{V_1 - V_2}{V_t}\right) - I_S\exp\left(\frac{V_{o2} - V_2}{V_t}\right) \tag{5.54}$$
$$y = I_S\exp\left(\frac{V_1}{V_t}\right)$$

In Eq. 5.54, a tuning current $I_o = C\omega_o V_t$ is defined, where C is a scaling factor which represents a capacitance. The linear state space equations have thus been transformed into a set of non-linear nodal equations, and the task for the designer is now to realize a circuit architecture that will implement these non-linear equations. Considering the first two equations in Eq. 5.54, the terms on the LHS can be implemented as currents flowing through grounded capacitors of value C connected at nodes V_1 and V_2, respectively. The expressions on the RHS can be implemented by constant current sources in conjunction with appropriately biased bipolar transistors to realize the exponential terms. The third equation in Eq. 5.54 indicates that the output y can be obtained as the collector current of a bipolar transistor biased with a base–emitter voltage V_1.

Figure 5.37 shows a possible circuit implementation, which is derived using Gilbert's translinear circuit principle.[64] The detailed circuit implementation is described in more detail in Ref. 63. The center frequency of this filter is given by $\omega_o = (I_o/CV_t)$ and can thus be tuned by varying the value of the bias current I_o.

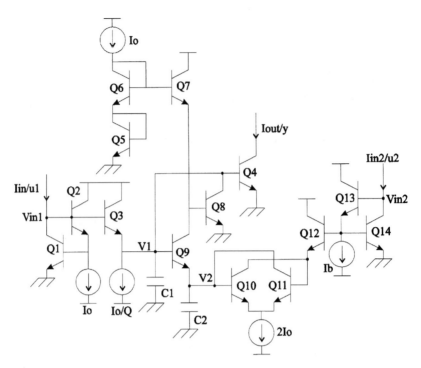

FIGURE 5.37   A log-domain biquadratic filter. A lowpass response is obtained by applying a signal at I_{u2} and keeping I_{u1} constant, while a bandpass response is obtained by applying a signal at I_{u1} and keeping I_{u2} constant.

Performance Aspects

Tuning Range

Both the quiescent bias current I_o and capacitance value C can be varied to alter the filter response. However, the allowable capacitor values are generally limited to within a certain range. On the lower side, C must not become smaller than the parasitic device capacitance. The base–emitter diffusion capacitance of the transistors is particularly important, since this is generally the largest device capacitance (up to the pF range) and is also non-linear. In addition, as the value of C decreases, it becomes more difficult to match capacitance values. The silicon area available limits the maximum value of C; for example, in a typical technology, a 50-pF poly-poly capacitor consumes 100 µm × 1000 µm of silicon area.

The range of allowable currents in a modern BJT is, in principle, fairly large (several decades); however, at very low and very high current levels, the current gain (β) is degraded. For high-frequency operation, particular attention must be paid to the actual f_t of the transistors, which is given by:[6]

$$f_t = \frac{g_m}{2\pi(C_\pi + C_\mu)} = \frac{g_m}{2\pi(C_d + C_{je} + C_{jc})} \tag{5.55}$$

C_{je} and C_{jc} represent the emitter and collector junction capacitance, respectively, and are only a weak function of the collector bias current. C_d represents the base–emitter diffusion capacitance and is proportional to the instantaneous collector current; $C_d = (\tau_f I_c/V_t)$, where τ_f is the effective base transit time. g_m represents the device transconductance, which is again dependent on the collector current; $g_m = (I_c/V_t)$. At high current levels, the diffusion capacitance is much larger than the junction capacitance, and thus:

$$f_t = \frac{g_m}{2\pi C_d} = \frac{1}{2\pi\tau_f} \tag{5.56}$$

At lower current levels, C_d and g_m decrease, whereas C_{je} and C_{jc} remain constant and f_t is reduced:

$$f_t = \frac{g_m}{2\pi(C_{je} + C_{jc})} \tag{5.57}$$

At very high current levels, f_t again reduces due to the effects of high-level injection.

Finite Current Gain

To verify the performance of the biquad filter, the circuit of Fig. 5.37 was simulated using HSPICE with transistor parameters from a typical bipolar process. Transient (large signal) simulations were carried out to confirm the circuit operation, and the lowpass characteristic (input at I_{u2}) is found to be very close to the ideal response. However, with an input signal applied at I_{u1} to obtain a bandpass response, the filter performance clearly deviates from the ideal characteristic as shown in Fig. 5.38. At low frequencies, the stop-band attenuation is only 25 dB. Re-simulating the circuit with ideal transistor models gives the required ideal bandpass characteristic as shown in Fig. 5.38; and by re-introducing the transistor parameters one at a time, the cause of the problem is found to be the finite current gain (β) of the bipolar transistors.

To increase the effective β, each transistor can be replaced by a Darlington pair, as shown in Fig. 5.39. This combination acts as a bipolar transistor with $\beta = \beta_a\beta_b + \beta_a + \beta_b \approx \beta^2$. Simulating the bandpass response of the circuit with Darlington pairs results in improved stopband attenuation as shown in Fig. 5.40, where the dc attenuation $H(0)$ is now approximately -50 dB. A disadvantage, however, is that a higher supply voltage is now required, since the effective base–emitter voltage V_{be} of the Darlington pair is double that of a single device. In addition, the current in device Q_a of Fig. 5.39 is now β times smaller than the design current I_o, resulting in a much lower f_t for this device.

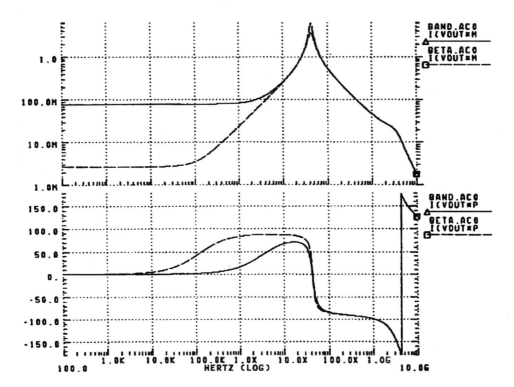

FIGURE 5.38 Bandpass response of the biquad of Fig. 5.37. Notice the poor low-frequency stop-band attenuation (approx. 25 dB) if real transistor models are used (solid line), versus the much better stop-band attenuation with β set to 1000 (dashed line).

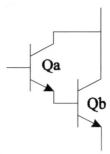

FIGURE 5.39   Darlington pair.

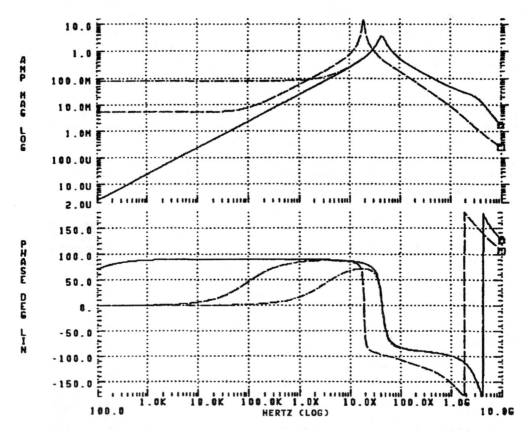

FIGURE 5.40   Possible solutions for the effect of finite β. The "dashed-dotted" line is the original response with a stopband attenuation of only 25 dB. The dashed line (–50 dB) is the result of replacing all BJTs by Darlington pairs. The solid line is the result of feeding a fraction of the input signal to the second (i.e., lowpass) input.

An alternative method to improve the bandpass characteristic is described below. The (non-ideal) transfer function from the bandpass input I_{u1} is described by:

$$H_1(s) = \frac{I_{out}(s)}{I_{u1}(s)} = \frac{\omega_o(s + \omega_z)}{s^2 + (\omega_o/Q)s + \omega_o^2} \tag{5.58}$$

ω_z is a parasitic zero, which describes the low-frequency "flattening-out" of the bandpass characteristic. The transfer function from the second input I_{u2} is a lowpass response:

$$H_2(s) = \frac{I_{out}(s)}{I_{u2}(s)} = \frac{\omega_o^2}{s^2 + (\omega_o/Q)s + \omega_o^2} \tag{5.59}$$

By applying a scaled version of the input signal I_{u1} to the lowpass input I_{u2}, the unwanted zero ω_z can thus be compensated. Setting $I_{u2} = -(\omega_z/\omega_o)I_{u1}$

$$I_{out}(s) = \frac{\omega_o(s + \omega_z)I_{u1}(s) + \omega_o^2\left(-\dfrac{\omega_z}{\omega_o}I_{u1}(s)\right)}{s^2 + (\omega_o/Q)s + \omega_o^2} = \frac{\omega_o s}{s^2 + (\omega_o/Q)s + \omega_o^2}I_{u1}(s) \tag{5.60}$$

This idea is confirmed by simulation as shown in Fig. 5.40. However, in practice, since this technique is very sensitive to the exact factor by which the input signal is scaled, active on-chip tuning would be required.

Frequency Performance

The log-domain filter operates in current mode, and all nodes within the circuit have an impedance of the order of $1/g_m$. Thus, any parasitic poles or zeros within the circuit are of the order of g_m/C_p, where C_p represents the parasitic capacitance at the given node. Typically, the parasitic capacitors are dominated by base–emitter diffusion capacitance $C\pi$, and the parasitic poles are situated close to the f_t of the technology. This underscores the potential of the log-domain technique for high-frequency operation. In practice, as with all high-frequency circuits, careful design and layout are required to achieve the maximum frequency potential.[76,77]

The Basic Log-Domain Integrator

The previous section has outlined some of the limitations of practical log-domain circuits that result from intrinsic device parasitics. This section analyzes these non-idealities in more detail by considering the simplest log-domain circuit, a first-order lowpass filter (lossy integrator).

A log-domain first-order lowpass filter is shown in Fig. 5.41. Referring to this circuit, and assuming an ideal exponential characteristic for each BJT, the following set of equations can be written (assuming matched components and neglecting the effect of base currents):

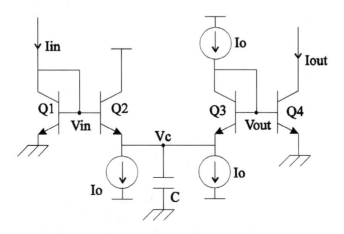

FIGURE 5.41 A first-order lowpass filter (lossy integrator).

$$I_{in} = I_S \exp\left(\frac{V_{in}}{V_t}\right) \qquad\qquad I_{out} = I_S \exp\left(\frac{V_{out}}{V_t}\right)$$

$$(5.61)$$

$$V_{out} = V_c + V_t \ln\left(\frac{I_o}{I_S}\right) \qquad I_o + C\frac{dV_c}{dt} = I_S \exp\left(\frac{V_{in} - V_c}{Vt}\right)$$

Combining these equations results in the following linear first-order differential equation:

$$\left(\frac{CV_t}{I_o}\right)\frac{dI_{out}}{dt} + I_{out} = I_{in} \qquad\qquad (5.62)$$

Taking the Laplace transform produces the following linear transfer function:

$$\frac{I_{out}(s)}{I_{in}(s)} = \frac{I_o}{I_o + sCV_t} \qquad\qquad (5.63)$$

Equation 5.63 describes a first-order lowpass circuit with −3 dB frequency $f_c = I_o/2\pi CV_t$. This transfer function has been derived assuming an ideal exponential BJT characteristic; however, in practice, device non-idealities will introduce deviations from this ideal exponential characteristic, resulting in transfer function errors and distortion. This is similar to the case of "conventional" linear system design, where any deviation from the ideal linear building block response will contribute to output distortion. A brief discussion of the performance limitations of log-domain filters due to device non-idealities is given below; further discussion of distortion and performance limitations can be found in Refs. 66 and 67.

Effects of Finite Current Gain

Assuming all transistors have equal β, Eq. 5.62 is modified to the following non-linear differential equation (neglecting terms with β^2 or higher in the denominator):

$$\left(\frac{CV_t}{I_o}\right)\frac{dI_{out}}{dt}\left(1 + \frac{I_o}{I_{out}(1+\beta)}\right) + I_{out}\left(1 + \frac{I_{out}}{\beta I_o}\right) + \frac{2I_o}{1+\beta} = I_{in}\left(1 + \frac{I_{in}}{\beta}\right) \qquad (5.64)$$

An analytical solution to Eq. 5.64 is difficult to obtain; thus, a qualitative discussion is given. At low frequencies, neglecting the differential terms, finite β causes a dc gain error and quadratic (even-order) distortion. At high frequencies, the differential term is modified, depending on the values of I_{out} and β; therefore, scalar error and modulation of the 3 dB frequency are expected. In practice, the effects of finite β are further complicated due to its dependence on frequency and device collector current.

Component Mismatch

Emitter area mismatches cause variations in the saturation current I_S between transistors. Taking the emitter area into account, Eq. 5.62 can be rewritten as:

$$\left(\frac{CV_t}{I_o}\right)\frac{dI_{out}}{dt} + I_{out} = \lambda I_{in} \qquad\qquad (5.65)$$

where $\lambda = (I_{S3}I_{S4}/I_{S1}I_{S2}) = (A_3A_4/A_1A_2)$. It is clear from Eq. 5.65 that area mismatches introduce only a change in the proportionality constant or dc gain of the integrator, and do not have any effect on the linearity or time constant of the integrator. The gain error can be compensated by easily adjusting one of the dc bias currents; thus, I_s mismatches do not seem to be a significant problem.

Ohmic Resistance

Ohmic resistances include the base and emitter diffusion resistance, and the resistance of interconnects and contact interfaces. For simplicity, all theses resistances can be referred to the base as an equivalent ohmic base resistance r_b. The device collector current can thus be defined as:

$$I_c = I_S \exp\left(\frac{V_{be} - I_b r_b}{V_t}\right) = I_S \exp\left(\frac{V_{be}}{V_t}\right) \exp(-\alpha I_c) \tag{5.66}$$

where V_{be} is the applied (extrinsic) base–emitter voltage and $\alpha = r_b/\beta V_t$. When the first-order filter of Fig. 5.41 is analyzed using the expression given in Eq. 5.66, a modified differential equation is obtained (assuming all base resistances are equal):

$$\left(\frac{CV_t}{I_o}\right)\frac{d[I_{out}e^{\alpha I_{out}}]}{dt} + I_{out}e^{\alpha I_{out}} = (e^{\partial\alpha I_o})I_{in}e^{\alpha I_{in}}e^{\left(-\alpha\frac{CV_t}{I_o}\frac{dI_{out}}{dt}(1+\alpha I_{out})\right)} \tag{5.67}$$

The term $\partial\alpha$ represents the difference between two α values and will be close to zero if all base resistances are assumed equal, and thus can be neglected. At frequencies well below ω_o, the time derivative in the exponent can also be neglected to give:

$$\left(\frac{CV_t}{I_o}\right)\frac{d[I_{out}\exp(\alpha I_{out})]}{dt} + I_{out}\exp(\alpha I_{out}) = I_{in}\exp(\alpha I_{in}) \tag{5.68}$$

Expanding the differential term in Eq. 5.68:

$$I_{out}\exp(\alpha I_{out}) + \tau\frac{dI_{out}}{dt}\exp(\alpha I_{out}) + \tau\alpha I_{out}\frac{dI_{out}}{dt}\exp(\alpha I_{out}) = I_{in}\exp(\alpha I_{in}) \tag{5.69}$$

Equation 5.69 is clearly no longer linear; and thus, distortion products will be present at the output. To quantify this distortion, we apply as an input signal $I_{in} = A\exp(j\omega t)$, and expect as output:

$$I_{out} = B\exp j(\omega w + \theta) + c\exp j(2\omega t + \phi) + D\exp j(3\omega t + \psi) \tag{5.70}$$

Expanding the exponential terms in Eq. 5.69 by the first few terms of their series expansion:

$$\exp(x) = 1 + x + \frac{x^2}{2} + \frac{x^3}{6} \tag{5.71}$$

and identifying the terms in $\exp(j\omega t)$, $\exp(2j\omega t)$, etc., results in:

$$B\exp j\theta = \frac{A}{1+s\tau} \qquad C\exp j\phi = \alpha A^2 p(\tau) \qquad D\exp j\psi = \alpha^2 A^3 q(s\tau) \tag{5.72}$$

The expression for B is (as expected) the first-order transfer function of the system. The expressions for C and D give the second and third harmonic terms:

$$HD_2 = \frac{|C\exp j\phi|}{|A|} \qquad HD_3 = \frac{|D\exp j\psi|}{|A|} \tag{5.73}$$

$p(s\tau)$ and $q(s\tau)$ are rational functions in $s\tau$; thus, the output distortion is frequency dependent. Distortion is low at low frequencies, and peaks around the cut-off frequency ($s\tau = 1$), where $p = 0.25$ and $q = 2$. The maximum distortion levels can be approximated as:

$$HD_{2(\text{max})} = 0.25\alpha A = 0.25\frac{r_b}{\beta V_t}I_{in}$$

$$HD_{3(\text{max})} = 2(\alpha A)^2 = 2\left(\frac{r_b}{\beta V_t}\right)^2 I_{in}^2$$

(5.74)

These expressions demonstrate that when the voltage drop across the base resistance becomes comparable to V_p, there is significant distortion. This analysis also predicts that the distortion is larger at frequencies closer to the cut-off frequency, which is confirmed by circuit simulation.

In practice, the base resistance can be minimized by good layout (multiple base contacts), or by connecting several transistors in parallel. The latter, however, decreases the current through each transistor, which may lead to a decrease in f_t. It should be noted that, from a technological point of view, a high f_t and a low r_b are opposing goals. To achieve a high f_t, the base should be as shallow as possible, reducing base transit time. At the same time, however, a shallow base increases the silicon (intrinsic) base resistance.

Early Effect

The Early effect (base-width modulation) causes the collector current to vary with the collector–emitter and base–collector voltages. Considering the variation of collector–emitter voltage, the collector current can be written as $I_c = (1 + V_{ce}/V_A)\exp(V_{be}/V_t)$, where V_A is the forward-biased Early voltage. An analysis of the circuit of Fig. 5.41 shows that the Early effect introduces a scalar error to the dc gain as in the case of emitter area (I_S) mismatch. In practice, the base-width modulation of the devices also introduces distortion because V_{ce} and V_{bc} are signal dependent. However, since voltage swings in current-mode circuits are minimized, this is not believed to be a major source of distortion.

Frequency Limitations

Each bipolar transistor has various intrinsic capacitors, the most important being C_μ (base–collector junction capacitance), C_{cs} (collector–substrate junction capacitance), and C_π (the sum of the base–emitter junction capacitance and the base–emitter diffusion capacitance). The junction capacitors depend only slightly on the operating point, while the base–emitter diffusion capacitance is proportional to the bias current, as given by $C_d = \tau_f I_c/V_t$. To determine the position of the parasitic poles and zeros, Fig. 5.41 should be analyzed using large-signal device models. Unfortunately, the complexity of the resulting expressions renders this approach impractical even for the simplest log-domain circuits. To gain an intuitive understanding of the high-frequency limitations of the circuit, a small-signal analysis can be performed. Although log-domain circuits are capable of large-signal operation, this small-signal approach is justified to some extent since small signals can be considered a "special case" of large signals. If the circuit fails to operate correctly for small signals, then it will almost certainly fail in large-signal operation (unfortunately, the opposite does not hold; if a circuit operates correctly for small signals, it does not necessarily work well for large signals). Analyzing the circuit of Fig. 5.41, replacing each transistor by a small-signal (hybrid-π) equivalent model (comprising g_m, r_b, r_π, C_π, C_μ),[65] results in the following expression:

$$\frac{I_{out}}{I_{in}} = \left(\frac{g_{m4}}{g_{m1}}\right)\frac{1 + \tau_z s}{(1 + \tau_{p1}s)(1 + \tau_{p2}s)}$$

(5.75)

τ_{p1} is the time constant of the first (dominant) pole, given by:

$$\tau_{p1} = \frac{C}{g_{m2}} + (C_{\mu4} + C_{\pi4})\left(r_{b4} + \frac{1}{g_{m2}} + \frac{1}{g_{m3}}\right) + \frac{C_{\pi1}}{g_{m1}} + \frac{C_{\pi2}}{g_{m2}}$$
$$+ \frac{C_{\pi3}}{g_{m3}} + C_{\mu1}r_{b1} + C_{\mu2}r_{b2} + C_{\mu3}r_{b3} \tag{5.76}$$

Ideally, this pole location should depend only on the design capacitance C, and not on the device parasitics. Therefore, $C_{\mu}r_b$, $C_{\pi}/g_m \ll C/g_{m2}$. Since $1/C_{\mu}r_b$ is typically much greater than f_t, this first constraint does not form a limit. The value of C_{π}/g_m depends on the operating point. For large currents, C_{π} is dominated by diffusion capacitance C_d so that $C_{\pi}/g_m = 1/f_t$. For smaller currents, g_m decreases while C_{π} becomes dominated by junction capacitance C_{je}, so that $C_{\pi}/g_m \gg 1/f_t$. Thus, it would seem that the usable cut-off frequency of the basic log-domain first-order filter is limited by the actual f_t of the transistors. The second pole time constant τ_{p2} (assuming that $\tau_{p1} = C/g_{m2}$) is:

$$\tau_{p2} = (C_{\mu4} + C_{\pi4})\left(r_{b4} + \frac{1}{g_{m3}}\right) + (C_{\mu2} + C_{\pi2})\left(r_{b2} + \frac{1}{g_{m1}}\right) + \frac{C_{\pi1} + C_{cs1}}{g_{m1}}$$
$$+ \frac{C_{\pi13} + C_{cs3}}{g_{m3}} + C_{\mu1}r_{b1} + C_{\mu4}r_{b4} \tag{5.77}$$

This corresponds approximately to the f_t of the transistors, although Eq. 5.77 shows that the collector–substrate capacitance also contributes toward limiting the maximum operating frequency. The zero time constant τ_z is given by:

$$\tau_z = (C_{\pi1} + C_{\mu1})r_{b1} + \frac{C_{\pi2}}{g_{m2}} + \frac{C_{\pi4}}{g_{m4}} + C_{\mu4}r_{b4} \tag{5.78}$$

This is of the same order of magnitude as the second pole. This means that the first zero and the second pole will be close together, and will compensate to a certain degree. However, in reality, there are more poles and zeros than Eq. 5.75 would suggest, and it is likely that others will also occur around the actual f_t of the transistors.

Noise

Noise in companding and log-domain circuits is discussed in some detail in Refs. 69 to 71, and a complete treatment is beyond the scope of this discussion. For linear (non-companding) circuits, noise is generally assumed to be independent of signal level, and the signal-to-noise ratio (SNR) will increase with increasing input signal level. This is not true for log-domain systems. At small input signal levels, the noise value can be assumed approximately constant, and an increase in signal level will give an increase in SNR. At high signal levels, the instantaneous value of noise will increase, and thus the SNR levels out at a constant value. This can be considered as an intermodulation of signal and noise power. For the Class A circuits discussed above, the peak signal level is limited by the dc bias current. In this case, the large-signal noise is found to be of the same order of magnitude as the quiescent noise level, and thus a linear approximation is generally acceptable (this is not the case for Class AB circuits).

Synthesis of Higher-Order Log-Domain Filters

The state-space synthesis technique outlined above proves difficult if implementation of high-order filters is required, since it becomes difficult to define and manipulate a large set of state equations. One solution is to use the signal flow graph (SFG) synthesis method proposed by Perry and Roberts[72] to simulate LC ladder filters using log-domain building blocks. The interested reader is also referred to Refs. 73 through 75, which present modular and transistor-level synthesis techniques that can be easily extended to higher-order filters.

References

1. A. Sedra and K. Smith, *Microelectronic Circuits*, Oxford, 1998.

2. P. R. Gray and R. G. Meyer, *Analysis and Design of Analog Integrated Circuits*, Wiley, 1993.

3. W. H. Gross, New high speed amplifier designs, design techniques and layout problems, in *Analog Circuit Design*, Ed. J. S. Huijsing, R. J. van der Plassche, W. Sansen, Kluwer Academic, 1993.

4. D.F. Bowers, The impact of new architectures on the ubiquitous operational amplifier, in *Analog Circuit Design*, Ed. J. S. Huijsing, R. J. van der Plassche, W. Sansen, Kluwer Academic, 1993.

5. J. Fonderie and J. H. Huijsing, Design of low-voltage bipolar opamps, in *Analog Circuit Design*, Ed. J. S. Huijsing, R. J. van der Plassche, and W. Sansen, Kluwer Academic, 1993.

6. M. Steyaert and W. Sansen, Opamp design towards maximum gain-bandwidth, in *Analog Circuit Design*, Ed. J. S. Huijsing, R. J. van der Plassche, W. Sansen, Kluwer Academic, 1993.

7. K. Bult and G. Geelen, The CMOS gain-boosting technique, in *Analog Circuit Design*, Ed. J. S. Huijsing, R. J. van der Plassche, W. Sansen, Kluwer Academic, 1993.

8. J. Bales, A low power, high-speed, current feedback opamp with a novel class AB high current output stage, *IEEE J. Solid-State Circuits*, vol. 32, no. 9, Sep. 1997, p. 1470.

9. C. Toumazou, Analogue signal processing: the 'current way' of thinking, *Int. J. High-Speed Electronics*, vol. 32, no. 3-4, p. 297, 1992.

10. K. Manetakis and C. Toumazou, A new CMOS CFOA suitable for VLSI technology, *Electron. Lett.*, vol. 32, no. 12, June 1996.

11. K. Manetakis, C. Toumazou, and C. Papavassiliou, A 120MHz, 12mW CMOS current feedback opamp, *Proc. IEEE Custom Int. Circuits Conf.*, p. 365, 1998.

12. D. A. Johns and K. Martin, *Analog Integrated Circuit Design*, Wiley, 1997.

13. C. Toumazou, J. Lidgey, and A. Payne, Emerging techniques for high-frequency BJT amplifier design: a current-mode perspective, *Parchment Press for Int. Conf. on Electron. Circuits Syst.*, Cairo, 1994.

14. M.C.H Cheng and C. Toumazou, 3V MOS current conveyor cell for VLSI technology, *Electron. Lett.*, vol. 29, p. 317, 1993.

15. K. Manetakis, Intermediate frequency CMOS analogue cells for wireless communications, Ph.D. thesis, Imperial College, London, 1998.

16. R. A. Johnson et al., A 2.4GHz silicon-on-sapphire CMOS low-noise amplifier, *IEEE Microwave and Guided Wave Lett.*, vol. 7, no. 10, pp. 350-352, Oct. 1997.

17. A. N. Karanicolas, A 2.7V 900MHz CMOS LNA and mixer, *Digest of IEEE Intl. Solid-State Circuits Conference*, pp. 50-51, 1996.

18. D. K. Shaffer and T. H. Lee, A 1.5-V, 1.5-GHz CMOS low noise amplifier, *IEEE J. Solid-State Circuits*, vol. 32, no. 5, pp. 745-759, May 1997.

19. J. C. Rudell et al., A 1.9GHz wide-band if double conversion CMOS integrated receiver for cordless telephone applications, *Digest of IEEE Intl. Solid-State Circuits Conference*, pp. 304-305, 1997.

20. E. Abou-Allam et al., CMOS front end RF amplifier with on-chip tuning, *Proc. IEEE Int. Symp. Circ. Syst. 1996*, pp. 148-151, 1996.

21. P. R. Gray and R. G. Meyer, *Analysis and Design of Analogue Integrated Circuits and Systems*, Chap. 11, 3rd ed., John Wiley & Sons, New York, 1993.

22. M. J. N. Sibley, *Optical Communications*, Chap. 4-6, Macmillan, 1995.

23. J. J. Morikuni et al., Improvements to the standard theory for photoreceiver noise, *J. Lightwave Tech.*, vol. 12, no. 4, pp. 1174-1184, Jul. 1994.

24. A. A. Abidi, Gigahertz transresistance amplifiers in fine line NMOS, *IEEE J. Solid-State Circuits*, vol. SC-19, no. 6, pp. 986-994, Dec. 1984.

25. M. B. Das, J. Chen, and E. John, Designing optoelectronic integrated circuit (OEIC) receivers for high sensitivity and maximally flat frequency response, *J. Lightwave Tech.*, vol. 13, no. 9, pp. 1876-1884, Sep. 1995.

26. B. Sklar, *Digital Communication: Fundamentals and Applications*, Prentice-Hall 1988.

27. S. D. Personick, Receiver design for optical fiber systems, *Proc. IEEE*, vol. 65, no. 12, pp. 1670-1678, Dec. 1977.

28. J. M. Senior, *Optical Fiber Communications: Principles and Practice*, Chap. 8-10, PHI, 1985.

29. N. Scheinberg et al, Monolithic GaAs transimpedance amplifiers for fiber-optic receivers, *IEEE J. Solid-State Circuits*, vol. 26, no. 12, pp. 1834-1839, Dec. 1991.

30. C. Toumazou and S. M. Park, Wide-band low noise CMOS transimpedance amplifier for gigahertz operation, *Electron. Lett.*, vol. 32, no. 13, pp. 1194-1196, Jun. 1996.

31. S. M. Park and C. Toumazou, Giga-hertz low noise CMOS transimpedance amplifier, *Proc. IEEE Int. Symp. Circ. Syst.*, vol. 1, pp. 209-212, June 1997.

32. D. M. Pietruszynski et al, A 50-Mbit/s CMOS monolithic optical receiver, *IEEE J. Solid-State Circuits*, vol. 23, no. 6, pp. 1426-1432, Dec. 1988.

33. Y. Akazawa et al., A design and packaging technique for a high-gain, gigahertz-band single-chip amplifier, *IEEE J. Solid-State Circuits*, vol. SC-21, no. 3, pp. 417-423, Jun. 1986.

34. N. Ishihara et. al., A Design technique for a high-gain, 10-GHz class-bandwidth GaAs MESFET amplifier IC module, *IEEE J. Solid-State Circuits*, vol. 27, no. 4, pp. 554-561, Apr. 1992.

35. M. Lee and M. A. Brooke, Design, fabrication, and test of a 125Mb/s transimpedance amplifier using MOSIS 1.2 μm standard digital CMOS process, *Proc. 37th Midwest Symp., Circ. and Syst.*, vol. 1, pp. 155-157, Aug. 1994.

36. R. P. Jindal, Gigahertz-band high-gain low-noise AGC amplifiers in fine-line NMOS, *IEEE J. Solid-State Circuits*, vol. SC-22, no. 4, pp. 512-520, Aug. 1987.

37. N. Takachio et al., A 10Gb/s optical heterodyne detection experiment using a 23GHz bandwidth balanced receiver, *IEEE Trans. M.T.T.*, vol. 38, no. 12, pp. 1900-1904, Dec. 1990.

38. K. Katsura et al., A novel flip-chip interconnection technique using solder bumps for high-speed photoreceivers, *J. Lightwave Tech.*, vol. 8, no. 9, pp. 1323-1326, Sep. 1990.

39. K. Murota and K. Hirade, GMSK modulation for digital mobile radio telephony, *IEEE Trans. Comm.*, vol. 29, pp. 1044-1050, 1981.

40. H. Krauss, C. W. Bostian, and F. H. Raab, *Solid State Radio Engineering*, New York, Wiley, 1980.

41. A. S. Sedra and K. C. Smith, *Microelectronic Circuits*, 4th Edition 1998.

42. N. O. Sokal and A. D. Sokal, Class E, A new class of high efficiency tuned single-ended switching power amplifiers, *IEEE J. Solid-State Circuits*, vol. SC-10, pp. 168-176, June 1975.

43. F. H. Raab, Effects of circuit variations on the class E tuned power amplifier, *IEEE J. Solid-State Circuits*, vol. SC-13, pp. 239-247, 1978.

44. T. Sowlati, C. A. T. Salama, J. Sitch, G. Robjohn, and D. Smith, Low voltage, high efficiency class E GaAs power amplifiers for mobile communications, in *IEEE GaAs IC Symp. Tech. Dig.*, pp. 171-174, 1994.

45. _____ Low voltage, high efficiency GaAs class E power amplifiers for wireless transmitters, *IEEE J. Solid-State Circuits*, vol. SC-13, no. 10, pp. 1074-1080, 1995.

46. A. Rofougaran et al., A single-chip 900 MHz spread-spectrum wireless transceiver in 1-μm CMOS. part I: architecture and transmitter design, *IEEE J. Solid-State Circuits*, vol. SC-33, no. 4, pp. 515-534.

47. J. Chang, A. A. Abidi, and M. Gaitan, Large suspended inductors on silicon and their use in a 2-μm CMOS RF amplifier, *IEEE Electron Device Letters*, vol. 14, no. 5, May 1993.

48. T. Sowlati et al., Linearized high efficiency class E power amplifier for wireless communications, *IEEE Custom Integrated Circuits Conf. Proc.*, pp. 201-204, 1996.

49. G. N. Henderson, M. F. O'Keefe, T. E. Boless, P. Noonan, et al., SiGe bipolar junction transistors for microwave power applications, *IEEE MTT-S Int. Microwave Symp. Dig.*, pp. 1299-1302, 1997

50. O. Shoaei and W. M. Snelgrove, A wide-range tunable 25MHz-110MHz BiCMOS continuous-time filter, *Proc. IEEE Int. Symp. Circ. Syst. (ISCAS)*, Atlanta, 1996.

51. P.-H. Lu, C.-Y. Wu, and M.-K. Tsai, Design techniques for VHF/UHF high-Q tunable bandpass filters using simple CMOS inverter-based transresistance amplifiers, *IEEE J. Solid-State Circuits*, vol. 31, no. 5, May 1996.

52. Y. Tsividis, Integrated continuous-time filter design — an overview, *IEEE J. Solid-State Circuits*, vol. 29, no. 3, Mar. 1994.

53. F. Rezzi, A. Baschirotto, and R. Castello, A 3V 12-55MHz BiCMOS Pseudo-Differential Continuous-Time Filter, *IEEE Trans. Circ. Syst.-I*, vol. 42, no. 11, Nov. 1995.

54. B. Nauta, *Analog CMOS Filters for Very High Frequencies*, Kluwer Academic Publishers.

55. C. Toumazou, F. Lidgey, and D. Haigh, *Analogue IC Design: The Current-Mode Approach*, Peter Peregrinus Ltd. for IEEE Press, 1990.

56. S. Szczepanski and R. Schauman, Nonlinearity-induced distortion of the transfer function shape in high-order filters, *Kluwer Journal of Analog Int. Circuits and Signal Processing*, vol. 3, p. 143-151, 1993.

57. S. Szczepanski, VHF fully-differential linearized CMOS transconductance element and its applications, *Proc. IEEE Int. Symp. Circ. Syst. (ISCAS)*, London 1994.

58. A. A. Abidi, Noise in active resonators and the available dynamic range, *IEEE Trans. Circ. Syst.-I*, vol. 39, no. 4, Apr. 1992.

59. S. Pipilos and Y. Tsividis, RLC active filters with electronically tunable center frequency and quality factor, *Electron. Lett.*, vol. 30, no. 6, Mar. 1994.

60. K. Manetakis, Intermediate frequency CMOS analogue cells for wireless communications, Ph.D. thesis, Imperial College, London, 1998.

61. K. Manetakis and C. Toumazou, A 50MHz high-Q bandpass CMOS filter, *Proc. IEEE Int. Symp. Circ. Syst. (ISCAS)*, Hong-Kong, 1997.

62. Y. Tsividis, Externally linear time invariant systems and their application to companding signal processors, *IEEE Trans. CAS-II*, vol. 44(2), pp. 65-85, Feb.1997.

63. D. Frey, Log-domain filtering: an approach to current-mode filtering, *IEE Proc.-G*, vol. 140, pp. 406-416, 1993.

64. B. Gilbert, Translinear circuits: a proposed classification, 1975, *Electron. Lett.*, vol. 11, no. 1, pp. 14-16, 1975.

65. P. Grey, and R. Meyer, *Analysis and Design of Analog Integrated Circuits*, John Wiley & Sons Inc., New York, 3rd Edition, 1993.

66. E. M. Drakakis, A. Payne, and C. Toumazou, Log-domain state-space: a systematic transistor-level approach for log-domain filtering, accepted for publication in *IEEE Trans. CAS-II*, 1998.

67. V. Leung, M. El-Gamal, and G. Roberts, Effects of transistor non-idealities on log-domain filters, *Proc. IEEE Int. Symp. Circ. Syst.*, Hong-Kong, pp. 109-112, 1997.

68. D. Perry and G. Roberts, Log domain filters based on LC-ladder synthesis, *Proc. 1997 IEEE Int. Symp. Circ. Syst. (ISCAS)*, pp. 311-314, Seattle, 1995.

69. J. Mulder, M. Kouwenhoven, and A. van Roermund, Signal × noise intermodulation in translinear filters, *Electron. Lett*, vol. 33(14), pp. 1205-1207.

70. M. Punzenberger and C. Enz, Noise in instantaneous companding filters, *Proc. 1997 IEEE Int. Symp. Circ. Syst.*, Hong Kong, pp. 337-340, June 1997.

71. M. Punzenberger and C. Enz, A 1.2V low-power BiCMOS class-AB log-domain filter, *IEEE J. Solid-State Circuits*, vol. SC-32(12), pp. 1968-1978, Dec. 1997.

72. D. Perry and G. Roberts, Log-domain filters based on LC ladder synthesis, *Proc. 1995 IEEE Int. Symp. Circ. Syst.*, Seattle, pp. 311-314, 1995.

73. E. Drakakis, A. Payne, and C. Toumazou, Bernoulli operator: a low-level approach to log-domain processing, *Electron. Lett.*, vol. 33(12), pp. 1008-1009, 1997.

74. F. Yang, C. Enz, and G. Ruymbeke, Design of low-power and low-voltage log-domain filters, *Proc. 1996 IEEE Int. Symp. Circ. Syst.*, Atlanta, pp. 125-128, 1996.

75. J. Mahattanakul and C. Toumazou, Modular log-domain filters, *Electron. Lett.*, vol. 33(12), pp. 1130-1131, 1997.

76. D. Frey, A 3.3 V electronically tuneable active filter useable to beyond 1 GHz, *Proc. 1994 IEEE Int. Symp. Circ. Syst.*, London, pp. 493-496, 1994.

77. M. El-Gamal, V. Leung, and G. Roberts, Balanced log-domain filters for VHF applications, *Proc. 1997 IEEE Int. Symp. Circ. Syst.*, Monterey, pp. 493-496, 1997.

6
Operational Transconductance Amplifiers

R.F. Wassenaar
University of Twente

Mohammed Ismail
The Ohio State University

Chi-Hung Lin
The Ohio State University

6.1 Introduction

In many analog or mixed analog/digital VLSI applications, an operational amplifier may not be appropriate to use for an active element. For example, when designing integrated high-frequency active filter circuitry, a much simpler building block, called an operational transconductance amplifier (OTA), is often used.[1] This type of amplifier is characterized as a voltage-driven current source and in its simplest form is a combination of a differential input pair with a current mirror as shown in Fig. 6.1. It is a simple circuit with a relatively small chip area. Further, it has a high bandwidth and also a good common-mode rejection ratio up to very high frequencies. The small signal transconductance, $g_m = \partial I_{out}/\partial V_{in}$, can be controlled by the tail current. This chapter discusses CMOS OTA design for modern VLSI applications. We begin the chapter with a brief study of noise in OTAs, followed by OTA design techniques.

6.2 Noise Behavior of the OTA

The noise behavior of the OTA is discussed here. Attention will be paid to thermal and flicker noise and to the fact that, for minimal noise, some voltage gain, from the input of the differential pair to the input of the current mirror, is required. Then, only the noise of the input pair becomes dominant and the other noise sources can be neglected to first order. The noise behavior of a single MOS transistor is modeled by a single noise voltage source. This noise voltage source is placed in series with the input (gate) of a "noiseless" transistor. Fig. 6.2(a) shows the simple OTA, including the noise sources, while Fig. 6.2(b) shows the same circuit with all the noise referred to the input of the stage.

All the noise sources indicated in Fig. 6.2(a) are converted to equivalent input noise voltages, which are then added to form a single noise source at the input (Fig. 6.2(b)). As a result, we obtain (assuming $g_{m1} = g_{m2}$ and $g_{m3} = g_{m4}$) the following mean-square input referred noise voltage.

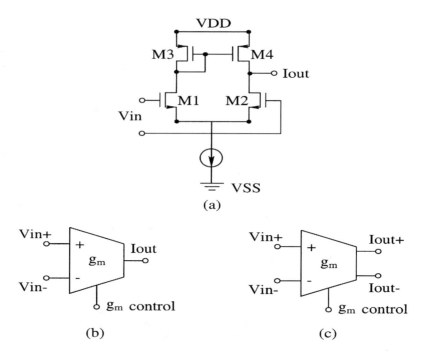

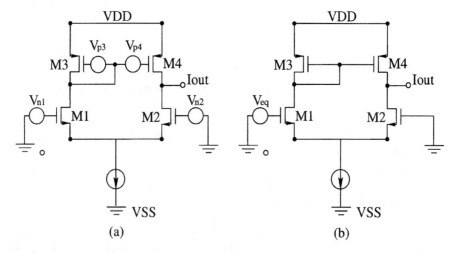

FIGURE 6.1 (a) An N-MOS differential pair with a P-MOS current mirror forming an OTA; (b) the symbol for a single-ended OTA; and (c) the symbol for a fully differential OTA.

FIGURE 6.2 (a) The OTA with its noise voltage sources, and (b) the same circuit with the noise voltage sources referred to one of the input nodes.

$$\overline{V_{eq}^{2}} = \overline{V_{n1}^{2}} + \overline{V_{n2}^{2}} + \left(\frac{g_{m3}}{g_{m1}}\right)^{2}(\overline{V_{p3}^{2}} + \overline{V_{p4}^{2}}) \qquad (6.1)$$

The thermal noise contribution of one transistor, over a band Δf, is written as:

$$\overline{V_{th}^{2}} = \frac{2}{3}4kT\frac{1}{g_{m}}\Delta f \qquad (6.2)$$

where k is the Boltzmann constant and T is the absolute temperature.

The equivalent noise voltage $\overline{V_{th_{eq}}^2}$ becomes:

$$\overline{V_{th_{eq}}^2} = \frac{2}{3} 4kT \left(\frac{1}{g_{m1}} + \frac{1}{g_{m2}} + \left(\frac{g_{m3}}{g_{m1}}\right)^2 \right) \left(\frac{1}{g_{m3}} + \frac{1}{g_{m4}} \right) \Delta f \tag{6.3}$$

and because $g_{m1} = g_{m2}$ and $g_{m3} = g_{m4}$, $\overline{V_{th_{eq}}^2}$ becomes:

$$\overline{V_{th_{eq}}^2} = \frac{16}{3} kT \left(\frac{1}{g_{m1}} + \left(\frac{g_{m3}}{g_{m1}}\right)^2 \frac{1}{g_{m3}} \right) \Delta f \tag{6.4}$$

or

$$\overline{V_{th_{eq}}^2} = \frac{16}{3} \frac{kT}{g_{m1}} \left(1 + \frac{g_{m3}}{g_{m1}} \right) \Delta f \tag{6.5}$$

Expressing g_m in physical parameters results in:

$$\overline{V_{th_{eq}}^2} = \frac{16kT}{3\sqrt{\mu_n C_{ox}(W/L)_1 I_0}} \left(1 + \sqrt{\frac{\mu_p (W/L)_3}{\mu_n (W/L)_1}} \right) \Delta f \tag{6.6}$$

In this equation, I_0 represents the tail current of the differential pair. Note that the term between brackets represents the relative noise contribution of the current mirror. This term can be neglected if M_3 and M_4 are chosen relatively long and narrow in comparison to M_1 and M_2.

It should be mentioned that the thermal noise of an N-MOS transistor and a P-MOS transistor with equal transconductance is the same. In most standard IC processes, a three to ten times lower $1/f$ noise is observed for P-MOS transistors in comparison to N-MOS transistors of the same size. However, in modern processes, the $1/f$ noise contribution of N- and P-MOS transistors tends to be equal.

For the $1/f$ noise, it is usually assumed for standard IC processes that:

$$\overline{V_{1/f}^2} = \frac{K'}{WLC_{ox}f} \Delta f \tag{6.7}$$

where K' is the flicker noise coefficient in the range of 10^{-24} J for N-MOS transistors and in the range of 3×10^{-25} to 10^{-25} J for P-MOS transistors. The equivalent $1/f$ input noise source of the OTA in Fig. 6.2(b) yields:

$$\overline{V_{eq(1/f)}^2} = \frac{2K'_n \Delta f}{W_1 L_1 C_{ox} f} \left(1 + \frac{K'_p \mu_p L_1^2}{K'_n \mu_n L_3^2} \right) \tag{6.8}$$

Here, the noise contributions of the current mirror (M_3, M_4) will be negligible if L_3 is chosen much larger than L_1.

The offset voltage of a differential pair is lowest when the transistors are in the weak-inversion mode; but on the contrary, the mismatch in the current transfer of a current mirror is lowest when the transistors are deep in strong inversion. Hence, the conditions that have to be fulfilled for both minimal equivalent input noise and minimal offset are easy to combine.

6.3   An OTA with an Improved Output Swing

A CMOS OTA with an output swing much higher than that in Fig. 6.1(a) is shown in Fig. 6.3. This configuration needs two extra current mirrors and consumes more current, but the output voltage "window" is, in the case when common-mode input voltage is zero, about doubled. The rules discussed earlier for sizing the input transistors and current-mirror transistors to reduce noise and offset still apply. However, there is still a tradeoff. On the one hand, a high voltage gain from the input nodes to the current mirror is good for reducing noise and mismatch effects; on the other hand, too much gain also reduces the upper limit of the common-mode input voltage range and the phase margin needed to ensure stability (this will be discussed later).[2] A voltage gain on the order of 3 to 10 is advised. The frequency behavior of the OTA in Fig. 6.3 is rather complex since there are two different signal paths in parallel, as shown in Fig. 6.4. In this scheme, r_p represents the parallel value of the output resistance of the stage $(r_{o6}\|r_{o8})$ and the load resistance (R_L); therefore,

$$r_p = r_{o6} \| r_{o8} \| R_L \tag{6.9}$$

The capacitor C_p represents the sum of the parasitic output capacitance and the load capacitance $C_p = C_o + C_L$. Using the half-circuit principle for the differential pair, a fast signal path can be seen from M_2 via current mirror M_7, M_8 to the output. This signal path contributes an extra high-frequency pole. The other signal path leads from transistor M_1 via both current mirrors M_3, M_4 and M_5, M_6 to the output. In this path, two extra poles are added. The transfer of both signal paths and their combination are

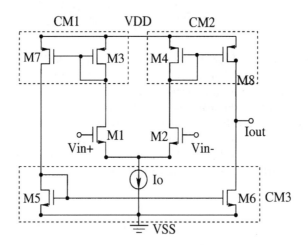

FIGURE 6.3   An OTA with an improved output window.

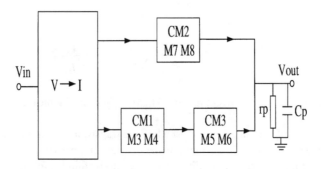

FIGURE 6.4   The signal paths of the OTA in Fig. 6.3.

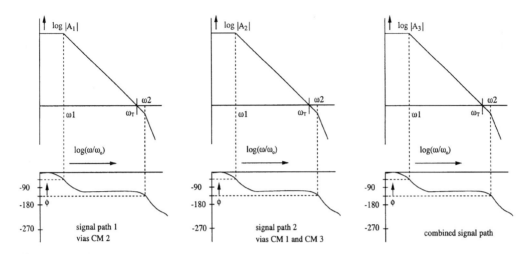

FIGURE 6.5 (a) The Bode plot belonging to signal path 1 in the OTA in Fig. 6.3 and 6.4, (b) signal path 2, and (c) to the combined signal path.

shown in the plots in Fig. 6.5, assuming equal pole positions of all three current mirrors. Note that the first (dominant) pole (ω_1) is determined by r_p and C_p.

$$\omega_1 = \frac{1}{r_p C_p} \tag{6.10}$$

The second pole (ω_2) is determined by the transconductance of M_3 and the sum of the gate-source capacitance of M_3 and M_4. If M_3 and M_4 are equal, the second pole is located at:

$$\omega_2 = \frac{g_{m3}}{2C_{gs3}} \tag{6.11}$$

The unity-gain corner frequency ω_T of the loaded OTA is at:

$$\omega_T = \frac{g_{m1}}{C_p} \tag{6.12}$$

Therefore, the ratio ω_2/ω_T is:

$$\frac{\omega_2}{\omega_T} = \frac{g_{m3}}{g_{m1}} \frac{C_p}{2C_{gs3}} \tag{6.13}$$

When the OTA is used for high-frequency filter design, an integrator behavior is required, that is, a constant 90° phase at least at frequencies around ω_T. Therefore, a high value of the ratio ω_2/ω_T is needed in order to have as little influence as possible from the second pole. It is obvious from Eq. 6.12 that the low-frequency voltage gain from the input nodes of the circuit to the input of the current mirrors ($= g_{m1}/g_{m3}$) must not be chosen too high. As mentioned, this is in contrast to the requirements for minimum noise and offset.

Sometimes, OTAs are used as unity gain voltage buffers; for example, in switched capacitor filters. In this case, the emphasis is placed on obtaining high open-loop voltage gain, improved output window, and good capability to drive capacitive loads efficiently (or small resistors); its integrator behavior is of less importance.

To increase the unloaded voltage gain, cascode transistors can be added in the output stage. This greatly increases the output impedance of the OTA and hardly decreases the phase margin. The penalty that has to be paid is an additional pole in the signal path and some reduction of the maximum possible output swing. This reduction can be very small if the cascode transistors are biased on the weak-inversion mode. The open-loop voltage gain can be in the order of 40 to 60 dB. A possible realization of such a configuration is shown in Fig. 6.6.[3]

6.4   OTAs with High Drive Capability

For driving capacitive loads (or small resistors), a large available output current is necessary. In the OTAs shown so far, the amount of output current available is equal to twice the quiescent current (i.e., the tail current I_0). In some situations, this current can be too small. There are several ways to increase the available current in an efficient way. To achieve this, four design principles will be discussed here:

1. Increasing the quiescent current by using current mirrors with a current transfer ratio greater than 1
2. Using a two-transistor level structure to drive the output transistors
3. Adaptive biasing techniques
4. Class AB techniques

OTAs with 1:B Current Mirrors

One way to increase available output current is to increase the transfer ratio of the current mirrors CM1 and CM2 by a factor B, as indicated in Fig. 6.7.[4] The amount of available output current and also the overall transconductance increase by the same factor. Unfortunately, the –3 dB frequency of the CM1-CM2 current mirrors will be reduced by a factor $(B + 1)/2$ due to the larger gate-source capacitance of the mirror output transistors. Moreover, ω_T will increase, ω_2 will decrease, and the ratio ω_2/ω_T will be strongly deteriorated. The amount of available output current though is B times the tail current. It is also possible to increase the current transfer ratio of current mirror CM3 instead of CM1. A better current efficiency then results, but at the expense of more asymmetry in the two signal paths. Although the amount of the maximum available output current is B times the tail current in both situations, the ratio between the maximum available current and quiescent current of the output stage remains equal to two, just as in the OTAs discussed previously.

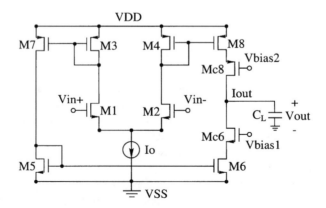

FIGURE 6.6   An OTA with improved output impedance.

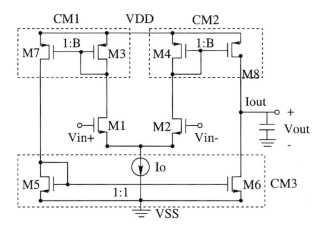

FIGURE 6.7 An OTA with improved load current using 1:B current mirrors.

OTA with Improved Output Stage

Another way of increasing the maximal available output current is illustrated in Fig. 6.8.[6] It improves upon the factor-two relationship between quiescent and maximal available current. Assuming equal K factors for all transistors shown in the circuit leads to the conclusion that the effective gate-source voltage of transistor M_{11} (= $V_{GS11} - V_{T11}$) equals that of transistor M_1 (=$V_{GS1} - V_{T1}$), since they carry the same current, assuming that transistors M_1, M_4, and M_6 are in saturation. Because the current drawn through transistor M_9 is equal to the current in transistor M_2, their effective source-gate voltages must also be equal assuming equal K factor for M_2 and M_9. Since the sum of the effective gate-source voltages transistors M_{11} and M_{12}, and also of M_9 and M_{10}, is fixed and equal to V_B, a situation exists which is equivalent to the two transistor level structure described in Ref. 5.

The ratio between the maximum available output current and the quiescent current of the output stage can be chosen by the designer. It is equal to: $(V_B/(V_B - V_{GS0}))^2$, where V_{GS0} is the quiescent gate-source voltage of transistor M_{11}.

If the OTA is used in an over-drive situation $|V_{in}| > \sqrt{(2I_0)/K}$, then either M_6 or M_5 will be cut off, while the other transistor carries its maximum current. As a result, one of the output transistors (M_{10} or M_{12}) carries its maximum current, while the other transistor is in a low-current stand-by situation. The maximum current that one of the output transistors carries is therefore proportional to V_B^2. With the

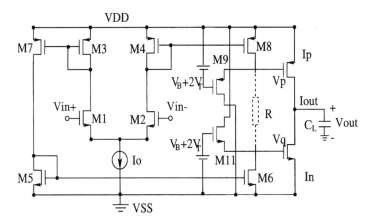

FIGURE 6.8 An OTA with an improved ratio between the maximum available current and the quiescent current of the output stage.

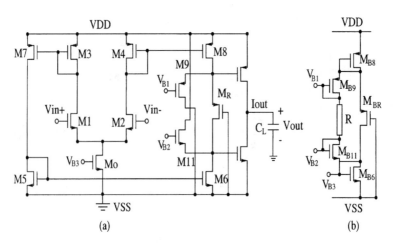

FIGURE 6.9   (a) The complete OTA, (b) and its bias stage.

high ohmic resistor R (indicated in Fig. 6.8 with dotted lines), this maximum current corresponds to either $(V_P - V_{SS} - V_{TN})^2$ or $(V_{DD} - V_Q - V_{TP})^2$, because in that situation no current flows through the resistor. Hence, with the extra resistor, it becomes possible to increase the maximum current in over-drive situations and therefore reduce the slewing time. Because resistor R is chosen to be high, it does not disturb the behavior of the circuit discussed previously. In practice, resistor R is replaced by transistor M_R working in the triode region, as shown in Fig. 6.9(a). Figure 6.9(b) shows the circuit which was used in Ref. 5 for biasing the gates of transistors M_0, M_9, and M_{11}. It is much like the so-called "replica biasing." The current in the circuit is strongly determined by the voltage across R (and its value) and is therefore very sensitive to variations in the supply voltage.

Adaptively Biased OTAs

Another combination of high available output current with low standby current can be realized by making the tail current of the differential input pair signal dependent. Figure 6.10 shows the basic idea of such an OTA with adaptive biasing.[7] The tail current I_0 of the differential pair is the sum of a fixed value I_R and an additional current equal to the absolute value of the difference between the drain currents multiplied by the current feedback factor B ($I_0 = I_R + B|I_1 - I_2|$). Therefore, with zero differential input voltage, only a low bias current I_R flows through the input pair. A differential input voltage, V_{ind}, will cause a difference in the drain currents which will increase the tail current. This, in turn, again gives rise to a greater difference in the drain current, and so on. This is the kind of positive feedback that can bring the differential input pair from the weak-inversion mode into the strong-inversion mode, depending on the input voltage and the chosen current feedback factor B.

Normally, when $V_{ind} = V_{in+} - V_{in-}$ is small, the input transistors are in weak inversion. The differential output current $(I_1 - I_2)$ of a differential pair operating in weak inversion equals the tail current times $\tanh((qV_{ind})/(2AkT))$. This leads to the following equation:

$$(I_1 - I_2) = I_R + B|I_1 - I_2| \tanh\left(\frac{qV_{ind}}{2AkT}\right) \tag{6.14}$$

or

$$(I_1 - I_2) = \frac{\tanh\left(\dfrac{qV_{ind}}{2AkT}\right)}{1 - B\left|\tanh\left(\dfrac{qV_{ind}}{2AkT}\right)\right|} I_R \tag{6.15}$$

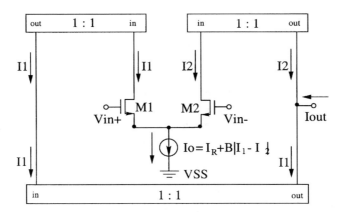

FIGURE 6.10 An OTA with an input dependent tail current.

and because $I_{out} = (I_1 - I_2)$:

$$I_{out} = \frac{\tanh\left(\dfrac{qV_{ind}}{2AkT}\right)}{1 - B\left|\tanh\left(\dfrac{qV_{ind}}{2AkT}\right)\right|} I_R \tag{6.16}$$

However, in the case of large currents, this expression will no longer be valid since $M_1 - M_2$ will leave the weak-inversion domain and enter the strong-inversion region. If that is the case, the output current becomes:

$$I_{out} = \begin{cases} \dfrac{k}{2}V_{ind}\sqrt{\dfrac{4I_R}{K} - (1 - B^2)V_{ind}^2} + B\dfrac{K}{2}V_{ind}^2 & \text{for } V_{ind} > 0 \\[3mm] \dfrac{k}{2}V_{ind}\sqrt{\dfrac{4I_R}{K} - (1 - B^2)V_{ind}^2} - B\dfrac{K}{2}V_{ind}^2 & \text{for } V_{ind} < 0 \end{cases} \tag{6.17}$$

In order to keep some control over the output current, a negative overall feedback must be applied, which is usually the case. For example, when an OTA is used as a unity-gain buffer with a load of C_L (see Fig. 6.11) and assuming a positive input step is applied, then the output current increases dramatically due to the positive feedback action described previously and, as a result, the output voltage will increase. This will lead to a decrease of the differential input voltage V_{ind} ($V_{ind} = V_s - V_{out}$). The result will be a very fast settling of the output voltage, and that is what we wanted to have. In order to realize current

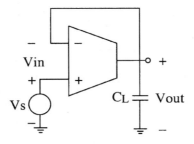

FIGURE 6.11 An OTA used as a unity gain buffer.

$|I_1 - I_2|$, two current-subtracter circuits can be combined (see Fig. 6.12). If the current I_2 is larger than current I_1, the output of current-subtracter circuit 1 (I_{out1}) will carry a current; otherwise, the output current will be zero. The opposite situation is found for the output current of subtracter circuit 2 because of the interchange of their input currents ($I_{out2} = B(I_1 - I_2)$). Consequently, either I_{out1} or I_{out2} will draw a current $B|I_1 - I_2|$ and the other current will be zero. It is for this reason that the upper current mirrors (in Fig. 6.13) have two extra outputs to support the currents for the circuit in Fig. 6.12. A practical realization of the adaptive biasing OTA is shown in Fig. 6.13. In order to avoid unwanted, relatively high stand-by currents due to transistors mismatches, the transfer ratio of the current mirrors (M_{12}, M_{13}) and (M_{19}, M_{18}) can be chosen somewhat larger than 1. This ensures an inactive region of the input voltage range whereby the feedback loop is deactivated.

Another example of an adaptive tail current circuit is shown in Fig. 6.14.[8] It has a normal OTA structure except that the input pair is realized in twofold, and the tail current transistor is used in a feedback loop. This feedback loop includes the inner differential pair and tail current transistor M_0 as well as a minimum current selector, the current source I_U, transistor M_R, and a current sink I_L. The minimum current selector[9] delivers an output current equal to the lowest value of its input currents ($I'_{out} = Min(I'_1, I'_2)$). The feedback loop ensures that the output current of the minimum current selector is equal to the difference in currents

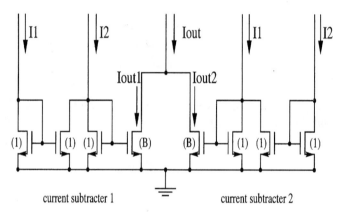

FIGURE 6.12   A combination of two current subtracters for realizing the adaptive biasing current for the circuit in Fig. 6.10.

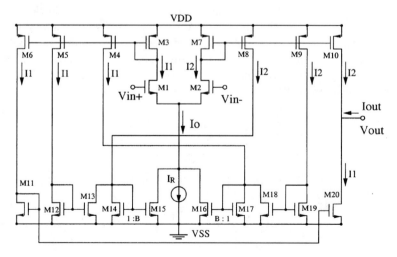

FIGURE 6.13   A practical realization of OTA with an adaptive biasing of its tail current.

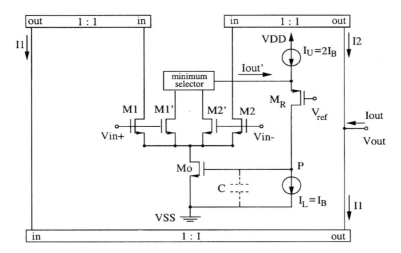

FIGURE 6.14   An OTA using a minimum selector for adapting the tail current.

between the upper and lower current sources. Assume that the upper current carries a current $2I_B$ and the lower current source carries I_B, then the feedback loop will bias the tail current in such a way that either I'_1 or I'_2 becomes equal to I_B; for positive values of V_{ind}, that will be I'_2. It should be realized that at $V_{ind} = 0$, all four input transistors are biased at the same gate-source voltage (V_{GS0}), corresponding to a drain current I_B. In the case of positive input voltages, the gate-source voltage of M_2/M'_2 will not change.

Therefore, all the input voltage will be added to the bias voltage of M_1/M'_1, that is,

$$V_{GS1} = V_{GS0} + V_{ind} \qquad (6.18)$$

Figure 6.15 shows the I_D vs. V_{GS} characteristic for both transistors M_1/M'_1 and M_2/M'_2. Accordingly, the relationship between $(I_1 - I_2)$ vs. V_{ind} (for $V_{ind} > 0$) follows the right side of the $I_D - V_{GS}$ curve of M_1, starting from the stand-by point (V_{GS0}, I_B) as indicated by the solid curve in Fig. 6.15. A similar view can be taken for negative values of the input voltage V_{ind}, resulting in an equal $(I_1 - I_2)$ vs. V_{in} curve rotated 180°. The result is shown in Fig. 6.16.

Note that this input stage has a relationship between $(I_1 - I_2)$ and V_{ind} that is different from that of a simple differential input stage. By increasing V_{ind}, the slope increases and, to a first-order approximation, there will not be a limit for the maximum value of $(I_1 - I_2)$.

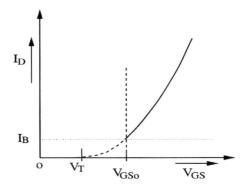

FIGURE 6.15   The I_D vs. V_{GS} characteristic for transistors M_1/M'_1 and M_2/M'_2, showing their standby point V_{GS0}, I_B.

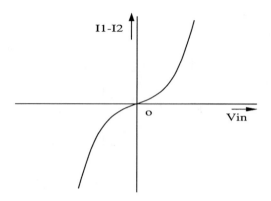

FIGURE 6.16 $I_1 - I_2$ vs. V_{ind}.

Note that there is an additional MOS transistor M_R in the circuit in Fig. 6.14 to fix the output voltage of the minimum current selector circuit. The lower current source I_L is necessary to be able to discharge the gate-source capacitor C of M_0 (indicated in Fig. 6.14 with dotted lines). The OTA in Fig. 6.14 is simpler than that in Fig. 6.13. However, its bandwidth is lower due to the high impedance of node P in the feedback loop.

Class AB OTAs

Another possibility in designing an OTA with a good current efficiency is to use an input stage exhibiting Class AB characteristics.[11] The input stage in Fig. 6.17 contains two CMOS pairs[12] connected as Class AB input transistors. They are driven by four source-followers. By applying a differential input voltage, the current through one of the input pairs will increase while the current through the other will decrease. The maximum current that can flow through the CMOS pair is, to first order, unlimited. In practice, it is limited by the supply voltage, the K_{eq} factor, the mobility reduction factor, and the series resistance. The currents are delivered to the output with the help of two current mirrors. In the OTA shown in Fig. 6.17, only one of the two outputs of each CMOS pair is used. The other output currents flow directly to the supply rails. Instead of wasting the other output currents, they can be used to supply an extra output. So with the addition of two current mirrors, an OTA with complementary outputs as shown in Fig. 6.18

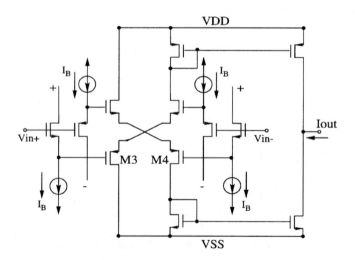

FIGURE 6.17   An OTA having a class AB input stage.

can be achieved.[10] An improvement of the output impedance and low-frequency voltage gain can be obtained by cascoding the output transistors of the current mirrors (Fig. 6.19). Usually, this reduces the output window. The function of transistors M_{41}-M_{44} is to control the dc output voltages. They form a part of a common-mode feedback system, which will be discussed next.

The relationship between the differential input voltage V_{in} and one of the output currents I_{out} is shown in Fig. 6.20. There is a linear relationship between V_{ind} and I_{out} for small to moderate values of V_{ind}. In the case of larger values of V_{ind}, one of the CMOS pairs becomes cut off, resulting in a quasi-quadratic relationship. At a further increase of V_{ind}, the output current will be somewhat saturated due to mobility reduction and to the fact that one of the transistors of the CMOS pair leaves saturation mode. The latter effect is, of course, also strongly dependent on the common input voltage.

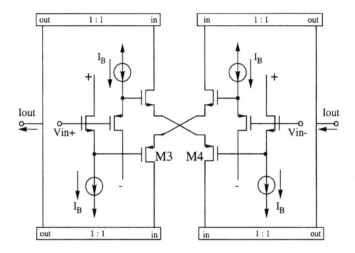

FIGURE 6.18   An OTA having a Class AB input stage and two complementary outputs.

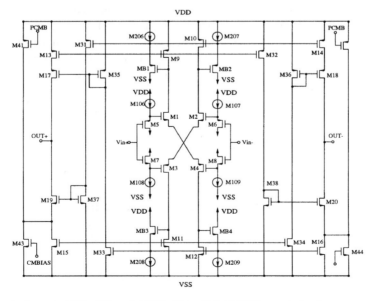

FIGURE 6.19   An improved fully differential OTA. (From Ref. 10. With permission.)

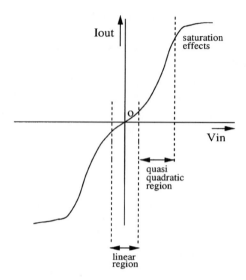

FIGURE 6.20   The $V_{in} \rightarrow I_{out}$ characteristic of the OTA in Fig. 6.19.

6.5   Common-Mode Feedback

A fully differential OTA circuit, as in Fig. 6.19, has many advantages compared with its single-ended counterpart. It is a basic building block in filter design. A fully differential approach, in general, leads to a more efficient current use, doubling of the maximum output-voltage swing, and an improvement of the power-supply rejection ratio (PSRR). It also leads to a significant reduction of the total harmonic distortion, since all even harmonics are canceled out due to the symmetrical structure. Even when there is a small imperfection in the symmetry, the reduction in distortion will be significant.

However, this type of symmetrical circuit needs an extra feedback loop. The feedback around a single-ended OTA usually only provides a differential-mode feedback and is ineffective for common-mode signals.

So, in the case of the fully differential OTA, a common-mode feedback (CMFB) circuit is needed to control the common output voltage. Without a CMFB, the common-mode output voltage of the OTA is not defined and it may drift out of its high-gain region. The general structure of a simple OTA circuit with a differential output and a CMFB circuit is shown in Fig. 6.21. The need for a CMFB circuit is a drawback since it counters many of the advantages of the fully differential approach. The CMFB circuit requires chip area and power, introduces noise, and limits the output-voltage swing.

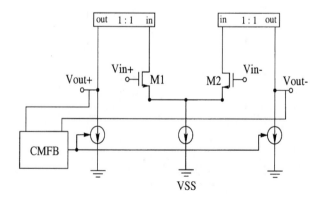

FIGURE 6.21   The general structure of a simple OTA circuit having a differential output and the required CMFB.

Figure 6.22(b) shows a simple implementation of a CMFB circuit. A differential pair (M_1, M_2) is used to sense the common-mode output voltage. So, the voltage at the common source of this differential pair (V_s) is used. Its voltage provides, with a level shift of one V_{GS}, the common-mode output voltage of the OTA. The voltage at this node is the first order insensitive to the differential input voltage. The relationship between the differential input voltage V_{in} of the differential pair, superimposed on a common-mode input voltage V_{CM}, and its common-source voltage V_s is shown in Fig. 6.22(a). The common-mode output voltage of the OTA is determined by the V_{GS} of M_1/M_2 and M_9/M_{10} and can be controlled by the voltage source V_0. There might be an offset in the dc value of the two output voltages due to a mismatch in transistors M_9 and M_{10}.

If the amplitude of the differential output voltage increases, the common-mode voltage will not remain constant, but will be slightly modulated by the differential output voltage, with a modulation frequency that is twice the differential input signal frequency. This modulation is caused by the "non-flat" characteristic of the V_s vs. V_{in} characteristic of the differential pair (M_1, M_2) (see Fig. 6.22(a)).

Another commonly used CMFB circuit is shown in the fully-differential folded cascode OTA in Fig. 6.23.[13] In this circuit, a similar high-output resistance and high unloaded voltage gain can be achieved as in the normal cascode circuits. An advantage of the folded cascode technique, however, is a higher accuracy in the signal-current transfer because current mirrors are avoided. In Fig. 6.23, all transistors are in saturation, with the exception of M_1, M_{11}, and M_{12}, which are in the triode region. The CMFB is provided with the help of M_{11} and M_{12}. These two transistors sense the output voltages V_P and V_Q. Since they operate in the triode region, their sum-current is insensitive to the differential output voltage $(V_P - V_Q)$ and depends only on the common output voltage $((V_P + V_Q)/2)$. Because the current that flows through M_{17} and M_{18} forces the value of the above-mentioned sum-current, they also determine, together with V_{bias4}, the common-mode output voltage. By choosing V_{bias1} in such a way that I_{M19} is twice I_{M17}, and making the width of transistor M_1 twice that of $M_{11} (= M_{12})$, the nominal common-mode output voltage will be equal to the gate voltage of M_1.

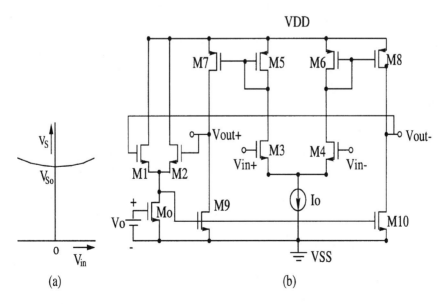

FIGURE 6.22 (a) The relationship between the differential input voltage, superimposed on a common-mode voltage V_{CM} of a differential pair (M_1, M_2) and its common-source voltage V_s; (b), a fully differential OTA with the differential pair (M_1, M_2) for providing a common-mode feedback.

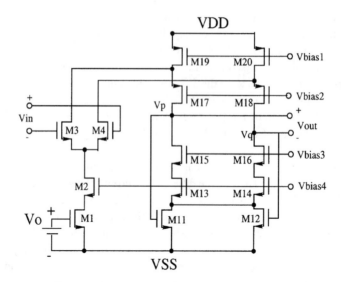

FIGURE 6.23   A fully differential folded cascode OTA with another commonly used CMFB circuit.

6.6   Filter Applications with Low-Voltage OTAs

Usually, g_m-C filters are considered suitable candidates for high-speed and low-power applications. Compared with the SC op-amp and RC op-amp techniques, the applicability of the g_m-C filter is limited by the low dynamic range and medium, even poor, linearity. The strategy of both simplifying the architecture and designing an ultra-low-voltage OTA[16] is used to meet low-power and dynamic range requirements. The filter topology is derived from a passive ladder form of 5th-order elliptic filtering. Using element replacement and sharing multiple inputs for gyrators, a fully differential 5th-order elliptic filter is shown in Fig. 6.24.[14] This multi-input sharing in the filter design reduces the numbers of OTAs from 11 to 6. Especially for wide bandwidth design, the method saves almost half of the die area and power dissipation. This design uses balanced signals to reduce even harmonics and to relax parasitic matching requirements. The capacitors realizing the filter poles are connected between the outputs of the transconductors and signal ground. This helps the stability of the common-mode feedback circuit because of the loading to both common-mode and fully-differential signal paths. The inherent 6 dB loss at low frequency is compensated for by the first transconductor with $2g_m$ gain. Fig. 6.25 shows the frequency response and

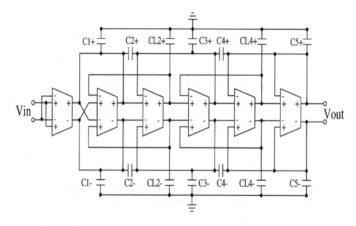

FIGURE 6.24   A g_m-C elliptic filter with low-voltage OTAs.

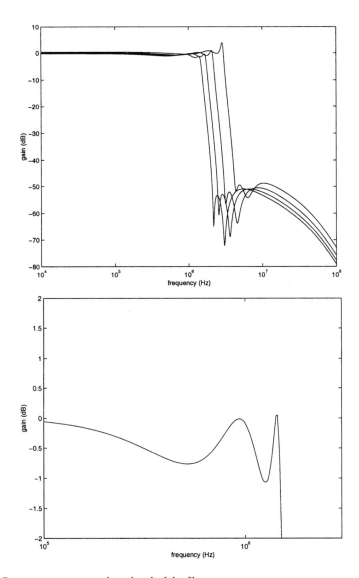

FIGURE 6.25   Frequency response and passband of the filter.

the passband of the filter. The filter has –3dB frequency tuning ranges of 1.2 MHz to 2.9 MHz and –60 dB stop band rejection. Obviously, the tuning range is limited by the low-power supply rail. This presents a problem for automatic tuning design at very low supply voltages. A possible application for this filter is for channel selection in wideband handy phones.[15]

References

1. M. Ismail and T. Fiez, *Analog VLSI Signal and Information Processing*, McGraw-Hill, 1994.
2. E. A. Vittoz, The design of high-performance analog circuits on digital CMOS chips, *IEEE J. Solid-State Circuits*, vol. SC-20, pp. 657-665, June 1985.
3. F. Krummenacher, High voltage gain CMOS OTA for micro-power SC filters, *Electronics Lett.*, vol. 17, pp. 160-162, 1981.

4. M. S. J. Steyaert, W. Bijker, P. Vorenkamp, and J. Sevenhans, ECL-CMOS and CMOS-ECL Interface in 1.27mm CMOS for 150 MHz Digital ECL Data Transmission Systems, *IEEE J. Solid-State Circuits*, vol. SC-26, pp. 15-24, Jan. 1991.

5. R. F. Wassenaar, Analysis of analog C-MOS circuits, Ph.D. thesis, University of Twente, The Netherlands, 1996.

6. S. L. Wong and C. A. T. Salama, An efficient CMOS buffer for driving large capacitive loads, *IEEE J. Solid-State Circuits*, vol. SC-21, pp. 464-469, June 1986.

7. M. G. Degrauwe, J. Rijmenants, E. A. Vittoz, and H. J. DeMan, Adaptive biasing CMOS amplifiers, *IEEE J. Solid-State Circuits*, vol. SC-17, pp. 522-528, June 1982.

8. E. Seevinck, R. F. Wassenaar, and W. de Jager, Universal adaptive biasing principle for micro-power amplifiers, *Digest of Technical Papers ESSCIRC'84*, pp. 59-62, Sept. 1984.

9. R. F. Wassenaar, Current-Mode Minimax Circuit, *IEEE Circuits and Devices*, vol. 8, pp. 47, Nov. 1992.

10. S. H. Lewis and P. R. Gray, A pipelined 5MHz 9b ADC, *Proc. IEEE Intl. Solid-State Circuits Conference*, pp. 210-211, 1987.

11. S. Dupuie and M. Ismail, High frequency CMOS transconductors, Ch. 5 in *Analog IC Design: The Current-Mode Approach*, Toumazou, Lidgey, and Haight, Eds., Peter Peregrinus, Ltd., London, 1990.

12. E. Seevinck and R. F. Wassenaar, A versatile CMOS linear transconductor/square-law function circuit, *IEEE J. Solid-State Circuits*, vol. SC-22, no. 3, pp. 366-377, June 1987.

13. T. C. Choi, R. T. Kaneshiro, R. Brodersen, and P. R. Gray, High-frequency CMOS switched capacitor filters for communication applications, *Proc. IEEE Intl. Solid-State Circuits Conference*, pp. 246-247, 314, 1983.

14. R. Schaumann, Continuous-Time Integrated Filters, Ch. 80, *The Circuits and Filters Handbook*, W.-K. Chen, Editor-in-Chief, CRC Press and IEEE Press, New York, 1995.

15. C.-C. Hung, K. A. I. Halonen, M. Ismail, V. Porra, and A. Hyogo, A low-voltage, low-power CMOS fifth-order elliptic GM-C filter for baseband mobile, wireless communication, *IEEE Trans. Circuits and Systems for Video Technology*, vol. 7, no. 4, pp. 584-592, August 1997.

16. C.-H. Lin and M. Ismail, Design and analysis of an ultra low-voltage CMOS class-AB V-I converter for dynamic range enhancement, *Int. Symp. Circ. Syst.*, Orlando, Florida, June 1999.

7

Nyquist-Rate ADC and DAC

Bang-Sup Song
*University of California
at San Diego*

7.1 Introduction

The rapidly growing electronics field has witnessed the digital revolution that started with the digital telephone switching system in the early 1970s. The trend continued with digital audio in the 1980s and with digital video in the 1990s. The digital technique is expected to prevail in the coming multimedia era and to influence even future wireless PCS/PCN systems. All electrical signals in the real world are analog in nature, and their waveforms are continuous in time. Since most signal processing is done numerically in discrete time, devices that convert an analog waveform into a stream of discrete digital numbers, or vice versa, have become technical necessities in implementing high-performance digital processing systems. The former is called an analog-to-digital converter (ADC or A/D converter), and the latter is called a digital-to-analog converter (DAC or D/A converter).

Typical systems in this digital era can be grouped and explained as in Fig. 7.1. The processed data are stored and recovered later using magnetic or optical media such as tape, magnetic disc, or optical disc. The system can also transmit or receive data through communication channels such as telephone switch,

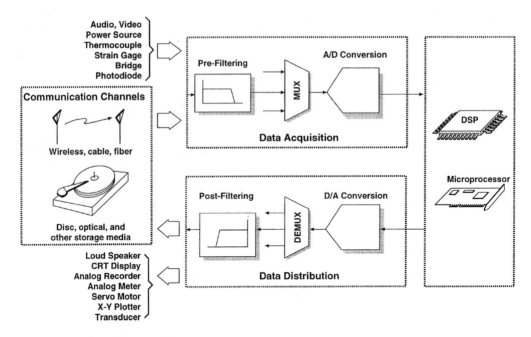

FIGURE 7.1   Information processing systems.

cable, optical fiber, and wireless RF media. Through the Internet computer networks, even compressed digital video images are now made accessible from anywhere and at any time.

Resolution

Resolution is a term used to describe a minimum voltage or current that an ADC/DAC can resolve. The fundamental limit is a quantization noise due to the finite number of bits used in the ADC/DAC. In an N-bit ADC, the minimum incremental input voltage of $V_{ref}/2^N$ can be resolved with a full-scale input range of V_{ref}. That is, limited 2^N digital codes are available to represent the continuous analog input. Similarly, in an N-bit DAC, 2^N input digital codes can generate distinct output levels separated by $V_{ref}/2^N$ with a full-scale output range of V_{ref}. The *signal-to-noise ratio* (SNR) is defined as the power ratio of the maximum signal to the in-band uncorrelated noise. The spectrum of the quantization noise is evenly distributed within the *Nyquist bandwidth* (half the sampling frequency). This inband rms noise decreases by 6 dB when the oversampling ratio is doubled. This implies that, when oversampled, the SNR within the signal band can be made higher. The SNR of an ideal N-bit ADC/DAC is approximated as

$$SNR = 1.5 \times 2^{2N} \approx 6.02N + 1.76 (dB) \qquad (7.1)$$

The resolution is usually characterized by the SNR, but the SNR accounts only for the uncorrelated noise. The real noise performance is better represented by the *signal-to-noise and distortion ratio* (SNDR, SINAD, or TSNR), which is the ratio of the signal power to the total inband noise including harmonic distortion. Also, a slightly different term is often used in place of the SNR. The useful signal range or *dynamic range* (DR) is defined as the power ratio of the maximum signal to the minimum signal. The minimum signal is defined as the smallest signal for which the SNDR is 0 dB, while the maximum signal is the full-scale signal. Therefore, the SNR of the non-ideal ADC/DAC can be lower than the ideal DR because the noise floor can be higher with a large signal present. In practice, performance is not only limited by the quantization noise but also by non-ideal factors such as noises from circuit components, power supply coupling, noisy substrate, timing jitter, settling, and nonlinearity, etc. An alternative definition of the resolution is the *effective number of bits* (ENOB), which is defined by

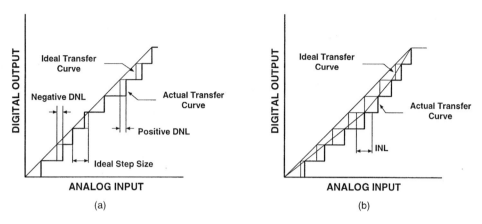

FIGURE 7.2   Definition of ADC nonlinearities: (a) DNL and (b) INL.

$$\text{ENOB} = \frac{\text{SNDR} - 1.76}{6.02} (\text{bits}) \tag{7.2}$$

Usually, the ENOB is defined for the signal at half the sampling frequency.

Linearity

The input/output ranges of an ideal N-bit ADC/DAC are equally divided into 2^N small units, and one least significant bit (LSB) in the digital code corresponds to the analog incremental voltage of $V_{ref}/2^N$. Static ADC/DAC performance is characterized by *differential nonlinearity* (DNL) and *integral nonlinearity* (INL). The DNL is a measure of deviation of the actual ADC/DAC step from the ideal step for one LSB, and the INL is a measure of deviation of the ADC/DAC output from the ideal straight line drawn between two end points of the transfer characteristic. Both DNL and INL are measured in the unit of an LSB. In practice, the largest positive and negative numbers are usually quoted to specify the static performance. The examples of these DNL and INL definitions for ADC are explained in Fig. 7.2.

However, several different definitions of INL may result, depending on how two end points are defined. In some architectures, the two end points are not exactly 0 and V_{ref}. The non-ideal reference point causes an offset error, while the non-ideal full-scale range gives rise to a gain error. In most applications, these offset and gain errors resulting from the non-ideal end points do not matter, and the integral linearity can be better defined in a relative measure using a straight-line linearity concept rather than the end-point linearity. The straight line can be defined as two end points of the actual transfer function, or as a theoretical straight line adjusted for best fit. The former definition is sometimes called end-point linearity, while the latter is called best-straight-line linearity.

Unlike ADC, the output of a DAC is a sampled-and-held step waveform held constant during a word clock period. Any deviation from the ideal step waveform causes an error in the DAC output. High-speed DACs which usually have a current output are either terminated with a 50 to 75-Ω low-impedance load or buffered by a wideband transresistance amplifier. The linearity of a DAC is often limited dynamically by the non-ideal settling of the output node. Anything other than ideal exponential settling results in linearity errors.

Monotonicity

In both the ADC and the DAC, the output should increase over its full range as the input increases. That is, the negative DNL should be smaller than one LSB for any ADC/DAC to be monotonic. Monotonicity is critical in most applications, in particular digital control or video applications. The source of non-monotonicity is an inaccuracy in binary weighting of a DAC. For example, the most significant bit (MSB) has a weight of half the full range. If the MSB weight is not accurate, the full range is divided into two non-ideal half ranges, and a major error occurs at the midpoint of the full scale. The similar non-monotonicity can take place at the quarter and one-eighth points. In DACs, monotonicity is inherently guaranteed if a DAC uses thermometer decoding. However, it is impractical to implement high-resolution DACs using thermometer codes since the number of elements grows exponentially as the number of bits increases. Therefore, to guarantee monotonicity in practical applications, DACs have been implemented using either a segmented DAC or an integrator-type DAC. Oversampling interpolative DACs also achieve monotonicity using a pulse-density modulated bitstream filtered by a lossy integrator or by a low-pass filter. Similarly, ADCs using slope-type, subranging, or oversampling architectures are monotonic.

Clock Jitter

Jitter is loosely defined as a timing error in analog-to-digital and digital-to-analog conversions. The clock jitter greatly affects the noise performance of both ADCs and DACs. For example, in ADCs, the right signal sampled at the wrong time is the same as the wrong signal sampled at the right time. Similarly, DACs need precise timing to correctly reproduce an analog output signal. If an analog waveform is not generated with the identical timing with which it is sampled, distortion will result because the output changes at the wrong time. This in turn introduces either spurious components related to the jitter frequency or a higher noise floor unless the jitter is periodic. If the jitter has a Gaussian distribution with an rms jitter of Δt, the worst-case SNR resulting from this random clock jitter is

$$\text{SNR} = -20 \times \log \frac{2\pi B \Delta t}{M^{1/2}} \tag{7.3}$$

where B is the signal bandwidth and M is the oversampling ratio. The oversampling ratio M is defined as

$$M = \frac{f_s}{2B} \tag{7.4}$$

where f_s is the sampling clock frequency. The timing jitter error is more critical in reproducing high-frequency components. In other words, for an N-bit ADC/DAC, an upper limit for the tolerable clock jitter is

$$\Delta t \le \frac{1}{2\pi B 2^N} \left(\frac{2M}{3} \right)^{1/2} \tag{7.5}$$

This implies that the error power induced in the baseband by clock jitter should be no larger than the quantization noise. For example, a Nyquist-rate 16-b ADC/DAC with a 22-kHz bandwidth should have a clock jitter of less than 90 ps.

Nyquist-Rate vs. Oversampling

In recent years, high-resolution ADCs and DACs at the low end of the spectrum such as for digital audio, voice, and instrumentation are dominantly implemented using oversampling techniques. Although Nyquist-rate techniques can achieve comparable resolution, such techniques are in general sensitive to non-ideal factors such as process, component matching, and even environmental changes. The inherent advantage of oversampling provides a unique solution in the digital VLSI environment. The oversampling technique achieves high resolution by trading speed for accuracy. The oversampling lessens the effect of quantization noise and clock jitter. However, the quantization or regeneration of a signal above MHz using oversampling techniques is costly even if possible. Therefore, typical applications for high-sampling rates require sampling at a Nyquist rate.

7.2 ADC Design Arts

The conversion speed of the ADC is limited by the time needed to complete all comparator decisions. Flash ADCs make all the decisions at once, while successive-approximation ADCs make one-bit decisions at a time. Although it is fast, the complexity of the flash ADC grows exponentially. On the other hand, the successive-approximation ADC is simple but slow since the bit decisions are made in sequence. Between these two extremes, there exist many architectures resolving a finite number of bits at a time, such as pipeline and multi-step ADCs. They balance complexity and speed. Figure 7.3 shows recent high-speed ADC applications in the resolution-versus-speed plot. ADC architecture depends on system requirements. For example, with IF (intermediate frequency) filters, wireless receivers need only 5 to 6 b ADC at a few MHz sampling rate. However, without IF filters, the dynamic range of 12 to 14 b is required for the IF sampling depending on IF as shown in Fig. 7.3.

State of the Art

Some architectures are preferred over others for certain applications. Three architectures stand out for three important areas of applications. For example, the oversampling converter is exclusively used to achieve high resolution above the 12-b level at low frequencies. The difficulty in achieving better than 12-b matching in conventional techniques gives a fair advantage to the oversampling technique. For medium speed with high resolution, pipeline or multi-step ADCs are promising. At extremely high frequencies, only flash and folding ADCs survive, but with low resolution. Figure 7.4 is a resolution-versus-speed plot showing this trend. As semiconductor process and design technologies advance, the performance envelope will be pushed further. The demand for higher resolution at higher sampling rates is a main driver of this trend.

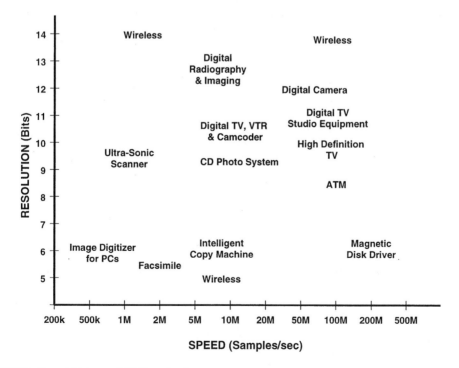

FIGURE 7.3 Recent high-speed ADC applications.

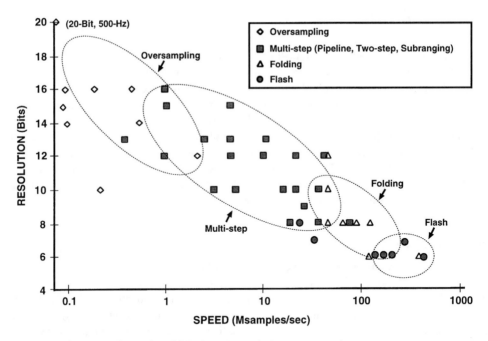

FIGURE 7.4   Performance of recently published ADCs: resolution versus speed.

Technical Challenge in Digital Wireless

In digital wireless systems, a need to quantize and to create a block of spectrum with low intermodulation has become the single most challenging problem. Implementing IF filters digitally has already become a necessity in wireless cell sites and base stations. Even in hand-held units, placing data conversion blocks closer to the RF (radio frequency) has many advantages. A substantial improvement in system cost and complexity of the RF circuitry can be realized by implementing high selectivity function digitally, and the digital IF can increase immunity to adjacent and alternate channel interferences. Furthermore, the RF transceiver architecture can be made independent of the system and can be adapted to different standards using software. Low-spurious, low-power data converters are key components in this new software radio environment.

The fundamental limit in quantizing IF spectrum is the crosstalk and overload, and the system performance heavily depends on the SFDR (spurious-free dynamic range) of the sampling ADC. To meet this growing demand, low-spurious data conversion blocks are being actively developed in ever-increasing numbers. For a 14b-level ideal dynamic range while sampling at 50 MHz, it is necessary to control the sampling jitter below 0.32 ps. Considering that the current state-of-the-art commercial bipolar chip exhibits the jitter range of 0.7 ps, the jitter on the order of a fraction of a picosecond is considered to be at the limit of CMOS capability. However, unlike nonlinearity that causes interchannel mixing, the random jitter in IF sampling increases only the random noise floor. As a result, the random jitter is not considered fundamental in this application.

This challenging new application for digital IF processing will lead to the implementation of data converters with very wide SFDR of more than 90 dB. Considering the current state of the art in CMOS ADCs, most architectures known to date are unlikely to achieve a high sampling rate of over 50 MHz, even with 0.2 to 0.3 μm technologies. Although a higher sampling rate of 65 MHz is reported using bipolar and BiCMOS (bipolar and CMOS), it has been implemented at a 12-b level. However, two high-speed candidate architectures, pipeline (or multi-step) and folding, are potential candidate architectures to challenge these limits with new system approaches.

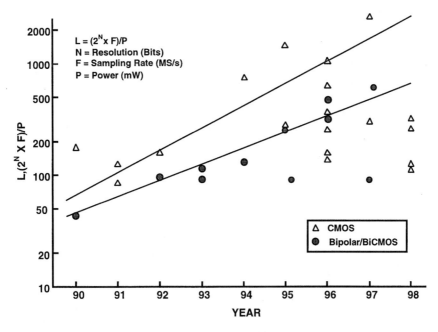

FIGURE 7.5 Figure of merit (L) versus year.

ADC Figure of Merit

The ADC performance is often represented by a figure of merit L which is defined as $L = 2^N \times f_s/P$, where N is the number of bits, f_s is the sampling rate in Msamples/s, and P is the power consumption in mW. The higher the L is, the more bits are obtained at higher speed with lower power. The plot of L versus year shown in Fig. 7.5 shows the low-power and high-speed trend both for leading integrated CMOS and bipolar/BiCMOS ADCs published in the last decade.

7.3 ADC Architectures

In general, the main criteria of choosing ADC architectures are resolution and speed, but auxiliary requirements such as power, chip area, supply voltage, latency, operating environment, or technology often limit the choices. The current trend is toward low-cost integration without using expensive discrete technologies such as thin film and laser trimming. Therefore, a growing number of ADCs are being implemented using mainstream VLSI technologies such as CMOS or BiCMOS.

Slope-Type ADC

Traditionally, slope-type ADCs have been used for multimeters or digital panel meters mainly because of their simplicity and inherent high linearity. There can be many variations, but dual- or triple-slope techniques are commonly used because the single-slope method is sensitive to the switching error. The resolution of this type of ADC depends on the accurate control of charge on the capacitor. The dual-slope technique in Fig. 7.6(a) starts with the initialization of the integrating capacitor by opening the switch S_1 with the input switch S_2 connected to V_{ref}. If V_{ref} is negative, V_x will increase linearly with a slope of V_{ref}/RC. After a time T_1, the switch S_2 is switched to V_{in}. Then, V_x will decrease with a new slope of $-V_{in}/RC$. The comparator detects the zero-crossing time T_2. From T_1 and T_2, the digital ratio of V_{in}/V_{ref} can be obtained as T_1/T_2. The triple-slope technique shown in Fig. 7.6(b) needs no op-amp to reduce the offset effect. Unlike the dual-slope method comparing two slopes, it measures three times T_1, T_2, and T_3 by charging the capacitor with V_{ref}, V_{in}, and ground with three switches S_1, S_2, and S_3, respectively.

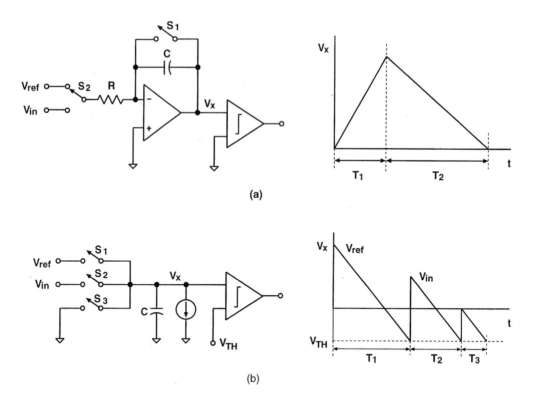

FIGURE 7.6   (a) Dual-slope and (b) triple-slope ADC techniques.

The comparator threshold can be set to negative V_{TH}. From three time measurements, the ratio of V_{in}/V_{ref} can be computed as $(T_2 - T_3)/(T_1 - T_3)$.

Successive-Approximation ADC

The simplest concept of A/D conversion is comparing analog input voltage with an output of a DAC. The comparator output is fed back through the DAC as explained in Fig. 7.7. The successive-approximation register (SAR) performs the most straightforward binary comparison. The sampled input is compared with the DAC output by progressively dividing the range by two as explained in the 4-b example. The conversion starts by sampling input, and the first MSB decision is made by comparing the sample-and-hold (S/H) output with $V_{ref}/2$ by setting the MSB of the DAC to 1. If the input is higher, the MSB stays as 1. Otherwise, it is reset to 0. In the second bit decision, the input is compared with $3V_{ref}/4$ in this example by setting the second bit to 1. Note that the previous decision set the MSB to 1. If the input is

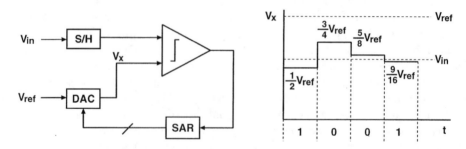

FIGURE 7.7   Successive-approximation ADC technique.

lower, as in the example shown, the second bit is set to 0, and the third bit decision is done by comparing the input with $5V_{ref}/8$. This comparison continues until all the bits are decided. Therefore, the N-bit successive-approximation ADC requires N+1 clock cycles to complete one sample conversion.

The performance of the successive-approximation ADC is limited by the DAC resolution and the comparator accuracy. The commonly used DACs for this architecture are a resistor-string DAC and a capacitor-array DAC. Although binary-weighted capacitors have a 10b-level matching in MOS,[1] diffused resistors have poor matching and high voltage coefficient. If differential resistor-string DACs are used, performance can be improved to the capacitor-array DAC level.[2] In general, the capacitor DAC exhibits poor DNL while the resistor-string DAC exhibits poor INL.

Flash ADC

The most straightforward way of making an ADC is to compare the input with all the divided levels of the reference simultaneously. Such a converter is called a flash ADC, and the conversion occurs in one step. The flash ADC is the fastest among all ADCs. The flash ADC concept is explained in Fig. 7.8, where divided reference voltages are compared to the input. The binary encoder is needed because the output of the comparator bank is thermometer-coded. The resolution is limited both by the accuracy of the divided reference voltages and by the comparator resolution. The metastability of the comparator produces a sparkle noise when the gain of the comparator is low. The reference division can be done using capacitor dividers[3,4] or transistor sizing[5] for small-scale flash ADCs. However, only resistor-string DACs can provide references as the number of bits grows.

In practical implementations, the limit is the exponential growth in the number of comparators and resistors. For example, an N-bit flash needs $2^N - 1$ comparators and 2^N resistors. Furthermore, for the Nyquist-rate sampling, the input needs an S/H to freeze the input for comparison. As the number of bits

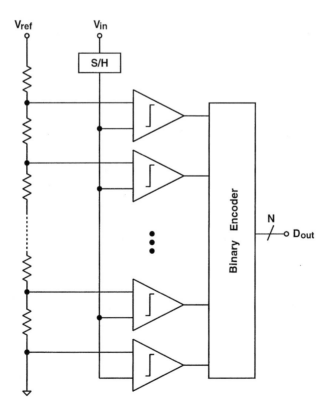

FIGURE 7.8 Flash ADC technique.

grows, the comparator bank presents a significant loading to the input S/H, diminishing the speed advantage of this architecture. Also, the control of the reference divider accuracy and the comparator resolution degrades, and the power consumption becomes prohibitively high. As a result, flash converters with more than 10-b resolution are rare. Flash ADCs are commonly used as coarse quantizers in the pipeline or multi-step ADCs. The folding/interpolation ADC, which is conceptually a derivative of the flash ADC, reduces the number of comparators by folding the input range.[6]

For high resolution, the flash ADC needs a low-offset comparator with high gain, and the comparator is often implemented in a multi-stage configuration with offset cancellation. The front-end of the multi-stage comparator is called a preamplifier. A technique called *interpolation* saves the number of preamplifiers by interpolating the adjacent preamplifier outputs as shown in Fig. 7.9(a), where two preamplifier outputs V_a and V_b are used to generate three more outputs V_1, V_2, and V_3 using a resistor divider. The interpolation can improve the DNL within the interpolated range, but the overall DNL and INL are not improved. Interpolating any arbitrary number of levels is possible by making more resistor taps. The interpolation is usually done using resistors, but the interpolation in the current domain is also possible. However, interpolating with independent current sources does not improve the DNL.

Another technique called *averaging*, as explained in Fig. 7.9(b) is often used to average out the offsets of the neighboring preamplifiers as well as to enhance the accuracy of the reference divider.[7] The idea is to couple the outputs of the preamplifier transconductance (G_m) stage so that the significant errors can be spread over the adjacent preamplifier outputs as explained. For example, if the coupling resistor value is infinite, there exists no averaging. As the coupling resistor value decreases, one preamplifier output becomes the weighted sum of the outputs of its neighboring preamplifiers. Therefore, the overall DNL and INL can improve significantly.[8] However, for the case in which errors to average have the same polarity, the averaging is not that effective. In practice, both the interpolation and the averaging concepts are often combined.

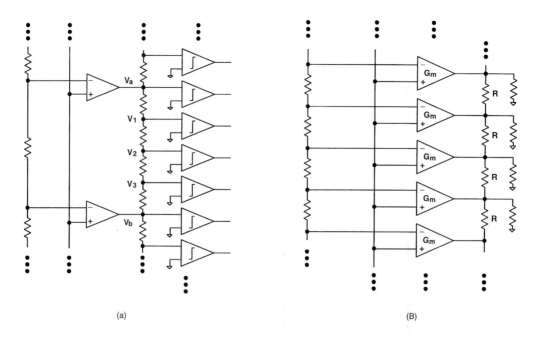

(a)   (B)

FIGURE 7.9   (a) Interpolation and (b) averaging techniques.

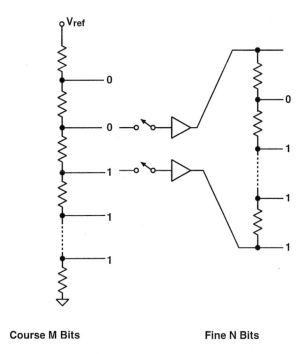

Course M Bits   **Fine N Bits**

FIGURE 7.10   Coarse and fine reference ladders for two-step subranging ADC.

Subranging ADC

Although the interpolation and averaging techniques simplify the flash ADC, the number of comparators stays the same. Instead of making all the bit decisions at once, resolving a few bits at a time makes the system simpler and more manageable. It also enables us to use a digital error correction concept. The simplest subranging ADC concept is explained in Fig. 7.10 for the two-step conversion case. It is a straightforward subranging since one subrange out of 2^M subranges is chosen in the coarse M-bit decision. Once one subrange is selected, the N-bit fine decision can be done using a fine reference ladder interpolating the selected subrange.

Note that the subrange after the coarse decision is $V_{ref}/2^M$ and the fine comparators should have a resolution of M+N bits. Unless the digital error correction with redundancy is used, the coarse comparators should also have a resolution of M+N bits.

Multi-Step ADC

The tactic of making a few bit decisions at a time as shown in the subranging case can be generalized. A slight modification of the subranging architecture shown in Fig. 7.11(a) to include a residue amplifier with a gain of 2^M results in Fig. 7.11(b). The residue is defined as the difference between the input and the nearest DAC output lower than the input. The difference between the two concepts is subtle, but including one residue amplifier drastically changes the system requirements. The obvious advantage of using the residue amplifier is that the fine comparators do not need to be accurate because the residue from the coarse decision is amplified by 2^M. That is, the subrange after the coarse decision is no longer $V_{ref}/2^M$. The disadvantage is the accuracy and settling of the high-gain residue amplifier.

Whether the residue is amplified or not, the subranging block consists of a coarse ADC, a DAC, a residue subtractor, and an amplifier. In theory, this block can be repeated as shown in Fig. 7.12. How many times it is repeated determines the number of steps. So, in general terms, the n-step ADC has n–1 subranging blocks. To complete a conversion in one cycle, usually poly-phase subdivided clocks are

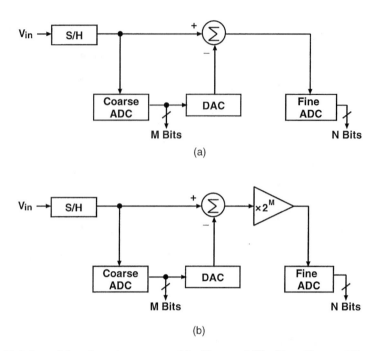

(a)

(b)

FIGURE 7.11   Variations of the subranging concepts: (a) without and (b) with residue amplifier.

needed. Due to the difficulty in clocking, the number of steps for the multi-step architecture is usually limited to two, which does not incur a speed penalty and needs the standard two-phase clocking.

There are many variations in the multi-step architecture. If no poly-phase clocking is used, it is called a *ripple ADC*. Also in the two-step ADC, if one ADC is repeatedly used both for the coarse and fine decisions, it is called a *recycling ADC*.[9] In this ADC example, the capacitor-array multiplying DAC (MDAC) also performs the S/H function in addition to the residue amplification. This MDAC, with either a binary-ratioed or thermometer-coded capacitor array, is a general form of the residue amplifier. The same capacitor array has been used with a comparator to implement a charge-redistribution successive-approximation ADC.[1] This MDAC is suited for MOS technologies, but other forms of the residue amplification are possible using resistor-strings or current DACs.

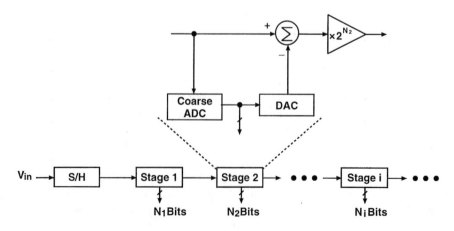

FIGURE 7.12   Multi-step ADC architecture.

Pipeline ADC

The complexity of the two-step ADC, although manageable and simpler than the flash ADC, still grows exponentially as the number of bits to resolve increases. Specifically, for high resolution above 12 b, the complexity reaches about the maximum, and a need to pipeline subranging blocks arises. The pipeline ADC architecture shown in Fig. 7.13 is the same as the subranging or multi-step ADC architecture shown in Fig. 7.12 except for the interstage S/H. Since the S/Hs are clocked by alternating clock phases, each stage needs to perform the decision and the residue amplification in each clock phase. Pipelining the residue greatly simplifies the ADC architecture. The complexity grows only linearly with the number of bits to resolve. Due to its simplicity, the pipeline ADCs have been gaining popularity in the digital VLSI environment.

In the pipeline ADC, each stage resolves a few bits quickly and transfers the residue to the following stage so that the residue can be resolved further in the subsequent stages. Therefore, the accuracy of the interstage residue amplifier limits the overall performance. The following four non-ideal error sources can affect the performance of the multi-step or pipeline ADCs: ADC resolution, DAC resolution, gain error of the residue amplifier, and inaccurate settling of the residue amplifier. The offset of the residue amplifier does not affect the linearity, but it appears as a system offset. Among these four error sources, the first three are static, but the residue amplifier settling is dynamic. If the residue amplifier is assumed to settle within one clock phase, three static error sources are limiting the linearity performance.

Figure 7.14 explains the residue from the 2-b stage in the systems shown in Figs. 7.12 and 7.13. In the ideal case, as the input is swept from 0 to the full range V_{ref}, the residue change from 0 to V_{ref} repeats each time V_{ref} is subtracted at the ideal locations of the 2-b ADC thresholds, which are $V_{ref}/4$ apart. In this case, the 2-b stage does not introduce any nonlinearity error. However, in the other cases with ADC, DAC, and gain errors, the residue ranges do not match with the ideal full-scale V_{ref}. If the residue range is smaller than the full range, missing codes are generated; and if the residue goes out of bounds, excessive codes are generated at the ADC thresholds. Unlike the DAC and gain errors, the ADC error appears as a shift of the residue either by the amounts of V_{ref} or $-V_{ref}$ as long as the DAC subtracts the ideal V_{ref} and the residue amplifier gain is ideal. This suggests that the error can be corrected digitally by adding or subtracting the full range.

Digital Error Correction

Any multi-step or pipeline ADC system can be made insensitive to the ADC error if the ADC error is digitally corrected. The residue normally going out of the full range can still be digitized by the following stage if the residue amplifier gain is reduced. That is, if the residue amplifier gain is set to 2^{N-1} instead of 2^N, the residue plots are as shown in Fig. 7.15. If the residue is bounded with the full range of 0 to

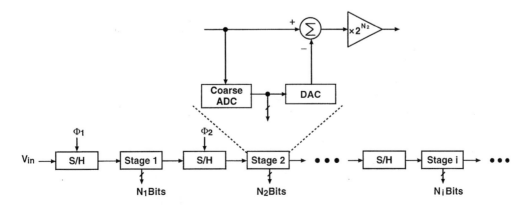

FIGURE 7.13   Pipeline ADC architecture.

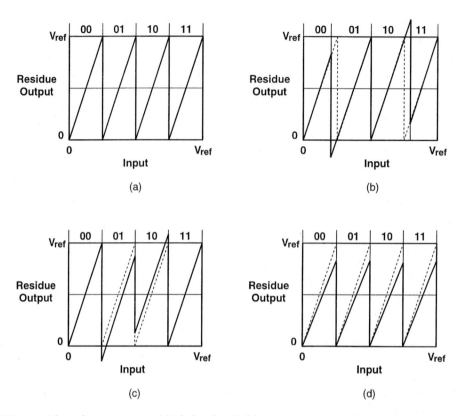

FIGURE 7.14   2-b residue versus input: (a) ideal, and with (b) ADC, (c) DAC, and (d) gain errors.

V_{ref}, the inner range from $V_{ref}/4$ to $3V_{ref}/4$ is the normal conversion range, and two redundant outer ranges are used to cover the residue error resulting from the inaccurate coarse conversion. Now the problem is that this redundancy requires extra resolution to cover the overrange. The half ranges on both sides are used to cover the residue error in this 2-b case. That is, one full bit of extra resolution is necessary for redundancy in the 2-b case. However, the amount of redundancy depends on the ADC error range to be corrected in the multi-bit cases.[10] In general, it is a tradeoff between comparator resolution and redundancy.

The range marked as B is the normal range, and A and C indicate the correction ranges in Fig. 7.15(b). The digital correction works by subtracting 1 from the previous decision if the residue exceeds $3V_{ref}/4$,

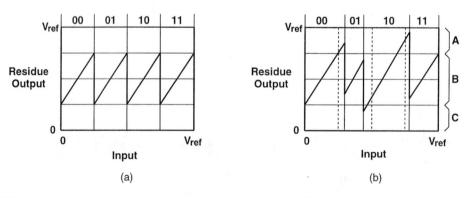

FIGURE 7.15   Over-ranged 2-b residue versus input: (a) ideal and with (b) ADC errors.

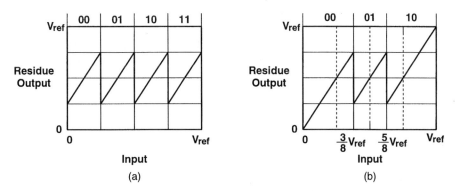

FIGURE 7.16 Half-bit shifted 2-b residue versus input: (a) with and (b) without ADC threshold-level shift.

as in the case marked as A. On the other hand, if the residue goes below $V_{ref}/4$, as in the case marked as C, 1 is added from the previous decision. Although the digital subtraction is simple, biasing the ADC threshold levels by half the ideal interval has an advantage. Figure 7.16 compares the residue plots for two cases with and without the ADC threshold shift by $V_{ref}/8$. This is to fully utilize the ADC conversion range from 0 to V_{ref}. The shift of $V_{ref}/8$ makes the residue start from 0 and end at V_{ref} in Fig. 7.16(b), contrary to the previous case where the residue starts from $V_{ref}/4$ and ends at $3V_{ref}/4$. This results in saving one comparator. The former case needs 2^N-1 comparators, while the latter case needs 2^N-2. The only minor issue is that the latter exhibits a half LSB systematic offset due to this shift.

This half-bit-level shift makes the ADC error occur only with the same polarity. As a result, only the addition is necessary for digital correction in the case of Fig. 7.16(b). This is explained in the 4-b ADC made of three stages using one-bit correction per stage in Fig. 7.17. The vertical axis marks the signal

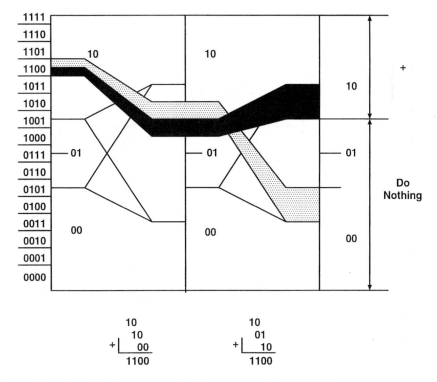

FIGURE 7.17 Example of digital error correction for three-stage 4b ADC (1100).

and residue levels as well as ADC decision levels. The dotted and shaded areas follow the residue paths when the ADC error occurs, but the end results are the same after digital correction. This half interval shift is valid for stages resolving any number of bits. Overall, the digital error correction enables fast data conversion using inaccurate comparators. However, the DAC resolution and the residue gain error still remain as the fundamental limits in multi-step and pipeline ADCs. The currently known ways to overcome these limits are either trimming or self-calibration.

One-Bit Pipeline ADC

The degenerate case of the pipeline ADC is when only one bit is resolved per stage as shown in Fig. 7.18. Each stage multiplies its input V_{in} by two and subtracts the reference voltage V_{ref} to generate the residue voltage. If the sign of $2V_{in} - V_{ref}$ is positive, the bit is 1 and the residue goes to the next stage. Otherwise, the bit is 0 and V_{ref} is added back to the residue before it goes to the next stage. However, in reality, it is more desirable if the reference restoring time is saved. In the non-restoring algorithm, the previous bit decision affects the polarity of the reference voltage to be used in the current bit decision. If the previous bit is 1, the residue voltage is $2V_{in} - V_{ref}$, as in the restoring algorithm. But if the previous bit is 0, the residue voltage is $2V_{in} + V_{ref}$.

The switched-capacitor implementation of the basic functional block performing $2V_{in} \pm V_{ref}$ is explained using two identical capacitors and one op-amp in Fig. 7.19.[11] During the sampling phase, the

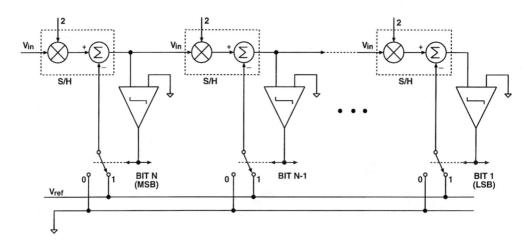

FIGURE 7.18   One-bit per stage pipeline ADC architecture.

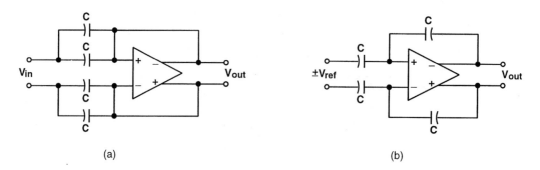

FIGURE 7.19   The simplest two-level MDAC: (a) sampling phase and (b) amplification phase.

bottom plates of two capacitors are switched to the input, and the top plate is connected either to the op-amp output or to the op-amp input common-mode voltage. During the amplification phase, one of the capacitors is connected to the output of the op-amp for feedback, but the other is connected to $\pm V_{ref}$. Then, the output of the op-amp will be $2V_{in} - V_{ref}$ and $2V_{in} + V_{ref}$, respectively, after the op-amp settles.

However, this simple one-bit pipeline ADC is of no use if the comparator resolution is limited. If any redundancy is used for digital correction, at least two bits should be resolved. A close look at Fig. 7.16(b) gives a clue to using the simple functional block shown in Fig. 7.19 for the 2-b residue amplification. The case explained in Fig. 7.16(b) is sometimes called 1.5-b rather than 2-b because it needs only three DAC levels rather than four. The functional block in Fig. 7.19 can have a two-level DAC subtracting $\pm V_{ref}$. However, in differential architecture, by shorting the input, one midpoint can be interpolated. Using the tri-level DAC, the 1.5-b pipeline ADC can be implemented with the following algorithm.[12] If the input V_{in} is lower than $-V_{ref}/4$, the residue output is $2V_{in} + V_{ref}$. If the input is higher than $V_{ref}/4$, the residue output is $2V_{in} - V_{ref}$. If the input is in the middle, the output is $2V_{in}$.

Algorithmic, Cyclic, or Recursive ADC

The interstage S/H used in the multi-step architecture provides a flexibility of the pipeline architecture. In the pipeline structure, the same hardware repeats as shown in Fig. 7.13. That is, the throughput rate of the pipeline is fast while the overall latency is limited by the number of stages. Instead of repeating the hardware, using the same stage repeatedly greatly saves hardware, as shown in Fig. 7.20. That is, the throughput rate of the pipeline is directly traded for hardware simplicity. Such a converter is called an *algorithmic, cyclic,* or *recursive ADC*. The functional blocks used for the algorithmic ADC are identical to the ones used in the pipeline ADC.

Time-Interleaved Parallel ADC

The algorithmic ADC just described sacrifices the throughput rate for small hardware. However, the *time-interleaved parallel ADC* takes quite the opposite direction. It duplicates more hardware in parallel for higher throughput rates. The system is shown in Fig. 7.21, where the throughput rate increases by the number of parallel paths strobed. Although it significantly improves the throughput rate and many refinements have been reported, it suffers from many problems.[13] Due to the multiplexing, even static non-linearity mismatch between paths appears as a fixed pattern noise. Also, it is difficult to generate clocks with exact delays, and inaccurate clocking increases the noise floor.

Folding ADC

The folding ADC is similar to the flash ADC except for using fewer comparators. This reduction in the number of comparators is achieved by replacing the comparator preamplifiers with folding amplifiers. In its original arrangements,[14] the folding ADC digitizes the folded signal with a flash ADC. The folded

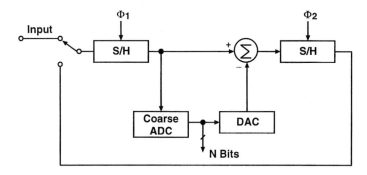

FIGURE 7.20   Algorithmic, cyclic, or recursive ADC architecture.

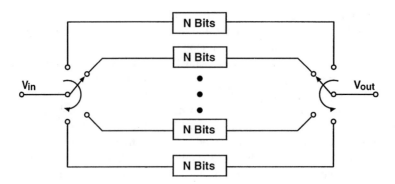

FIGURE 7.21   Time-interleaved parallel ADC architecture.

signal is equivalent in concept to the residue of the subranging, multi-step, or pipeline ADC, but the difference is that the generation of the folding signal is done solely in the analog domain. Since the digitized code from the folding amplifier output repeats over the whole input range, a coarse coding is required, as in all subranging-type ADCs.

Consider a system configured as a 4-b folding ADC as shown in Fig. 7.22. Four folding amplifiers can be placed in parallel to produce four folded signals. Comparators check the outputs of the folding amplifiers for zero crossing. If the input is swept, the outputs of the fine comparators show a repeating pattern, and eight different codes can be obtained by the four comparators. Because there are two identical fine code patterns, one comparator is needed to distinguish them. However, if this coarse comparator is misaligned with the fine quantizer, missing codes will result. A digital correction similar to that for the multi-step or pipeline ADC can be employed to correct the coarse quantizer error. For this system example, one-bit redundancy is used by adding two extra comparators in the coarse quantizer. The shaded region in the figure is where errors occur.

Having several folded signals instead of one has many advantages in designing high-speed ADCs. The folding amplifier requires neither linear output nor accurate settling. This is because in the folding ADC, the zero-crossings of the folded signals matter, but not their absolute values. Therefore, the offset of the folding amplifiers becomes the most important design issue. The resolution of the folding ADC can be further improved using the interpolation concept. When the adjacent folded signals are interpolated by I times, the number of zero-crossing points are also increased by I times. So, the resolution of the final ADC is improved by $\log_2 I$ bits. The higher bound for the degree of interpolation is set by the comparator resolution, the gain of the folding amplifiers, the linearity of folded signals, and the interpolation accuracy. Since the folding process increases the internal signal bandwidth by the number of foldings, the folding ADC performance is limited by the folding amplifier bandwidth. To increase the number of foldings while maintaining the reasonable comparator resolution, the folding amplifier's gain should be high. Since the higher gain limits the amplifier bandwidth, it is necessary to cascade the folding stages.[6,8]

7.4   ADC Design Considerations

In general, multi-step ADCs are made of cascaded low-resolution ADCs. Each low-resolution ADC stage provides a residue voltage for the subsequent stage, and the accuracy of the residue voltage limits the resolution of the converter. One of the residue amplifiers commonly used in CMOS is a switched-capacitor MDAC, whose connections during two clock phases are illustrated in Fig. 7.23 for an N-bit case. An extra capacitor C is usually added to double the feedback capacitor size so that the residue voltage may remain within the full-scale range for digital correction.

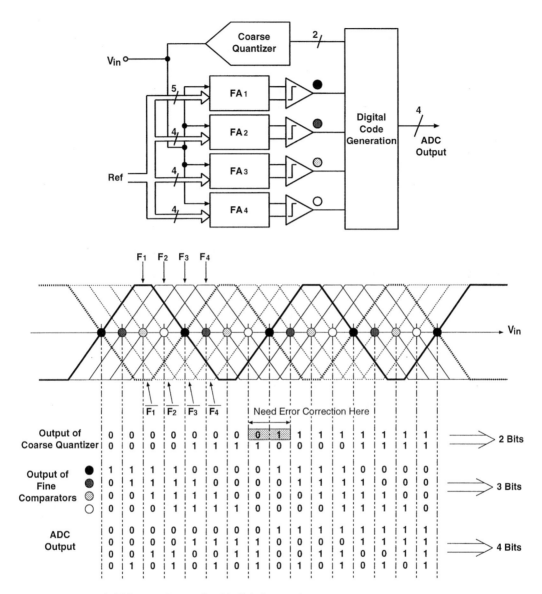

FIGURE 7.22 A 4-b folding ADC example with digital correction.

Sampling Error Considerations

Since the ADC works on a sampled signal, the accuracy in sampling fundamentally limits the system performance. It is well known that the noise power to be sampled on a capacitor along with the signal is KT/C. It is inversely proportional to the sampling capacitor size. The sampled rms voltage noise is 64 μV with 1 pF, but decreases to 20 μV with 10 pF. For accurate sampling, sampling capacitors should be large, but sampling on large capacitors takes time. The speed of the ADC is fundamentally limited by the sampling KT/C noise.

In sampling, there exists another important error source. Direct sampling on a capacitor suffers from switch feedthrough error due to the charge injection when switches are opened. A common way to reduce this charge feedthrough error is to turn off the switches connected to the sampling capacitor top plate slightly earlier than the switches connected to the bottom plate. This is explained in Fig. 7.24. Usually, the top plate is connected to the op-amp summing or comparator input node. The top plate switch is

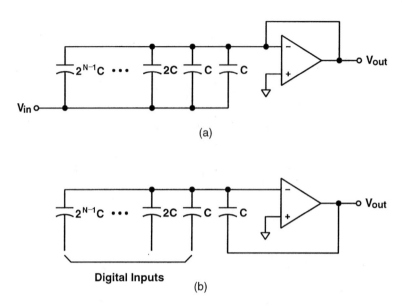

(a)

Digital Inputs

(b)

FIGURE 7.23   General N-bit residue amplifier: (a) sampling phase and (b) amplification phase.

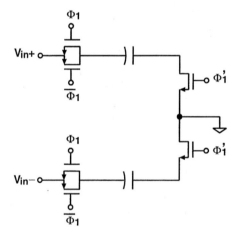

FIGURE 7.24   Open-loop bottom-plate differential sampling on capacitors.

switched with one MOS transistor with a clock phase marked as Φ'_1 which makes a falling transition earlier than other clocks. The bottom plate is switched with a CMOS switch (both NMOS and PMOS) with clocks marked as Φ_1 and $\overline{\Phi}_1$. These clocks make falling transitions after the prime clock does. The net effect is that the feedthrough voltage stays constant because the top plate samples the same voltage repeatedly. The differential sampling using two capacitors symmetrically is known to provide the most accurate sampling known to date.

Unless limited by speed, the sampling error as well as the low-end spectrum of the sampled noise can be eliminated using a correlated double sampling (CDS) technique. The system has been used to remove the flicker noise or slowly-varying offset such as in charge-coupled device (CCD). The CDS needs two sampling clocks. The idea is to subtract the previously sampled sampling error from the new sample after one clock delay. The result is to null the sampling error spectrum at every multiple of the sampling frequency f_s. The CDS is effective only for the low-frequency spectrum.

TABLE 7.1 Three Dominant ADC Architectures

	Interpolated Flash	Multi-step	Pipeline
Matching	Least	Medium	Most critical
Feedthrough	Most critical	Medium	Least
Bandwidth	Least	Most critical	Medium
Settling	Least	Medium	Most critical
Gain	Least	Medium	Most critical
Speed	Fast	Slow	Medium
Complexity	Complex	Medium	Simple
Problems	Clock jitter	Low loop gain	Matching
	Time skew		Wide bandwidth
	Sampling error		High gain

Techniques for High-Resolution and High-Speed ADCs

Considering typical requirements, three representative ADC architectures are compared in Table 7.1. To date, all techniques known to improve ADC resolution are as follows: trimming, dynamic matching, ratio-independent technique, capacitor-error averaging, walking reference, and self-calibration. However, the trimming is irreversible and expensive. It is only possible at the factory or with precision equipment. The dynamic matching technique is effective, but it generates high-frequency noise. The ratio-independent techniques either require many clock cycles or are limited to improve differential linearity. The latter case is good for monotonicity, but it also requires accurate comparison. The capacitor-error averaging technique requires three clock cycles, and the walking reference is sensitive to clock sampling error. The self-calibration technique requires extra hardware for calibration and digital storage, but its compatibility with existing proven architectures may provide potential solutions both for high resolution and for high speed.

The ADC self-calibration concepts originated from the direct code-mapping concept using memory. The calibration is to predistort the digital input to the DAC so that the DAC output can match the ideal level from the calibration equipment. Due to the precision equipment needed, this method has limited use. The first self-calibration concept applied to the binary-ratioed successive-approximation ADC is to internally measure capacitor DAC ratio errors using a resistor-string calibration DAC as shown in Fig. 7.25.[15] Later, an improved concept of the digital-domain calibration was developed for the multi-step or pipeline ADCs.[16]

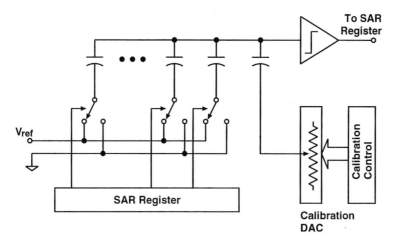

FIGURE 7.25 Self-calibrated successive-approximation ADC.

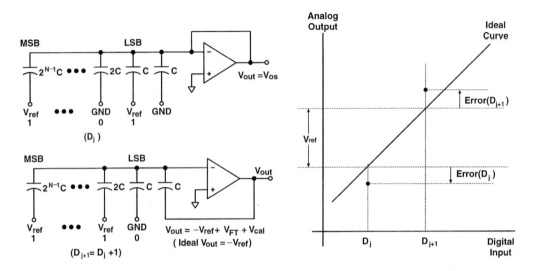

FIGURE 7.26 Segment-error measuring technique for digital self-calibration.

The general concept of the digital-domain calibration is to measure code or bit errors, to store them in the memory, and to subtract them during the normal operation. The concept is explained in Fig. 7.26 using a generalized N-bit MDAC with a capacitor array. If the DAC code increases by 1, the MDAC output should increase by V_{ref} or $V_{ref}/2$ with digital correction. Any deviation from this ideal step is defined as a segment error. Code errors are obtained by accumulating segment errors. This segment-error measurement needs two cycles. The first cycle is to measure the switch feedthrough error, and the next cycle is to measure the segment error. The segment error is simply measured as shown in Fig. 7.26 using the LSB-side of the ADC by increasing the digital code by 1. In the case of $N = 1$, the segment error becomes a bit error. If binary bit errors are measured and stored, code errors should be calculated for subtraction during the normal operation. How to store DAC errors is a tradeoff issue. Examples of the digital calibration are well documented for the cases of segment-error[17] and bit-error[18] measurements, respectively.

7.5 DAC Design Arts

There are many different circuit techniques used to implement DACs, but the popular ones widely used today are of the parallel type in which all bits change simultaneously upon the application of an input code word. Serial DACs, on the other hand, produce an analog output only after receiving all digital input data in a sequential form. When DACs are used as stand-alone devices, their output transient behaviors limited by glitch, slew rate, word clock jitter, settling, etc., are of paramount importance, but used as subblocks of ADCs, DACs need only to settle within a given time interval. An output S/H, usually called a *deglitcher*, is often used for better transient performance. Three of the most popular architectures of DACs are resistor string, ratioed current sources, and a capacitor array. The current-ratioed DAC finds most applications as a stand-alone DAC, while the resistor-string and capacitor-array DACs are mainly used as ADC subblocks.

For speeds over 100 MHz, most state-of-the-art DACs employ current sources switched directly to output resistors.[19-23] Furthermore, owing to the high bit counts (12 to 16 b), segmented architectures are employed, with the current sources broken into two or three segments. The CMOS design has the advantages of lower power, smaller area, and lower manufacturing costs. In all cases, it is of interest to note that the dynamic performance of the DACs degrades rapidly as input frequencies increase, and true dynamic performance is not attained except at low frequencies. Since a major application of wide-bandwidth, high-resolution DACs is in communications, poor dynamic performance is undesirable, owing to the noise leakage from frequency multiplexed channels into other channels. The goal of better dynamic performance continues to be a target of ongoing research and development.

7.6 DAC Architectures

An N-bit DAC provides a discrete analog output level, either voltage or current, for every level of 2^N digital words that is applied to the input. Therefore, an ideal voltage DAC generates 2^N discrete analog output voltages for digital inputs varying from 000...00 to 111...11. In the unipolar case, the reference point is 0 when the digital input is 000...00; but in bipolar or differential DACs, the reference point is the midpoint of the full scale when the digital input is 100...00. Although purely current-output DACs are possible, voltage-output DACs are common in most applications.

Resistor-String DAC

The simplest voltage divider is a resistor string. Reference levels can be generated by connecting 2^N identical resistors in series between V_{ref} and ground. Switches to connect the divided reference voltages to the output can be either 1-out-of-2^N decoder or binary tree decoder as shown in Fig. 7.27 for the 3-b example. Since it requires a good switch, the stand-alone resistor-string DAC is easier to implement using CMOS. However, the lack of switches does not limit the application of the resistor string as a voltage reference divider subblock for ADCs in other process technologies.

Resistor strings are widely used as an integral part of the flash ADC as a reference divider. All resistor-string DACs are inherently monotonic and exhibit good differential linearity. However, they suffer from poor integral linearity and also have the drawback that the output resistance depends on the digital input

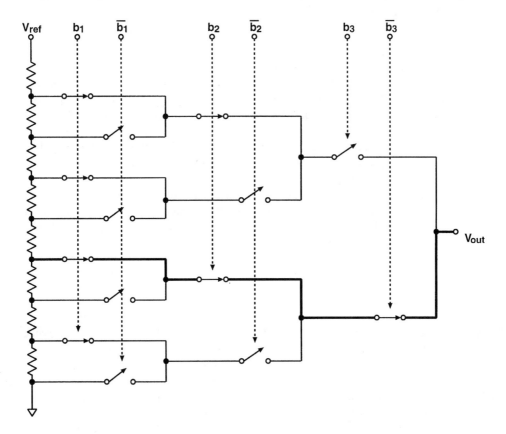

FIGURE 7.27   Resistor-string DAC.

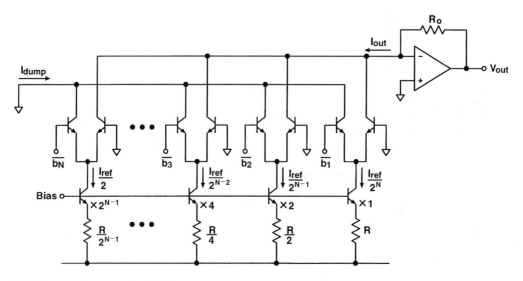

FIGURE 7.28   Current-ratioed DAC.

code. This causes a code-dependent settling time when charging the capacitive load. This non-uniform settling time problem can be alleviated by adding low-resistance parallel resistors or by compensating the MOS switch overdrive voltages.

Current-Ratioed DAC

The most popular stand-alone DACs in use today are *current-ratioed DACs*. There are two types: one is a weighted-current DAC and the other is an R-2R DAC. The weighted-current DAC shown in Fig. 7.28 is made of an array of switched binary-weighted current sources and the current summing network. In bipolar technology, the binary weighting is achieved by ratioed transistors and emitter resistors with binary related values of R, R/2, R/4, etc., while in MOS technology, only ratioed transistors are used. DACs relying on active device matching can achieve an 8b-level performance with a 0.2 to 0.5% matching accuracy using a 10- to 20-μm device feature size, while degeneration with thin-film resistors gives a 10b-level performance. The current sources are switched on or off by means of switching diodes or emitter-coupled differential pairs (source-coupled pairs in CMOS). The output current summing is done by a wideband transresistance amplifier; but in high-speed DACs, the output current directly drives a resistor load for maximum speed. The weighted-current design has the advantage of simplicity and high speed, but it is difficult to implement a high-resolution DAC because a wide range of emitter resistors and transistor sizes are used, and very large resistors cause problems with both temperature stability and speed.

R-2R Ladder DAC

This large resistor ratio problem is alleviated by using a resistor divider known as an *R-2R ladder*, as shown in Fig. 7.29. The R-2R network consists of series resistors of value R and shunt resistors of value 2R. The top of each shunt resistor of value 2R has a single-pole double-throw electronic switch that connects the resistor either to ground or to the current summing node. The operation of the R-2R ladder network is based on the binary division of current as it flows down the ladder. At any junction of series resistor of value R, the resistance looking to the right side is 2R. Therefore, the input resistance at any junction is R, and the current splits into two equal parts at the junction since it sees equal resistances in both directions. As a result, binary-weighted currents flow into shunt resistors in the ladder. The digitally controlled switches direct the currents either to ground or to the summing node. The advantage of the R-2R ladder method is that only two values of resistors are used, greatly simplifying the task of matching

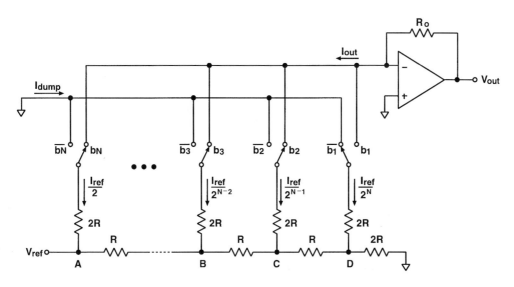

FIGURE 7.29 R-2R DAC.

or trimming and temperature tracking. In addition, for high-speed applications, relatively low resistor values can be used. Excellent results can be obtained using laser-trimmed thin-film resistor networks. Since the output of the R-2R DAC is the product of the reference voltage and the digital input word, the R-2R ladder DAC is often called an MDAC.

Capacitor-Array DAC

Capacitors made of double-poly or poly-diffusion in MOS technology are considered one of the most accurate passive components comparable to thin-film resistors in the bipolar process, both in the matching accuracy and voltage and temperature coefficients.[1] The only disadvantage in the capacitor-array DAC implementation is the use of a dynamic charge redistribution principle. A switched-capacitor counterpart of the resistor-string DAC is a parallel capacitor array of 2^N unit capacitors with a common top plate. The capacitor-array DAC is not appropriate for stand-alone applications without a feedback amplifier virtually grounding the top plate and an output S/H or deglitcher. The operation of the capacitor-array DAC shown in Fig. 7.30(a) is based on the thermometer-coded DAC principle and has the distinct advantage of monotonicity. However, due to the complexity of handling the thermometer-coded capacitor array, a binary-weighted capacitor array is often used, as shown in Fig. 7.30(b) by grouping unit capacitors in binary ratio values. One important application of the capacitor-array DAC is as a reference DAC for ADCs. As in the case of the R-2R MDAC, the capacitor-array DAC can be used as an MDAC to amplify residue voltages for multi-step or pipeline ADCs.

Thermometer-Coded Segmented DAC

Applying a two-step conversion concept, a DAC can be made in two levels using coarse and fine DACs. The fine DAC divides one coarse MSB segment into fine LSBs. If one fixed MSB segment is subdivided to generate LSBs, matching among MSB segments creates a non-monotonicity problem. However, if the next MSB segment is subdivided instead of the fixed segment, the segmented DAC can maintain monotonicity regardless of the MSB matching. This is called the next-segment approach. The most widely used segmented DAC is a current-ratioed DAC, whose MSB DAC is made of identical elements for the next-segment approach, except that the LSB DAC is a current divider as shown in Fig. 7.31. To implement a segmented DAC using two resistor-string DACs, voltage buffers are needed to drive the LSB DAC without loading the MSB DAC. Although the resistor-string MSB DAC is monotonic, overall monotonicity is not

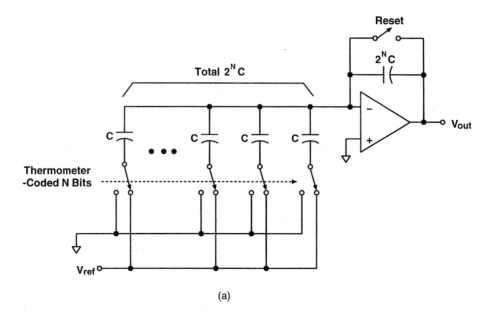

(a)

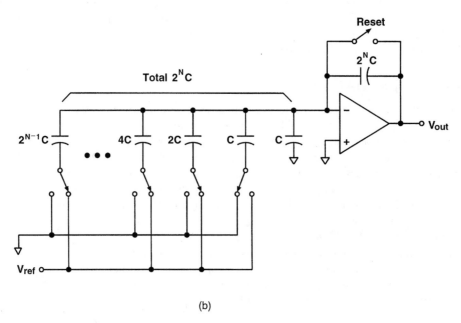

(b)

FIGURE 7.30 Capacitor-array DACs: (a) thermometer-coded and (b) binary-weighted.

guaranteed due to the offsets of the voltage buffers. The use of a capacitor-array LSB DAC eliminates the need for voltage buffers.

Integrator-Type DAC

As mentioned, monotonicity is guaranteed only in a thermometer-coded DAC. The thermometer coding of a DAC output can be implemented either by repeating identical DAC elements many times or by using the same element over and over. The former requires more hardware, but the latter requires more time. In the continuous-time integrator-type DAC, the integrator output is a linear ramp and the time to stop

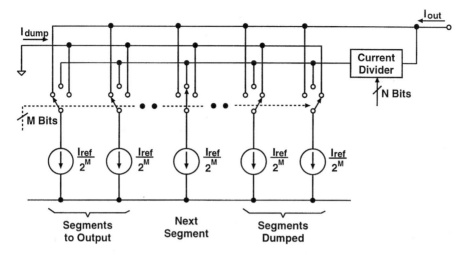

FIGURE 7.31 Thermometer-coded segmented DAC.

integration can be controlled digitally. Therefore, monotonicity can be maintained. Similarly, the discrete-time integrator can integrate a constant amount of charge repeatedly and the number of integrations can be controlled digitally. The integration approach can give high accuracy, but its disadvantage is that its slow speed limits its applications.

7.7 DAC Design Considerations

Figure 7.32 illustrates two step responses of a DAC when it settles with a time constant τ and when it slews with a slew rate S. The transient errors given by the shaded areas are h/τ and $h^2/2S$, respectively. This implies that a single time-constant settling of the former case only generates a linear error in the output, which does not affect the DAC linearity, but the slew-limited settling generates a nonlinear error. Even in the single-time constant case, the code-dependent settling time constant can introduce a non-linearity error because the settling error is a function of the time constant t. This is true for a resistor-string DAC, which exhibits a code-dependent settling time because the output resistance of the DAC depends on the digital input.

Effect of Limited Slew Rate

The slew-rate limit is a significant source of nonlinearity since the error is proportional to the square of the signal, as shown in Fig. 7.32(b). The height and width of the error term change with the input. The

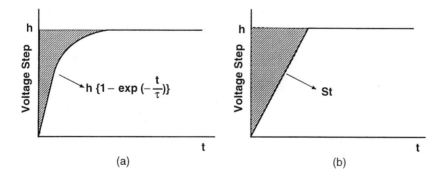

FIGURE 7.32 DAC settling cases: (a) exponential and (b) slew-limited case.

worst-case harmonic distortion (HD) when generating a sinusoidal signal with a magnitude V_o with a limited slew rate of S is:[24]

$$\text{HD}_k = 8\frac{\sin^2\frac{\omega T_c}{2}}{\pi k\left(k^2 - 4\right)} \times \frac{V_o}{ST_c}, k = 1,3,5,7\cdots \qquad (7.6)$$

where T_c is the clock period. For a given distortion level, the minimum slew rate is given. Any exponential system with a bandwidth of ω_o gives rise to signals with the maximum slew rate of $2\omega_o V_o$. Therefore, by making $2\omega_o V_o > S$, the DAC system will exhibit no distortion due to the limited slew rate.

Glitch

Glitches are caused by small time differences between some current sources turning off and others turning on. Take, for example, the major code transition at half-scale from 011...11 to 100...00. Here, the MSB current source turns on while all other current sources turn off. The small difference in switching times results in a narrow half-scale glitch, as shown in Fig. 7.33. Such a glitch, for example, can produce distorted characters in CRT display applications. To alleviate both glitch and slew-rate problems related to transients, a DAC is followed by a deglitcher. The deglitcher stays in the hold mode while the DAC changes its output value. After the switching transients have settled, the deglitcher is changed to the sampling mode. By making the hold time suitably long, the output of the deglitcher can be made independent of the DAC transient response. However, the slew rate of the deglitcher is on the same order as that of the DAC, and the transient distortion will still be present — now as an artifact of the deglitcher.

Techniques for High-Resolution DACs

The following methods are often used to improve the linearity of DACs: Laser trimming, off-chip adjustment, common-centroid layout technique, dynamic element matching technique, voltage or current sampling, and electronic calibration techniques. The trend is toward more sophisticated and intelligent electronic solutions that overcome and compensate for some of the limitations of conventional trimming techniques. *Electronic calibration* is a general term to describe various circuit techniques, which usually predistort the DAC transfer characteristic so that the DAC linearity can be improved. The self-calibration is to incorporate all the calibration mechanisms and hardware on the DAC as a built-in function so that users can recalibrate whenever necessary.

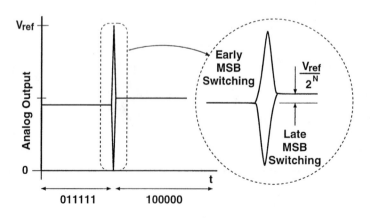

FIGURE 7.33   DAC output glitch.

The application of *dynamic element matching* to the binary-weighted current DAC is a straightforward switching of two complementary currents.[25] Its application to the binary voltage divider using two identical resistors or capacitors requires exchanging resistors or capacitors. This can be easily achieved by reversing the polarity of the reference voltage for the divide-by-two case. However, in the general case of N-element matching, the current division is inherently simpler than the voltage division. In general, to match the N independent elements, a switching network with N inputs and N outputs is required. The function of the switching network is to connect any input out of N inputs to one output with an average duty cycle of 1/N. The simplest one is a barrel shifter rotating the input-output connections in a predetermined manner. This barrel shifter generates a low-frequency modulated error when N gets larger because the same pattern repeats every N clocks. A more sophisticated randomizer with the same average duty cycle can distribute the mismatch error over the wider frequency range.

The *voltage or current sampling concept* is an electronic alternative to direct mechanical trimming. The voltage sampler is usually called an S/H, while the current sampler is called a current copier. The voltage is usually sampled on the input capacitor of a buffer amplifier, and the current is usually sampled on the input capacitor of a transconductance amplifier such as a MOS transistor gate. Therefore, both voltage and current sampling techniques are ultimately limited by their sampling accuracy.

The idea behind the voltage or current sampling DAC is to use one voltage or current element repeatedly. One example of the voltage sampling DAC is a discrete-time integrating DAC. The integrator integrates a constant charge repeatedly, and its output is sampled. This is equivalent to generating equally spaced reference voltages by stacking identical unit voltages.[26] The fundamental problem associated with this sampling voltage DAC approach is the accumulation of the sampling error and noise in generating larger voltages. Similarly, the current sampling DAC can sample a constant current on current sources made of MOS transistors.[27] Since one reference current is copied on other identical current samplers, the matching accuracy can be maintained as long as the sampling errors are kept constant. Since it is not practical to make a high-resolution DAC using voltage or current sampling alone, this approach is limited to generating MSB DACs for the segmented DAC or for the subranging ADCs.

Self-calibration is based on an assumption that the segmented DAC linearity is limited by the MSB DAC so that only errors of MSBs can be measured, stored in memory, and recalled during normal operation. There are two different ways of measuring the MSB errors. In one method, individual-bit non-linearities, usually appearing as component mismatch errors, are measured digitally,[15,18] and a total error, which is called a code error, is computed from individual-bit errors depending on the output code during normal conversion. On the other hand, the other method measures and stores digital code errors directly and eliminates the digital code-error computation during normal operation.[16,17] The former requires less digital memory, while the latter requires fewer digital computations.

References

1. J. McCreary and P. Gray, All-MOS charge redistribution analog-to-digital conversion techniques-part I, *IEEE J. Solid-State Circuits*, vol. SC-10, pp. 371-379, Dec. 1975.
2. S. Ramet, A 13-bit 160kHz differential analog to digital converter, *ISSCC Dig. Tech. Papers*, pp. 20-21, Feb. 1989.
3. C. Mangelsdorf, H. Malik, S. Lee, S. Hisano, and M. Martin, A two-residue architecture for multistage ADCs, *ISSCC Dig. Tech. Papers*, pp. 64-65, Feb. 1993.
4. W. Song, H. Choi, S. Kwak, and B. Song, A 10-b 20-Msamples/s low power CMOS ADC, *IEEE J. Solid-State Circuits*, vol. 30, pp. 514-521, May 1995.
5. T. Cho and P. Gray, A 10-bit, 20-Msamples/s, 35-mW pipeline A/D converter, *IEEE J. Solid-State Circuits*, vol. 30, pp. 166-172, Mar. 1995.
6. P. Vorenkamp and R. Roovers, A 12-bits, 60MSPS cascaded folding & interpolating ADC, *IEEE J. Solid-State Circuits*, vol. 32, pp. 1876-1886, Dec. 1997.
7. K. Kattmann and J. Barrow, A technique for reducing differential nonlinearity errors in flash A/D converters, *ISSCC Dig. Tech. Papers*, pp. 170-171, Feb. 1991.

8. K. Bult and A. Buchwald, An embedded 240-mW 10-bit 50MS/s CMOS ADC in 1-mm^2, *IEEE J. Solid-State Circuits*, vol. 32, pp. 1887-1895, Dec. 1997.

9. B. Song, S. Lee, and M. Tompsett, A 10b 15-MHz CMOS recycling two-step A/D converter, *IEEE J. Solid-State Circuits*, vol. SC-25, pp. 1328-1338, Dec. 1990.

10. S. Lewis and P. Gray, A pipelined 5-Msamples/s 9-bit analog-to-digital converter, *IEEE J. Solid-State Circuits*, vol. SC-22, pp. 954-961, Dec. 1987.

11. B. Song, M. Tompsett, and K. Lakshmikumar, A 12-bit 1-Msample/s capacitor error-averaging pipelined A/D converter, *IEEE J. Solid-State Circuits*, vol. SC-23, pp. 1324-1333, Dec. 1988.

12. S. Lewis, S. Fetterman, G. Gross Jr., R. Ramachandran, and T. Viswanathan, A 10-b 20-Msample/s analog-to-digital converter, *IEEE J. Solid-State Circuits*, vol. SC-27, pp. 351-358, Mar. 1992.

13. C. Conroy, D. Cline, and P. Gray, An 8-b 85-MS/s parallel pipeline A/D converter in 1-μm CMOS, *IEEE J. Solid-State Circuits*, vol. SC-28, pp. 447-454, Apr. 1993.

14. R. Plassche and R. Grift, A high speed 7 bit A/D converter, *IEEE J. Solid-State Circuits*, vol. SC-14, pp. 938 – 943, Dec. 1979.

15. H. Lee, D. Hodges, and P. Gray, A self-calibrating 15-bit CMOS A/D converter, *IEEE J. Solid-State Circuits*, vol. SC-19, pp. 813-819, Dec. 1984.

16. S. Lee and B. Song, Digital-domain calibration of multistep analog-to-digital converters, *IEEE J. Solid-State Circuits*, vol. SC-27, pp. 1679-1688, Dec.1992.

17. S. Kwak, B. Song, and K. Bacrania, A 15-b, 5-Msamples/s low spurious CMOS ADC, *IEEE J. Solid-State Circuits*, vol. 32, pp. 1866-1875, Dec. 1997.

18. A. Karanicolas, H. Lee, and K. Bacrania, A 15-b 1-Msample/s digitally self calibrated pipeline ADC, *IEEE J. Solid-State Circuits*, vol. 28, pp. 1207-1215, Dec. 1993.

19. D. Mercer, A 16-b D/A converter with increased spurious free dynamic range, *IEEE J. Solid-State Circuits*, vol. 29, pp. 1180-1185, Oct. 1994.

20. B. Tesch and J. Garcia, A low glitch 14-b 100-MHz D/A converter, *IEEE J. Solid-State Circuits*, vol. 32, pp. 1465-1469, Sept. 1997.

21. D. Mercer and L. Singer, 12-b 125 MSPS CMOS D/A designed for special performance, *Proc. IEEE Int. Symp. Low Power Electronics and Design*, pp. 243-246, Aug. 1996.

22. C. Lin and K. Bult, A 10b 250MSample/s CMOS DAC in 1mm^2, *ISSCC Dig. Tech. Papers*, pp. 214-215, Feb. 1998.

23. A. Marques, J. Bastos, A. Bosch, J. Vandenbusche, M. Steyaert, and W. Sansen, A 12b accuracy 300M sample/s update rate CMOS DAC, *ISSCC Dig. Tech. Papers*, pp. 216-217, Feb. 1998.

24. D. Freeman, Slewing distortion in digital-to-analog conversion, *J. Audio Eng. Soc.*, vol. 25, pp. 178-183, Apr. 1977.

25. R. Plassche, Dynamic element matching for high accuracy monolithic D/A converters, *IEEE J. Solid-State Circuits*, vol. SC-11, pp. 795-800, Dec. 1976.

26. D. Kerth, N. Sooch, and E. Swanson, A 12-bit 1-MHz two-step flash ADC, *IEEE J. Solid-State Circuits*, vol. SC-24, pp. 250-255, Apr. 1989.

27. D. Groeneveld, H. Schouwenaars, H. Termeer, and C. Bastiaansen, A self-calibration technique for monolithic high-resolution D/A converters, *IEEE J. Solid-State Circuits*, vol. SC-24, pp. 1517-1522, Dec. 1989.

8

Oversampled Analog-to-Digital and Digital-to-Analog Converters

John W. Fattaruso
Louis A. Williams, III
Texas Instruments, Incorporated

8.1 Introduction

In the absence of some form of calibration or trimming, the precision of the Nyquist rate converters described Chapter 7 is strictly dependent on the precision of the VLSI components that comprise the converter circuits. Oversampled data converters are a means of exchanging the speed and data processing capability of modern sub-micron integrated circuits for precision that would otherwise not be readily attainable.[1,2] The precision of an oversampled data converter can exceed the precision of its circuit components by several orders of magnitude.

In this chapter, the basic operation and design techniques of the most widely used class of oversampled data converters — *sigma-delta*[*] *modulators* — are described. In Section 8.2, the basic theory of sigma-delta

---

[*]The reader will find functionally identical modulator blocks in the literature named either "sigma-delta" modulators or "delta-sigma" modulators, with the choice of terminology largely up to personal preference. We use the former term here.

modulators is presented, using both time-domain and frequency-domain approaches. The issue of non-harmonic tones is also discussed. In Section 8.3, more complex sigma-delta architectures are described, including higher-order architectures, cascaded architectures, and bandpass architectures. Filtering techniques unique to sigma-delta modulators are presented in Section 8.4. In Section 8.5, the basic circuit building blocks for sigma-delta modulators are described; and in Section 8.6, circuit design issues specific to sigma-delta-based data converters are discussed.

8.2   Basic Theory of Operation

Oversampled data conversion techniques have their roots in the design of signal coders for communication systems.[3-6] Oversampling techniques differ from Nyquist techniques in that their comprehension and design procedures draw equally from time-domain and frequency-domain representations of signals, whereas Nyquist techniques are readily understood in just the time domain.

In general, the function of data conversion by oversampling is typically performed by a serial connection of a modulator and various filter blocks. In analog-to-digital (A/D) conversion, shown in Fig. 8.1, the analog input signal $x_i(t)$ is first bandlimited by an anti-alias filter, then sampled at a rate f_s. This sampling rate is M times faster than a comparable Nyquist rate converter with the same signal bandwidth; the value of M is the oversampling ratio. The sampled signal $x[n]$ is coded by a modulator block that quantizes the data into a finite number of discrete levels. The resulting coded signal $y[n]$ is down-sampled, or decimated, by a factor of M to produce an output that is comparable to a Nyquist rate converter. Digital-to-analog oversampled data conversion is basically the reverse of analog-to-digital conversion. As shown in Fig. 8.2, the Nyquist-rate digital samples are oversampled by an interpolation filter, coded by a modulator, and then reconstructed in the analog domain by an analog filter.

In both analog-to-digital and digital-to-analog data conversion, the block with the most unique signal processing properties is the modulator. The remainder of this section and the subsequent sections focus on the properties and architectures for oversampled data modulators.

Time-Domain Representation

The simplest modulator that would perform the requisite conversion to discrete output levels is the quantization function $Q(x)$ shown in Fig. 8.3. This quantization can be thought of as merely the sum of

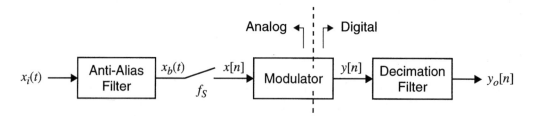

FIGURE 8.1   Oversampled A/D conversion.

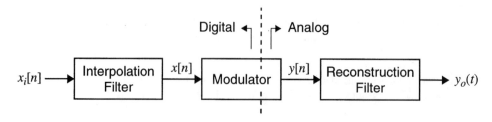

FIGURE 8.2   Oversampled D/A conversion.

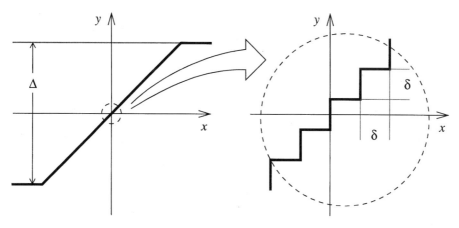

FIGURE 8.3 Quantizer transfer function.

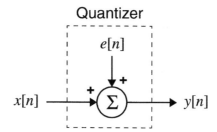

FIGURE 8.4 Quantization error.

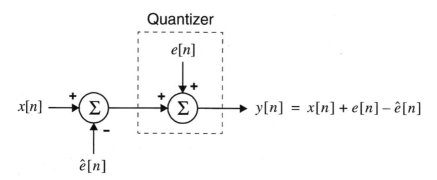

FIGURE 8.5 Error correction feedback.

the original signal $x[n]$ with a sampled error signal $e[n]$, as illustrated in Fig. 8.4. In Nyquist rate converters, the error is reduced by using a large number of small steps in the quantizer characteristic. In oversampled data converters, specifically sigma-delta modulators, the error is corrected by a feedback network. This correction is made by estimating the error in advance and subtracting it from the input, as shown in Fig. 8.5, where is the error estimate. If this estimate were perfect, $\hat{e}[n]$ would equal $e[n]$ and the output $y[n]$ would equal the input $x[n]$. However, since the error is not known until it is made, $e[n]$ is not known when $\hat{e}[n]$ is needed. Therefore, some means must be found to estimate the error. In the case of sigma-delta converters, the error can be estimated by exploiting some knowledge of the frequency domain behavior of the input signal. Specifically, it is assumed that the signal is changing very slowly from sample to sample, or equivalently, its bandwidth is much less than the sampling rate.

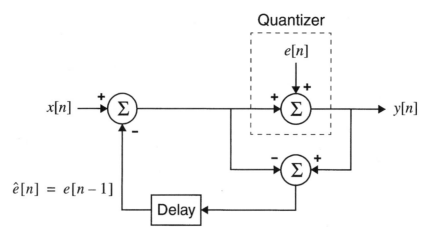

FIGURE 8.6   First-order error estimation.

For exceedingly slow signals, a first-order estimate of the error to be committed in quantization can be formed. The first-order estimate of the current error $e[n]$ is simply the previous error $e[n-1]$. This error may be found simply by a subtraction across the quantization block as shown in Fig. 8.6, and the output $y[n]$ is

$$y[n] = x[n] + e[n] - e[n-1] \qquad (8.1)$$

The essential property of this structure is that if an error is committed that is not large enough to be corrected by a displacement to another quantizer level on the next sample, then the history of successive errors accumulate in the feedback loop until they eventually push the quantizer into another level. In this manner, the output of the quantizer will, over time, correct the errors committed in previous samples, increasing the precision of the information being generated as a time sequence of samples.

As will be shown in Section 8.5, the most convenient and accurate sampled-data circuit building block in practice is an integrator. With a few straightforward steps, the system of Fig. 8.6 can be transformed into that of Fig. 8.7, where the delay element is now immersed in an integrator feedback loop. The output of this transformed modulator is

$$y[n] = x[n-1] + e[n] - e[n-1] \qquad (8.2)$$

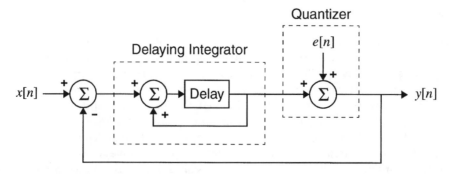

FIGURE 8.7   First-order equivalent modulator.

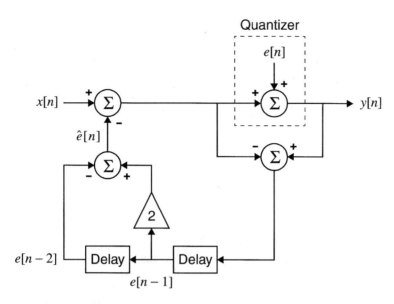

FIGURE 8.8 Second-order error estimation.

Comparing Eqs. 8.1 and 8.2, it is evident that this transformation does require the addition of a single clock delay block in the input path $x[n]$, but this extra clock cycle of latency has no effect on the precision or frequency response of the modulator. The structure in Fig. 8.7 is generally known as a first-order sigma-delta modulator.[7-9]

An increase in precision can be obtained by using more accurate estimates of the expected quantizer error.[6] A second-order estimate of $e[n]$ may be formed by assuming that the error $e[n]$ varies linearly with time. In this case, an estimate of the current error $e[n]$ may be computed by changing the previous error $e[n-1]$ by an amount equal to the change between $e[n-2]$ and $e[n-1]$. The second order error estimate is thus

$$\hat{e}[n] = e[n-1] + \left(e[n-1] - e[n-2]\right) = 2e[n-1] - e[n-2] \tag{8.3}$$

and is illustrated in Fig. 8.8. The output of the second-order estimation modulator is

$$y[n] = x[n] + e[n] - 2e[n-1] + e[n-2] \tag{8.4}$$

It can be shown, after a number of steps, that the modulator in Fig. 8.8 can be transformed into a modulator in which the feedback loop delays are again immersed in practical integrator blocks. This second-order sigma-delta modulator[10-12] is shown in Fig. 8.9; the output of this transformed modulator is

$$y[n] = x[n-2] + e[n] - 2e[n-1] + e[n-2] \tag{8.5}$$

which is entirely equivalent to that given by Eq. 8.4, except for the addition of two inconsequential delays of the input signal $x[n]$.

A further increase in precision can be obtained using even higher-order estimates of the quantizer error, such as quadratic or cubic. These high-order error estimate modulators can also be transformed into a series of delaying integrators in a feedback loop. Unfortunately, as discussed in Section 8.3, practical difficulties emerge for orders greater than two, and alternative architectures are generally needed.

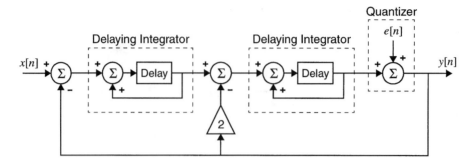

FIGURE 8.9   Second-order equivalent modulator.

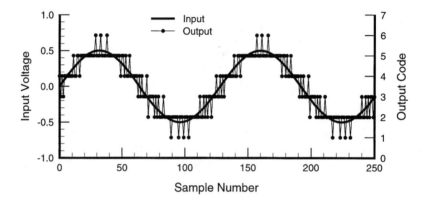

FIGURE 8.10   First-order sample output.

Computer simulation of modulator systems is straightforward, and Fig. 8.10 shows the simulated output of the first-order modulator of Fig. 8.7 when fed with a simple sinusoidal input. The resolution of the quantizer in the modulator loop was assumed to be eight levels. (The modulator output is drawn with continuous lines to emphasize the oscillatory nature of the modulator output, but the quantities plotted have meaning only at each sample time.) The coarsely quantized output code generally follows the input, but with occasional transitions that track intermediate values over local sample regions.

A second-order modulator with an eight-level quantizer exhibits the simulated behavior shown in Fig. 8.11. Note that the oscillations by which the loop attempts to minimize quantization error appear

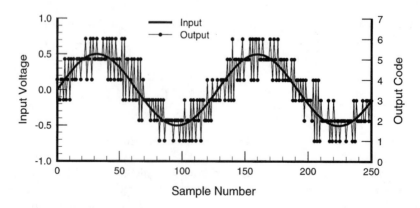

FIGURE 8.11   Second-order sample output.

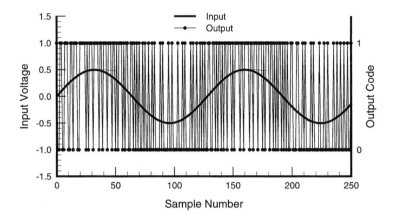

FIGURE 8.12   Second-order one-bit modulator sample output.

"busier" than in the first-order case of Fig. 8.10. It will be shown in the frequency domain that, for a given signal bandwidth, the more vibrant output code oscillations in Fig. 8.11 actually represent the input signal with higher precision than the first-order case in Fig. 8.10.

A special case that is of practical significance is the second-order modulator with a two-level quantizer, that is, simply a comparator closing the feedback loop. A simulation of such a modulator is shown in Fig. 8.12. Although the quantized representation at the output appears crude, the continuous nature of the input level is expressed in the density of output codes. When the input level is high (around sample numbers 32 and 160), there is a greater preponderance of '1' output codes, and at low swings of the input (around sample numbers 96 and 224), the '0' output code dominates.

The examples in Figs. 8.10 to 8.12 demonstrate that information generated from the modulator expresses, in a high-speed coded form, the coarsely quantized input signal and the deviation between the signal and the quantization levels. Although the time-domain-coded modulator output looks somewhat unintelligible, the output characteristics are clearer in the frequency domain.

Frequency-Domain Representation

The modulators in Figs. 8.7 and 8.9 can be generalized as the sampled-data system shown in Fig. 8.13, where the time domain signals $x[n]$, $y[n]$, and $e[n]$ are written as their frequency-domain equivalents $X(z)$, $Y(z)$, and $E(z)$. The modulator output in Fig. 8.13 can be written in terms of the input $X(z)$ and the quantizer error $E(z)$ as

$$Y\left(z\right) = H_x\left(z\right)X\left(z\right) + H_e\left(z\right)E\left(z\right) \tag{8.6}$$

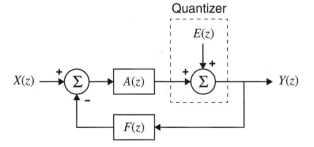

FIGURE 8.13   Generalized sigma-delta modulator.

where the input transfer function, $H_x(z)$, is

$$H_x(z) = \frac{A(z)}{1 + A(z)F(z)} \tag{8.7}$$

and the error transfer function, $H_e(z)$, is

$$H_e(z) = \frac{1}{1 + A(z)F(z)} \tag{8.8}$$

Strictly speaking, the error $E(z)$ is directly dependent on the input $X(z)$. Nonetheless, if the input to the quantizer is sufficiently busy; that is, the input to the quantizer crosses through several quantization levels, the quantizer error approaches having the behavior of a random value that is uniformly distributed between $\pm\delta/2$ and is uncorrelated with the input, where δ is the quantization level separation illustrated in Fig. 8.3. In the frequency domain, the error noise power spectrum is uniform with a total error power of $\delta^2/12$.[13,14] The error power between the frequencies f_L and f_H is

$$S_{ee} \approx \frac{\delta^2}{6f_S} \int_{f_L}^{f_H} \left| H_e\left(e^{j\pi f/f_s}\right)\right|^2 df \tag{8.9}$$

where f_s is the sampling rate. The error power between f_L and f_H can be reduced independent of the quantizer error level separation δ by having a small error transfer function $H_e(z)$ in that frequency band. Sigma-delta modulators have this property.

A sigma-delta modulator is a system such as that in Fig. 8.13 in which the error transfer function $H_e(z)$ is small and the input transfer function $H_x(z)$ is about unity for some band of frequencies. That is,

$$\left| H_e\left(e^{j2\pi f/f_s}\right)\right| \ll 1; \; f_L \le f \le f_H \tag{8.10}$$

$$\left| H_x\left(e^{j2\pi f/f_s}\right)\right| \approx 1; \; f_L \le f \le f_H \tag{8.11}$$

The requirements in Eqs. 8.10 and 8.11 are equivalent to requiring that the loop gain be large and the feedback gain be unity; that is

$$\left| A\left(e^{j2\pi f/f_s}\right)\right| \gg 1; \; f_L \le f \le f_H \tag{8.12}$$

$$\left| F\left(e^{j2\pi f/f_s}\right)\right| \approx 1; \; f_L \le f \le f_H \tag{8.13}$$

There are many system designs that have the sigma-delta properties of high-loop gain and unity feedback gain. The previous examples in Figs. 8.7 and 8.9 are part of an important class of modulator architectures called *noise-differencing modulators* that are particularly well suited to VLSI implementation. The forward path in a noise-differencing sigma-delta modulator consists of a series of delaying integrators. The *order* of the modulator is defined as the number of integrators. The forward gain of an L-th order modulator is

$$A(z) = \left(\frac{z^{-1}}{1 - z^{-1}} \right)^L \tag{8.14}$$

The feedback gain in a noise-differencing modulator is designed such that the modulator open-loop gain is

$$A(z)F(z) = \frac{1}{\left(1 - z^{-1}\right)^L} - 1 \tag{8.15}$$

From Eqs. 8.14 and 8.15, it follows that the output for an *L*-th order noise-differencing sigma-delta modulator is

$$Y(z) = z^{-L} X(z) + \left(1 - z^{-1}\right)^L E(z) \tag{8.16}$$

The simulated frequency response for a second-order noise-differencing sigma-delta modulator with a sinusoidal input is shown in Fig. 8.14. The large spike in the center is the original input signal. It is clear from the plot that the noise energy is lowest at low frequencies. Noise-differencing modulators are designed to reduce the quantization noise in the baseband; that is, $f_L = 0$ and $f_H \ll f_S$. The oversampling ratio, *M*, is

$$M = \frac{f_S}{2 f_H} \tag{8.17}$$

(In a Nyquist rate converter, $M = 1$.) The baseband noise power for a noise-differencing modulator is

$$S_{ee} \approx \frac{\delta^2}{6 f_S} \int_0^{f_H} \left| \left(1 - e^{-j2\pi f/f_S}\right)^L \right|^2 df \approx \frac{\delta^2}{12} \frac{\pi^{2L}}{2L+1} \frac{1}{M^{2L+1}} \tag{8.18}$$

One important measure of a sigma-delta modulator is its dynamic range, defined here as the ratio of the maximum sinusoidal input power to the noise power. With the quantizer output limited, as shown in Fig. 8.3, to a range of Δ, the maximum sinusoidal signal power is $\Delta^2/8$. The quantizer range Δ is related to the quantizer level separation δ by the number of quantization levels *K*, where

$$\delta = \frac{\Delta}{K-1} \tag{8.19}$$

The dynamic range of a noise-differencing sigma-delta modulator is then

$$DR = \frac{3}{2} \frac{2L+1}{\pi^{2L}} M^{2L+1} (K-1)^2 \tag{8.20}$$

Because the dynamic range is such a strong function of the oversampling ratio, the number of bits required to achieve a given dynamic range is substantially less in a sigma-delta modulator than in a Nyquist-rate converter. To illustrate this, the dynamic range, as given by Eq. 8.20, is shown in Fig. 8.15

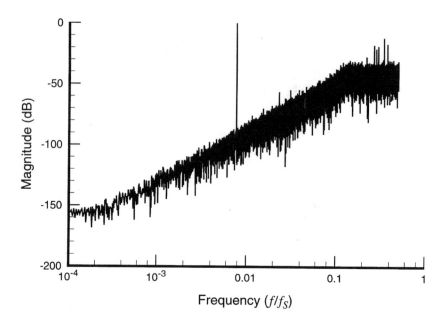

FIGURE 8.14   Second-order simulated frequency response.

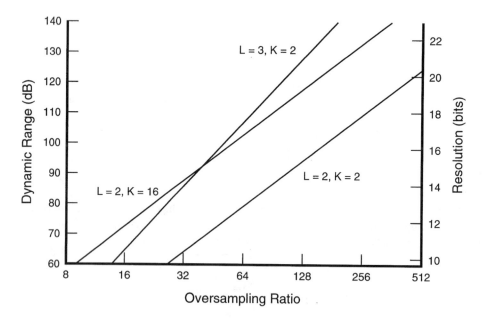

FIGURE 8.15   Calculated dynamic range vs. oversampling ratio.

as a function of the oversampling ratio, M, for three combinations of modulator order, L, and number of quantization levels, K. The equivalent resolution in bits that would be required of a Nyquist-rate converter to achieve the same dynamic range is shown in the right-hand axis of this figure. It can be inferred from Eq. 8.20 that a large dynamic range can be obtained even with only two quantization levels. This is important when circuit imperfections in actual sigma-delta data converter implementations are considered.

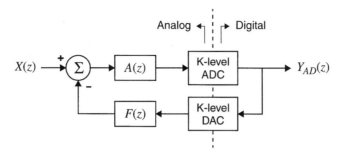

FIGURE 8.16 Sigma-delta modulator for A/D conversion.

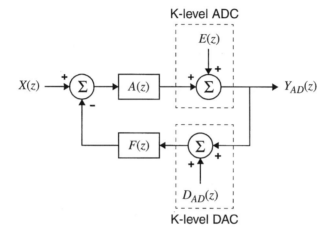

FIGURE 8.17 *K*-level DAC error in sigma-delta A/D converter.

Sigma-Delta Modulators in Data Converters

The generalized modulator shown in Fig. 8.13 must be subtly modified when applied to the A/D and D/A converters in Figs. 8.1 and 8.2. In an A/D converter, the quantizer is actually a coarse *K*-level A/D converter (ADC), having an analog input and a digital output. (Since *K* is generally small, the *K*-level ADC is usually just a small flash converter, as described in Chapter 7.) The quantized digital code is fed back into the analog *F(z)* through a *K*-level D/A converter (DAC), as shown in Fig. 8.16. Imperfections in this *K*-level DAC will introduce an additional error term $D_{AD}(z)$ as shown in Fig. 8.17. With the addition of this DAC error term, the modulator output is

$$Y_{AD}(z) = H_x(z)\big[X(z) - F(z)D_{AD}(z)\big] + H_e(z)E(z) \tag{8.21}$$

Since the feedback transfer function, *F(z)*, is unity in the band of interest (see Eq. 8.13), the DAC error is indistinguishable from the input. If there are more than two quantization levels, any mismatch between the level separations in the DAC will manifest itself as distortion because the DAC input is signal dependent. On the other hand, if there are only two quantization levels, there is only one level separation, and errors in the DAC levels will not cause distortion. (At worst, DAC errors in a two-level modulator will introduce a dc offset and a gain error.) Thus, with two-level sigma-delta modulators, it is possible to achieve low distortion and low-noise performance without precise component matching.

Unfortunately, most, if not all, of the statistical conditions that led to Eq. 8.9 and the subsequent equations are violated when a two-level single-threshold quantizer is used in a sigma-delta modulator.[15] Furthermore, the effective gain of the quantizer, which in Fig. 8.3 is implied to be unity, is undefined for

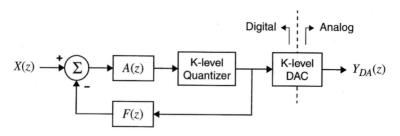

FIGURE 8.18   Sigma-delta modulator for D/A conversion.

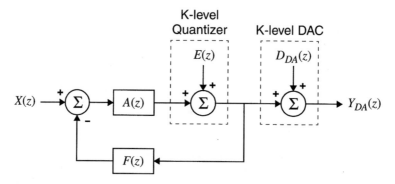

FIGURE 8.19   *K*-level DAC error in sigma-delta D/A converter.

a single-threshold quantizer. Nonetheless, empirical evidence has indicated that Eq. 8.20 is still a reasonable approximation for two-level noise-differencing sigma-delta modulators, and is useful as a design guideline for the amount of oversampling needed to achieve a specific dynamic range for a given modulator order.[16]

As in sigma-delta ADCs, there is also a DAC error term in sigma-delta based DACs. In a sigma-delta DAC, the modulator loop is implemented digitally, and the output of that loop is applied to a coarse *K*-level DAC that provides the analog input for the reconstruction filter, as shown in Fig. 8.18. Imperfections in the *K*-level DAC will introduce an error term $D_{DA}(z)$ as shown in Fig. 8.19. With the addition of this error term, the modulator output is

$$Y_{DA}(z) = H_x(z)X(z) + D_{DA}(z) + H_e(z)E(z) \tag{8.22}$$

Since the input transfer function $H_x(z)$ is unity in the band of interest (see Eq. 8.11), the DAC error is indistinguishable from the input, just as in the A/D case. Once again, two-level quantization can be used to avoid DAC-introduced distortion.

Tones

One problem with the simplified noise model of sigma-delta modulators that led to Eq. 8.20 is the failure to predict non-harmonic tones. This is especially true for two-level modulators. Repetitive patterns in the coded modulator output that cause discrete spectral peaks at frequencies not harmonically related to any input frequency can occur in sigma-delta modulators.[10,16-18] These tones can manifest themselves as odd "chirps" or "pops," and they exist even in ideal sigma-delta modulators; they are not caused by circuit imperfections.[2]

The origin of sigma-delta tones is illustrated in the following example. Consider a first-order sigma-delta modulator, such as that in Fig. 8.7, with a dc input of 0.0005. Let the quantizer have two output

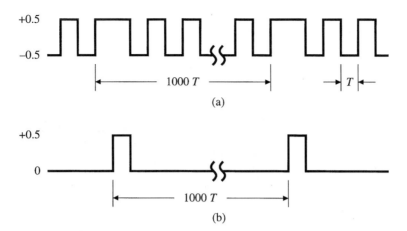

FIGURE 8.20 (a) Output sequence with average of 0.0005 and (b) running average of (a).

levels, +0.5 and –0.5. The output of such a modulator will be a sequence of +0.5's and –0.5's such that the time average of the outputs is 0.0005. To achieve this average, the output of the first-order modulator will be a stream of alternating +0.5's and –0.5's, with an extra +0.5 every 1000 clock cycles. This is illustrated in Fig. 8.20(a), where T is the clock period ($T = 1/f_S$). The two-cycle running average of this output is shown in Fig. 8.20(b). For the most part, this running average is zero, except that at every 1000 clock cycles there is a one clock cycle pulse. This repetitive pulse produces a tone in the output spectrum at a frequency of

$$f_P = \frac{f_S}{1000} = \frac{M}{500} f_H \tag{8.23}$$

If the oversampling ratio M is less than 500, this tone will appear in the baseband spectrum.

In sigma-delta modulators with more active input signals, the output sequence is typically more complex than that illustrated in Fig. 8.20. Nonetheless, the concept underlying tone behavior is that repeating patterns in the quantizer output cause non-uniformity in the quantizer error spectrum, which in the worst case is a discrete spectral peak. A measured tone for a second-order modulator with a dc input is shown in Fig. 8.21.[19]

Several means of mitigating sigma-delta tones have been used. The first rule is to avoid using first-order sigma-delta modulators. Aside from having inferior noise-shaping properties compared to other modulator architectures, first-order modulators have terrible tone properties.[15,16] The situation improves dramatically with second- and higher-order modulators. In fact, the presence of tones may only be a perceptual or marketing concern, as the total tone power is usually less than the broadband quantization noise power.[20]

When tone magnitudes must be reduced, several techniques have proven effective. These include dither, cascaded architectures, and multi-level quantization. Of these three, dither is the only technique whose sole benefit is the reduction of tones. The simplest type of dither is to add a moderate amplitude out-of-band signal, such as a square wave, to the input.[9,21,22] This dither signal is attenuated by the same filter that attenuates the quantization noise. The purpose of this dither is to keep the sigma-delta modulator out of modes that produce patterns, and for some types of tones this technique is effective. A more rigorously effective technique is to add a large amplitude pseudo-random noise signal at the quantizer input.[23] This noise is spectrally shaped just like the quantization noise, and is the most effective dither scheme for eliminating tones. Its drawbacks are the expense in silicon area of the random noise generator and the 2- to 3-dB reduction in dynamic range caused by the dither noise.

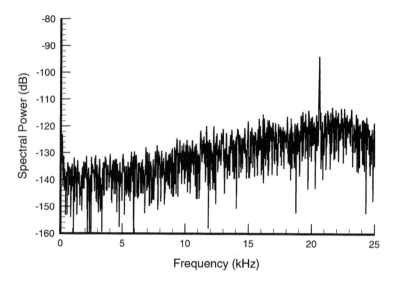

FIGURE 8.21   Measured tone in second-order modulator with dc input.

The other two tone mitigation techniques, cascaded architectures and multi-level quantization, are simply more complex sigma-delta architectures that happen to have improved tone properties over simple noise-differencing two-level sigma-delta modulators. These techniques are covered in Sections 8.3 and 8.4, respectively.

8.3   Alternative Sigma-Delta Architectures

Equation 8.20 appears to indicate that the order of the modulator, L, can be any value, and that increasing L would be beneficial. However, one further problem with two-level sigma-delta modulators is that two-level noise-differencing modulators of order greater than two can exhibit unstable behavior.[10] For this reason, only first- and second-order modulators were discussed in the Section 8.2. Nonetheless, there have been acceptably stable practical alternative architectures that achieve quantization noise shaping that is superior to a second-order modulator. Two such architectures, high-order and cascaded modulators, are discussed in this section.

Another assumption in the previous section was that the noise-shaped region in a sigma-delta modulator is centered around dc. This is not necessarily the case; sigma-delta modulators with noise-shaped regions at frequencies other than near dc are called *bandpass modulators* and are discussed at the end of this section.

High-Order Modulators

A high-order modulator is a modulator such as that depicted in Fig. 8.13 in which there are more than two zeros in the noise transfer function. As stated earlier, if two-level quantization is employed, a simple noise-differencing series of integrators cannot be used, as such architectures produce unstable oscillations with large inputs that do not recover when the input is removed. To overcome this problem, high-order modulators use forward and feedback transfer functions that are more complex than the noise-differencing functions in Eqs. 8.14 and 8.15.[24-26]

The general rule of thumb in the design of high-order modulators is that the modulator can be made stable if

$$\lim_{z \to \infty} H_e(z) = 1 \qquad (8.24)$$

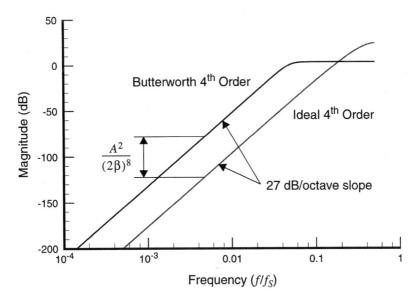

FIGURE 8.22   Fourth-order error spectrum.

$$\left|H_e(z)\right| \leq A, \text{ for } |z| = 1 \tag{8.25}$$

and the integrator outputs are clipped and/or scaled to prevent self-sustaining instability.[26,27] The maximum error gain A is about 1.5, but the value used represents a tradeoff between noise attenuation and modulator stability. These rules cover a broad class of filter types and modulator architectures, and the type of filter used generally follows the traditions of the previous designers in an organization.

As an example, consider a fourth-order modulator with a highpass Butterworth error transfer function having a maximum gain, A, of 1.5, and a cutoff frequency set such that Eq. 8.24 is satisfied. The error spectrum of the Butterworth filter is shown in Fig. 8.22, along with the error transfer function of an ideal fourth-order difference. While the Butterworth filter holds the maximum gain to 1.5 (3.5 dB), and while both filters have a fourth-order noise shaping slope in the baseband (27 dB/octave), the error power in the baseband is 44 dB higher with the Butterworth filter than with the ideal noise-differencing filter. This error penalty is typical of high-order designs; there is usually a direct tradeoff between stability and noise reduction.

Consider the more general case of an L-th order highpass Butterworth error transfer function. The error transfer function of such a filter around the unit circle is

$$\left|H_e\left(e^{j\omega}\right)\right|^2 = \frac{A^2\left(\dfrac{1}{\beta}\tan\dfrac{\omega}{2}\right)^{2L}}{1+\left(\dfrac{1}{\beta}\tan\dfrac{\omega}{2}\right)^{2L}} \tag{8.26}$$

The filter coefficients $H_e(z)$ for needed to satisfy Eq. 8.26 can be computed using standard digital filter design techniques.[28] For a given filter order, L, and gain, A, the parameter β must be chosen to satisfy Eq. 8.24. (The condition in Eq. 8.25 is always satisfied when Eq. 8.26 is true.) These solutions can be computed numerically, and it is found empirically that

TABLE 8.1 High-Order Butterworth Gain Factors and Dynamic Range (DR) Loss

L	β_N	b	Loss in DR (dB)	Zero Placement DR Improvement (dB)	Zero Placement Net Loss in DR (dB)
3	0.052134	0.1570	33.7	8.0	25.8
4	0.051709	0.1557	44.1	12.8	31.2
5	0.033866	0.1020	72.6	17.9	54.7
6	0.034903	0.1051	84.8	23.2	61.6
7	0.025390	0.0764	117.7	28.6	89.1

Note: Except for β_N, all values are calculated for $A = 1.5$.

$$\beta = A^e \beta_N \qquad (8.27)$$

where the values for β_N are tabulated in Table 8.1. The loss in dynamic range relative to an ideal noise-differencing modulator, given by, $A^2/(2\beta)^{2L}$, is also tabulated. In spite of this loss, high-order modulators can still achieve better noise performance than second-order modulators. However, because of the compromise in dynamic range required to stabilize high-order modulators, third-order modulators are generally not worth the effort. More common are fourth- and fifth-order modulators.

The noise penalty required to stabilize high-order modulators can be mitigated to some extent by alternate zero placement.[25] Classic noise-differencing modulators place all of the zeros of the error transfer function at dc ($z = 1$). This causes most of the noise power to be concentrated at the highest baseband frequencies. If, instead, the zeros are distributed throughout the baseband, the total noise in the baseband can be reduced, as illustrated in Fig. 8.23. The amount by which zero placement can improve the noise transfer function is summarized in Table 8.1. Also tabulated is the net loss in dynamic range of a high-order Butterworth modulator that uses zero placement relative to an ideal noise-differencing modulator that has zeros at dc.

Cascaded Modulators

Cascaded, or multi-stage, architectures are an alternative means of achieving higher-order noise shaping without the stability problems of the high-order modulators described in the previous section.[29,30] In a cascaded modulator, two or more stable first- or second-order modulators are connected in series, with the input of each stage being the error from the previous stage, as illustrated in Fig. 8.24. Referring to this illustration, the first stage of the cascade has two outputs, y_1 and e_1. The output y_1 is an estimate of the input x. The error in this estimate is e_1. The second stage has as its input the error from the first stage, e_1, and its outputs are y_2 and e_2. The second stage output y_2 is an estimate of the first stage error e_1. By subtracting this estimate of the first stage error from the output of the first stage, y_1, only the second stage error remains. Thus, the error cancellation network uses the output of one stage to cancel the error in the previous stage.

For example, in a cascaded architecture comprising a second-order noise-differencing modulator followed by a first-order noise-differencing modulator, the transforms of the output of the two stages, as given by Eq. 8.16, are

$$Y_1(z) = z^{-2}X(z) + \left(1 - z^{-1}\right)^2 E_1(z) \qquad (8.28)$$

$$Y_2(z) = z^{-1}E_1(z) + \left(1 - z^{-1}\right)E_2(z) \qquad (8.29)$$

If the error cancellation network combines the two outputs such that

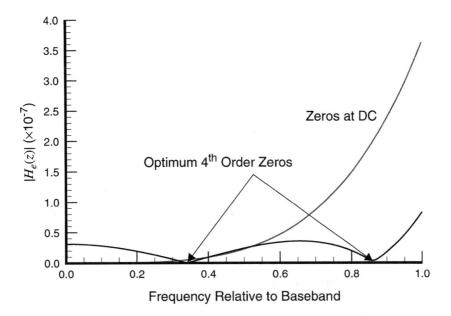

FIGURE 8.23 Fourth-order distributed zeros.

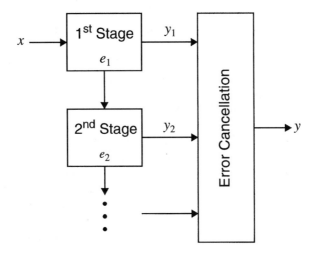

FIGURE 8.24 Cascaded sigma-delta modulators.

$$Y(z) = z^{-1} Y_1(z) - (1 - z^{-1})^2 Y_2(z)$$ (8.30)

then the error in the first stage will be cancelled and the output will be

$$Y(z) = z^{-3} X(z) - (1 - z^{-1})^3 E_2(z)$$ (8.31)

The final output of this cascaded modulator is third-order noise shaped. As a general rule, the noise shaping of a cascaded architecture is comparable to a single-stage modulator whose order is the sum of all the orders in the cascade.

The extent to which the errors in a cascaded modulator can be cancelled depends on the matching between the stages. The earliest multi-stage modulators were cascades of three first-order stages, often called the MASH architecture.[29] The disadvantage of this structure is that in order to achieve third-order performance, the error in the first stage, which is only first-order shaped, must be cancelled. Cancelling this relatively large error places a stringent requirement on inter-stage matching. An alternative architecture that has much more relaxed matching requirements is the cascade of a second-order modulator followed by a first-order modulator. This architecture, like the MASH, ideally achieves third-order noise shaping. Its advantage is that the matching can be 100 times worse than a MASH and still achieve better noise shaping performance.[31]

An additional benefit of cascaded modulators is improved tone performance. It has been shown both analytically and experimentally that the error spectra of the second and subsequent stages in a cascade are not plagued by the spectral tones that can exist in single-stage modulators.[19,32] To the extent that the first-stage error is cancelled, any tones in the first-stage error spectrum are attenuated, and the final output of the cascaded modulator is nearly tone-free.

Bandpass Modulators

The aforementioned sigma-delta architectures, called herein *baseband modulators*, all have zeros at or near dc; that is, at frequencies much less than the modulator sampling rate. It is also possible to group these zeros at some other point in the sampling spectrum; such architectures are called *bandpass modulators*. Bandpass architectures are useful in systems that need to quantize a narrow band signal that is centered at some frequency other than dc. A common example of such a signal is the intermediate frequency (IF) signal in a communications receiver.

The simplest method for designing a bandpass modulator is by applying a transformation to an existing baseband modulator architecture. The most common transformation is to replace occurrences of z with $-z^2$.[2] Such an architecture has zeros at $f_S/4$ and is stable if the baseband modulator is stable.[33] A comparison of the error transfer function of baseband and bandpass modulators is shown in Fig. 8.25. Note that a bandpass modulator generated through this transformation has twice the order of its equivalent baseband counterpart. For example, a fourth-order bandpass modulator is comparable to a second-order baseband modulator.

The noise shaping properties of a bandpass modulator generated through the $-z^2$ transformation are equivalent to the baseband modulator that was transformed. Thus, the approximation in Eq. 8.20 can be used where L is the order of the baseband modulator that was transformed and M is the effective oversampling ratio, which in a bandpass modulator is the sampling rate divided by the signal bandwidth.

There are advantages and disadvantages to bandpass modulators when compared with traditional down-conversion and baseband modulation. One advantage of the bandpass modulator is its insensitivity to $1/f$ noise. Since the signal of interest is far from dc, $1/f$ noise is often insignificant. Another advantage of bandpass modulation applies specifically to bandpass modulators having zeros at $f_S/4$ that are used in quadrature I and Q demodulation systems. If the narrowband IF signal is to be demodulated by a cosine and sine waveform, as shown in Fig. 8.26, the demodulation operation becomes simple multiplication by 1, −1, or 0 when the demodulation frequency is $f_S/4$.[34] Furthermore, because a single modulator is used, the bandpass modulator is free of the I/Q path mismatch problems that can exist in baseband demodulation approaches.

Two disadvantages of bandpass modulators involve the sampling operation. Sampling in a bandpass modulator has linearity requirements that are comparable to a Nyquist-rate converter sampling at the same IF frequency; this is much more severe than the linearity requirements of the sampling operation in a baseband converter with the same signal bandwidth. Also, because of the higher signal frequencies, the sampling in bandpass modulators is much more sensitive to clock jitter. To date, the state-of-the-art in bandpass modulators has about 20 dB less in dynamic range than comparable baseband modulators.[2] While the remainder of this chapter focuses once again on baseband modulators, many of the techniques are applicable to bandpass modulators as well.

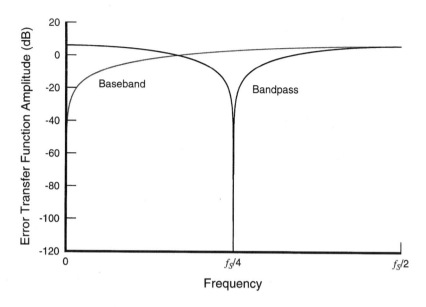

FIGURE 8.25 Bandpass noise transfer function.

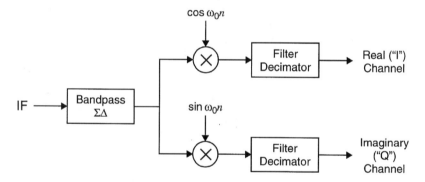

FIGURE 8.26 IQ demodulation with a bandpass modulator.

8.4 Filtering for Sigma-Delta Modulators

In Sections 8.2 and 8.3 of this chapter, the discussion focused on the operation of the sigma-delta modulator core. While this core is the most unique aspect of sigma-delta data conversion, there are also filtering blocks that constitute an important part of sigma-delta A/D and D/A converters. In this section, the non-modulator components in baseband sigma-delta converters, namely the analog and digital filters, are described. First, the requirements of the analog anti-alias and reconstruction filters are described. Second, typical architectures for the decimation and interpolation filters are discussed. While much of the design of these filters use standard techniques covered elsewhere in this volume, there are aspects of these filters that are specific to sigma-delta modulator applications.

Anti-Alias and Reconstruction Filters

The purpose of the anti-alias filter, shown in Fig. 8.1 at the input of the sigma-delta A/D converter, is, as the name would indicate, to prevent aliasing. The sampling operation maps, or aliases, all frequencies into the range bounded by $\pm f_S/2$.[28] Specifically, all signals within a baseband bandwidth of multiples of the sampling rate are mapped into the baseband. This is generally undesirable, so the anti-alias filter is

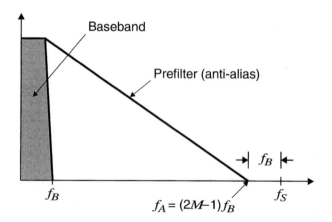

FIGURE 8.27   Anti-alias filter for sigma-delta A/D converters.

designed to attenuate this aliasing to some tolerable level. One advantage of sigma-delta converters over Nyquist-rate converters is that this anti-aliasing filter has a relatively wide transition region. As illustrated in Fig. 8.27, the passband region for this filter is the signal bandwidth f_B, while the stopband region for this filter is only within f_B of the sampling rate. Thus, the transition region is $2(M - 1)\,f_B$, and since $M \gg 1$, the transition region is relatively wide. A wide transition region generally means a simple filter design. The precise nature of the anti-alias filter is application dependent, and can be designed using any number of standard analog filter techniques.[35]

The reconstruction filter, shown in Fig. 8.2 at the output of the sigma-delta D/A converter, is also an analog filter. Its primary purpose is to remove unwanted out-of-band quantization noise. The extent to which this noise must be removed varies widely from system to system. If the analog output is to be applied to an element that is naturally bandlimited, such as a speaker, then very little attenuation may be necessary. On the other hand, if the output is applied to additional analog circuitry, care must be taken lest the high-frequency noise distort and map itself into the baseband. Circuit techniques for this filter are addressed further in Section 8.5.

Decimation and Interpolation Filters

In general, the filter characteristics of the decimation filter, shown in Fig. 8.1 at the output of the sigma-delta A/D converter, are much sharper than those of the anti-alias filter; that is, the transition region is narrower. The saving grace is that the filter is implemented digitally, and modern sub-micron processes have made complex digital filters economically feasible. Nonetheless, care must be taken or the filter architecture will become more computationally complex than is necessary.

The basic purpose of the decimation filter is to attenuate quantization noise and unwanted signals outside the baseband so that the output of the decimation filter can be down-sampled, or decimated, without significant aliasing. Normally, the most efficient means of accomplishing this is to apply a multi-rate filter architecture, such as that illustrated in Fig. 8.28.[36,37] The comb filter is a relatively crude, but easy to implement, filter that has zeros equally spaced throughout the sampled spectrum. The frequency response of an N-th order comb filter, $H_C(z)$, is

$$H_C\!\left(z\right) = \left(\frac{1}{R}\frac{1 - z^{-R}}{1 - z^{-1}}\right)^{\!N} \tag{8.32}$$

where R is the impulse response length of the comb filter. If R is set equal to the decimation ratio of the comb filter (the comb filter input rate divided its output rate), then the filter zeros will occur at every

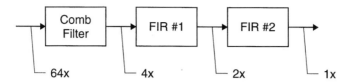

FIGURE 8.28   Typical decimation filter architecture.

point that would alias to dc.[38,39] If the filter order N is one more than the modulator order, then the comb filter will be adequate to attenuate the out-of-band quantization noise to the point where it does not adversely increase the baseband noise after decimation.[40]

Following the comb filter is typically a series of one or more FIR filters. Since the sample rates of these FIR filters are much slower than the oversampled clock rate, each filter output can be computed over many clock cycles. Also, since the output of each filter is decimated, only the samples that will be output need to be calculated. These properties can be exploited to devise computationally efficient structures for decimation filtering.[41]

In the example in Fig. 8.28, the first FIR filter is decimating from 4× to 2× oversampling. Since the output of this filter is still oversampled, the transition region is relatively wide and the attenuation at midband need not be very high. Thus, an economical half-band filter (a filter in which every other coefficient is zero) can be used.[37]

The final FIR filter is by far the most complex. It usually has to have a very sharp transition region, and for strict anti-alias performance, it cannot be a halfband filter. In high-performance sigma-delta modulators, this filter is often in the range of 50 to 200 taps in length. Standard digital filter design techniques can be used to select that tap weights for this filter.[28] Since it is going to be a complex filter anyway, it can also be used to compensate for any frequency drop in the previous filter stages.

The interpolation filter, shown in Fig. 8.2 at the input of the sigma-delta D/A converter, up-samples the input digital words to the oversampling rate. In many ways, this filter is the inverse of a decimation filter, typically comprising a complex up-sampling FIR filter, optionally followed by one or more simple FIR filters, followed by an up-sampling comb filter. The up-sampling operation, without this filter, would produce images of the baseband spectrum at multiples of the baseband frequency. The purpose of the interpolation is to attenuate these images to a tolerable level. What constitutes tolerable is very much a system-dependent criterion. Design techniques for the interpolation filter parallel those of the decimation filter discussed above.

8.5   Circuit Building Blocks

For analog-to-digital conversion, the modulator is implemented primarily in the analog domain as shown in Fig. 8.16. In digital-to-analog conversion, the modulator output if filtered by an analog reconstruction filter as depicted in Fig. 8.2. The basic analog circuit building blocks for these data converters are described in this section. These building blocks include switched-capacitor integrators, the amplifiers that are imbedded in the integrators, comparators, and circuits for sigma-delta based D/A conversion. At the end of this section, the techniques for continuous-time sigma-delta modulation are briefly discussed.

Switched-Capacitor Integrators

Switched-capacitor integration stages are commonly used to perform the signal processing functions of integration and summation required for realization of the discrete time transfer functions $A(z)$ and $F(z)$ in Fig. 8.16. The circuit techniques outlined herein are drawn from a rich literature of switched-capacitor filters[42-45] that is detailed elsewhere in this volume.

Figure 8.29 is a typical integrator stage for the case of single bit feedback,[11,20] and is designed to perform the discrete time computation

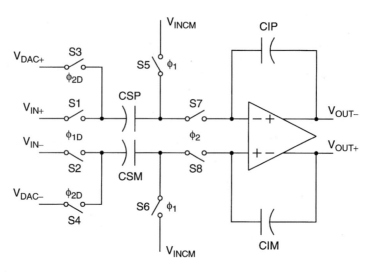

FIGURE 8.29 Typical integrator stage.

$$V_{OUT}(z) = K \frac{z^{-1}}{1 - z^{-1}} \left(V_{IN}(z) - V_{DAC}(z) \right) \qquad (8.33)$$

independent of the parasitic capacitances associated with the capacitive devices shown. The curved line in the capacitor symbol is the device terminal with which the preponderance of the parasitic capacitance is associated. For example, this will be the bottom plate of a stacked planar capacitance structure, where the parasitic capacitance is that between the bottom plate and the IC substrate. The circuit's precision stems from the conservation of charge at the two input nodes of the operational amplifier, and the cyclic return of the potential at those nodes to constant voltages. More details may be found in the chapter on switched capacitor filters (Chapter 12).

Fully differential circuits will be shown here, as these are almost universally preferred over single-ended circuits in monolithic implementations owing to their greatly improved power supply rejection, MOS switch feedthrough rejection, and suppression of even-order non-linearities. The switches shown in Fig. 8.29 are generally full CMOS switches, as detailed in Fig. 8.30. However, integrators with very low power supply voltages may necessitate the use of only one polarity of switch device, possibly with a switch gate voltage-boosting arrangement.[46] Sampling capacitors CSP and CSM are designed with the same capacitance C_S, and the effect of slight fabrication mismatches between the two will be mitigated by the common-mode rejection of the amplifier. Similarly, integration capacitors CIP and CIM are designed to be identical with capacitance C_I.

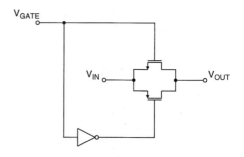

FIGURE 8.30 Full CMOS switch.

The discrete-time signal to be integrated is applied between the input terminals V_{IN+} and V_{IN-}, and the output is taken between V_{OUT+} and V_{OUT-}. V_{INCM} is the common mode input voltage required by the amplifier. The single-bit DAC feedback voltage is applied between V_{DAC+} and V_{DAC-}. The stage must be clocked by two non-overlapping signals, ϕ_1 and ϕ_2. During the ϕ_1 phase, the differential input voltage is sampled on the bottom plates of CSP and CSM, while their top plates are held at the amplifier common-mode input level. During this phase, the amplifier summing nodes are isolated from the capacitor network, and the amplifier output will remain constant at its previously integrated value. During the ϕ_2 phase, the bottom plates of the sampling capacitors CSP and CSM experience a differential potential shift of $(V_{DAC} - V_{IN})$, while the top plates are routed into the amplifier summing nodes. By forcing its differential input voltage to a small level, the amplifier will effect a transfer of a charge of $C_S(V_{IN} - V_{DAC})$ to the integration capacitors, and therefore the differential output voltage will shift to a new value by an increment of $(C_S/C_I)(V_{IN} - V_{DAC})$. Since this output voltage shift will accumulate from cycle to cycle, the discrete-time transfer function will be that of Eq. 8.33, with

$$K = \frac{C_S}{C_I} \qquad (8.34)$$

Over several cycles of initial operation, the amplifier input terminals will be driven to the common-mode level that is precharged onto the top plates of the sampling capacitors.

In order to suppress any signal-dependent clock feedthrough from the switches, it is helpful to slightly delay the clock phases that switch variable signal voltages with respect to the phases that switch current into constant potentials. The channel charge in each turned-on switch device can potentially dissipate onto the sampling capacitors when the switches are turned off, producing an error in the sampled charge. This channel charge is dependent on the difference between the switch gate-to-source voltage and its threshold voltage; and as the source voltage varies with signal voltage, the clock feedthrough charge will vary with the signal. By first turning off the switches that see constant potentials at the end of each cycle, and thus floating the sampling capacitor, the only clock feedthrough is a charge that is to the first order independent of signal level, and results only in a common-mode shift that is suppressed by the amplifier. This acts to reduce the non-linearity of the integrator and the harmonic distortion generated by the modulator.

The timing for the delayed and undelayed clocks is illustrated in Fig. 8.31, where the clock phases ϕ_{1D} and ϕ_{2D} represent phases that are slightly delayed versions of ϕ_1 and ϕ_2, respectively. The delayed clocks drive the switches that are subject to full-signal voltage swings, the analog and reference voltage inputs, as shown in Fig. 8.29. The undelayed clocks drive the switches associated with the amplifier summing node and common-mode input bias voltage, which will always be driven to the same potential by the end of each clock cycle. A typical clock generator circuit to produce these phase relationships is shown in Fig. 8.32. The delay time Δt is generated by the propagation delay through two CMOS inverters.

Other more complex integration circuits are used in some sigma-delta implementations, for example, to suppress errors due to limited amplifier gain[47,48] or to effectively double the sampling rate of the integrators.[49,50] For the modulator structures discussed in Section 8.3 that are more elaborate than a second-order loop, more complex switched-capacitor filtering is required. These may still, however, be designed with the same basic integrator architecture as in Fig. 8.29, but with extra sampling capacitors feeding the amplifier summing node to implement additional signal paths.[26,33,51] Consult Chapter 12 in this volume on switched-capacitor filtering for more information.

Operational Amplifiers

Embedded in the switched-capacitor integrator shown in Fig. 8.29 is an operational amplifier. There are three major types of operational amplifiers typically used in switched-capacitor integrators:[52] the folded cascode amplifier,[42] shown in Fig. 8.33; the two-stage amplifier,[43] shown in Fig. 8.34; and the class AB amplifier,[45] shown in Fig. 8.35.

When the available supply voltage is high enough to permit stacking of cascode devices to develop high gain, a *folded cascode amplifier* is commonly used. A typical topology is shown in Fig. 8.33. The input devices are PMOS, since most IC processes feature PMOS devices that exhibit lower $1/f$ noise than their NMOS counterparts.[53] The input differential pair M1 and M2 is biased with the drain current of M3. FETs M5-M8 function as current sources, and M9-M12 form cascode devices that boost the output

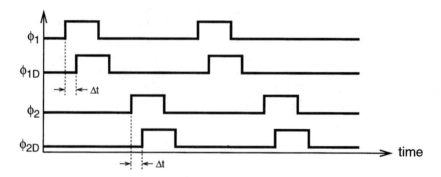

FIGURE 8.31   Delayed clock timing.

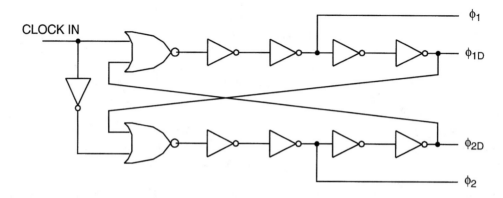

FIGURE 8.32   Non-overlapping clock generator with delayed clocks.

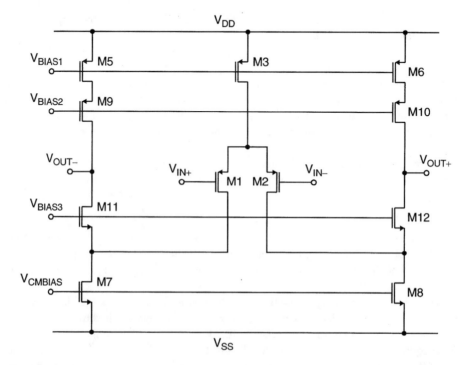

FIGURE 8.33   Folded cascode amplifier.

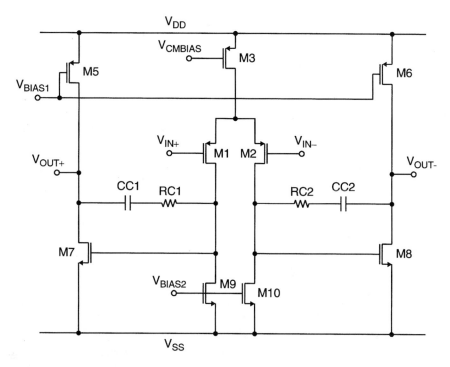

FIGURE 8.34 Two-stage amplifier.

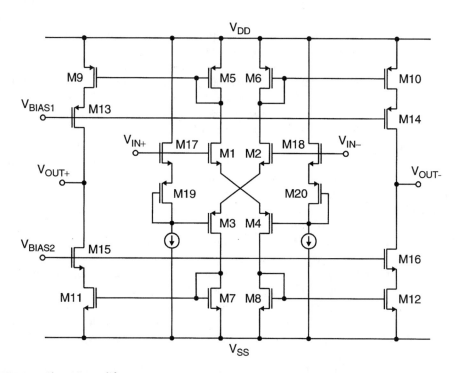

FIGURE 8.35 Class AB amplifier.

impedance. The amplifier is compensated for stability in the integrator feedback loop by the dominant pole that is formed at its output node with the high output impedance and the load capacitance. In an integrator stage, the amplifier will be loaded with the load capacitance of the following stage sampling capacitance as well as its own integration capacitance. The non-dominant pole at the drains of M1 and M2 limit the unity-gain frequency, which can be quite high.

When the power supply voltage is limited and cascode devices cannot be stacked and still preserve adequate signal swing, a *two-stage amplifier* is a common alternative to the folded cascode amplifier. As shown in Fig. 8.34, the input differential pair of M1 and M2 now feed the active load current sources of M9 and M10 to form the first stage. The second stage comprises common-source amplifiers M7 and M8, loaded with current sources M5 and M6. Due to the presence of two poles from the two stages of roughly comparable frequencies, compensation is generally achieved with a pole-splitting RC local feedback network as shown.[52] Often, the resistors RC1 and RC2 are actually implemented as NMOS devices biased into their ohmic region by tying their gates to V_{DD}. In this arrangement, the effective resistance of RC1 and RC2 will approximately track any drift in mobility of M7 and M8 over temperature and processing variations, preserving the compensated phase margin. For a given process, the bandwidth of a two-stage amplifier is less than what can be achieved than by a folded cascode design; but because the two-stage has no stacked cascode devices, the signal swing is higher.

In the case of modulators with higher clock speeds, both folded cascode and two-stage amplifiers may have unacceptably long settling times; in these amplifiers, the maximum slewing current that can be applied to charge or discharge the load capacitance is limited by fixed current sources. This slewing limitation can be overcome by a *class AB amplifier* topology that can supply a variable amount of output current and is capable of providing a large pulse of current early in the settling cycle when the differential input error voltage is high. A typical class AB amplifier topology is shown in Fig. 8.35. The input differential pair from the folded cascode and two-stage designs is replaced by M1 through M4, and their drain currents are mirrored to the output current sources M9–M12 by diode-connected devices M5–M8. Cascode devices M13–M16 enhance the output impedance and gain. As with the folded cascode design, frequency compensation is accomplished by a dominant pole at the output node. The input voltage is fed directly to the NMOS input devices and to the PMOS input devices through the level-shifting source follower and diode combination M17–M20. This establishes the quiescent bias current through the input network M1–M4, and therefore through the output devices as well.

In each of the three amplifier topologies discussed above, there is either one or a set of two matched current sources driving both differential outputs. These current sources are controlled by a gate bias line labeled V_{CMBIAS}. The current output of these devices will determine the common-mode output voltage of the amplifier independent, to the first order, of the amplified differential signal. The appropriate potential for V_{CMBIAS} is determined by a feedback loop that is only operable in the common mode and is separate from the differential feedback instrumental in the charge integration process.

Since a discrete time modulator is, by its nature, clocked periodically, a natural choice for the implementation of this common-mode feedback loop is the switched-capacitor network of Fig. 8.36.[44,45] Capacitors CCM1 and CCM2 act as a voltage divider for transient voltages that derives the average, or common-mode, voltage of the amplifier output terminals. This applies corrective negative feedback transients to the V_{CMBIAS} node to stabilize the feedback loop during each clock period while the amplifier is differentially settling.

A dc bias is then maintained on CCM1 and CCM2 by the switched-capacitor network on the left side of the figure. This will slowly transfer the charge necessary to establish and maintain a dc level shift that makes up the difference between the common-mode level desired at the amplifier output terminals (V_{CMDES}) and the approximate gate bias required by the common mode current devices ($V_{BAPPROX}$). The former is usually set at mid-supply by a voltage divider, and the latter can be derived from a matched diode-connected device. Since the clocking of this switching network is done synchronously to the amplifier integrator clocking, no charge injection will occur during the sensitive settling process of the amplifier. In order to minimize the charge injection at the clock transitions, capacitors CS1 and CS2 are usually made very small, and therefore dozens of clock cycles may be required for the common-mode bias to settle and the modulator to become operable.

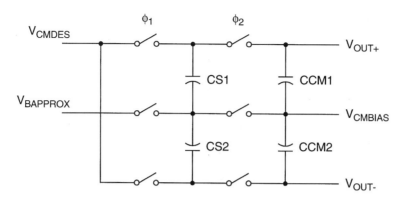

FIGURE 8.36 Switched-capacitor common-mode feedback.

Comparators

The noise shaping mechanism of the modulator feedback loop allows the loop behavior to be tolerant of large errors in circuit behavior at locations closer to the output end of the network. Modulators are generously tolerant of large offset errors in the comparators used in the A/D converter forming the feedback path. For this reason, almost all modulators use simple regenerative latches as comparators. No preamp is generally needed, as the small error from clock kickback can easily be tolerated. Simulations show that offset errors that are even as large as 10% of the reference level will not degrade modulator performance significantly.

The circuit of Fig. 8.37 is typical.[54] This is essentially a latch composed of two cross-connected CMOS inverters, M1–M4. Switch devices M5–M8 will disconnect this network when the clock input is low, and throw the network into a regenerative mode with the rising edge of the clock. The state in which the regeneration will settle may be steered by the relative strengths of the bias current output by devices M9 and M10, which in turn depend on the differential input voltage.

Complete Modulator

Figure 8.38 illustrates a complete second-order, single-bit feedback modulator assembled from the components discussed above.[11] The discrete time integrator gain factors that are derived in Sections 8.2 and 8.3 are realized by appropriate ratios between the integration and sampling capacitors in each stage. Since the single-bit feedback DAC is only responsible for generating two output levels, it may be implemented by simply switching an applied differential reference voltage V_{REF+} to V_{REF-} in a direct or reversed sense to the sampling capacitor bottom plates during the amplifier integration phase, ϕ_2.

D/A Circuits

For the DAC system shown in Fig. 8.2, the oversampled bit stream is generated by straightforward digital implementations of the modulator signal flow graphs discussed in Section 8.2. The remaining analog components are the low-resolution DAC block and the reconstruction filter. Integrated sigma-delta D/A implementations often employ two-level quantization, and the DAC block may either be designed as charge-based[55] or current-based.[56] Multi-level DAC approaches are also used, but for harmonic content less than about –60 dB below the reference, some form of dynamic element matching must be added, as discussed in Section 8.6.

The charge-based approach for sigma-delta D/A conversion is illustrated in Fig. 8.39, which is similar to the switched-capacitor integrator of Fig. 8.29, but without an analog signal input. As in Fig. 8.38, V_{DAC} may be either polarity of V_{REF} according to the bit value to be converted. Figure 8.40 shows a typical topology for current-based converters. In both cases, the leaky integration function around the amplifier contributes to the first pole of the reconstruction filtering. An efficient combination of the current-based approach and a digital delay line realizing an FIR reconstruction filter is also possible.[57]

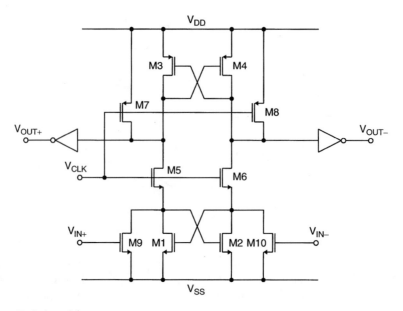

FIGURE 8.37 Typical modulator comparator.

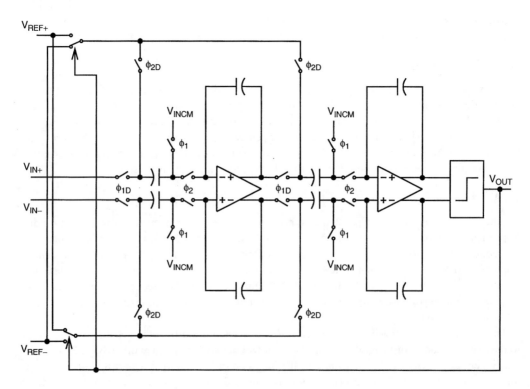

FIGURE 8.38 Complete second-order sigma-delta modulator.

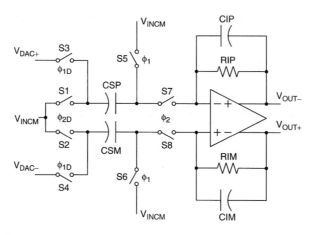

FIGURE 8.39   Charge-based DAC.

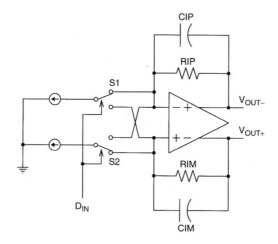

FIGURE 8.40   Current-based DAC.

Additional reconstruction filtering beyond that provided by in the DAC may also be necessary. This is accomplished using the appropriate analog sampled-data filtering techniques described in Chapter 12 of this section.

Continuous-Time Modulators

In general, the amplifiers contained in the switched-capacitor integrators in a sampled-data sigma-delta data converter dissipate the majority of the analog circuit power. Since the integrator sections must settle accurately within each clock period at the oversampled rate, the amplifiers must often be designed with a unity-gain frequency much higher than the oversampled rate; typical unity-gain frequencies are in the hundreds of MHz.

In applications in which dissipating the lowest possible power is important, sigma-delta modulators may also be implemented using continuous-time integrators. In these continuous-time modulators, the analog signal is not sampled until the quantizer at the back of the modulator loop.[58] Owing to the typical means employed for the DAC feedback, continuous-time modulators tend to be more sensitive to sampling clock jitter, but the influences of any aliasing distortion and non-linearity at the sampler take place late in the loop where noise shaping is steepest, and as a consequence the anti-aliasing filter of Fig. 8.1 may often be omitted.[59] The power advantage comes from the relaxed speed requirement of the

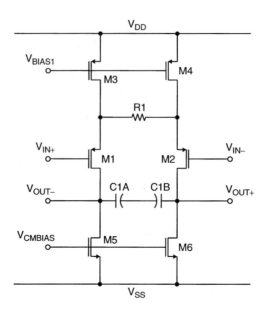

FIGURE 8.41   Gm-C integrator.

integrator stages, which now need only have unity-gain frequencies on the order of the oversampled clock frequency.

Instead of switched-capacitor discrete-time integrators, the continuous-time modulators generally use active Gm-C integrators. Circuits like that shown in Fig. 8.41 are typical.[59] The input differential pair M1 and M2 is degenerated by source resistance R1 to improve linearity. The output analog voltage is developed across capacitor C1, which may be split as shown to place the bottom plate parasitic capacitance at a common-mode node. As the integrator is now unclocked, continuous-time common-mode feedback must be used, as discussed in the literature for continuous time filtering.[60]

8.6   Practical Design Issues

As with any design involving analog components, there are a number of circuit limitations and tradeoffs in sigma-delta data converter design. The design considerations discussed in this section include kT/C noise, integrator scaling, amplifier gain, and sampling non-linearity. Also discussed in this section are the techniques of integrator reset and multi-level feedback.

kT/C Noise

In switched-capacitor-based modulators, one fundamental non-ideality associated with using a MOS device to sample a voltage on a capacitor is the presence of a random variation of the sampled voltage after the MOS switch opens.[61-63] This random component has a Gaussian distribution with a variance of kT/C, where k is Boltzmann's constant, C is the capacitance, and T is the absolute temperature. The variation stems from thermal noise in the resistance of the MOS channel as it is opening. The noise voltage has a mean power of $4kTRB$, where R is the channel resistance and B is the bandwidth. It is low-pass filtered by its characteristic resistance and the sampling capacitor to an equivalent noise bandwidth of $1/RC$. The total integrated variance thus will be kT/C, independent of the resistance of the switch.

If, in the process of developing the integrated signal, a sampling operation on n capacitors is used, then, since we assume Gaussian noise distribution, the variance of the eventual integrated value will be nkT/C. For the case of a fully differential integrator, where a differential signal is sampled onto two sampling capacitors and then transferred to two integration capacitors, n is 4. This effect, along with the

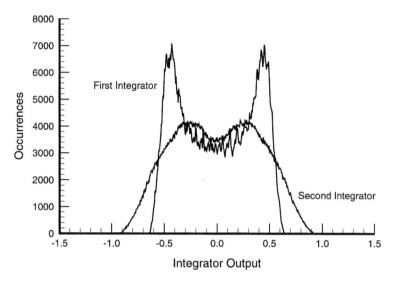

FIGURE 8.42   Integrator output distribution for an eight-level modulator.

input referred noise of the amplifier, will limit the achievable noise floor of the modulator. The first stage sampling capacitors must be sized so as to limit this noise contribution to an acceptable level. From this starting point, and the capacitive ratios required for realizing the various integrator gains, the remaining capacitor sizes can be determined. The modulator will be much less sensitive to kT/C noise generated in integrators past the first, and the capacitors in these integrators can be made considerably smaller.

Integrator Gain Scaling

The integration stages in Section 8.2 were discussed as ideal elements, capable of developing any real output voltage. In practice, the output voltage of real integrators is limited to at most the supply voltage of the embedded amplifier. To ensure that this limitation does not adversely affect the modulator performance, a survey of the likely limit of integrator output voltages must be made for a given value of the DAC reference voltage. The modulator may be simulated over a large number of samples with a representative sinusoidal input, and a histogram of all encountered output voltages tabulated. These histograms may be expected to scale linearly with the reference voltage level. In general, this statistical survey will show that a modulator designed to realize the integrator gain constants in the ideal topologies of Sections 8.2 and 8.3 will have different ranges of expected output voltages from each of its integrators. For example, Figs. 8.42 and 8.43 show the simulated output voltages at the two integrators in a second-order modulator with eight-level and two-level feedback, respectively. Since the largest value possible of reference level will generally mean the best available signal-to-noise ratio for a given circuit power consumption, the integrator gain constants may be adjusted from their straightforward values so that the overall modulator transfer function remains the same, but the output voltages are scaled so that no integrator limits the signal swing markedly before the other.[11] Figures 8.44 and 8.45 illustrate the properly scaled second-order modulator examples.

Amplifier Gain

Another mechanism by which the actual characteristic of the integrator circuits fall short of the ideal is the limitation of finite amplifier gain. A study of many simulations of modulators with various amplifier gains[11] has shown that a modulator needs amplifiers with gains about numerically equal to the decimation ratio of the filter that follows it in order to avoid significant noise shaping errors. At least this is the result with perfectly linear amplifiers, and, in practice, amplifier gains often need to be at least 10 times this

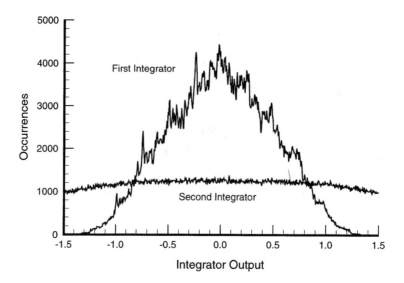

FIGURE 8.43   Integrator output distribution for a two-level modulator.

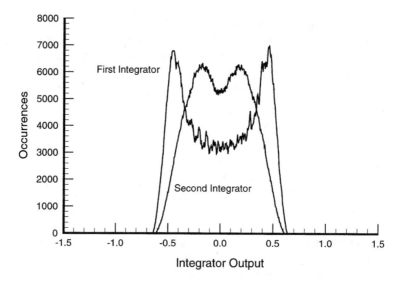

FIGURE 8.44   Integrator output distribution for a scaled eight-level modulator.

high to avoid distortion in the integrator characteristic due to the non-linearity of the amplifier gain characteristic.

One approach used when the simple circuits of Section 8.5 (Operational Amplifiers) do not develop enough gain in a given process is the regulated cascode gain enhancement.[64,65,68] Figure 8.46 illustrates a typical circuit topology. This subcircuit may be substituted for the output common-source amplifier stages in the amplifiers of Figs. 8.33 to 8.35 if the power supply voltage can accommodate its somewhat increased requirement for headroom.

Sampling-Non-linearity and Reference Corruption

The sigma-delta modulator is remarkably tolerant of most circuit non-idealities past the input sampling network. However, the linearity of the sampling process at the very first input sampling capacitor will be the upper bound for the linearity for the entire modulator. Care must be exercised to ensure the

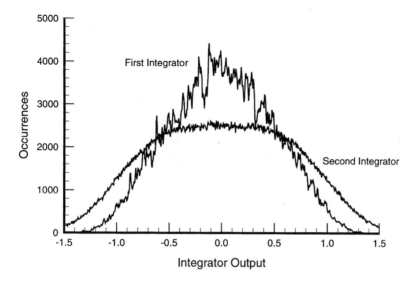

FIGURE 8.45 Integrator output distribution for a scaled two-level modulator.

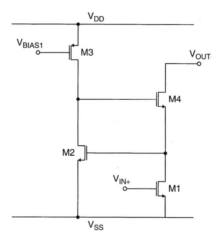

FIGURE 8.46 Regulated cascode topology.

switches are sufficiently large so that the sampled voltage will be completely settled through their non-linear resistance, but not so large that any residual signal dependent clock feedthrough is significant.

Another susceptibility of modulators is to non-linear corruption of the reference voltage. If the digital bit stream output, through a parasitic feedback path either on- or off-chip, can affect the reference voltage sampled during clock phase ϕ_2 in Fig. 8.29, then there will be a term in the output signal dependent on the square of the input voltage. This will distort the noise shaping properties of the modulator and generate second harmonic distortion, even with fully differential circuitry. This is illustrated in the spectrum in Fig. 8.47, which is the output of a modulator having the same conditions as Fig. 8.14, except that a parasitic feedback path is assumed that would change the reference voltage by 1% for the "1" output bits on the previous cycle, relative to its value with "0" output bits. As can be seen by comparison with Fig. 8.14, that the ability of the modulator to shift quantization noise out of the baseband has been greatly compromised, and a prominent second harmonic has been generated. Care must be taken in chip and printed circuit board application design so that the reference voltage remains isolated from the signals carrying the output bit stream.

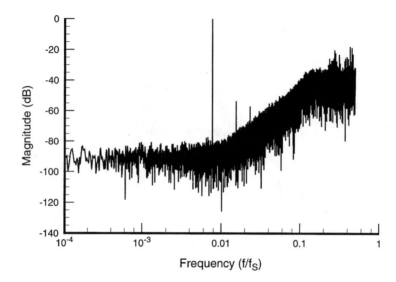

FIGURE 8.47 Output spectrum with reference corruption.

Fully differential circuitry is almost universally employed in integrated VLSI modulators to reduce sampling non-linearity and reference contamination. Even-order non-linearities and common-mode switch feedthrough are cancelled with fully differential circuits, and power supply rejection is greatly improved, leading to more isolated reference potentials. For high-precision modulators, the integrator topology is often changed from that of Fig. 8.29 to Fig. 8.48.[26,51] The input signal and the DAC output voltage are sampled independently during phase ϕ_1, and then both are discharged together into the summing node during ϕ_2. At the expense of additional area for capacitors and higher kT/C noise, this

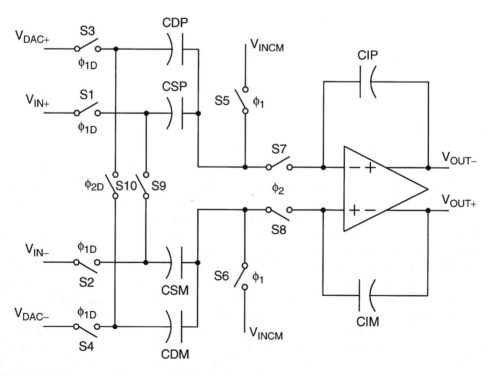

FIGURE 8.48 Integrator with separate DAC feedback capacitor.

arrangement ensures that the same charge is drawn from the reference supply onto the DAC sampling capacitors CDP and CDM and then discharged into the summing node each cycle. Thus, a potentially undesirable mechanism for reference supply loading that is dependent on the output bit history is eliminated.[66]

High-Order Integrator Reset

Although careful design of the loop filter for higher-order modulators, as discussed above in Section 8.3 (High-Order Modulators), will yield a generally stable design, their stability cannot be mathematically guaranteed, as in the case of second-order loops. To protect against the highly undesirable state of low frequency limit cycle oscillations due to an occasional, but improbable, input overload condition, some form of forced integrator reset is sometimes used.[26,51] Generally, these count the consecutive '1' or '0' bits out of the modulator, and close a resetting switch to discharge integration capacitors for a short time if the modulator generates a longer consecutive sequence than normal operation allows. This will naturally interrupt the operation of the modulator, but will only be triggered in the case of pathological input patterns for which linear operation would not necessarily be expected.

Another approach to a stability safety mechanism for higher-order loops is to arrange the scaling of the integrator gains so that they clip against their maximum output voltage swings in a prescribed sequence as the input level rises. The sequence is designed to gradually lower the effective order of the modulator[59] and return operation to a stable mechanism.

Multi-level Feedback

Expanding the second-order modulator to more than two-level feedback can be accomplished by the circuit in Fig. 8.49. For K-level feedback, $K - 1$ latch comparators are arranged in a flash structure as

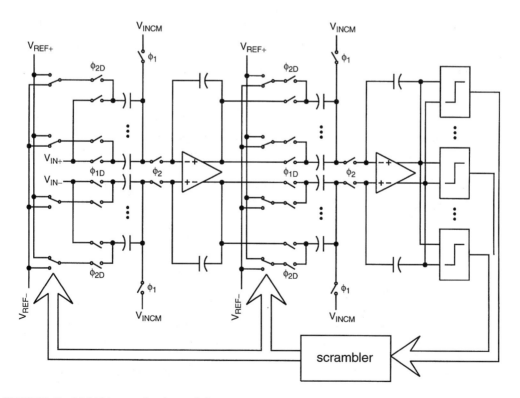

FIGURE 8.49 Multi-bit second-order modulator.

shown on the right. There must be a different offset voltage designed into each of the comparators so that they detect when each quantization level is crossed by the second integrator output. This can be implemented by a resistor string[54,67] or an input capacitive sampling network.[68] The output of the $K - 1$ comparators is then a thermometer code representation of the integrator output. This may be translated into binary for the modulator output, but the raw thermometer code is the most convenient to use as a set of feedback signals. They each will drive a switch that will select either V_{REF+} or V_{REF-} to be used as the bottom plate potential for the integrator sampling capacitors. If all sampling capacitors are of equal value, the net charge being integrated will have a term that varies linearly with the quantization level. Each comparator output drives two switches, so there are $2(K - 1)$ switches and capacitors in the sampling array.

In any practical integrated structure, even if careful common-centroid layout techniques are used, the precision with which the various capacitors forming the sampling array will be matched is typically limited to 0.1 or 0.2%. As discussed in Section 8.2, this will limit the harmonic distortion that is inherent in the modulator to about –60 dB or higher. However, by varying the assignment of which sampling switch is driven by which comparator output dynamically as the modulator is clocked, much of this distortion may be traded off for white or frequency-shaped noise at the modulator output. This technique is referred to as *dynamic element matching*.

One simple way of implementing dynamic element matching is indicated in Fig. 8.49 with the block labeled "scrambler." This block typically comprises an array of switches that provide a large number of permutations in the way the comparator output lines can be mapped onto sampling switch lines. A multiple-stage butterfly network is one relatively simple approach.[69] The mapping permutation is then changed every modulator clock cycle. Assuming each comparator output will, over a time period less than the final baseband period, end up mapped to all the sampling capacitors, all of the capacitor mismatches will be averaged out. The energy that would be found in input signal harmonics without scrambling will be spread out in some fashion into a higher modulator noise output.

There have been various algorithms published in the literature on how best to control the sequence of mapping perturbations. A simple random sequence will render the mismatch into white noise, increasing the baseband output noise floor.[68,69] More complex sequences relying on a knowledge of the history of quantizer levels are capable of coloring the spread noise so that much of it appears outside the baseband and is suppressed by the following decimation filtering.[70-73]

8.7   Summary

In this chapter, a brief overview of sigma-delta data converters has been presented. Sigma-delta data conversion is a technique that effectively trades speed for resolution. High-linearity data conversion can be accomplished in modern IC processes without expensive device trimming or calibration. For a far more detailed treatment of this topic, refer to Norsworthy, Schreier, and Temes.[2] For a compilation of some of the seminal papers that helped establish sigma-delta modulation as a mainstream technique, refer to Candy and Temes.[1]

References

1. J. Candy and G. Temes, *Oversampling Delta-Sigma Data Converters*, IEEE Press, 1992.
2. S. Norsworthy, R. Schreier, and G. Temes, *Delta-Sigma Data Converters: Theory, Design, and Simulation*, IEEE Press, 1996.
3. C. Cutler, Transmission systems employing quantization, U.S. Patent 2,927,962, Mar. 8, 1960.
4. H. Spang III and P. Schultheiss, Reduction of quantizing noise by use of feedback, *IRE Trans. Communication Systems*, pp. 373–380, Dec. 1962.
5. H. Inose and Y. Yasuda, A unity bit coding method by negative feedback, *Proc. IEEE*, vol. 51, pp. 1524–1535, Nov. 1963.

6. S. Tewksbury and R. Hallock, Oversampled, linear predictive and noise-shaping coders of order N>1, *IEEE Trans. Circ. Syst.*, vol. CAS-25, pp. 436–447, July 1978.

7. J. Candy, A use of limit cycle oscillations to obtain robust analog-to-digital converters, *IEEE Trans. Communications*, vol. COM-22, pp. 298–305, Mar. 1974.

8. H. Fiedler and B. Hoefflinger, A CMOS pulse density modulator for high-resolution A/D converters, *IEEE J. Solid-State Circuits*, vol. SC-19, pp. 995–996, Dec. 1984.

9. B. Leung, R. Neff, P. Gray, and R. Brodersen, Area-efficient multichannel oversampled PCM voice-band coder, *IEEE J. Solid-State Circuits*, vol. SC-23, pp. 1351–1357, Dec. 1988.

10. J. Candy, A use of double integration in sigma-delta modulation, *IEEE Trans. Communications*, vol. COM-33, pp. 249–258, Mar. 1985.

11. B. Boser and B. Wooley, The design of sigma-delta modulation analog-to-digital converters, *IEEE J. Solid-State Circuits*, vol. 23, pp. 1298–1308, Dec. 1988.

12. V. Friedman, D. Brinthaupt, D. Chen, T. Deppa, J. Elward, E. Fields, J. Scott, and T. Viswanathan, A dual-channel voice-band PCM codec using ΣΔ modulation technique, *IEEE J. Solid-State Circuits*, vol. 24, pp. 274–280, Apr. 1989.

13. W. Bennett, Spectra of quantized signals, *Bell System Tech. Journal*, vol. 27, pp. 446–472, 1948.

14. B. Widrow, Statistical analysis of amplitude quantized sampled-data systems, *Trans. AIEE*, vol. 79, pp. 555–568, Jan. 1961.

15. R. Gray, Oversampled sigma-delta modulation, *IEEE Trans. Communications*, vol. COM-35, pp. 481–489, May 1987.

16. J. Candy and O. Benjamin, The structure of quantization noise from sigma-delta modulation, *IEEE Trans. Communications*, vol. COM-29, pp. 1316–1323, Sept. 1981.

17. B. Boser and B. Wooley, Quantization error spectrum of sigma-delta modulators, *1988 IEEE Int. Symp. Circ. Syst.*, pp. 2331–2334, 1988.

18. R. Gray, Quantization noise spectra, *IEEE Trans. Information Theory*, vol. 36, pp. 1220–1244, Nov. 1990.

19. L. Williams and B. Wooley, A third-order sigma-delta modulator with extended dynamic range, *IEEE J. Solid-State Circuits*, vol. 29, Mar. 1994.

20. B. Brandt, D. Wingard, and B. Wooley, Second-order sigma-delta modulation for digital-audio signal acquisition, *IEEE J. Solid-State Circuits*, vol. 26, pp. 618–627, Apr. 1991.

21. J. Everard, A single-channel PCM codec, *IEEE J. Solid-State Circuits*, vol. SC-14, pp. 25–37, Feb. 1979.

22. M. Hauser, P. Hurst, and R. Brodersen, MOS ADC-filter combination that does not require precision analog components, *ISCC Dig. Tech. Papers*, pp. 80–82, Feb. 1985.

23. S. Norsworthy, Effective dithering of sigma-delta modulators, *Proc. of the 1992 IEEE Int. Symp. Circ. Syst.*, pp. 1304–1307, May 1992.

24. D. Welland, B. Del Signore, E. Swanson, T. Tanaka, K. Hamashita, S. Hara, and K. Takasuka, A stereo 16-bit delta-sigma A/D converter for digital audio, *J. Audio Engineering Society*, vol. 37, pp. 476–486, June 1989.

25. K. Chao, S. Nadeem, W. Lee, and C. Sodini, A higher order topology for interpolative modulators for oversampling A/D converters, *IEEE Trans. Circ. Syst.*, vol. 37, pp. 309–318, Mar. 1990.

26. R. Adams, P. Ferguson, A. Ganesan, S. Vincelette, A. Volpe, and R. Libert, Theory and practical implementation of a fifth-order sigma-delta A/D converter, *J. Audio Eng. Soc.*, vol. 39, pp. 515–528, July/Aug. 1991.

27. R. Schreier, An empirical study of high-order single-bit delta-sigma modulators, *IEEE Trans. Circ. Syst. II. Analog and Digital Signal Processing*, vol. 40, Aug. 1993.

28. A. Oppenheim and R. Schafer, *Discrete-Time Signal Processing*, Prentice-Hall, Englewood Cliffs, NJ, 1989.

29. Y. Matsuya, K. Uchimura, A. Iwata, T. Kobayashi, M. Ishikawa, and T. Yoshitome, A 16-bit over-sampling A-to-D conversion technology using triple-integration noise shaping, *IEEE J. Solid-State Circuits*, vol. SC-22, pp. 921–929, Dec. 1987.

30. L. Longo and M. Copeland, A 13 bit ISDN-band oversampling ADC using two-stage third order noise shaping, *IEEE 1988 Custom Integrated Circuits Conference,* pp. 21.2.1–4, 1988.
31. L. Williams and B. Wooley, Third-order cascaded sigma-delta modulators, *IEEE Trans. Circ. Syst.,* vol. 38, pp. 489–498, May 1991.
32. P.-W. Wong and R. Gray, Two stage sigma-delta modulation, *IEEE Trans. Acoustics, Speech, and Signal Processing,* vol. 38, pp. 1937–1952, Nov. 1990.
33. L. Longo and B.-R. Horng, A 15b 30kHz bandpass sigma-delta modulators, *1993 IEEE Intl. Solid-State Circuits Conf.,* pp. 226–227, Feb. 1993.
34. R. Schreier and W. M. Snelgrove, Decimation for bandpass sigma-delta analog-to-digital conversion, *1990 IEEE Intl. Symp. Circ. Syst.,* vol. 3, pp. 1801–1804, May 1990.
35. R. Gregorian and G. Temes, *Analog MOS Integrated Circuits for Signal Processing,* John Wiley & Sons, 1986.
36. D. Goodman and M. Carey, Nine digital filters for decimation and interpolation, *IEEE Trans. Acoustics, Speech, and Signal Processing,* vol. ASSP-25, pp. 126–126, Apr. 1977.
37. R. Crochiere and L. Rabiner, *Multirate Digital Signal Processing,* Prentice-Hall, Englewood Cliffs, NJ, 1983.
38. E. Hogenauer, An economical class of digital filters for decimation and interpolation, *IEEE Trans. Acoustics, Speech, and Signal Processing,* vol. ASSP-29, pp. 155–162, Apr. 1981.
39. S. Chu and C. Burrus, Multirate filters designs using comb filters, *IEEE Trans. Circ. Syst.,* vol. CAS-31, pp. 913–924, Nov. 1984.
40. J. Candy, Decimation for sigma delta modulation, *IEEE Trans. Communications,* vol. COM-34, pp. 72–76, Jan. 1986.
41. B. Brandt and B. Wooley, A low-power, area-efficient digital filter for decimation and interpolation, *IEEE J. Solid-State Circuits,* vol. 29, pp. 679–687, June 1994.
42. T. Choi, et al., High-frequency CMOS switched-capacitor filters for communications applications, *IEEE J. Solid-State Circuits,* vol. 15, pp. 929–938, Dec. 1980.
43. W. C. Black, et al., A high-performance low power CMOS channel filter, *IEEE J. Solid-State Circuits,* vol. 15, pp. 929–938, Dec. 1980.
44. D. Senderowitz, et al., A family of differential NMOS analog circuits for a PCM codec filter chip, *IEEE J. Solid-State Circuits,* vol. 17, pp. 1014–1023, Dec. 1982.
45. R. Castello and P. R. Gray, A high-performance micropower switched-capacitor filter, *IEEE J. Solid-State Circuits,* vol. 20, pp. 1122–1132, Dec. 1985.
46. T. B. Cho and P. R. Gray, A 10b, 20 Msample/s, 35 mW pipeline A/D converter, *IEEE J. Solid-State Circuits,* vol. 30, pp. 166–172, Mar. 1995.
47. K. Nagaraj et al., Switched-capacitor integrator with reduced sensitivity to amplifier gain, *Electronics Letters,* vol. 22, p. 1103, Oct. 1986.
48. K. Huag, G. C. Temes, and L. Martin, Improved offset-compensation scheme for SC circuits, *1984 IEEE Intl. Symp. Circ. Syst.,* pp. 1054–1057, 1984.
49. P. J. Hurst and W. J. McIntyre, Double sampling on switched-capacitor delta-sigma A/D converters, *1990 IEEE Symp. Circ. Syst.,* pp. 902–905, May 1990.
50. D. Senderowitz et al., Low voltage double-sampled sigma-delta converters, *IEEE J. Solid-State Circuits,* vol. 32, pp. 1907–1919, Dec. 1997.
51. P. Furguson et al., An 18b, 20kHz dual sigma-delta A/D converter, *1991 IEEE Intl. Solid-State Circuits Conf.,* pp. 68–69, Feb. 1991.
52. P. R. Gray and R. G. Meyer, MOS operational amplifier design — A tutorial overview, *IEEE J. Solid-State Circuits,* vol. 17, pp. 969–981, Dec. 1982.
53. A. Abidi, C. Viswanathan, J. Wu, and J. Wikstrom, Flicker noise in CMOS: a unified model for VLSI processes, *1987 Symp. VLSI Technology,* pp. 85–86, May 1987.
54. A. Yukawa, A CMOS 18-bit high speed A/D converter IC, *IEEE J. Solid-State Circuits,* vol. 20, pp. 775–779, June 1985.

55. B. Kup, E. Dijkmans, P. Naus, and J. Sneep, A bit-stream digital-to-analog converter with 18-b resolution, *IEEE J. Solid-State Circuits,* vol. 26, pp. 1757–1763, Dec. 1991.
56. R. Adams, K. Q. Nguyen, and K. Sweetland, A 113-dB SNR oversampled DAC with segmented noise-shaped scrambling, *IEEE J. Solid-State Circuits,* vol. 33, pp. 1871–1878, Dec. 1998.
57. D. Su and B. Wooley, A CMOS oversampling D/A converter with a current-mode semidigital reconstruction filter, *IEEE J. Solid-State Circuits,* vol. 28, pp. 1224–1233, Dec. 1993.
58. R. Schreier and B. Zhang, Delta-sigma modulators employing continuous-time circuitry, *IEEE Trans. on Circuits and Systems. I. Fundamental Theory and Applications,* vol. 44, pp. 324–332, Apr. 1996.
59. E. J. van der Zwan and E. C. Dijkmans, A 0.2-mW CMOS sigma-delta modulator for speech coding with 80dB dynamic range, *IEEE J. Solid-State Circuits,* vol. 31, pp. 1873–1880, Dec. 1996.
60. Y.P. Tsividis, Integrated continuous-time filter design — an overview, *IEEE J. Solid-State Circuits,* vol. 29, pp. 166–176, Mar. 1994.
61. K. C. Hsieh, Noise limitations in switched-capacitor filters, Ph.D. dissertation, Univ. California, Berkeley, Dec. 1981.
62. C. Gobet and A. Knob, Noise analysis of switched-capacitor networks, *1981 Intl. Symp. Circ. Syst.,* Apr. 1981.
63. C. Gobet and A. Knob, Noise generated in switched-capacitor networks, *Electronics Letters,* vol. 19, no. 19, 1980.
64. E. Säckinger and W. Guggenbühl, A high-swing, high-impedance MOS cascode circuit, *IEEE J. Solid-State Circuits,* vol. SC-25, pp. 289–298, Feb. 1990.
65. K. Bult and G. J. G. M. Geelen, A fast-settling CMOS opamp for SC circuits with 90-dB DC gain, *IEEE J. Solid-State Circuits,* vol. SC-25, no. 6, pp. 1379–1384, Dec. 1990.
66. D. Ribner, R. Baertsch, S. Garverick, D. McGrath, J. Krisciunas, and T. Fuji, A third-order multistage sigma-delta modulator with reduced sensitivity to nonidealities, *IEEE J. Solid-State Circuits,* vol. 26, pp. 1764–1774, Dec. 1991.
67. B. Brandt and B. Wooley, A 50-MHz multibit sigma-delta modulator for 12-b 2-MHz A/D conversion, *IEEE J. Solid-State Circuits,* vol. 26, pp. 1746–1756, Dec. 1991.
68. J. Fattaruso et al., Self-calibration techniques for a second-order multibit sigma-delta modulator, *IEEE J. Solid-State Circuits,* vol. 28, pp. 1216–1223, Dec. 1993.
69. L. Carley, A noise-shaping coder topology for 15+ bit converters, *IEEE J. Solid-State Circuits,* vol. 24, pp. 267–273, Apr. 1989.
70. F. Chen and B. Leung, A high resolution multibit sigma-delta modulator with individual level averaging, *IEEE J. Solid-State Circuits,* vol. 30, pp. 453–460, Apr. 1995.
71. T. Kwan, R. Adams, and R. Libert, A stereo multi-bit $\Sigma\Delta$ D/A with asynchronous master-clock interface, *IEEE J. Solid-State Circuits,* vol. 31, pp. 1881–1887, Dec. 1996.
72. B. Leung and S. Sutarja, Multibit Σ-Δ A/D converter incorporating a novel class of dynamic element matching techniques, *IEEE Trans. Circ. Syst. II. Analog and Digital Signal Processing,* vol. 39, pp. 35–51, Jan. 1992.
73. L. Williams III, An audio DAC with 90 dB linearity using MOS to metal-metal charge transfer, *1998 IEEE Intl. Solid-State Circuits Conf.,* pp. 58–59, Feb. 1998.

9

RF Communication Circuits

Michiel Steyaert
Marc Borremans
Johan Janssens
Bram De Muer
Katholieke Universiteit Leuven,
ESAT-MICAS

9.1 Introduction

A few years ago, the world of wireless communications and its applications started to grow rapidly. The main cause for this event was the introduction of digital coding and digital signal processing in wireless communications. This digital revolution is driven by the development of high-performance, low-cost, CMOS technologies that allow for the integration of an enormous numbers of digital functions on a single die. This allows, in turn, for the use of sophisticated modulation schemes, complex demodulation algorithms, and high-quality error detection and correction systems, resulting in high-performance lossless communication channels.

Today, the digital revolution and the high growth of the wireless market also bring many changes to the analog transceiver front-ends. The front-ends are the interface between the antenna and the digital modem of the wireless transceiver. They have to detect very weak signals (µV) which come in at a very high frequency (1 to 2 GHz) and, at the same time, they have to transmit at the same high-frequency high power levels (up to 2 W). This requires high-performance analog circuits, like filters, amplifiers, and mixers which translate the frequency bands between the antenna and the A/D-conversion and digital signal processing. Low cost and a low power consumption are the driving forces and they make the analog front-ends the bottleneck for future RF design. Both low cost and low power are closely linked to the trend toward full integration. An even further level of integration renders significant space, cost, and power reductions. Many different techniques to obtain a higher degree of integration for receivers, transmitters, and synthesizers have been presented over the past years.[1-3] This chapter introduces and analyzes some advantages and disadvantages and their fundamental limitations.

Parallel to the trend to further integration, there is the trend to the integration of RF circuitry in CMOS technologies. The mainstream use for CMOS technologies is the integration of digital circuitry. The use of these CMOS technologies for high-performance analog circuits yields, however, many benefits. The technology is, of course — if used without any special adaptations toward analog design — cheap. This is especially true if one wants to achieve the ultimate goal of full integration: the complete transceiver system on a single chip, with both the analog front-end and the digital demodulator implemented on the same die. This can only be achieved in either a CMOS or a BiCMOS process. BiCMOS has better devices for analog design, but its cost will be higher, not only due to the higher cost per area, but also due to the larger area that will be needed for the digital part. Plain CMOS has the extra advantage that the performance gap between devices in BiCMOS and nMOS devices in deep sub-micron CMOS, and even nMOS devices in the same BiCMOS process, is becoming smaller and smaller due to the much higher investments in the development of CMOS than bipolar. The f_T's of the nMOS devices are getting close to the f_T's of npn devices.

Although some research had been done in the past on the design of RF in CMOS technologies,[4] it is only in the last few years that real attention has been given to its possibilities.[5,6] Today several research groups at universities and in industry are researching this topic.[2,3,7,9] As bipolar devices are inherently better than CMOS devices, RF CMOS is seen by some as a possibility only for low-performance systems, with reduced specification (like ISM),[8,10] feeling that the CMOS processes need adaptations, like substrate etching under inductors.[7] Others feel, however, that the benefits of RF CMOS can be much bigger and that it will be possible to use plain deep sub-micron CMOS for the full integration of transceivers for high-performance applications, like GSM, DECT, and DCS 1800.[2,3] First, this chapter analyzes some trends, limitations, and problems in technologies for high-frequency design. Second, the down-converter topologies and implementation problems are addressed. Third, the design and trends toward fully integrated low-phase noise PLL circuits are discussed. Finally, the design of fully integrated up-converters is studied.

9.2 Technology

Active Devices

Due to the never-ending progress in technology and the requirement to achieve a higher degree of integration for DSP circuits, sub-micron technologies are nowadays considered standard CMOS technologies. The trend is even toward deep sub-micron technologies (e.g., transistor lengths of 0.1 μm. Using the square law relationship for MOS transistors to calculate the f_t of a MOS device no longer holds, due to the high electrical fields. Using a more accurate model, which includes the mobility degradation due to the high electrical fields, results in

$$f_t = \frac{g_m}{2\pi\,C_{gs}}$$

$$= \frac{\mu}{2\pi 2/3L^2}\;\frac{\left(V_{gs}-V_t\right)}{\left(1+2\left(q+\dfrac{\mu}{v_{max}L}\right)\left(V_{gs}-V_t\right)\right)} \qquad (9.1)$$

Hence, by going to deep sub-micron technologies, the square law benefit in L for speed improvement drastically reduces due to the second term in the denominator of Eq. 9.1. Even for very deep sub-micron technologies, the small signal parameter g_m has no square law relationship anymore:

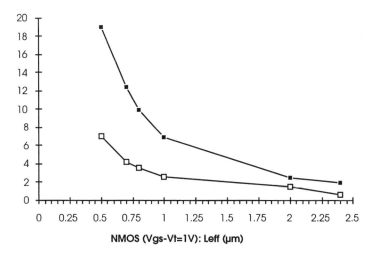

NMOS (Vgs-Vt=1V): Leff (µm)

FIGURE 9.1   Comparison of f_t and f_{max}.

$$g_m = \frac{\mu C_{ox} W \left(V_{gs} - V_t \right)}{L \left[1 + 2 \left(\theta + \dfrac{\mu}{v_{max} L} \right) \left(V_{gs} - V_t \right) \right]} \quad (9.2)$$

with transistor lengths smaller than approximately

$$L < \frac{\mu}{v_{max}} \frac{1}{\dfrac{1}{2 \left(V_{gs} - V_t \right)} - q} \approx 0.12 \, \mu m \quad (9.3)$$

with $\mu/v_{max} = 0.3$, $V_{gs} - V_t = 0.2$ (boundary of strong inversion), and $\theta = 0.06$, the transistor has only the weak inversion and the velocity saturation area. This will result in even higher biasing currents in order to achieve the required g_m and will result in higher distortion and intermodulation components, which will be further discussed in the tradeoff of low-noise amplifier designs.

Furthermore, the speed increase of deep sub-micron technologies is reduced by the parasitic capacitance of the transistor, meaning the gate-drain overlap capacitances and drain-bulk junction capacitances. This can clearly be seen in Fig. 9.1 in the comparison for different technologies of the f_t and the f_{max} defined as the 3-dB point of a diode-connected transistor.[11] The f_{max} is more important because it reflects the speed limitation of a transistor in a practical configuration. As can be seen, the f_t rapidly increases; but for real circuit designs (f_{max}), the speed improvement is only moderate.

Passive Devices

The usability of a process for RF design depends not only on the quality of the active devices, but also, more and more, on the availability of good passive devices. The three passive devices (resistors, capacitors, and inductors) will be discussed.

Low-ohmic resistors are available today in all CMOS technologies and their parasitic capacitance is such that they allow for more than high enough bandwidth (i.e., more than 2 to 3 GHz). A more important passive device is the capacitor. In RF circuits, capacitors can be used for ac-coupling. This allows dc-level shifting between different stages, resulting in a more optimal design of each stage and in the ability to use lower power supply voltages. The quality of a capacitor is mainly determined by the ratio between the capacitance value and the value of the parasitic capacitance to the substrate. Too high a parasitic capacitor loads the transistor stages, thus reducing their bandwidth, and it causes an inherent signal loss due to a capacitive division. Capacitors with ratios lower than 8 are, as a result, difficult to use in RF circuit design as coupling devices.

The third passive device, the inductor, is gaining more and more interest in RF circuit design on silicon. The use of inductors allows for a further reduction of the power supply voltage and for compensation of parasitic capacitances by means of resonance, resulting in higher operating frequencies. The problem is that the conducting silicon substrate under a spiral inductor reduces the quality of the inductor. Losses occur due to capacitive coupling to the substrates, and eddy currents induced in the substrate will also result in losses and in a reduction of the effective inductance value. This problem can be circumvented by using extra processing steps that etch away the substrate under the spiral inductor,[23] having the large disadvantage that it eliminates all the benefits of using a standard CMOS process. It is therefore important that in CMOS, spiral inductors are used without any process changes and that their losses are accurately modeled. In Ref. 12, it is shown that spiral inductors can be accurately modeled and that they can be used in CMOS RF circuit design. As an example, Section 9.4 discusses all the different possibilities for the use of inductors in the design of VCOs. It shows that high-performance VCOs can be integrated with spiral inductors, even on lossy substrates, without requiring any external component.

9.3   The Receiver

Receiver Topologies

The heterodyne or IF receiver is the best known and most frequently used receiver topology. In the IF receiver, the wanted signal is down-converted to a relatively high intermediate frequency. A high quality passive bandpass filter is used to prevent a mirror signal to be folded upon the wanted signal on the IF frequency. Very high performances can be achieved with the IF receiver topology, especially when several IF stages are used (e.g., 900 MHz to 300 MHz, 300 MHz to 70 MHz, 70 MHz to 30 MHz, 30 MHz to 10 MHz). The main disadvantage of the IF receiver is the poor degree of integration that can be achieved as every stage requires going off-chip and requires the use of a discrete bandpass filter. This is both costly (the cost of the discrete filters and the high pin-count for the receiver chip) and power consuming (often the discrete filters have to be driven by a 50-Ω signal source).

The homodyne or zero-IF receiver, introduced as an alternative to the IF receiver, can achieve a much higher degree of integration. The zero-IF receiver uses a direct, quadrature down-conversion of the wanted signal to the baseband. In this case, the wanted signal has itself as mirror signal and sufficient mirror signal suppression can therefore be achieved, even with a limited quadrature accuracy (e.g., 3 degrees phase accuracy and 1-dB amplitude accuracy). Theoretically, there is thus no discrete high-frequency bandpass filter required in the zero-IF receiver, allowing in this way the realization of a fully integrated receiver. The limited performance of the LNA and the mixers reveals, however, that — although not for mirror signal suppression — a high-frequency bandpass filter is still required. The reason why LNAs and mixers require bandpass filtering and how this can be prevented, is explained later.

In the zero-IF receiver, down-conversion can be performed in a single stage (e.g., directly from 900 MHz to the baseband), giving large benefits toward full integration, low cost, and low power consumption.[13] The problem with the zero-IF receiver, however, is its poor performance compared to IF receivers. The zero-IF receiver is intrinsically very sensitive to parasitic baseband signals like dc-offset voltages and crosstalk products caused by RF and LO self-mixing. It is precisely these drawbacks that have kept the zero-IF receiver from being used on a large scale in new wireless applications. The use of the zero-IF receiver has therefore been limited to either low-performance applications like pagers and ISM[10] or as a second stage in a combined IF–zero-IF receiver topology.[14,15] It has, however, been shown that by using dynamic non-linear dc-correction algorithms, implemented in the DSP, the zero-IF topology can be used for high-performance applications like GSM and DECT.[1,16]

In recent years, new receiver topologies, like the quasi-IF receiver[3] and the low-IF receiver[2] have been introduced for use in high-performance applications. The quasi-IF receiver uses a quadrature down-conversion to an IF frequency, followed by a further quadrature down-conversion to the baseband. The channel selection is done with the second local oscillator on the IF frequency, giving the advantage that a fixed-frequency first local oscillator can be used. The disadvantages of the quasi-IF receiver are that, with a limited accuracy of the first quadrature down-converter (e.g., a phase error of 3 degrees), the mirror signal suppression is not good enough and an HF filter that improves the mirror signal suppression is still necessary. A second disadvantage is that a high IF is required in order to obtain a high enough ratio between the IF frequency and the full band of the application. Otherwise, the tunability of the second VCO has to be too large. Unfortunately, a high IF requires a higher power consumption. Moreover, the first stage of mixers cannot be true down-conversion mixers in the sense that they still need to have a relatively high output bandwidth. To conclude, multi-stage topologies inherently require more power.

The low-IF receiver performs a down-conversion from the antenna frequency directly down to — as the name already indicates — a low IF (i.e., in the range a few 100 kHz).[2] Down-conversion is done in quadrature and the mirror signal suppression is performed at low frequency, after down-conversion, in the DSP. The low-IF receiver topology is thus closely related to the zero-IF receiver. It can be fully integrated (it does not require an HF mirror signal suppression filter) and uses a single-stage direct-down-conversion. The difference is that the low-IF does not use baseband operation, resulting in a total immunity to parasitic baseband signals, resolving in this way the main disadvantage of the zero-IF receiver. The drawback is that the mirror signal is different from the wanted signal in the low-IF receiver topology; but by carefully choosing the IF frequency, an adjacent channel with low signal levels can be selected for which the typical mirror signal suppression (i.e., a phase accuracy of 3 degrees) is sufficient.

Full Integration

With newly developed receiver topologies such as the zero-IF receiver and the low-IF receiver, the need disappears for the use of external filters that suppress the mirror signal (see previous section). This does not mean, however, that there would not be any HF filtering required anymore. Filtering before the LNA is, although not for mirror signal suppression, still necessary to suppress the blocking signals. Moreover, between the LNA and the mixer, filtering may be necessary in order to suppress second and third harmonic distortion products that are introduced by the LNA. Due to the use of a switching down-converter or the non-linearities of the mixer and local oscillator harmonics, these distortion products will be down-converted to the same frequency as the wanted signal. The latter problem can be eliminated by using either a very good blocking filter before the LNA (resulting in small signals after the LNA) or by using a highly linear LNA. The use of linear down-converters (i.e., based on the multiplication with a sinusoidal local oscillator signal) reduces the problem as well.

In mobile communications systems, very high, out-of-band signals may be present. In order to prevent saturation of the LNA, these signals must be suppressed with an HF filter that passes only the signals in the band of the application. In the GSM system, for example, the ratio between the largest possible out-of-band signal and the lowest detectable signal is 107 dB. Without a blocking filter, the LNA and the mixer must be able to handle this dynamic range. For the LNA, this means that the input should be able to handle an input signal of 0 dBm (i.e., the −1 dB compression point P_{-1dB} should be about 0 dBm), while having a noise figure of 6 dB. Consequently, this means that the IP3 value should be +10 dBm (IP3 ≈ P_{-1dB} + 10.66 dB). The IMFDR3 (intermodulation free dynamic range) of an LNA or mixer for a given channel bandwidth is given by:

$$\text{IMFDR3} = \frac{2}{3}\left[\text{IP3} + 174\text{dB} - 10\log(\text{BW}) - \text{NF}\right] \tag{9.4}$$

The required IMFDR3 for an LNA is thus (for a 200-kHz bandwidth) 80 dB. CMOS down-converters can be made very linear by using MOS transistors in the linear region,[2,6,17] much more linear than the bipolar cross-coupled multipliers. IP3 values of +45 dBm and noise figures of 18 dB have been demonstrated for CMOS realizations.[2,6] This results in an IMFDR3 for a 200-kHz bandwidth of more than 95 dB. The consequence is that the IMFDR3 spec of 80 dB (i.e., without blocking filter) is achievable for the mixer. In this manner, CMOS opens the way to the development of a true fully integrated single-chip receiver for wireless systems that does not require a single external component, not even a blocking filter. In order to achieve this goal, highly linear mixers that multiply with a single sine must be used. However, the noise performance of mixers is intrinsically worse than the noise of an amplifier, and the use of an LNA is still necessary. In order to be able to cope with the blocking levels, the LNA will have to be highly linear and its gain will have to be reduced from a typical value of, for example, 18 dB to 12 dB. The mixers' noise figure will then have to be lowered by about 6 dB too. This will require a higher power consumption from the down-conversion mixer, but the benefit would be that the receiver can then be fully integrated.

The Down-Converter

The most-often used topology for a multiplier is the multiplier with cross-coupled variable transconductance differential stages. The use of this topology or related topologies (e.g., based on the square law) in CMOS is limited for high-frequency applications. Two techniques are used in CMOS: the use of the MOS transistor as a switch and the use of the MOS transistor in the linear region.

The technique often used in CMOS down-conversion for its ease of implementation is subsampling on a switched-capacitor amplifier.[5,18,19] Here, the MOS transistor is used as a switch with a high input bandwidth. The wanted signal is commutated via these switches. Subsampling is used in order to be able to implement these structures with a low-frequency op-amp. The switches and the switched capacitor circuit run at a much lower frequency (comparable to an IF frequency or even lower). The clock jitter must, however, be low so that the high-frequency signals can be sampled with a high enough accuracy. The disadvantage of subsampling is that all signals and noise on multiples of the sampling frequency are folded upon the wanted signal. The use of a high-quality HF filter in combination with the switched-capacitor subsampling topology is therefore absolutely necessary.

Figure 9.2 shows the block diagram of a fully integrated quadrature down-converter realized in a 0.7-μm CMOS process.[2] The proposed down-converter does not require any external components, nor does it require any tuning or trimming. It uses a newly developed double-quadrature structure, which renders a very high performance in quadrature accuracy (less than 0.3 degrees in a very large passband). The down-converter topology is based on the use of nMOS transistors in the linear region.[2,6] By using capacitors on the virtual ground, a low-frequency op-amp can be used for down-conversion. The MOS transistors in the linear region result in very high linearity (input-referred IP3 is +27 dBm) for both the RF and the LO input. The advantages of such high linearity on both inputs are, as explained in the previous section, that the mixer can handle a very high IMFDR3, resulting in no need for any kind of HF filtering. This opens the way to the implementation of a fully integrated receiver.

The LNA

As denoted in the previous section, the HF down-conversion mixer tends to have a high noise floor; if the mixer is positioned directly behind the antenna, small antenna signals will be drowned in noise and the overall receiver sensitivity will be low. In order to increase the receiver sensitivity and the SNR at minimum antenna input power, the antenna signal has to be pushed above the noise floor of the mixer by means of a low noise amplifier (LNA). As long as the output noise of the LNA is greater than the input noise of the mixer, the sensitivity is fully determined by the LNA noise figure (NF). This is illustrated in Fig. 9.3.

The noise figure (NF) of a low-noise amplifier embedded in a 50-Ω system is defined as:

$$NF = 10\log_{10}\left(\frac{LNA\ output\ noise}{LNA\ output\ noise\ if\ the\ LNA\ itself\ was\ noiseless}\right) \quad (9.5)$$

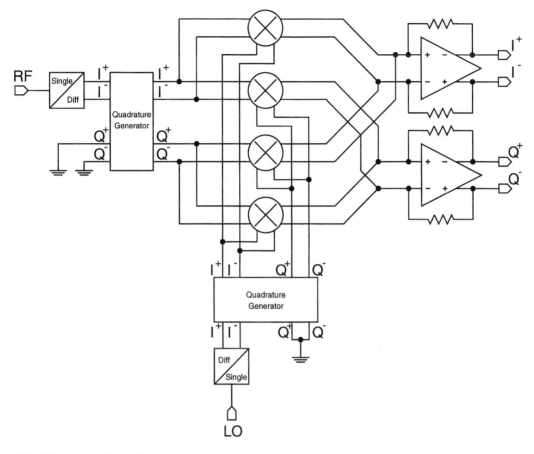

FIGURE 9.2   A double-quadrature down-conversion mixer.

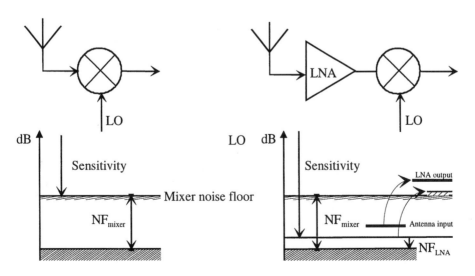

FIGURE 9.3   The benefit of using a low-noise amplifier.

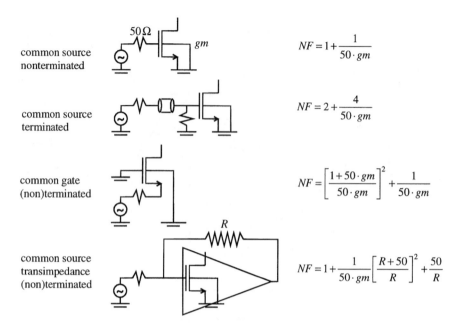

FIGURE 9.4 Noise figure of some common input structures.

that is, the real output noise power (dv²/Hz) of the LNA (consisting of the amplified noise of the 50-Ω source and including all the noise contributions of the amplifier itself to the output noise), divided by the amplified noise of the 50-Ω source only (dv²/Hz). In this way, the noise figure can be seen as the deterioration of the SNR due to insertion of the non-ideal amplifier. The noise figure is generally dominated by the noise of the first device in the amplifier.

Figures 9.4 and 9.5 compare some common input structures regarding noise. As can be seen from the NF equations and the plotted noise figure as function of the g_m of the transistor for the different topologies,

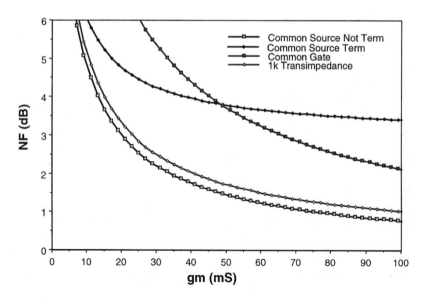

FIGURE 9.5 Performance comparison.

the non-terminated common-source input stage and the (terminated) transimpedance stage are superior as far as noise is concerned. For those circuits, the NF can be approximated as:

$$\left(NF-1\right)=\frac{1}{50\cdot g_m}=\frac{\left(V_{gs}-V_t\right)}{50\cdot 2\cdot I} \tag{9.6}$$

indicating that a low noise figure needs a high transconductance in the first stage. In order to generate this transconductance with high power efficiency, a low $V_{gs}-V_t$ is preferred. However, this will result in a large gate capacitance. Together with the 50-Ω source resistance in a 50-Ω antenna system, the achievable bandwidth is limited by:

$$f_{3dB}\cong\frac{1}{2\pi\cdot 50\Omega\cdot C_{gs}} \tag{9.7}$$

Together with Eq. 9.6, this results in (f_T is the cutoff frequency of the input transistor)

$$\left(NF-1\right)=\frac{f_{3dB}}{f_T} \tag{9.8}$$

Due to overlap capacitances and Miller effect, this relationship becomes approximately (f_d is the 3-dB point of a transistor in diode configuration):[11]

$$\left(NF-1\right)=\frac{f_{3dB}}{f_d} \tag{9.9}$$

This means that a low noise figure can only be achieved by making a large ratio between the frequency performance of the technology (f_d) and the working frequency (f_{3dB}). Because for a given technology, f_d is proportional to $V_{gs}-V_t$, this requires a large $V_{gs}-V_t$ and, associated with it, a large power drain. Only by going to real deep sub-micron technologies, will the f_d be high enough to achieve GHz working frequencies with low $V_{gs}-V_t$ values. Only then can the power drain be reduced to an acceptable value.

In practice, the noise figure and the power transfer from the antenna to the LNA is further optimized by doing, respectively, noise and source impedance matching. These matching techniques often rely on inductors to cancel out parasitics by a resonance phenomenon to boost up the maximum working frequency; the LNA works in "overdrive" mode. Although these aspects are not discussed in this chapter, they are very important when designing LNAs for practical boundary conditions like antenna termination, etc.

In contrast to what one might think, there are still some drawbacks in using short-channel devices for low noise. The large electric field at the drain of a sub-micron transistor may produce hot carriers, having a noise temperature significantly above the lattice temperature.[20] This indicates that a good LDD (lightly doped drain) is as crucial for low noise as it is for device reliability.

At high antenna input powers, the signal quality mainly degrades due to in-band distortion components that are generated by third-order intermodulation in the active elements. The linearity performance of LNAs is generally described by the input-referred third-order intercept point (IIP3), as can be seen in Fig. 9.6. IIP3 specifies the extrapolated input signal where third-order intermodulation products start to dominate the output.

As out-of-band signals are, in general, orders of magnitude larger than the wanted signal ("blocking levels"), the mixing of out-of-band signals toward an in-band intermodulation product must be avoided by all means. Therefore, it is very important to know the limiting factors and the dynamics of the most

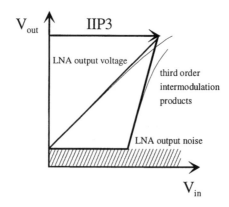

FIGURE 9.6   Linearity performance of an LNA.

important linearity spec, that is, IIP3. Because the core of an amplifier always contains one or more active elements, we will focus on their internal distortion mechanisms.

Long-channel transistors are generally described by a quadratic model. Consequently, a one-transistor common-source amplifier ideally suffers only from second-order distortion and produces no third-order intermodulation products. As a result, high IP3 values should easily be achieved. In fact,

$$IM2 = \frac{1}{2} \frac{v}{V_{gs} - V_t} \quad \text{and} \quad IM3 = 0 \tag{9.10}$$

where v denotes the input amplitude of the amplifier.

However, when the channel length decreases toward deep sub-micron, this story no longer holds; third-order intermodulation starts to become important. To understand the main mechanism behind third-order distortion in a sub-micron CMOS transistor, we start from the equation for the drain current of a short-channel transistor,

$$I_{ds} = \frac{\mu_0 C_{ox}}{2n} \cdot \frac{W}{L} \cdot \frac{\left(V_{gs} - V_t\right)^2}{1 + \Theta \cdot \left(V_{gs} - V_t\right)} \tag{9.11}$$

with

$$\Theta = \theta + \frac{\mu_0}{L_{eff} \cdot v_{max} \cdot n} \tag{9.12}$$

where θ stands for the mobility degradation due to the transversal electrical field (surface scattering at the oxide–silicon interface) and the $\frac{\mu}{L_{eff} \cdot v_{max} \cdot n}$ -term models the degradation caused by the longitudinal electric field (electrons reaching the thermal saturation speed). As the $_\theta$-term is small in today's technologies (increasingly better quality of the oxide–silicon interface), it can often be neglected relative to the longitudinal term. For a typical 0.5 µm CMOS technology, the Θ-parameter equals about 0.9.

It can be seen from Eq. 9.11 that for large values of $V_{gs} - V_t$, the current becomes a linear function of $V_{gs} - V_t$. The transistor is then operating in the velocity saturation region. For smaller values of $V_{gs} - V_t$, the effect of Θ consists apparently of "linearizing" the quadratic characteristic, but in reality, the effect results in an intermodulation behavior that is worse than in the case of quadratic transistors. Indeed, we will have a slightly lower amount of second-order intermodulation, but it comes at the cost of third-order intermodulation.

The following equations for the intermodulation ratios *IMx* can be found[13] by calculating the Taylor expansion of the drain current around a certain $V_{gs} - V_t$ value:

$$IM2 = \frac{v}{V_{gs} - V_t} \cdot \frac{1}{(1+r) \cdot (2+r)} \tag{9.13}$$

and

$$IM3 = \frac{3}{4} \frac{v}{(V_{gs} - V_t)} \cdot \frac{v}{V_{sv}} \cdot \frac{1}{(1+r)^2 \cdot (2+r)} \tag{9.14}$$

where

$$V_{sv} = \frac{1}{\Theta} \tag{9.15}$$

represents the transit voltage between strong inversion and velocity saturation and

$$r = \frac{V_{gs} - V_t}{V_{sv}} \equiv \Theta \cdot (V_{gs} - V_t) \tag{9.16}$$

denotes the relative amount of velocity saturation. The transit voltage V_{sv} depends only on technology parameters and is about 2 V for a 0.7-µm CMOS technology. For deep sub-micron processes (e.g., a 0.1-micron technology), this voltage becomes even smaller than 300 mV, which is very close to the $V_{gs} - V_t$ at the boundary of strong inversion.

Based on Eq. 9.14, one can directly derive an expression for IIP3:

$$IIP3 \cong 11.25 + 10 \cdot \text{Log}_{10} \left((V_{gs} - V_t) \cdot V_{sv} \cdot (1+r)^2 \cdot (2+r) \right) \tag{9.17}$$

This value is normalized to 0 VdBm, the voltage that corresponds to a power of 0 dBm in a 50-Ω resistor. For the 0.5-µm technology that was mentioned before and a $V_{gs} - V_t$ value of 0.2 V, IIP3 equals +9.5 VdBm. It is worth noting that for a given L_{eff}, the intrinsic IIP3-value of a transistor is only a function of the gate overdrive.

In Figure 9.7, the formula for IP3 is evaluated for a minimum-length transistor in three different technologies. As can be seen from the figure, for a given L_{eff}, the linearity becomes better with increasing gate overdrive. For small gate overdrives, the increase in IIP3 is proportional to the square root of $V_{gs} - V_t$. At high $V_{gs} - V_t$ values (near velocity saturation), the increase in IIP3 becomes even more pronounced. However, this region of operation exhibits a very low transconductance efficiency (g_m/I_{ds}), particularly for sub-micron transistors, where this parameter is given by

$$\frac{g_m}{I_{ds}} = \frac{2}{V_{gs} - V_t} \cdot \frac{1 + \Theta \cdot (V_{gs} - V_t)}{1 + 2\Theta \cdot (V_{gs} - V_t)} \tag{9.18}$$

The influence of L_{eff} on IIP3 can be seen in Fig. 9.7; For practical values of the gate overdrive, the linearity gets worse with decreasing gate length, because V_{sv} is proportional to L_{eff}. This may pose a problem when very small transistor lengths are required to reduce the power drain and a high IP3 is necessary to avoid the blocking filters. For large values of the gate overdrive, there is a point where the intermodulation performance of a short-channel transistor gets better compared to a large-channel one,

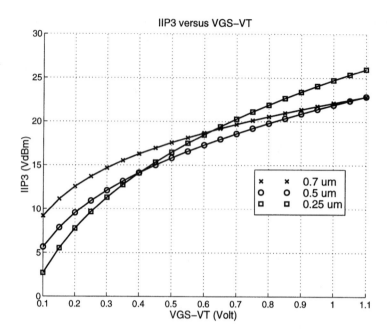

FIGURE 9.7   IIP3 versus $V_{gs} - V_t$ for three different technologies.

because the first already enters the velocity saturation region. As mentioned before, this region of operation is not highly recommended.

Nevertheless, when a certain IIP3 is required, there are basically two methods to ensure this: using a high enough $V_{gs} - V_t$ or using some kind of feedback mechanism (e.g., source degeneration). It can be shown that for the same equivalent g_m and the same distortion performance, the required dc current is lower when local feedback at the source is applied. It comes, however, at the cost of a larger transistor width, eventually compromising the amplifier bandwidth.

9.4   The Synthesizer

Synthesizer Topology

The *frequency synthesizer* generates the local oscillator signal, responsible for the correct frequency selection in the up- and down-converters. Since the frequency spectrum in modern wireless communication systems must be used as efficiently as possible, channels are placed very close together. The signal level of the desired receiving channel can be made very small, whereas adjacent channels can have very large power levels. Therefore, the phase noise specifications for the LO signal are very high, which makes the design of the frequency synthesizer very critical.

Meanwhile, mobile communication means low power consumption, low cost, and low weight. This implies that a completely integrated synthesizer is desirable, where integrated means a standard CMOS technology without any external components or processing steps. Usually, the frequency synthesizer is realized as a phase-locked loop (PLL) as shown in Fig. 9.8. The most critical building blocks of a PLL for integration in CMOS are the building blocks that operate at high frequency: the voltage-controlled oscillator (VCO) and the prescaler. Both will be discussed in the following sections.

As stated above, the most important specification of a frequency synthesizer is low phase noise. The following formulae indicate how the noise of the different building blocks is transferred toward the synthesizer's output.

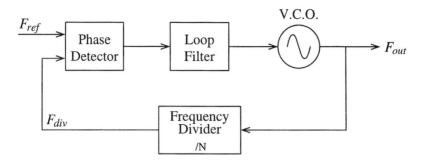

FIGURE 9.8   PLL-based frequency synthesizer.

$$\frac{\theta_{out}(s)}{\theta_{vco}(s)} = \frac{N \cdot s}{N \cdot s + K_{pd} \cdot G_{lf}(s) \cdot K_{vco}} \tag{9.19}$$

$$\frac{\theta_{out}(s)}{\theta_{other}(s)} = \frac{N \cdot K_{pd} \cdot G_{lf}(s) \cdot K_{vco}}{N \cdot s + K_{pd} \cdot G_{lf}(s) \cdot K_{vco}} \tag{9.20}$$

with θ_{out} the synthesizer's output noise, θ_{vco} the phase noise of the VCO, θ_{other} the phase noise of the reference signal, the prescaler, the phase detector, and the loop filter, N the prescaler division factor, K_{pd} the phase detector gain, $G_{lf}(s)$ the loop filter transfer function, and K_{vco} the VCO gain.

With $G_{lf}(s)$ as the transfer function of a low-pass filter, the phase noise of the VCO is high-passed toward the output. The phase noise of the other components is low-passed. In other words, the VCO is the main contributor for out-of-band phase noise, while the other building blocks account for the in-band noise.

As can be seen, the choice of the loop bandwidth is critical for phase noise performance. In order to have enough suppression of phase noise at frequency offsets, important for communication standards (e.g., 600kHz for GSM and DCS-1800) and of spurious value due to the reference signal, the loop bandwidth cannot be very large (a few kHz typically). Also, for stability reasons, the loop bandwidth has to be small compared to the PLL reference frequency. To realize relatively small loop bandwidths, large capacitor and resistor values are necessary to implement the large time constant. In current commercial designs, the capacitors are often implemented off-chip, to limit the chip area. To go to full integration of the frequency synthesizer, a way must be found to implement the loop filter without the need for large amounts of capacitance. Several possibilities exist.

The first possibility is to increase the resistance needed to create the large time constant of a narrow low-pass filter, which means a decrease of capacitance. The disadvantage of this approach is the increase of loop filter phase noise. Hence, a tradeoff between integrated capacitance and phase noise exists.

A more appealing approach is the one used in Ref. 31. Here, a type-II fourth-order PLL is integrated, using a dual-path filter topology. The loop filter consists of two filter paths: one active filter path and one passive filter path. By combining the signals of both paths, a zero is realized without the need for an extra capacitor and resistor. The zero is necessary to provide enough phase margin for loop stability. The principle is explained in Figure 9.9. In this way, the integrated capacitance is small enough to be integrated on-chip, without degrading the phase noise performance. The total chip area was only 1.9 μm by 1.6 μm.

A third possibility is increasing the reference frequency. As a consequence, the loop bandwidth can be made larger, with still enough suppression of the reference spurious noise. This means less integrated capacitance and a better PLL settling time. In addition, the prescaler's division factor N will be decreased. As can be seen in Eq. 9.20, the phase noise of the phase detector, the loop filter, the reference signal, and

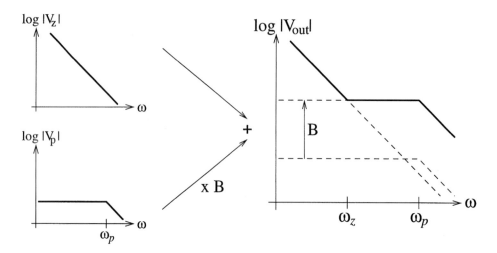

FIGURE 9.9   Dual-path filter principle.

the prescaler is multiplied by *N*. In other words, increasing the reference frequency also results in better in-band phase noise. One disadvantage exists. The prescaler can only divide by integer values, so the smallest frequency step that can be synthesized is equal to the reference frequency. In GSM and DCS-1800 systems, this would mean that the reference frequency must be 200 kHz in order to select all possible channels. This problem can be circumvented by the use of the fractional-N technique. This technique allows the use of fractional division factors by very fast switching between different integer division factors. By combining fractional-N with delta-sigma modulation,[32] the spurs generated by the switching action are shaped to high-frequency noise. This noise can then be filtered by the PLL loop filter.

The Oscillator

As stated above, the *oscillator* will be the main source of out-of-band phase noise. Therefore, its design is one of the most critical parts in the integration of a frequency synthesizer for high-quality communication standards. For the realization of a gigahertz VCO in a sub-micron CMOS technology, two options exist: ring oscillators or oscillators based on the resonance frequency of an LC-tank. The inductor in this LC-tank can be implemented as an active inductor or a passive one. It has been shown that for ring oscillators[21] as well as active LC-oscillators,[22] the phase noise is inversely related to the power consumption.

$$\text{Ring osc. (Ref. 21):}\quad L\{\Delta\omega\} \sim kTR\left(\frac{\omega}{\Delta\omega}\right)^2 \quad \text{with } g_m = \frac{1}{R}$$

$$\text{Active-LC (Ref. 22):}\quad L\{\Delta\omega\} \sim \frac{kT}{2\omega C}\cdot\left(\frac{\omega}{\Delta\omega}\right)^2 \quad \text{with } g_m = 2\omega C \tag{9.21}$$

Therefore, the only viable solution to a low-power, low-phase-noise VCO is an LC-oscillator with a passive inductor. In this case, the phase noise changes proportionally with the power consumption:

$$\text{Passive-LC (Ref. 22):}\quad L\{\Delta\omega\} \sim kTR\left(\frac{\omega}{\Delta\omega}\right)^2 \quad \text{with } g_m = R\left(\omega C\right)^2 \tag{9.22}$$

As could be expected, the only limitation in this oscillator is the integrated passive inductor. Equation 9.22 shows that for low phase noise, the resistance R (i.e., the equivalent series resistance in the LC-loop) must be as small as possible. A low resistance also means low losses in the circuit and thus low power needed to compensate for these losses. Capacitors are readily available in most technologies. But since the resistance R will be dominated by the contribution of the inductors' series resistance, the inductor design is critical. Three solutions exist.

Spiral inductors on a silicon substrate usually suffer from high losses in this substrate, which limit the obtainable Q-factor. Recently, techniques have been developed to etch this substrate away underneath the spiral coil in a post-processing step.[7,23] The cavity created by such an etching step can clearly be seen in Fig. 9.10. However, since there is an extra etching step required after normal processing of the ICs, this technique is not allowed for mass production.

For extremely low phase noise requirements, the concept of bondwire inductors has been investigated. Since a bondwire has a parasitic inductance of approximately 1 nH/mm and a very low series resistance, very-high-Q inductors can be created. Bondwires are always available in IC technology, and can therefore be regarded as being standard CMOS components. Two inductors, formed by four bondwires, can be combined in an enhanced LC-tank[22] to allow a noise/power tradeoff. A microphotograph of the VCO is shown in Fig. 9.11.[25] The measured phase noise is as low as −115 dBc/Hz at an offset frequency of 200 kHz from the 1.8-GHz carrier. The power consumption is only 8 mA at 3 V supply. Although chip-to-chip bonds are used in mass commercial products,[28] they are not characterized on yield performance for mass production. Therefore, the industry is reluctant with regard to this solution.

The most elegant solution is the use of a spiral coil on a standard silicon substrate, without any modifications. Bipolar implementations do not suffer from substrate losses because they usually have a high-ohmic substrate.[24] Most sub-micron CMOS technologies use a highly doped substrate, and therefore have large induced currents in the substrate, which is responsible for the high losses. The effects present in these low-ohmic substrates can be investigated with finite-element simulations. The finite-element simulations also take the losses in the metal conductors into account. Four phenomena contribute to these losses. The first is the dc series resistance of the metal. The others are high frequency effects, the

FIGURE 9.10   Etched spiral inductor.

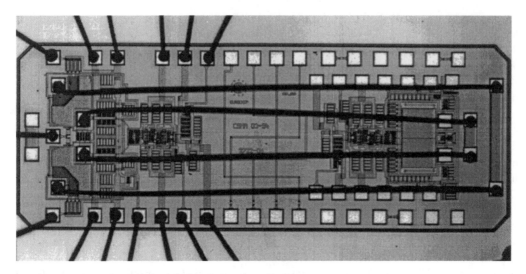

FIGURE 9.11 Microphotograph of the bondwire LC-oscillator.

skin effect, and eddy currents. Due to the skin effect, the metal turns are only partially used for conduction of the high frequency current. Eddy currents are generated by the changing magnetic field that crosses the metal lines, resulting in an increased resistance, especially in the inner turns of the inductors. This analysis can lead to a coil design optimized for a certain technology. A coil was implemented, using the above analysis, in a spiral-inductor LC-oscillator. The technology is a standard two-metal layer, 0.4-μm CMOS technology. With a power consumption of only 11 mW, a phase noise of −122.5 dBc/Hz at 600 kHz offset of the 1.8 GHz carrier has been obtained.[29] A microphotograph is shown in Fig. 9.12.

The Prescaler

To design a high-speed dual-modulus prescaler, a new architecture has been developed that is based on the 90-degrees phase relationship between the master and the slave outputs of an M/S toggle-flip-flop.[26] This architecture is shown in Fig. 9.13. No additional logic is present in the high frequency path to realize the dual-modulus division, as is the case in classic prescalers, based on synchronous counters. Here, the dual-modulus prescaler is as fast as an asynchronous fixed divider. Using this new principle, a 1.75-GHz input frequency has been obtained at a power consumption 24 mW and 3 V power supply. At 5 V power supply, input frequencies above 2.5 GHz can even be processed in a standard 0.7-μm CMOS technology. By going to sub-micron technologies, even higher frequencies can be obtained at low power consumption.

Fully Integrated Synthesizer

The fully integrated VCO and dual-modulus prescaler make it possible to integrate a complete LO synthesizer in a standard CMOS technology, without tuning, trimming, or post-processing, that achieves modern telecom specs. Using the dual-path filter topology for minimizing the necessary integrated capacitance, a type-II, fourth-order, fully integrated PLL frequency synthesizer for DCS-1800 applications has been realized. The PLL is implemented in a standard 0.4-μm CMOS, achieving a phase noise of −121 dBc/Hz at 600 kHz offset frequency, while consuming only 51 mW from a 3-V power supply. The integrated capacitance could be decreased to less than 1 nF, for a loop bandwidth of 45 kHz, resulting in a chip area of 1.7 μm by 1.9 μm. A chip microphotograph is shown in Fig. 9.14.

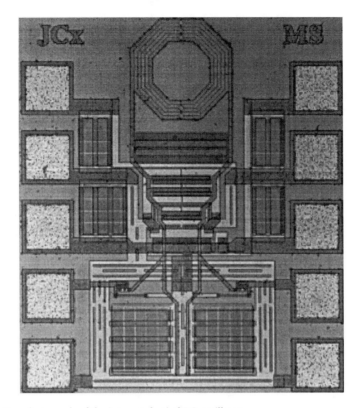

FIGURE 9.12   Microphotograph of the integrated spiral LC-oscillator.

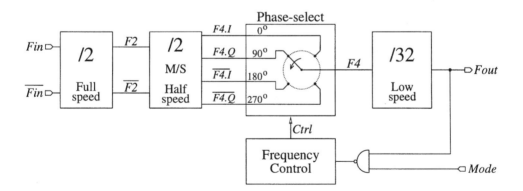

FIGURE 9.13   New dual-modulus prescaler architecture.

9.5   The Transmitter

For communication systems like GSM, two-way communication is required and a transmitter circuit must be implemented to achieve a full transceiver system. In the open literature, most reported mixer circuits in CMOS are down-conversion mixers. However, as will be explained in the first section below, there is a huge difference between receivers. This implies that a lot of research for the development of CMOS transmitter circuits still needs to be done.

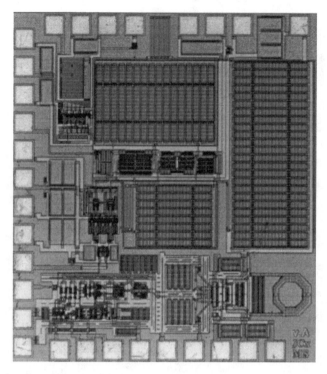

FIGURE 9.14   A fully integrated frequency synthesizer.

Down-Conversion vs. Up-Conversion

The modulation of the baseband signals on the local oscillator carrier frequency requires an up-conversion mixer topology. In classical bipolar transceiver implementations, the up- and down-converter mixer use typically the same four-quadrant topology. There are, however, some fundamental differences between up- and down-converters, which can be exploited to derive optimal dedicated mixer topologies.

In a down-conversion topology, the two input signals are at a high frequency (e.g., 900 MHz for GSM systems) and the output signal is a low-frequency signal of maximum a few MHz for low-IF or zero-IF receiver systems. This low-frequency output signal can easily be processed making optimal use of the advantages of feedback circuits. Also, high-frequency spurious signals (e.g., LO feedthrough) can be filtered.

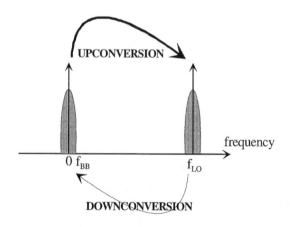

FIGURE 9.15   The difference between up-conversion and down-conversion.

For up-conversion mixers, the situation is totally different. The high frequent local oscillator (LO) and the low frequent baseband (BB) input signal are multiplied to form a high frequent (RF) output signal. All further signal processing has to be performed at high frequencies, which is very difficult and power consuming when using current sub-micron CMOS technologies. Furthermore, all unwanted signals (like the intermodulation products and LO-leakage) have to be limited to a specified level (e.g., below –30 dBc) as they cannot be filtered.

Also, the specifications have to be interpreted differently for both kinds of mixer circuits: for example, an important specification for down-conversion mixers is the conversion gain G_c, defined as the ratio between the output signal and the RF input signal. As for a down-conversion mixer, the RF input signal is fixed — and usually very small; G_c is a good measure of the circuit performance. For an up-conversion mixer, the situation is different. G_c would be the ratio between the RF output signal and the baseband signal. However, in this case, both the input ports, the baseband signal, and the LO signal are free design parameters. As it is easier and more power-friendly to amplify the low-frequency signal, a large baseband signal is preferred. Because of this extra design parameter, it should be better to compare up-conversion circuits based on, for example, the same distortion level or on the same output power level.

CMOS Mixer Topologies

Switching Modulators

Many published CMOS mixer topologies are based on the traditional variable transconductance multiplier with cross-coupled differential modulator stages. Since the operation of the classical bipolar cross-coupled differential modulator stages is based on the translinear behavior of the bipolar transistor, the MOS counterpart can only be effectively used in the modulator or switching mode. Large LO-signals have to be used to drive the gates and this results in a huge LO-feedthrough. In CMOS down-converters, this is already a problem; in Ref. 9, for example, the output signal level is –23 dBm with an LO-feedthrough signal of –30 dBm, which represents a suppression of only –7 dB. This gives rise to very severe problems in direct up-conversion topologies. Moreover, by using a square wave modulating LO signal, 30% of the signal power is present at the third-order harmonic. This unwanted signal can only be filtered with an extra external output blocking filter.

In CMOS, the variable transconductance stage is typically implemented using a differential pair biased in the saturation region. To avoid distortion problems, large $V_{gs} - V_t$ values or a large source degeneration resistance have to be used, which results in large power drain and noise problems, especially compared to the bipolar converter circuits. This can be avoided by replacing the bottom differential pair by a pseudo-differential topology with MOS transistors in the linear region.[17]

Linear MOS Mixers

Next, an intrinsically linear CMOS mixer topology is presented. The modulation is performed by biasing MOS transistors in the linear region. The circuit is focused on a real single-ended output topology, which avoids the use of external combining and the circuits have been optimized based on the analysis of the non-linearities of the mixer structure. The understanding of the origins of the distortion, results in a better compromise between the linearity, the output signal power, and the power consumption. Therefore, the results of the linearity analysis and some guidelines to optimize the circuit performance are also presented in this section. The general design ideas will be illustrated by numerical data from realized chips.[33,34]

Figure 9.16 shows the four up-conversion mixers M1a, M1b, M1c, and M1d and the single-ended output current buffer. The realized mixer topology is based on an intrinsic linear mixer circuit that converts the baseband and LO voltages into modulated currents.

Each mixer transistor converts a quadrature LO voltage and baseband voltage to a linearly modulated current. The expression for the drain-source current for an MOS transistor in the linear region is given by:

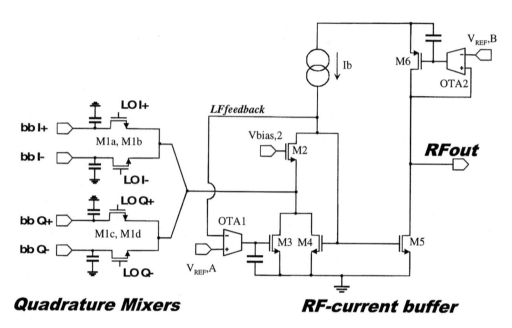

FIGURE 9.16 Schematic of the quadrature up-conversion mixers and output driver.

$$I_{ds} = \beta \left[\left(V_{gs} - V_t \right) V_{ds} - \frac{V_{ds}^2}{2} \right] \tag{9.23}$$

This equation is rewritten in terms of dc and ac terms as:

$$I_{ds} = \beta \left(V_{ds} + v_{ds} \right) \left(V_{gs} - V_t - \frac{V_d - V_s}{2} + v_g - \frac{v_d + v_s}{2} \right) \tag{9.24}$$

The differential baseband voltage is applied at the drain/source of the mixer transistors M1a and M1b, and the LO signal is applied at the gates of the mixers. Equation 9.24 shows that only two high-frequency components (products with the LO signal v_g) occur in the signal current of each mixer transistor: $\beta V_{DS} v_g$ and $\beta v_{ds} v_g$. The last term is the wanted mixed signal. The first term is proportional to the product of the dc drain-source voltage and the gate (LO) signal. Hence, it is situated at the oscillator frequency. This unwanted signal has been eliminated by applying zero dc drain-source voltage over the mixer transistor. In this way, only the wanted frequency component is formed by each mixer transistor.

The voltage-to-current conversion is performed balanced. The currents are immediately added at the common node, which is made virtual ground by the very low input impedance of the buffer stage (Fig. 9.16).

Quadrature modulation is performed by summing the four modulated currents of each mixer transistor. The resulting single-ended signal current is given by Eq. (9.25).

$$I_{mixer} = \delta \beta \left(v_{bb,I}^2 + v_{bb,Q}^2 + 2 v_{lo} v_{bb} \left(SSB \right) \right) \tag{9.25}$$

where:

δ = A reduction factor due to the degeneration by the finite input conductance of the output stage

$$\beta = \mu \, C_{ox} \, W/L$$

$v_{bb,I}$ and $v_{bb,Q}$ = the baseband I and Q voltage signals, respectively

v_{lo} = the local oscillator voltage

The modulated current contains a low-frequency square baseband signal and the wanted RF single side band (SSB) mixing product.

The mixer has been designed to have all distortion and intermodulation components, including LO-feedthrough, lower than −30 dBc, which is typical for wireless GSM, DECT, and DCS applications.

An essential improvement in the RF mixer design is the implementation of large on-chip capacitors to the ground at the baseband nodes of the mixer transistors. In this way, the modulated RF current flows through the capacitor and not through the bonding wires, which form a considerable impedance at GHz frequency. The RF signal linearity and amplitude become independent of the bonding wire matching and length as the RF signal remains on-chip. Also, a degeneration of the high-frequency signal current by the output impedance of the baseband signal source is prevented by this low impedance.

The Low-Frequency Feedback Loop

The low-frequency feedback loop, which consists of OTA1 and transistors M2 and M3, suppresses the low-frequency (square baseband) signals to enlarge the dynamic range of the output stage and to prohibit intermodulation products of the unwanted low frequent square baseband mixer current with the high-frequency signal. It also lowers the input impedance of the output stage at low frequencies. The operation can be further explained and illustrated by the numerical data given in Ref. 33 and Fig. 9.17. The LF loop has a gain bandwidth of 500 MHz and a high dc gain, which is obtained by using a cascoded OTA structure. The large gain results in a very small (<1 Ω) LF input impedance. In this way, no LF voltage signal is formed at the summing node. This is absolutely essential to obtain a linear operation of the mixer as an additional voltage signal is also up-converted. This structure offers the advantage that the LF and HF modulated current components are separated, and only the high-frequency component of the signal current is mirrored to the output. The dc current through M3 needs to be sufficiently large to reject all the LF current generated in the mixer transistors.

The High-Frequency Current Buffer

The high-frequency current buffer (M2, M4) ensures a very low input impedance at high frequencies to realize the virtual ground node and mirrors the RF current component (~vlo·vbb) to the output. In Ref. 33, a 2.0-GHz GBW was obtained by designing gm_4 = 14 mS, consuming 5 mA dc current, and the parasitic capacitance at node 2 = 1.1 pF. As the loop gain is unity at 2 GHz, the input impedance of the output driver is determined by $1/g_m2$ = 12 Ω (Fig. 9.18). This low impedance also ensures the stability

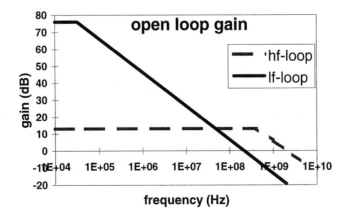

FIGURE 9.17   Open loop conversion gain of the feedback loops.

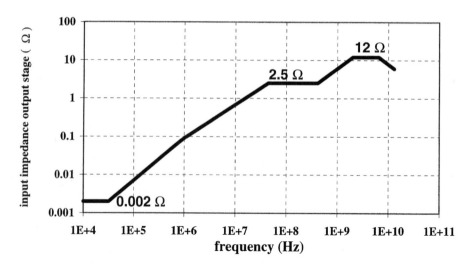

FIGURE 9.18 Input impedance on the virtual ground summing node.

of the HF feedback loop. The pole at node 1, determined by the parasitic capacitance at the summing node and this input impedance, is at more than 3 times the GBW of HF feedback loop. For the second-order feedback system, this results in a phase margin of more than 68 degrees.

Analysis of the Non-Linearities and the LO-Feedthrough

To obtain an optimal tradeoff between distortion, power consumption, frequency performance, signal power, and distortion, a profound analysis of the causes of the non-linearities is indispensable.

First, the intrinsic non-linearity of the voltage-to-current conversion by the mixer transistors is examined. As the LO feedthrough is also an unwanted signal in the signal band, it is also covered in this section.

A non-ideal virtual ground at the summing node has severe consequences for the modulated signal. A finite output conductance of the baseband signal has similar influences on the linearity of the output signal. These elements are investigated.

Intrinsic Non-Linearity of the Mixer Transistors

Equations 9.24 and 9.25, which describe the generated current, are only valid if a very low impedance is seen at the drain and source nodes of the mixer transistors. If this condition is fulfilled, no unwanted high-frequency mixing components seem to appear in the modulated current. However, both in measurements and in simulations, a significant unwanted signal is noticed at oscillator frequency (f_{lo}) $\pm$ 3 times the baseband frequency (f_{bb}). As this signal originates from a product $v_{lo} \cdot v_{bb}^3$, the magnitude of this distortion component is expected to have the same third-order relationship to the baseband signal. However, only a second-order magnitude relationship is perceived if the virtual ground condition is fulfilled. An explanation for this effect has been searched and the result is described in the paragraphs that follow.

The observed phenomenon is a result of the short-channel effects in an MOS transistor. Both the mobility and the threshold voltage are affected by the gate-source and drain-source voltage. However, the calculated impact of the V_t dependency on the signal is an order of magnitude lower than the observed effect.

The distortion component can be explained by exploring the drain-source voltage dependency of μ_{eff} in the β factor. The expressions for β and μ_{eff} in the linear region are given by:

$$\beta = \mu_{eff} C_{ox} W/L, \text{ where } \mu_{eff} = \frac{\mu_0}{1 + \vartheta\left(V_{gs} - V_t\right) + \dfrac{\mu_0}{V_{max}L}V_{ds}} \tag{9.26}$$

where: μ_{eff} = The effective mobility

μ_0 = The surface mobility

ϑ = The empirical mobility modulation factor

V_{max} = The maximum drift velocity

If the summing node is made virtual ground and zero dc drain-source voltage is applied, drain and source of each mixer transistor alternate when a sinusoidal signal is applied. If the applied baseband signal is positive, the virtual ground node is the source of that mixer transistor; if the signal is negative, this node is the drain. Mathematically, the source and drain voltages of a mixer transistor can be described as:

$$v_s = \frac{v_{bb} - |v_{bb}|}{2} \qquad v_d = \frac{v_{bb} + |v_{bb}|}{2} \tag{9.27}$$

Equation 9.27 can be transformed to:

$$\mu_{eff} = \frac{\mu_0}{1 + \vartheta\left(V_{gs} - V_t\right)_{dc} + \vartheta\left(v_{lo} - \dfrac{v_{bb}}{2}\right) + \left(\dfrac{\mu_0}{V_{max}L} + \dfrac{\theta}{2}\right)|v_{bb}|} \tag{9.28}$$

Substituting v_{bb} by $A\sin(\omega_{bb}t)$ and making the Fourier series expansion of $|v_{bb}|$ results in Eq. 9.29:

$$\mu_{eff} = \frac{\mu_0}{B\left(1 + \dfrac{\vartheta}{B}v_{lo} - \dfrac{\vartheta A}{2B}\sin(\omega_{bb}t) + C\cos(2\omega_{bb}t) + D\cos(4\omega_{bb}t) + \ldots\right)}$$

where $A\sin(\omega_{bb}t) = $ The baseband signal

$$B = 1 + \left(V_{gs} - V_t\right)_{DC} + \frac{2}{\pi}A\left(\frac{\mu_0}{V_{max}L} + \frac{\theta}{2}\right)$$

$$\approx 1 + \left(V_{gs} - V_t\right)_{DC} \tag{9.29}$$

$$C = \frac{A}{B}\frac{4}{\pi 3}\left(\frac{\mu_0}{V_{max}L} + \frac{\theta}{2}\right)$$

$$D = \frac{A}{B}\frac{4}{\pi \cdot 3 \cdot 5}\left(\frac{\mu_0}{V_{max}L} + \frac{\theta}{2}\right)$$

The equation for μ_{eff} shows that a second-order baseband frequency component ($\cos(2\omega_{bb}t)$) appears. In the dc reduction factor B, the third term is an order of magnitude smaller than 1. Hence, it appears that the magnitude C of the second-order component has only a first-order relationship to the baseband signal amplitude A. In the mixer voltage-to-current relationship, μ_{eff} is multiplied with $v_{lo} \cdot v_{bb}$. As a result, a mixing component at $f_{lo} \pm 3f_{bb}$ occurs. The magnitude of this signal is only second-order related to the amplitude of the baseband signal. This explains the observed effect.

In the amplitude C of this distortion component, $\frac{\mu_0}{V_{max}L}$ is dominant to $\frac{\theta}{2}$ for most sub-micron CMOS technologies. It is important to notice that the distortion is inversely proportional to the gate length. This implies that this effect will become even more important when going to deeper sub-micron technologies.

Oscillator Feedthrough

As a direct up-conversion topology is used, the oscillator feedthrough cannot be filtered and, consequently, it has to be as low as the other in-band unwanted frequency components. A signal at oscillator frequency can have several origins: for example, capacitive feedthrough of the gate voltage or a component due to mixing of a dc drain-source voltage with the oscillator gate signal (see Eq. 9.23). The last is avoided by applying zero dc drain-source voltage over the mixer transistor, as stated earlier.

Due to the differential input voltages, the capacitive LO-feedthrough is canceled at the virtual ground node. However, this cancellation is never perfect, due to technology mismatch. The capacitive LO-feedthrough of one mixer transistor is given by (Eq. 9.30).

$$i_{lo} = \left(\frac{C_{ox}}{2} + \frac{C_{ov}}{L}\right) WLv_{lo} 2\pi f \tag{9.30}$$

where C_{ox} = The oxide capacitance
C_{ov} = The gate-drain overlap capacitance
v_{lo} = The amplitude of the LO signal
f = The LO frequency

When the asymmetry caused by mismatches in the differential mixer structure is taken into account, the expression for the capacitive LO-feedthrough current is given by:

$$\Delta\left(i_{lo}\right) = \delta\left(i_{lo}\right) i_{lo} \tag{9.31}$$

where $\delta(i_{lo})$ is the relative difference in the LO-feedthrough for the differential mixer transistors.

Based on Eqs. 9.25, 9.30, and 9.31, the ratio between the LO feedthrough current and the modulated signal is given in Eq. 9.32.

$$\frac{i_{signal}}{\Delta\left(i_{lo}\right)} = \frac{2\mu C_{ox} \frac{W}{L} v_{bb} v_{lo}}{\delta\left(i_{lo}\right)\left(\frac{C_{ox}}{2} + \frac{C_{ov}}{L}\right) WLv_{lo} 2\pi f} \tag{9.32}$$

This formula can be simplified to:

$$\frac{i_{signal}}{\Delta\left(i_{lo}\right)} = \frac{\mu C_{ox} v_{bb}}{\delta\left(i_{lo}\right)\left(\frac{C_{ox}}{2} + \frac{C_{ov}}{L}\right) L^2 \pi f} \tag{9.33}$$

Even in the case where the two LO-feedthrough currents are added instead of canceled ($\delta(i_{lo}) = 1$), a ratio of 30 dBc signal to LO-feedthrough is achieved with a 1-GHz oscillator signal and a 316 mV baseband signal. If a relative inaccuracy ($\delta(i_{lo})$) of, for example, 10% on the parameters is taken into account, 50 dBc can easily be accomplished. This analysis illustrates the strategy not to rely on cancelling of RF signals to achieve the requirements.

9.6 Toward Fully Integrated Transceivers

Combining all of the above techniques has resulted in the recent development of single-chip CMOS transceiver circuits.[30] Figure 9.19 provides a microphotograph of the 0.25-μm CMOS transceiver test-circuit. The chip does not require a single external component, nor does it require any tuning or trimming. The objective of this design is to develop a complete system for wireless communications at 1.8 GHz that can be built with a minimum of surrounding components: only an antenna, a duplexer, a power amplifier, and a baseband signal processing chip. The high level of integration is achieved using a low-IF topology for reception, a direct quadrature up-conversion topology for transmission, and an oscillator with on-chip integrated inductor. The presented chip has been designed for the DCS-1800 system, but the broadband nature of the LNA, the down-converter, the up-converter, and the output amplifier makes the presented techniques equally suited for use at other frequencies, for example, for use in a DCS-1900 or a DECT system. The presented circuit consumes 240 mW from a 2.5-V supply and occupies a die area of 8.6 mm².

9.7 Conclusion

The trends toward deep sub-micron technologies have resulted in the exploration by several research groups of the possible use of CMOS technologies for the design of RF circuits. Especially the development of new receiver topologies, such as quasi-IF and low-IF topologies, in combination with highly linear down-converters, has opened the way to fully integrated down-converters with no external filters or components. However, due to the moderate speed performance of the present sub-micron technologies, lower noise circuits in combination with less power drain have to be worked out. The trends toward deep sub-micron technologies will allow us to achieve those goals as long as the short-channel effects do not limit the performance concerning linearity and intermodulation problems.

Concerning synthesizers, in the last few years, high-performance, low phase noise, low power drain, fully integrated VCO circuits have been demonstrated. Starting with difficult post-processing techniques, research has resulted in the use of standard CMOS technologies using bondwires as inductors. Today, even low phase noise performances with optimized integrated spiral inductors in standard CMOS technologies without any post-processing, tuning, trimming, or external components have been announced. This opens the way toward fully integrated receiver circuits.

However, telecommunication systems are usually two-way systems, requiring transmitter circuits as well. It is only recently that CMOS up-converters with moderate output power have been announced in open literature. Again, thanks to the trends toward deep sub-micron technologies, fully integrated CMOS transmitter circuits with an acceptable power consumption will be feasible. This opens the way for fully integrated transceiver circuits in standard CMOS technologies.

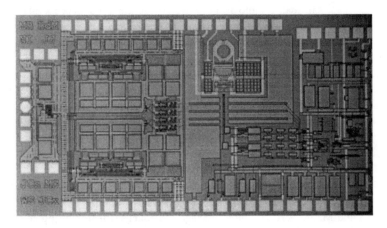

FIGURE 9.19   Microphotograph of a 0.25-μm CMOS transceiver test-circuit.

References

1. J. Sevenhans, A. Vanwelsenaers, J. Wenin, and J. Baro, An integrated Si bipolar transceiver for a zero IF 900 MHz GSM digital mobile radio front-end of a hand portable phone, *Proc. CICC*, pp.7.7.1-7.7.4, May 1991.
2. J. Crols and M. Steyaert, A single-chip 900 MHz CMOS receiver front-end with a high performance low-IF topology, *IEEE J. Solid-State Circuits*, vol. 30, no.12, pp.1483-1492, Dec. 1995.
3. P. R. Gray and R. G. Meyer, Future directions in silicon ICs for RF personal communications, *Proc. CICC*, May 1995.
4. B.-S. Song, CMOS RF circuits for data communications applications, *IEEE J. Solid-State Circuits*, vol. SC-21, no. 2, pp. 310-317, Apr. 1986.
5. P. Y. Chan, A. Rofougaran, K. A. Ahmed, and A. A. Abidi, A highly linear 1-GHz CMOS down-conversion mixer, *Proc. ESSCIRC*, pp.210-213, Sevilla, Sept. 1993.
6. J. Crols and M. Steyaert, A 1.5 GHz highly linear CMOS down-conversion mixer, *IEEE J. Solid-State Circuits*, vol. 30, no. 7, pp. 736-742, July 1995.
7. J. Y.-C. Chang, A. A. Abidi, and M. Gaitan, Large suspended inductors on silicon and their use in a 2-μm CMOS RF amplifier, *IEEE Electron Device Letters*, vol. 14, no. 5, pp. 246-248, May 1993.
8. C. H. Hull, R. R. Chu, and J. L. Tham, A Direct-Conversion Receiver for 900 MHz (ISM Band) Spread-Spectrum Digital Cordless Telephone, *Proc. IEEE Intl. Solid-State Circuits Conference*, pp. 344-345, San Francisco, Feb. 1996.
9. A. N. Karanicolas, A 2.7 V 900 MHz CMOS LNA and Mixer, *Proc. IEEE Intl. Solid-State Circuits Conference*, pp. 50-51, San Francisco, Feb. 1996.
10. A. A. Abidi, Radio frequency integrated circuits for portable communications, *Proc. CICC*, San Diego, pp. 151-158, May 1994.
11. M. Steyaert and W. Sansen, Opamp design toward maximum gain-bandwidth, *Proc. AACD Workshop*, Delft, pp. 63-85, March 1993.
12. J. Crols, P. Kinget, J. Craninckx, and M. Steyaert, An analytical model of planar inductors on lowly doped silicon substrates for high frequency analog design up to 3 GHz, *Proc. VLSI Circuits Symposium*, June 1996.
13. D. Rabaey and J. Sevenhans, The challenges for analog circuit design in mobile radio VLSI chips, *Proc. AACD Workshop*, vol. 2, Leuven, pp. 225-236, Mar. 1993.
14. T. Stetzler, I. Post, J. Havens, and M. Koyama, A 2.7V to 4.5V single-chip GSM transceiver RF integrated circuit, *Proc. IEEE Intl. Solid-State Circuits Conference*, San Francisco, pp. 150-151, Feb. 1995.
15. C. Marshall et al., A 2.7V GSM Transceiver ICs with On-chip Filtering, *Proc. IEEE Intl. Solid-State Circuits Conference*, San Francisco, pp. 148-149, Feb. 1995.
16. J. Sevenhans et al., An analog radio front-end chip set for a 1.9 GHz mobile radio telephone application, *Proc. IEEE Intl. Solid-State Circuits Conference*, San Francisco, pp. 44-45, Feb. 1994.
17. A. Rofougaran et al., A 1GHz CMOS RF front-end IC with wide dynamic range, *Proc. ESSCIRC*, Lille, pp. 250-253, Sept. 1995.
18. D. H. Shen, C.-M. Hwang, B. Lusignan, and B. A. Wooley, A 900 MHz integrated discrete-time filtering RF front-end, *Proc. IEEE Intl. Solid-State Circuits Conference*, San Francisco, pp. 54-55, Feb. 1996.
19. S. Sheng et al., A low-power CMOS chipset for spread spectrum communications, *Proc. IEEE Intl. Solid-State Circuits Conference*, San Francisco, pp. 346-347, Feb. 1996.
20. A. A. Abidi, High-frequency noise measurements on FETs with small dimensions, *IEEE Trans. Electron Devices*, vol. 33, no. 11, pp. 1801-1805, Nov. 1986.
21. B. Razavi, Analysis, modeling, and simulation of phase noise in monolithic voltage-controlled oscillators, *Proc. CICC*, pp. 323-326, May 1995.
22. J. Craninckx and M. Steyaert, Low-noise voltage controlled oscillators using enhanced LC-tanks, *IEEE Trans. Circ. Syst. II. Analog and Digital Signal Processing*, vol. 42, no. 12, pp. 794-804, Dec. 1995.

23. A. Rofourgan, J. Rael, M. Rofourgan, and A. Abidi, A 900-MHz CMOS LC-oscillator with quadrature outputs, *Proc. IEEE Intl. Solid-State Circuits Conference*, pp. 392-393, Feb. 1996.

24. N. M. Nguyen and R. G. Meyer, A 1.8-GHz monolithic LC voltage-controlled oscillator, *IEEE J. Solid-State Circuits*, vol. 27, no. 3, pp. 444-450, March 1992.

25. J. Craninckx and M. Steyaert, A 1.8-GHz low-phase-noise voltage-controlled oscillator with prescaler, *IEEE J. Solid-State Circuits*, vol. 30, no. 12, pp. 1474-1482, Dec. 1995.

26. J. Craninckx and M. Steyaert, A 1.75-GHz/3-V dual modulus divide-by-128/129 prescaler in 0.7-μm CMOS, *Proc. ESSCIRC*, pp. 254-257, Sept. 1995.

27. P. Kinget and M. Steyaert, A 1 GHz CMOS upconversion mixer, *Proc. CICC*, session 10.4, May 1996.

28. —, AD 7886, a 12-bit, 750 kHz, sampling ADC, *Analog Devices Data Sheet*, Apr. 1991.

29. J. Craninckx and M. Steyaert, A fully integrated spiral-LC CMOS VCO set with prescaler for GSM and DCS-1800 systems, *Proc. CICC*, pp. 403-406, May 1997.

30. M. Steyaert et al. A single chip CMOS transceiver for DCS1800 wireless communications, *Proc. IEEE Intl. Solid-State Circuits Conference*, Feb. 1998.

31. J. Craninckx and M. Steyaert, A fully integrated CMOS DCS-1800 frequency synthesizer, *Proc. IEEE Intl. Solid-State Circuits Conference*, San Francisco, pp. 372-373, Feb. 1998.

32. T. A. D. Riley, M. A. Copeland, and T. A. Kwasniewski, Delta-sigma modulation in fractional-N frequency synthesis, *IEEE J. Solid-State Circuits*, vol. 28, no. 5, pp. 553-559, May 1993.

33. M. Borremans, M. Steyaert, and T. Yoshitomi, A 1.5V wide band 3GHz CMOS quadrature direct up-converter for multi-mode wireless communications, *Proc. CICC '98*, pp. 79-82.

34. M. Borremans and M. Steyaert, A 2V low distortion 1 GHz CMOS up-converter mixer, *IEEE J. Solid-State Circuits*, vol. 33, no. 3, pp. 359-366, Mar. 1998.

10

PLL Circuits

Min-shueh Yuan
Chorng-kuang Wang
National Taiwan University

10.1 Introduction

What Is Phase-Locked?

Phase-locked loop (PLL) is a circuit architecture that causes a particular system to track with another one. More precisely, PLL synchronizes a signal (usually a local oscillator output) with a reference or an input signal in frequency as well as in phase.

Phase locking is a useful technique that can provide effective synchronization solutions in many data transmission systems such as optical communications, telecommunications, disk drive systems, and local networks, in which data are transmitted in baseband or passband. In general, only data signals are transmitted in most of these applications; namely, clock signals are not transmitted in order to save hardware cost. Therefore, the receiver should have some schemes to extract the clock information from the received data stream in order to recover transmitted data. The scheme is called a *timing recovery* or *clock recovery* circuit.

The cost of electronic interfaces in communication systems is higher in conjunction with the higher data rate. Hence, high-speed circuits are the practical issue of the high data rate systems implementation, and advanced VLSI technology plays an important role in cost reduction for the high-speed communication systems.

Basic Operation Concepts of Phase-Locked Loops (PLLs)

Typically, a PLL consists of three basic functional blocks: a phase detector (PD), a loop filter (LF), and a voltage-controlled oscillator (VCO). The PD detects the phase difference between the VCO output and the input signal, and generates a signal proportional to the phase error. The PD output contains a dc component and an ac component; the former is accumulated and the latter is filtered by the loop filter. The loop filter output that is near a dc signal is applied to the VCO. This almost dc control voltage changes the VCO frequency toward a direction to reduce the phase error between the input signal and the VCO. Depending on the type of loop filter used, the steady-state phase error will be reduced to zero or to a finite value.

PLL has an important feature: the ability to suppress both the noises superimposed on the input signal and generated by the VCO. The narrower the bandwidth of the PLL, the more effective jitter filtering the PLL can achieve. However, the error of the VCO frequency cannot be reduced rapidly. Although a narrow bandwidth is better for rejecting large amounts of input noise, it also prolongs the settling time in the acquisition process. So, there is a tradeoff between jitter filtering and fast acquisition.

Classification of PLL Types

Different PLL types are built from different classes of building blocks. The first PLL ICs appeared around 1965 and were purely analog devices. In the so-called "*linear PLLs*" (LPLLs), an analog multiplier (four-quadrant) is used as the phase detector, the loop filter is built from a passive or an active RC filter, and the voltage-controlled oscillator is used to generate the output signal of the PLL. In most cases, the input signal to this linear PLL is a sine wave, whereas the VCO output signal is a symmetrical square wave.

The classical *digital PLL* (DPLL) uses a digital phase detector such as an XOR gate, a JK-flipflop, or a phase-frequency detector (PFD), with the remaining blocks still being the same as LPLL. In many aspects, the DPLL performance is similar to the LPLL.

The function blocks of the *all-digital PLL* (ADPLL) are implemented by purely digital circuits, and the signals within the loop are digital too. Digital versions of the phase detector are the same as DPLL. The digital loop filter is built from an ordinary up/down counter, N-before-M counter, or K-counter.[1] The digital counterpart of the VCO is the digital-controlled oscillator (DCO).[2,3]

Analogous to filter designs, PLLs can be implemented by software such as a microcontroller, micro-computer, or digital signal processing (DSP); this type of PLL is called *software PLL* (SPLL).

10.2 PLL Techniques

Basic Topology

A phase-locked loop is a feedback system that operates and minimizes the phase difference between two signals. Figure 10.1 is the basic function block of a PLL, which consists of a phase detector (PD), a loop filter, and a VCO. The PD works as a phase error amplifier. It compares the phase of the VCO output signal $u_o(t)$ with the phase of the reference signal $u_i(t)$ and develops an output signal $u_d(t)$, that is proportional to the phase error θ_e. Within a limited range

$$u_d(t) = k_d \theta_e \qquad (10.1)$$

where k_d represents the gain of the PD and the unit of k_d is volt/rad.

The output signal $u_d(t)$ of the PD consists of a dc component and a superimposed ac component. The latter is undesired and removed by the loop filter (LPF). Thus, the LPF generates an almost-dc control voltage for the VCO to oscillate at a frequency equal to the input frequency.

How the building blocks of the basic PLL work together will be explained in the following. First, the angular frequency ω_i of the input signal $u_i(t)$ is assumed equivalent to the central frequency ω_o of the VCO signal $u_o(t)$. Now, a small positive frequency step is applied to $u_i(t)$ at $t = t_o$ (shown in Fig. 10.2). $u_i(t)$ accumulates a phase faster than $u_o(t)$ of the VCO does. The PD then generates wider pulses increasingly, which results in a higher dc voltage at the LPF output to increase the VCO frequency. Depending on the type of the loop filter, the final phase error will be reduced to zero or to a finite value.

It is important to note from the descriptions in the above that the loop locks only after two conditions are satisfied: (1) ω_i and ω_o are equal and (2) the phase difference between the input $u_i(t)$ and the VCO

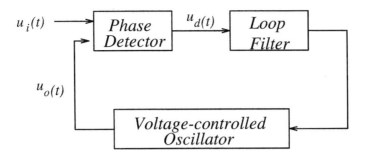

FIGURE 10.1   Basic block diagram of the PLL.

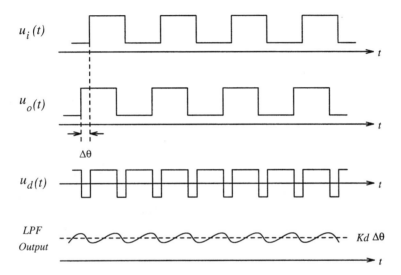

FIGURE 10.2   Waveforms in a PLL.

output $u_o(t)$ settles to a steady-state value. If the phase error varies with time so fast that the loop is unlocked, the loop must keep on the transient process, which involves both *frequency acquisition* and *phase acquisition*.

To design a practical PLL system, it is required to know the status of the responses of the loop if (1) the input frequency is varied slowly (tracking process), (2) the input frequency is varied abruptly (lock-in process), and (3) the input and the output frequencies are not equal initially (acquisition process). Using linear PLL as an example, these responses will be shown in the following subsections.

Loop Order of the PLL

Second-Order Loop

The loop order of the PLL depends on the characteristics of the loop filter; therefore, the loop filter is a key component that affects the PLL dynamic behavior. Figure 10.3 shows three types of loop filters that are widely used. Figure 10.3(a) is a passive lead-lag filter; the associate transfer function $F(s)$ is given by

$$F(s) = \frac{1 + s\tau_2}{1 + s(\tau_1 + \tau_2)} \tag{10.2}$$

where $\tau_1 = R_1 C$ and $\tau_2 = R_2 C$. Figure 10.3(b) shows an active lead-lag filter; its transfer function is

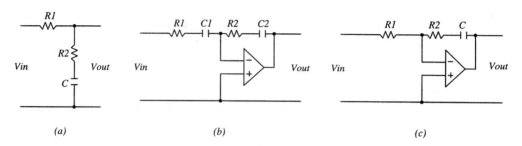

FIGURE 10.3 (a) Passive lead-lag filter, (b) active lead-lag filter, and (c) active PI filter.

$$F(s) = k_a \frac{1+s\tau_2}{1+s\tau_1} \qquad (10.3)$$

where $\tau_1 = R_1 C_1$, $\tau_2 = R_2 C_2$ and $k_a = -\frac{C_1}{C_2}$. A PI filter is shown in Fig. 10.3(c), where PI stands for *Proportional and Integral* action. The transfer function is given by

$$F(s) = \frac{1+s\tau_2}{s\tau_1} \qquad (10.4)$$

where $\tau_1 = R_1 C$ and $\tau_2 = R_2 C$. Their Bode diagrams are shown in Fig. 10.4. High-order filters could be used in some applications, but additional filter poles introduce a phase shift. High-order systems are not trivial generally to maintain a stable system.

Figure 10.5 shows the linear block diagram of the PLL. According to control theory, the closed loop transfer function of PLL can be derived as

$$H(s) \triangleq \frac{\theta_o(s)}{\theta_i(s)} = \frac{k_d k_o F(s)}{s + k_d k_o F(s)} \qquad (10.5)$$

where k_d with units V/rad is called the phase-detector gain, k_o is the VCO gain factor and has units rad/s-V. In addition to the phase transfer function, a phase-error transfer function $H_e(s)$ is defined as follows:

$$H_e(s) \triangleq \frac{\theta_e(s)}{\theta_i(s)} = \frac{s}{s + k_d k_o F(s)} \qquad (10.6)$$

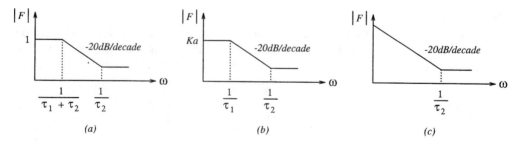

FIGURE 10.4 Bode diagrams of (a) passive lead-lag filter, (b) active lead-lag filter, and (c) active PI filter.

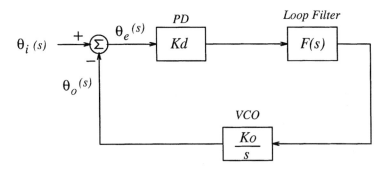

FIGURE 10.5 Linear model of PLL.

The transfer functions of the loop filters shown in Fig. 10.3 are substituted for $F(s)$ in Eq. 10.5 in order to analyze the phase transfer function. We obtain the phase transfer functions as follows:

For the passive lead-lag filter,

$$H(s) = \frac{k_d k_o \dfrac{1 + s\tau_2}{\tau_1 + \tau_2}}{s^2 + s\dfrac{1 + k_d k_o \tau_2}{\tau_1 + \tau_2} + \dfrac{k_d k_o}{\tau_1 + \tau_2}} = \frac{\omega_n\left(2\zeta - \dfrac{\omega_n}{k_d k_o}\right)s + \omega_n^2}{s^2 + 2s\zeta\omega_n + \omega_n^2} \tag{10.7}$$

For the active lead-lag filter,

$$H(s) = \frac{k_d k_a k_o \dfrac{1 + s\tau_2}{\tau_1}}{s^2 + s\dfrac{1 + k_d k_a k_o \tau_2}{\tau_1} + \dfrac{k_d k_a k_o}{\tau_1}} = \frac{\omega_n\left(2\zeta - \dfrac{\omega_n}{k_d k_o k_a}\right)s + \omega}{s^2 + 2s\zeta\omega_n + \omega_n^2} \tag{10.8}$$

For the active PI filter,

$$H(s) = \frac{k_d k_o \dfrac{1 + s\tau_2}{\tau_1}}{s^2 + s\dfrac{k_d k_o \tau_2}{\tau_1} + \dfrac{k_d k_o}{\tau_1}} = \frac{2\zeta\omega_n s + \omega_n^2}{s^2 + 2s\zeta\omega_n + \omega_n^2} \tag{10.9}$$

where ω_n is the *natural frequency* and ξ is the *damping factor*. These two parameters are important to characterize a PLL. If the condition $k_d k_o \gg \omega_n$ or $k_d k_o k_o \gg \omega_n$ is true, this PLL system is called a *high-gain loop*. If the reverse is true, the system is a *low-gain loop*. Most practical PLLs are high-gain loops for good tracking performance. For a high-gain loop, Eqs. 10.7 to 10.9 become approximately

$$H(s) \approx \frac{2\zeta\omega_n s + \omega_n^2}{s^2 + 2s\zeta\omega_n + \omega_n^2} \tag{10.10}$$

Similarly, assuming a high-gain loop, the approximate expression of the phase-error transfer function $H_e(s)$ for all three loop filter types becomes

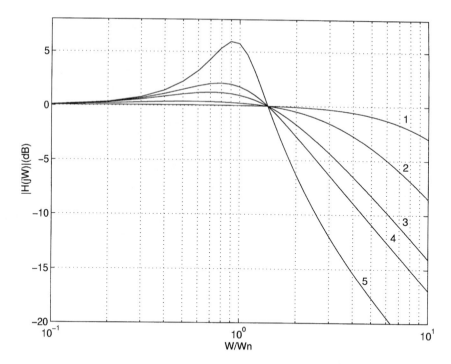

FIGURE 10.6 Frequency responses of the phase transfer function $H(j\omega)$ for different damping factors. Trace 1: $\zeta = 5$, Trace 2: $\zeta = 2$, Trace 3: $\zeta = 1$, Trace 4: $5 = 0.707$, Trace 5: $3 = 0.3$.

$$H_e(s) \approx \frac{s^2}{s^2 + 2s\zeta\omega_n + \omega_n^2} \tag{10.11}$$

Because the highest power of s in the denominator of the transfer function is 2, the loop is known as the second-order loop.

The magnitude of the frequency responses of $H_e(s)$ for a high-gain loop with several values of damping factor are plotted in Fig. 10.6. It shows that the loop performs a low-pass filtering on the input phase signal. That is, the second-order PLL is able to track both phase and frequency modulations of the input signal as long as the modulation frequency remains within the frequency band roughly between zero and ω_n.

The magnitude of the frequency responses of $H_e(s)$ are plotted in Fig. 10.7. A high-pass characteristic is derived. It indicates that the second-order PLL tracks the low-frequency phase error but cannot track high-frequency phase error.

The transfer function $H(s)$ has a -3dB frequency, ω_{-3db}. ω_{-3db} stands for the closed-loop bandwidth of the PLL. The relationship between ω_{-3db} and ω_n is presented here to provide a comparison with the familiar concept of bandwidth. In the high-gain loop case, by setting $|H(j\omega)| = \frac{1}{\sqrt{2}}$ and solving for ω, we find that

$$\omega_{-3db} = \omega_n \left[2\zeta^2 + 1 + \sqrt{\left(2\zeta^2 + 1\right)^2} \right]^{\frac{1}{2}} \tag{10.12}$$

The relationship between ω_{-3db} and ω_n for different damping factors is plotted in Fig. 10.8.[4]

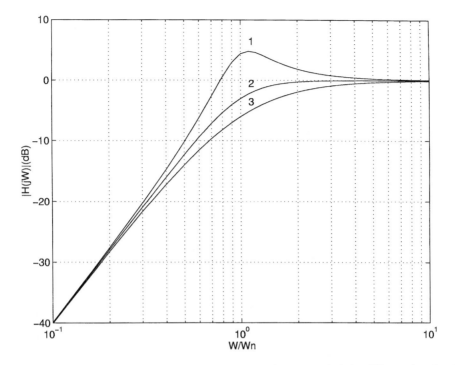

FIGURE 10.7 Frequency responses of the phase-error transfer function $H_e(j\omega)$ for different damping factors. Trace 1: $\zeta = 0.3$, Trace 2: $\zeta = 0.707$, Trace 3: $\zeta = 1$.

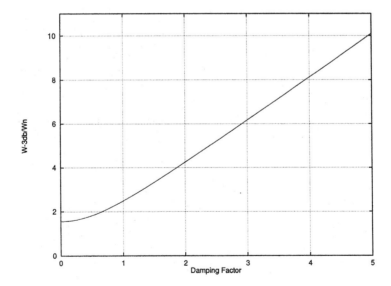

FIGURE 10.8 ω_{-3db} bandwidth of a second-order loop versus different damping factors.

Other-Order Loop

The second-order PLL with a lead-lag loop filter is commonly used because the lead-lag filter has two time constants to determine the natural frequency and the damping factor independently. Therefore, a high dc loop gain for good tracking, while maintaining an overdamped PLL system, can be achieved. No loop filter and single pole loop filter cases are discussed briefly in the following for completeness.

If the loop filter is left out, a first-order loop is obtained. As shown in Fig. 10.5, set $F(s) = 1$ and the closed-loop transfer function can be derived as

$$H(s) = \frac{k}{s+k} \tag{10.13}$$

where $k = k_d k_o$. If it is necessary to design a high dc gain loop for fast tracking, then the bandwidth of the PLL must be wide enough because the dc loop gain k is the only parameter available, which is not suitable for noise suppression. Therefore, fast tracking and narrow bandwidth are incompatible in a first-order loop.

Another commonly used loop filter is the passive lag filter. The transfer function is

$$F(s) = \frac{1}{1+s\tau} \tag{10.14}$$

The closed-loop transfer function can be derived as

$$H(s) = \frac{\dfrac{k_d k_o}{\tau}}{s^2 + \dfrac{1}{\tau}s + \dfrac{k_d k_o}{\tau}} = \frac{\tau \omega_n^2}{s^2 + 2\zeta\omega_n s + \omega_n^2} \tag{10.15}$$

where $\omega_n = \sqrt{\dfrac{k_d k_o}{\tau}}$ and $\zeta = \dfrac{1}{2\sqrt{\tau k_d k_o}}$. Although there are two parameters (τ and $k = k_o k_d$) available, three loop parameters must be determined (ω_n, ξ, k). That is if it is necessary to have a large dc loop gain and a narrow bandwidth, the loop will be severely underdamped and the transient response will be poor.

A high-order filter is used for critical applications because it provides better noise filtering. However, it is difficult to design a high-order loop due to some problems such as loop stability.

Tracking Process

The linear model of a PLL shown in Fig. 10.5 is suitable for analyzing the tracking performance of a PLL that is initially locked, but with small phase error. If the phase error changes so abruptly that the PLL loses lock, a large phase error will be induced, even if it happens only momentarily. The unlock condition is a non-linear process that cannot be analyzed by a linear model. The acquisition process will be described later.

At first, consider a step phase error applied to the input. The input phase signal is $\theta_i(t) = \Delta\theta u(t)$, the Laplace transform of the input is $\theta_i(s) = \dfrac{\Delta\theta}{s}$, which is substituted into Eq. 10.11 to get

$$\theta_e(s) = \frac{\Delta\theta}{s} \frac{s^2}{s^2 + 2\zeta\omega_n s + \omega_n^2} \tag{10.16}$$

According to the final value theorem of the Laplace transform,

$$\lim_{t\to\infty} \theta_e(t) = \lim_{s\to 0} s\theta_e(s) = 0$$

In other words, the loop will eventually track on the step phase change without steady-state phase error.

Applying a step change of frequency $\Delta\omega$ to the input, the input phase change is a ramp (i.e., $\theta_i(t) = \Delta\omega t$); therefore, $\theta_i(s) = \dfrac{\Delta\omega}{s^2}$. Substituting $\theta_i(s)$ in Eq. 10.6 and applying the final value theorem, one obtains

$$\theta_v = \lim_{t \to \infty} \theta_e(t) = \lim_{s \to 0} s\theta_e(s)$$

$$= \lim_{s \to 0} \frac{\Delta\omega}{s + k_d k_o F(s)}$$

(10.17)

$$= \frac{\Delta\omega}{k_d k_o F(0)}$$

$$= \frac{\Delta\omega}{k_v}$$

θ_v is called the *velocity error* or *static phase error*.[4] Because the input frequency almost never agrees exactly with the VCO free-running frequency, there is a frequency difference $\Delta\omega$ between the two. From Eq. 10.17, if the PLL has a high dc loop gain, (i.e., $k_d k_o F(0) \gg \Delta\omega$), the steady-state phase error to a step frequency error input approaches zero. This is the reason why a high-gain loop has a good tracking performance. Now the advantage of a second-order loop using an active loop filter of high dc gain is evident. The active lead-lag loop filter with a high dc gain will make the steady-state phase error approach zero and the noise bandwidth narrow simultaneously, which is impossible in a first-order loop.

If the input frequency is changed linearly with time at a rate of $\Delta\dot{\omega}$, which is $\theta_i(t) = \frac{1}{2} \Delta\dot{\omega} t^2$, so $\theta_i(s) = \frac{\Delta\dot{\omega}}{s^3}$. According to a high-gain loop and applying the final value theorem of Laplace transform, then

$$\theta_a = \lim_{t \to \infty} \theta_e(t) = \lim_{s \to 0} s\theta_e(s)$$

$$= \frac{\Delta\dot{\omega}}{\omega_n^2}$$

(10.18)

θ_a is called an *acceleration error* (sometimes called *dynamic tracking error* or *dynamic lag*).[4]

In some applications, PLL needs to track an accelerating phase error without static tracking error. The expression of the static phase error when frequency ramp is applied

$$\theta_e(s) = \lim_{s \to 0} \frac{\Delta\dot{\omega}}{s(s + k_d k_o F(s))}$$

(10.19)

For θ_e to be zero, it is necessary to make $F(s)$ be a form of $\frac{G(s)}{s^2}$, where $G(0) \neq 0$. $\frac{G(s)}{s^2}$ implies that the loop filter has two cascade integrators. This is a third-order loop. In order to eliminate the static acceleration error, a third-order loop is very useful for some special applications, such as satellites and missile systems.

Based on Eq. 10.18, a large natural frequency ω_n is used to reduce the static tracking phase error in a second-order loop; however, a wide natural frequency has an undesired noise filtering performance. In contrast, the zero tracking phase error for a frequency ramp error is concordant with a small loop bandwidth in a third-order loop.

The preceding analysis on tracking process is under the assumption that the phase error is relatively small and the loop is linear. If the phase error is large enough to make the loop drop out of lock, the linear assumption is invalid. For a sinusoidal-characteristic phase detector, the exact phase expression of Eq. 10.17 should be

$$\sin\theta_v = \frac{\Delta\omega}{k_v} \tag{10.20}$$

The sine function has solutions when $\Delta\omega \leq k_v$. Therefore, there is no solution if $\Delta\omega > k_v$. This is the case when the loop loses lock and the output of the phase detector will be beat notes signal rather than a dc control voltage. Therefore, k_v can be used to define the *hold range* of the PLL; that is

$$\Delta\omega_H = \pm k_v = k_o k_d F(0) \tag{10.21}$$

The hold range is the frequency range in which a PLL is able to maintain lock *statically*. Namely, if input frequency offset exceeds the hold range statically, the steady-state phase error would drop out of the linear range of the phase detector and the loop loses lock. The hold range expressed in Eq. 10.21 is not correct when some other components in PLL are saturated earlier than the phase detector. k_v is a function of k_o, k_d, and $F(0)$. The dc gain $F(0)$ of the loop filter depends on the filter type. Then hold range $\Delta\omega_H$ can be $k_o k_d$, $k_o k_d k_a$, and ∞ for passive lead-lag filter, active lead-lag filter, and active PI filter, respectively. When the PI filter is used, the real hold range is actually determined by the control range of the VCO.

Considering the dynamic phase error θ_a in a second-order loop, the exact expression for a sinusoidal characteristic phase detector is

$$\sin\theta_a = \frac{\Delta\dot{\omega}}{\omega_n^2} \tag{10.22}$$

which implies that the maximum change rate of the input frequency is ω_n^2. If the rate exceeds ω_n^2, the loop will fall out of lock.

Lock-in Process

The *lock-in* process is defined as PLL locks within one single beat note between the input and the output (VCO output) frequency. The maximum frequency difference between the input and the output that the PLL can lock within one single beat note is called the *lock-in range* of the PLL.

Figure 10.9 is the case in which a frequency offset $\Delta\omega$ is less than the lock-in range. Then, PLL will lock within one single beat note between ω_i and ω_o, and the lock-in process happens. In Fig. 10.10(b), the frequency offset $\Delta\omega$ between input (ω_i) and output (ω_o) is larger than the lock-in range; hence, the lock-in process will not take place, at least not instantaneously.

The magnitude of the lock-in range can be derived approximately in the following. Suppose the PLL is unlocked initially. The input frequency ω_i is $\omega_o + \Delta\omega$. If the input signal $v_i(t)$ is a sine wave and given by

$$v_i(t) = A_i \sin(\omega_o t + \Delta\omega t) \tag{10.23}$$

and the VCO output signal $v_o(t)$ is usually a square wave written as a Walsh function[5]

$$v_o(t) = A_o W(\omega_o t) \tag{10.24}$$

$v_o(t)$ can be replaced by the Fourier Series,

$$v_o(t) = A_o \left[\frac{4}{\pi}\cos(\omega_o t) + \frac{4}{3\pi}\cos(3\omega_o t) + \ldots \right] \tag{10.25}$$

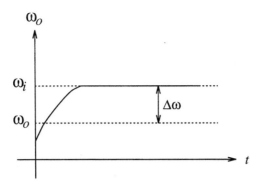

FIGURE 10.9   Lock-in process of the PLL.

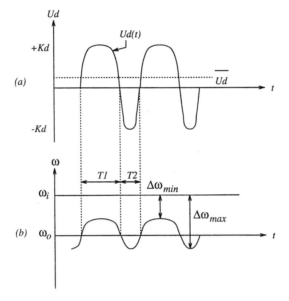

FIGURE 10.10   Pull-in process of the PLL.

So, the phase detector output v_d is

$$v_d(t) = v_i(t)v_o(t) = A_i A_o \left[\frac{2}{\pi} \sin(\Delta\omega t) + \ldots \right]$$ (10.26)

$$= k_d \sin(\Delta\omega t) + high\text{-}frequency\ terms$$

The high-frequency components can be filtered out by the loop filter. The output of the loop filter is given by

$$v_f(t) \approx k_d \left| F(\Delta\omega) \right| \sin(\Delta\omega t)$$ (10.27)

The peak frequency deviation based on Eq. 10.27 is equal to $k_d k_o |F(\Delta\omega)|$. If the peak deviation is larger than the frequency error between ω_i and ω_o, the lock-in process will take place. Hence, the lock-in range is given by

$$\Delta\omega_L = k_d k_o \left| F(\Delta\omega_L) \right| \qquad (10.28)$$

The lock-in range is always larger than the corner frequency $\frac{1}{\tau_1}$ and $\frac{1}{\tau_2}$ of the loop filter in practical cases. An approximation of the loop filter gain $F(\Delta\omega_L)$ is shown as follows:

For the passive lead-lag filter,

$$F(\Delta\omega_L) \approx \frac{\tau_2}{\tau_1 + \tau_2}$$

For the active lead-lag filter,

$$F(\Delta\omega_L) \approx k_a \frac{\tau_2}{\tau_1}$$

For the active PI filter,

$$F(\Delta\omega_L) \approx \frac{\tau_2}{\tau_1}$$

τ_2 is usually much smaller than τ_1, and the $F(\Delta\omega_L)$ can be further approximated as follows:

For the passive lead-lag filter,

$$F(\Delta\omega_L) \approx \frac{\tau_2}{\tau_1}$$

For the active lead-lag filter,

$$F(\Delta\omega_L) \approx k_a \frac{\tau_2}{\tau_1}$$

For the active PI filter,

$$F(\Delta\omega_L) \approx \frac{\tau_2}{\tau_1}$$

Substituting the above equations into Eq. 10.28 and assuming a high-gain loop,

$$\Delta\omega_L = 2\zeta\omega_n \qquad (10.29)$$

can be obtained for all three types of loop filters shown in Fig. 10.3.

Acquisition Process

Suppose that the PLL does not lock initially, and the input frequency is $\omega_i = \omega_o + \Delta\omega$, where ω_o is the initial frequency of VCO. If the frequency error $\Delta\omega$ is larger than the lock-in range, the lock-in process will not happen. Consequently, the output signal $u_d(t)$ of the phase detector shown in Fig. 10.10(a) is a sine wave that has a frequency $\Delta\omega$. The ac phase detector output signal $u_d(t)$ passes through the loop filter. Then the output $u_f(t)$ of the loop filter modulates the VCO frequency. As shown in Fig. 10.10(b), when ω_o increases, the frequency difference between ω_i and ω_o becomes smaller and vice versa. Therefore, the phase detector output $u_d(t)$ becomes asymmetric. That is, the duration of positive half-periods of the phase detector output is larger than the negative ones. The average value $\overline{u_d(t)}$ of the phase detector output therefore goes to slightly positive. Then the frequency of VCO will be pulled up until it reaches the input frequency. This phenomenon is called a *pull-in process*.

Because the pull-in process is a non-linear behavior, the mathematical analysis is quite complicated. According to the results in Ref. 1, the pull-in range and the pull-in time depend on the type of loop filter. For an active lead-lag filter with a high-gain loop, the pull-in range is

$$\Delta\omega_P \approx \frac{4\sqrt{2}}{\pi}\sqrt{\zeta\omega_n k_o k_d} \tag{10.30}$$

and the pull-in time is

$$T_P \approx \frac{\pi^2}{16}\frac{\Delta\omega_0^2 k_a}{\zeta\omega_n^3} \tag{10.31}$$

where $\Delta\omega0$ is the initial frequency error. Equations 10.30 and 10.31 will be modified for different types of phase detectors.[1]

Aided Acquisition

The PLL bandwidth is always too narrow to lock a signal of large frequency error. Furthermore, the frequency acquisition is slow and impractical. Therefore, there are aided frequency-acquisition techniques to solve this problem, such as the frequency-locked loop (FLL) and bandwidth-widening methods.

The frequency-locked loop, which is very similar to a PLL, is composed of a frequency discriminator, a loop filter, and a VCO. PLL is a coherent mechanism to recover a signal buried in noise. FLL, in contrast, is a non-coherent scheme that cannot distinguish between signal and noise. Therefore, an FLL can only be useful to provide signal frequency, which usually implies that the input signal power must exceed the noise.

The major difference between PLL and FLL is the phase detector and the frequency discriminator. The frequency discriminator is the frequency detector in the FLL. It generates a voltage proportional to the frequency difference between the input and the VCO. The frequency difference will be driven to zero in a negative feedback fashion. If a linear frequency detector is employed, it can be shown that the frequency-acquisition time is proportional to the logarithm of the frequency error.[6] In the literature, some frequency detectors like the quadricorrelator,[7] balance quadricorrelator,[8] rotational frequency detector,[9] and frequency delimiter[10] are disclosed.

Delay-Locked Loop

Two major approaches for adjustable timing elements are VCO and voltage-controlled delay line (VCDL). Figure 10.11 shows a typical delay-locked loop (DLL)[11,12] that replaces the VCO of a PLL with a VCDL. The input signal is delayed by an integer multiple of the signal period because the phase error is zero when the phase difference between V_{in} and V_o approaches a multiple of the signal periods. The VCDL usually consists of a number of cascaded gain stages with variable delay. Delay lines, unlike ring oscillators, cannot generate a signal; therefore, it is difficult to make frequency multiplication in a DLL.

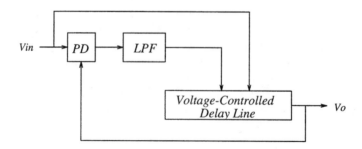

FIGURE 10.11 DLL block diagram.

In a VCO, the output "frequency" is proportional to the input control voltage. The phase transfer function contains a pole, which is $H(s) = \frac{k_o}{s}$ (k_o is the VCO gain). In a VCDL, the output "phase" is proportional to the control voltage, and the phase transfer function is $H(s) = k_{VCDL}$. So, the DLL can be easily stabilized with a simple first-order loop filter. Consequently, DLLs have much more relaxed tradeoffs among gain, bandwidth, and stability. This is one of the two important advantages over PLLs. Another advantage is that delay lines typically introduce much less jitter than a VCO.[13] Because a delay chain is not configured as a ring oscillator, there is no jitter accumulation because the noise does not contribute to the starting point of the next clock cycle.

Charge-Pump Phase-Locked Loop

A charge-pump PLL usually consists of four major blocks as shown in Fig. 10.12. The phase detector is a purely phase-frequency detector. The charge-pump circuit converts the digital signals UP, DN, and null (neither up nor down) generated by the phase detector into a corresponding charge-pump current I_p, $-I_p$, and zero. The loop filter is usually a passive RC circuit converting the charge-pump current into an analog voltage to control the VCO. So, the purpose of the "charge-pump" is to convert the logic state of the phase-frequency detector output into an analog signal suitable for controlling the voltage-controlled oscillator. The linear model of a charge-pump PLL is shown in Fig. 10.13. The k_d is the equivalent gain of a charge-pump circuit and a loop filter, which is shown in Fig. 10.14. If the loop bandwidth is much smaller than the input frequency, the detailed behavior within a single cycle can be ignored. Then, the state of a PLL can be assumed to be only changed by a small amount during each input cycle. Actually, the "average" behavior over many cycles is interesting. Then, the average current charging the capacitor is given by

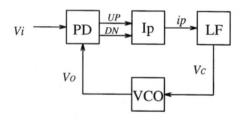

FIGURE 10.12 Charge-pump PLL diagram.

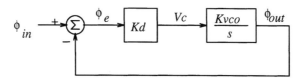

FIGURE 10.13   The linear model of charge-pump PLL.

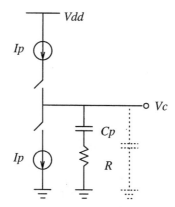

FIGURE 10.14   The schematic of a loop filter.

$$I_{avg} = \frac{Q}{T} = \frac{I\Delta t}{T}$$

$$= \frac{I\left(\frac{\phi_e}{2\pi}\right)T}{T} \tag{10.32}$$

$$= \frac{I\phi_e}{2\pi}$$

And the average k_d in Fig. 10.13 is

$$k_d \triangleq \frac{v_c}{\phi_e} = \frac{I}{2\pi}\left(R + \frac{1}{C_p s}\right) \tag{10.33}$$

The closed-loop transfer function can be obtained as

$$H(s) \triangleq \frac{\phi_{out}}{\phi_{in}} = \frac{k_d \dfrac{k_{vco}}{s}}{1 + \dfrac{k_d k_{vco}}{s}}$$

$$= \frac{\dfrac{I}{2\pi C_p}\left(RC_p s + 1\right)k_{vco}}{s^2 + \dfrac{I}{2\pi}k_{vco}Rs + \dfrac{Ik_{vco}}{2\pi C_p}} \tag{10.34}$$

Generally, a second-order system is characterized by the natural frequency $f_n = \dfrac{\omega_n}{2\pi}$ and the damping factor ζ. So,

$$\omega_n = \sqrt{\frac{I}{2\pi C_p} k_{vco}} \; rad \; sec$$

$$\zeta = \frac{RC_p}{2} \omega_n \tag{10.35}$$

For the stability consideration, there is a limitation of a normalized natural frequency F_N:[14]

$$F_N \triangleq \frac{f_n}{f_i} < \frac{\sqrt{1+\zeta^2}-\zeta}{\pi} \tag{10.36}$$

In the single-ended charge pump, the resistor added in series with the capacitor shown in Fig. 10.14 can introduce "ripple" in the control voltage V_c even when the loop is locked.[15] The ripple control voltage modulates the VCO frequency and results in phase noise. This effect is especially undesired in frequency synthesizers. In order to suppress the ripple, a second-order loop filter, as shown in Fig. 10.14 with a shunt capacitor in dotted lines, is used. This configuration introduces a third pole in the PLL. Stability issues must be taken care of. Gardner[15] provides criteria for the stability of the third-order PLL.

An important property of any PLL is the static phase error that arises from a frequency offset $\Delta\omega$ between the input signal and the free-running frequency of the VCO. According to the analysis of Ref. 15, the static phase error is

$$\theta_v = \frac{2\pi\Delta\omega}{k_o I_p F(0)} \; rad \tag{10.37}$$

To eliminate the static phase error in conventional PLLs, an active loop filter with a high dc gain ($F(0)$ is large) is preferred. But the charge-pump PLL allows zero static phase error without the need for a large dc gain of the loop filter. This effect arises from the input open circuit during the "null" state (charge-pump current is zero). Real circuits will impose some resistive loading R_s in parallel to the loop filter. Therefore, the static phase error, from Eq. 10.37 will be

$$\theta_v = \frac{2\pi\Delta w}{k_o I_p R_s} \; rad \tag{10.38}$$

The shunt resistive loading most likely comes from the input of a VCO control terminal. Compared with the static phase error of a conventional PLL as expressed in Eq. 10.17, the same performance can be obtained from a charge-pump PLL without a high dc-gain loop filter.[16]

PLL Noise Performance

In high-speed data recovery applications, better performance of the VCO and the overall phase-locked loop itself is desired. The random variations of the sampling clock, so-called jitter, is the critical performance parameter.

Jitter sources of PLL, when using a ring voltage-controlled oscillator, mainly come from the input and the VCO itself. The ring oscillator jitter is associated with the power supply noise, the substrate noise, $1/f$ noise, and the thermal noise. The former two noise sources can be reduced by fully differential circuit structure. $1/f$ noise, on the other hand, can be rejected by the tracking capability of the PLL. Therefore, the thermal noise is the worst noise source. From the analysis in Ref. 17, the one-state RMS timing jitter error of the ring oscillator normalized to the time delay per stage can be shown as

$$\frac{\Delta\tau_{rms}}{t_d} \approx \sqrt{\frac{2KT}{C_L}} \left(\sqrt{1+\frac{2}{3}a_v}\right) \frac{1}{V_{pp}} \tag{10.39}$$

where C_L is the load capacitance, $\sqrt{1+\frac{2}{3}a_v}$ is called the noise contribution factor ς, a_v is the small-signal gain of the delay cell, and Vpp is the VCO output swing. From Eq. 10.39, for a fixed output bandwidth, higher gain contributes larger noise.

Because the ring oscillator is a feedback architecture, the noise contribution of a single delay cell may be amplified and filtered by the following stage. To consider two successive stages, Eq. 10.39 can be rearranged as:[17]

$$\frac{\Delta\tau_{rms}}{t_d} \approx \sqrt{\frac{2KT}{C_L}\frac{1}{\left(V_{gs}-V_t\right)}}\varsigma \tag{10.40}$$

Therefore, the cycle-to-cycle jitter of the ring oscillator in a PLL can be predicted[17] by

$$\overline{\left(\Delta\tau_N\right)^2} = \frac{KT}{I_{ss}}\frac{a_v\varsigma^2}{\left(V_{gs}-V_t\right)}T_o \tag{10.41}$$

where I_{ss} is the current of the delay cell, and T_o is the output period of the VCO. Based on Eq. 10.41, designing a low jitter VCO, $(V_{gs} - V_t)$ should be as large as possible. For fixed delay and fixed current, a lower gain of each stage is better for jitter performance, but the loop gain must satisfy the Barkhausen criterion. From the viewpoint of VCO jitter, a wide bandwidth of the PLL can correct the timing error of the VCO rapidly.[13] If the bandwidth is too wide, the input noise jitter may be so large that it dominates the jitter performance of the PLL. Actually, this is a tradeoff.

For a phase-locked loop design, the natural frequency and the damping factor are the key parameters to be determined by designers. If the signal-to-noise ratio $(SNR)_i$ is defined, then the output signal-to-noise ratio $(SNR)_o$ can be obtained:[4]

$$\left(SNR\right)_o = \left(SNR\right)_i\frac{B_i}{2B_L} \tag{10.42}$$

where B_i is the bandwidth of the prefilter and B_L is the noise bandwidth. Hence, the B_L can be derived using Eq. 10.42. And the relationship of B_L with ω_n and ζ is

$$B_L = \frac{\omega_n}{2}\left(\zeta+\frac{1}{4\zeta}\right) \tag{10.43}$$

Therefore, the ω_n and ζ can be designed to satisfy the $(SNR)_o$ requirement.

Besides the system and circuit designs, jitter can be reduced in the board level design. Board jitter can be alleviated by better layout and noise decoupling schemes like such as appending proper decouple and bypass capacitances.

PLL Design Considerations

A PLL design starts with specifying the key parameters such as natural frequency ω_n, lock-in range $\Delta\omega_L$, damping factor ζ, and the frequency control range which depend significantly on applications. Design procedures based on a practical example will be described as follows:

Step 1. Specify the damping factor ζ. The damping factor of the PLL determines the responses of phase or frequency error steps applied to the input. ζ should be considered to achieve fast response, small overshoot, and minimum noise bandwidth B_L. If ζ is very small, large overshoot will occur and the overshoot causes phase jitter.[18] If ζ is too large, the response will become sluggish.

Step 2. Specify the lock-in range $\Delta\omega_L$ or the noise bandwidth B_L. As shown in Eqs. 10.29 and 10.43, the natural frequency ω_n depends on $\Delta\omega_L$ and ζ (or B_L and ζ). If the noise is not the key issue of the PLL, one can ignore the noise bandwidth and specify the lock-in range. Where noise is of concern, one should specify B_L first, and keep the lock-in range of the PLL.

Step 3. Calculate the ω_n according to Step 2. If the lock-in range has been specified, Eq. 10.29 indicates that

$$\omega_n = \frac{\Delta\omega_L}{2\zeta} \tag{10.44}$$

If the noise bandwidth has been specified, Eq. 10.43 indicates the natural frequency as

$$\omega_n = \frac{2B_L}{\zeta + \frac{1}{4}\zeta} \tag{10.45}$$

Step 4. Determine the VCO gain factor k_o and the phase detector gain k_d. k_o and k_d are both characterized by circuit architectures and they must achieve the requirement of the lock-in range specified in Step 2. For example, if k_o or k_d is too small, the PLL will fail to achieve the desired lock-in range.

Step 5. Choose the loop filter. Different types of loop filters are available, as shown in Fig. 10.3. Eqs. 10.7 to 10.9, ω_n and ζ (specified above) are used to derive the time constants of the loop filter.

10.3 Building Blocks of the PLL Circuit

Voltage-Controlled Oscillators

The function of a voltage-controlled oscillator (VCO) is to generate a stable and periodic waveform whose frequency can be varied by an applied control voltage. The relationship between the control voltage and the oscillation frequency depends upon the circuit architecture. A linear characteristic is generally preferred because of its wider applications. As a general classification, VCOs can be roughly categorized into two types by the output waveforms: (1) *harmonic oscillators* that generate nearly sinusoidal outputs, and (2) *relaxation oscillators* that provide square or triangle outputs. In general, a harmonic oscillator is composed of an amplifier that provides an adequate gain and a frequency-selective network that feeds a certain output frequency range back to the input. LC-tank oscillators and crystal oscillators belong to this type.

Relaxation oscillators are the most commonly used oscillator configuration in monolithic IC design because they can operate in a wide frequency range with a minimum number of external components. According to the mechanism of the oscillator topology employed, relaxation oscillators can be further categorized into three types: (1) grounded capacitor VCO,[19] (2) emitter-coupled VCO, and (3) delay-based ring VCO.[20] The operation of the first two oscillators is similar in the sense that time duration spent in each state is determined by the timing components and charge/discharge currents. The delay-

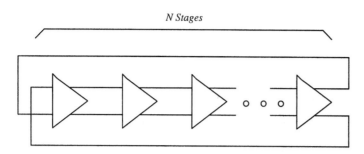

N Stages

FIGURE 10.15   Ring oscillator.

based ring VCO operates quite differently since the timing relies on the delay in each of the gain stages that are connected in a ring configuration.

Generally, harmonic oscillators have the following advantages: (1) superior frequency stability, which includes the frequency stability with temperature, power supply, and noise; and (2) good frequency accuracy control, because the oscillation frequency is determined by a tank circuit or a crystal.

Nevertheless, harmonic oscillators are not compatible with monolithic IC technology and their frequency turning range is limited. On the contrary, relaxation oscillators are easy to implement in monolithic ICs. Since frequency is normally proportional to a controlled-current or -voltage and inversely proportional to timing capacitors, the frequency of oscillation can be varied linearly over a very wide range. Coming from the ease of frequency tuning, the drawbacks of such oscillators are poor frequency stability and frequency inaccuracy.

Recently, the ring oscillator has received considerable attentions in high-frequency PLL applications for clock synchronization and timing recovery. Since they can provide high-frequency oscillation with simple digital-like circuits that are compatible with digital technology, they are suitable for VLSI implementations.

In order to achieve high rejection of power supply and substrate noises, both the signal path and the control path of a VCO must be fully differential. A common ring oscillator topology in monolithic PLLs is shown in Figure 10.15. The loop oscillates with a period equal to $2NT_d$ where T_d is the delay of each stage. The oscillation can be obtained when the total phase shift is zero and the loop gain is greater or equal to unity at a certain frequency. To vary the frequency of oscillation, the effective number of stages or the delay of each stage must be changed. The first approach is called "delay interpolating" VCO,[20] where a shorter delay path and a longer delay path are used in parallel. The total delay is tuned by increasing the gain of one path and decreasing the other, and the total delay is a weighted sum of the two delay paths. The second approach is to vary the delay time of each stage to adjust the oscillation frequency. The delay of each stage is tuned by varying the capacitance or the resistance seen at the output node of each stage. Because the tuning range of the capacitor is small and the maximum oscillation frequency is limited by the minimum value of the load capacitor, it makes the "resistive tuning" a better alternative technique. The resistive tuning method provides a large, uniform frequency tuning range and lends itself easily to differential control. In Figure 10.16(a), the on-resistance of the triode PMOS loads are adjusted by V_{cont}. The more V_{cont} decreases, the more the delay of the stage drops; because the time constant at the output node decreases, the small-signal gain decreases too. The circuit eventually fails to oscillate when the loop gain at the oscillation frequency is less than unity. In Fig. 10.16(b), the delay of the gain stage is tuned by adjusting the tail current, but the small-signal gain remains constant. So, the circuit is better than Fig. 10.16(a). As shown in Fig. 10.16(c),[21] the PMOS current source with a pair of cross-coupled diode loads provides a differential load impedance that is independent of common-mode voltage. This makes the cell delay insensitive to common-mode noise. Figure 10.16(d) is a poor delay cell for a ring oscillator because the tuning range is very small.

The minimum number of stages that can be used while maintaining reliable operation is an important issue in a ring oscillator design. When the number of stages decreases, the required phase shift and dc

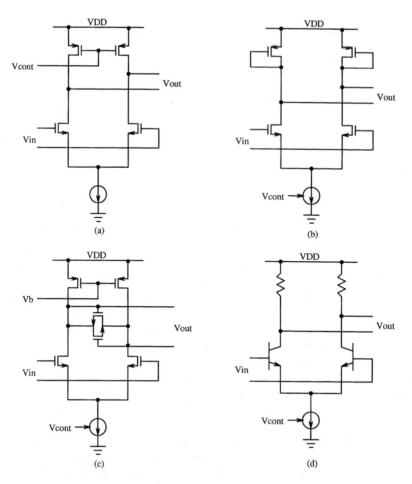

FIGURE 10.16   The gain stages using resistive tuning.

gain per stage increases. Two-stage bipolar ring oscillators can be designed reliably,[22] but CMOS implementations cannot. Thus, CMOS ring oscillators typically utilize three or more stages.

Phase and Frequency Detectors

The phase detector type has influence on the dynamic range of PLLs. Hold range, lock-in range, and pull-in range are analyzed in Section 10.2, based on the multiplier phase detector. Most other types of phase detectors have a greater linear output span and a larger maximum output swing than a sinusoidal characteristic phase detector. A larger tracking range and a larger lock limit are available if the linear output range of the PD increases. The three widely used phase detectors are XOR PD, edge-triggered JK-flipflop, and PFD (phase-frequency detector). The characteristics of these phase detectors are plotted in Fig. 10.17.

The XOR phase detector can maintain phase tracking when the phase error θ_e is confined in the range of

$$\frac{-\pi}{2} < \theta_e < \frac{\pi}{2}$$

as shown in Fig. 10.17(a). The zero phase error takes place when the input signal and the VCO output are quadrature in phase as shown in Fig. 10.18(a). As the phase difference deviates from $\frac{\pi}{2}$, the output duty cycle is no longer 50%, which provides a dc value proportional to the phase difference as shown in

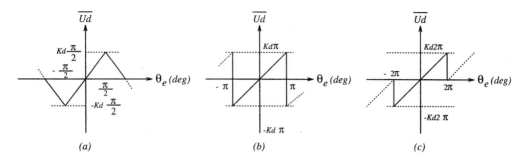

FIGURE 10.17   Phase detector characteristics of (a) XOR, (b) JK-flipflop, and (c) PFD.

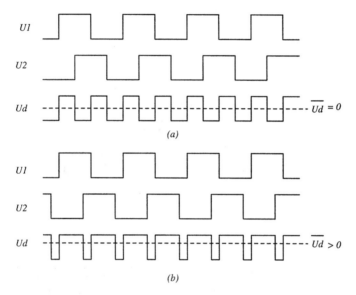

FIGURE 10.18   Waveforms of the signals for the XOR phase detector: (a) waveforms at zero phase error, and (b) waveforms at positive phase error.

Fig. 10.18(b). But the XOR phase detector has a steady-state phase error if the input signal or the VCO output are asymmetric.

The JK-flipflop phase detector shown in Fig. 10.19, also called a two-state PD, is barely influenced by the asymmetric waveform because it is edge-triggered. The zero phase error happens when the input signal and the VCO output are out-of phase as illustrated in Fig. 10.19(a). As shown in Fig. 10.17(b), the JK-flipflop phase detector can maintain phase tracking when the phase error is within the range of

$$-\pi < \theta_e < \pi$$

Here, a positive edge appearing at the "J" input triggers the flipflop into "high" state ($Q = 1$), and the rising edge of u_2 drives Q to zero. Figure 10.19(b) shows the output waveforms of the JK-flipflop phase detector for $\theta_e > 0$.

The PFD output depends not only on the phase error, but also on the frequency error. The characteristic is shown in Fig. 10.17(c). When the phase error is greater than 2π, the PFD works as a frequency detector. The operation of a typical PFD is as follows, and the waveforms are shown in Fig. 10.20. If the frequency of input A, ω_A, is less than the frequency of input B, ω_B, then the PFD produces positive pulses at Q_A, while Q_B remains at zero. Conversely, if $\omega_A > \omega_B$, the positive pulses appear at Q_B while $Q_A = 0$. If $\omega_A = \omega_B$, then the PFD generates pulses at either Q_A or Q_B with a width equal to the phase difference between

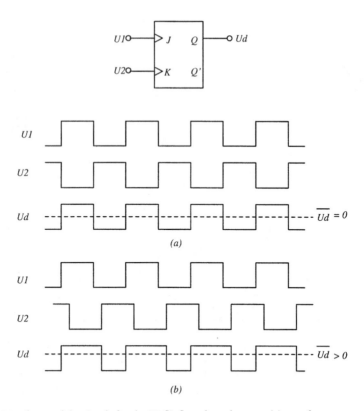

FIGURE 10.19 Waveforms of the signals for the JK-flipflop phase detector: (a) waveforms at zero phase error, (b) waveforms at positive phase error.

the two inputs. The outputs Q_A and Q_B are usually called the "up" and "down" signals, respectively. If the input signal fails, which usually happens at the NRZ data recovery applications during missing or extra transmissions, the output of the PFD would stick on the high state (or low state). This condition may cause the VCO to oscillate rapidly or slowly abruptly, which results in noise jitter or even losing lock. This problem can be remedied by additional control logic circuits to make the PFD output toggle back and forth between the two logic levels with 50% duty cycle,[18] the loop is interpreted as zero phase error. The "rotational FD" described by Messerschmitt can also solve this issue.[9] The output of a PFD can be converted to a dc control voltage by driving a three-state charge-pump, as described in Section 10.2.

10.4 PLL Applications

Clock and Data Recovery

In data transmission systems such as optical communications, telecommunications, disk drive systems, and local networks, data are transmitted on baseband or passband. In most of these applications, only data signals are transmitted by the transmitter; clock signals are not transmitted in order to save hardware cost. Therefore, the receiver should have some scheme to extract the clock information from the received data stream and regenerate transmitted data using the recovery clock. This scheme is called *timing recovery* or *clock recovery*.

To recover the data correctly, the receiver must generate a synchronous clock from the input data stream, and the recovered clock must synchronize with the bit rate (the baud of data). The PLL can be used to recover the clock from the data stream, but there are some special design considerations. For

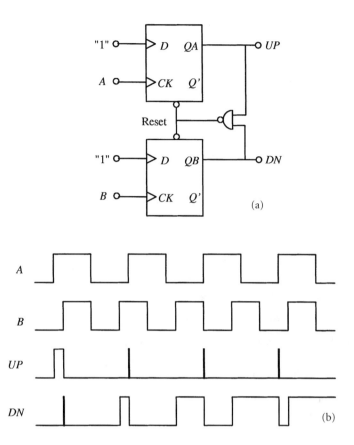

FIGURE 10.20 (a) PFD diagram and (b) input and output waveforms of PFD.

example, because of the random nature of data, the choice of phase-frequency detectors is restricted. In particular, a three-state PD is not proper; because of missing data transitions, the PD will interpret the VCO frequency to be higher than the data frequency, and the PD output stays on "down" state to make the PLL lose lock, as shown in Fig. 10.21. Thus, the choice of phase-frequency detector for random binary data requires a careful examination of their responses when some transitions are absent. One useful method is the rotational frequency detector described in Ref. 9. The random data also causes the PLL to introduce undesired phase variation in the recovered clock; this is called *timing jitter* and is an important issue of the clock recovery.

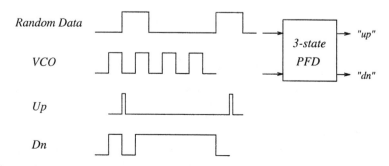

FIGURE 10.21 Response of a three-state PD to random data.

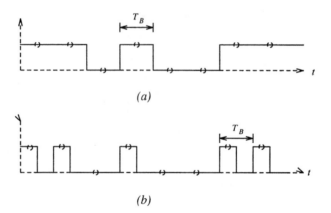

(a)

(b)

FIGURE 10.22   (a) NRZ data and (b) RZ data.

Data Format

Binary data are usually transmitted in an NRZ (Non-Return-to-Zero) format, as shown in Fig. 10.22(a), because of the consideration of bandwidth efficiency. In NRZ format, each bit has a duration of T_B (bit period). The signal does not go to zero between adjacent pulses representing 1's. It can be shown[23] in that the corresponding spectrum has no line component at $f_B = \dfrac{1}{T_B}$; most of the spectrum of this signal lies below $\dfrac{f_B}{2}$. The term "non-return-to-zero" distinguishes from another data type called "return-to-zero" (RZ), as shown in Fig. 10.22(b), in which the signal goes to zero between consecutive bits. Therefore, the spectrum of RZ data has a frequency component at f_B. For a given bit rate, RZ data needs wider transmitting bandwidth; therefore, NRZ data are preferable when channel or circuits bandwidth is a concern.

Due to the lack of a spectral component at the bit rate of NRZ format, a clock recovery circuit may lock to spurious signals or fail to lock at all. Thus, a non-linear process at NRZ data is essential to create a frequency component at the baud rate.

Data Conversion

One way to recover a clock from NRZ data is to convert it to RZ-like data that has a frequency component at bit rate, and then recover clock from data using a PLL. Transition detection is one of the methods to convert NRZ data to RZ-like data. As illustrated in Fig. 10.23(a), the edge detection requires a mechanism to sense both positive and negative data transitions. In Fig. 10.23(b), NRZ data are delayed and compared with itself by an exclusive-OR gate; therefore, the transition edges are detected. In Fig. 10.24, the NRZ data V_i are first differentiated to generate pulses corresponding to each transition. These pulses are made to be all positive by squaring the differentiated signal $\dot{v}_i$. The result is the signal V_i' that looks just like RZ data, where pulses are spaced at an interval of T_B.

Clock Recovery Architecture

Based on different PLL topologies, there are several clock recovery approaches. Here, the early-late and the edge-detector based methods are described.

Figure 10.25 shows the block diagram of the early-late method. If the input lags the VCO output, Fig. 10.26 shows the waveforms for this case. In Fig. 10.26, the early integrator integrates the input signal for the early-half period of the clock signal and holds it for the remainder of the clock signal. On the other hand, the late integrator integrates the input signal for the late-half period of the clock signal and holds it for the next early-half period. The average difference between the absolute values of the late hold and the early hold voltage generated from a low-pass filter gives the control signal to adjust the frequency of the VCO. As mentioned above, this method is popular for rectangular pulses. However, there are some

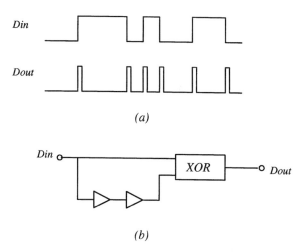

FIGURE 10.23   Edge detection of NRZ data.

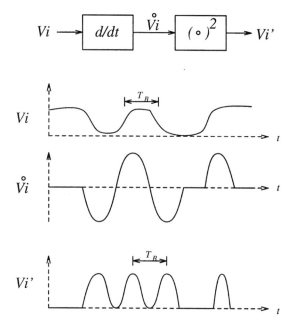

FIGURE 10.24   Converting NRZ to RZ-like signal.

drawbacks to this method. Since this method relies on the shape of pulses, a static phase error can be introduced if the pulse shape is not symmetric. In high-speed applications, this approach requires a fast settling integrator that limits the operating speed of the clock recovery circuit and the acquisition time cannot be easily controlled.

The most widely used technique for clock recovery in high-performance, wide-band data transmission applications is the edge-detection based method. The edge-detection method is used to convert data format such that the PLL can lock the correct band frequency. More details were given in the previous subsection. There are many variations of this method, depending on the exact implementation of each PLL loop component. The "quadricorrelator" introduced by Richman[7] and modified by Bellisio[24] is a frequency-difference discriminator and has been implemented in a clock recovery architecture. Figure 10.27 is a phase-recovery locked loop using edge-detection method and quadricorrelator to recover timing information from NRZ data.[25] As shown in Fig. 10.27, the quadricorrelator follows the edge-

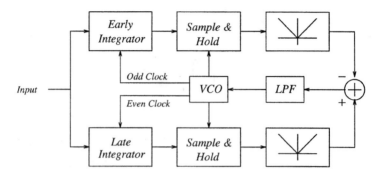

FIGURE 10.25 Early-late block diagram.

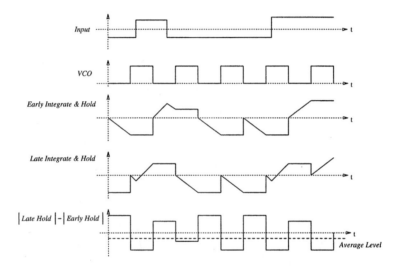

FIGURE 10.26 Clock waveforms for early-late architecture.

detector with a combination of three loops sharing the same VCO. Loop I and II form a frequency-locked loop that contains the quadricorrelator for frequency detection. Loop III is a typical phase-locked loop for phase alignment. The phase- and frequency-locked loops share the same VCO; the interaction between two loops is a very important issue. As described in Ref. 25, when $\omega_1 \approx \omega_2$, the dc feedback signal produced by loop I and II approaches zero, and loop III dominates the loop performance. A composite frequency- and phase-locked loop is a good method to achieve fast acquisition and a narrow PLL loop bandwidth to minimize the VCO drift. Nevertheless, because the wide band frequency-locked loop can respond to noise and spurious components, it is essential to disable the frequency-locked loop when the frequency error gets into the lock-in range of the PLL to minimize the interaction. More clock recovery architectures are described in Refs. 18, 20, 22, and 26–28.

Frequency Synthesizer

A frequency synthesizer generates any of a number of frequencies by locking a VCO to an accurate frequency source such as a crystal oscillator. For example, RF systems usually require a high-frequency local oscillator whose frequency can be changed in small and precise steps. The ability to multiply a reference frequency makes PLLs attractive for synthesizing frequencies.

The basic configuration used for frequency synthesis is shown in Fig. 10.28(a). This system is capable of generating an integer multiple frequency of a reference frequency. A quartz crystal is usually used as the reference clock source because of its low jitter characteristic. Due to the limited speed of a CMOS

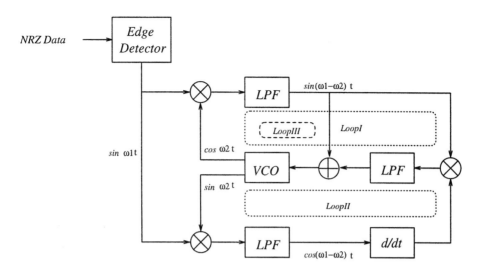

FIGURE 10.27 Quadricorrelator.

device, it is difficult to generate frequency directly in the range of GH_z or more. To generate higher frequencies, prescalers are used; they are implemented with other IC technologies such as ECL. Figure 10.28(b) shows a synthesizer structure using a prescaler V; the output frequency becomes

$$f_{out} = \frac{NVf_i}{M} \qquad (10.46)$$

Because the scaling factor V is much greater than one, obviously, it is no longer possible to generate any desired integer multiple of the reference frequency. This drawback can be circumvented by using a so-called dual-modulus prescaler, as shown in Fig. 10.29. A dual-modulus prescaler is a divider whose division can be switched from one value to another by a control signal. The following shows that the dual-modulus prescaler makes it possible to generate a number of output frequencies that are spaced only by one reference frequency. The VCO output is divided by V/V+1 dual-modulus prescaler. The output of the prescaler is fed into a "program counter" $\frac{1}{N}$ and a "swallow counter" $\frac{1}{A}$. The dual-modulus prescaler is set to divide by V+1 initially. After "A" pulses out of the prescaler, the swallow counter is full and changes the prescaler modulus to V. After additional "N-A" pulses out of the prescaler, the program counter changes the prescaler modulus back to V+1, restarts the swallow counter, and the cycle is repeated. In this way, the VCO frequency is equal to $(V + 1) A + V (N - A) = VN + A$ times the reference frequency. Note that N must be larger than A. If this is not the case, the program counter would be full earlier than $\frac{1}{A}$, and both counters would be reset. Therefore, the dual-modulus prescaler would never be switched from $V + 1$ to V. For example, if V = 64, then A must be in the range of 0 to 63 such that $N_{min} = 64$. The smallest realizable division ratio is

$$\left(N_{tot}\right)_{min} = N_{min}V = 4096 \qquad (10.47)$$

The synthesizer of Fig. 10.29 is able to generate all integer multiples of the reference frequency, starting from $N_{tot} = 4096$. For extending the upper frequency range of frequency synthesizers, but still allowing the synthesis of lower frequency, the four-modulus prescaler is a solution.[1]

Based on the above discussions, the synthesized frequency is an integer multiple of a reference frequency. In RF applications, the reference frequency is usually larger than the channel spacing for loop dynamic performance considerations, in which the wider loop bandwidth for a given channel spacing

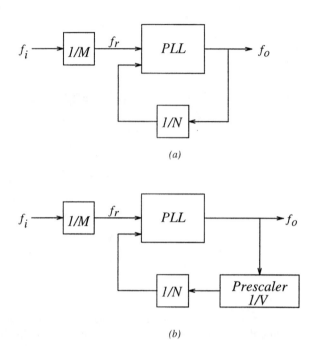

(a)

(b)

FIGURE 10.28 Frequency-synthesizer block diagrams: (a) basic frequency-synthesizer system; (b) system extends the upper frequency range by using an additional high-speed prescaler.

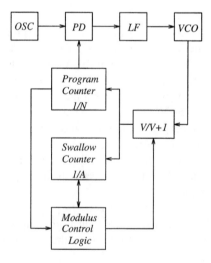

FIGURE 10.29 The block diagram of dual-modulus frequency synthesizer.

allows faster settling time and reduces the phase jitter requirements to be imposed on the VCO. Therefore, a "fractional" scaling factor is needed. Fractional division ratios of any complexity can be realized. For example, a ratio of 3.7 is obtained if a counter is forced to divide by 4 in seven cycles of each group of ten cycles and by 3 in the remaining three cycles. On the average, this counter effectively divides the input frequency by 3.7.

References

1. R. E. Best, *Phase-Locked Loops Theory, Design, Applications*, McGraw-Hill, New York, 1984.
2. D. G. Troha and J. D. Gallia, Digital phase-locked loop design using S-N54/74LS297, Application Note AN 3216, Texas Instruments Inc., Dallas, TX.
3. W. B. Rosink, All-digital phase-locked loops using the 74HC/HCT297, *Philips Components*, 1989.
4. F. M. Gardner, *Phaselock Techniques*, 2nd ed.
5. S. G. Tzafestas, *Walsh Functions in Signal and Systems Analysis and Design*, Van Nostrand, 1985.
6. F. M. Gardner, Acquisition of phaselock, *Conference Record of the International Conference on Communications*, vol. I, pp. 10-1 to 10-5, June 1976.
7. D. Richman, Color carrier reference phase synchronization accuracy in NTSC color television, *Proc. IRE*, vol. 42, pp. 106-133, Jan. 1954.
8. F. M. Gardner, Properties of frequency difference detector, *IEEE Trans. Comm.*, vol. COM-33, no. 2, pp. 131-138, Feb. 1985.
9. D. G. Messerschmitt, Frequency detectors for PLL acquisition in timing and carrier recovery, *IEEE Trans. Comm.*, vol. COM-27, no. 9, pp. 1288-1295, Sept. 1979.
10. R. B. Lee, Timing recovery architecture for high speed data communication system, Masters thesis, 1993.
11. M. Bazes, A novel precision MOS synchronous delay lines, *IEEE J. Solid-State Circuits*, vol. 20, pp. 1265-1271, Dec. 1985.
12. M. G. Johnson and E. L. Hudson, A variable delay line PLL for CPU-coprocessor synchronization, *IEEE J. Solid-State Circuits*, vol. 23, pp. 1218-1223, Oct. 1988.
13. B. Kim, T. C. Weigandt, and P. R. Gray, PLL/DLL systems noise analysis for low jitter clock synthesizer design, *ISCAS Proceedings*, pp. 31-35, 1994.
14. M. V. Paemel, Analysis of a charge-pump PLL: a new model, *IEEE Trans. Comm.*, vol. 42, no. 7, pp. 131-138, Feb. 1994.
15. F. M. Gardner, Charge-pump phase-locked loops, *IEEE Trans. Comm.*, vol. COM-28, pp. 1849-1858, Nov. 1980.
16. F. M. Gardner, Phase accuracy of charge pump PLL's, *IEEE Trans. Comm.*, vol. COM-30, pp. 2362-2363, Oct. 1982.
17. T. C. Weigandt, B. Kim, and P. R. Gray, Analysis of timing recovery jitter in CMOS ring oscillator, *ISCAS Proceedings*, pp. 27-30, 1994.
18. T. H. Lee and J. F. Bulzacchelli, A 155-MHz clock recovery delay- and phase-locked loop, *IEEE J. Solid-State Circuits*, vol. 27, no. 12, pp. 1736-1746, Dec. 1992.
19. M. P. Flyun and S. U. Lidholm, A 1.2 μm CMOS current-controlled oscillator, *IEEE J. Solid-State Circuits*, vol. 27, no. 7, pp. 982-987, July 1992.
20. S. K. Enam and A. A. Abidi, NMOS IC's for clock and data regeneration in gigabit-per-second optical-fiber receivers, *IEEE J. Solid-State Circuits*, vol. 27, no. 12, pp. 1763-1774, Dec. 1992.
21. M. Horowitz et al., PLL design for a 500MB/s interface, *ISSCC Digest Technical Paper*, pp. 160-161, Feb. 1993.
22. A. Pottbacker and U. Langmann, An 8GHz silicon bipolar clock-recovery and data-regenerator IC, *IEEE J. Solid-State Circuits*, vol. 29, no. 12, pp. 1572-1751, Dec. 1994.
23. B. P. Lathi, *Modern Digital and Analog Communication System*, HRW, Philadelphia, 1989.
24. J. S. Bellisio, A new phase-locked loop timing recovery method for digital regenerators, *IEEE Int. Comm.Conf. Rec.*, vol. 1, pp. 10-17-10-20, June 1976.
25. B. Razavi, A 2.5-Gb/s 15-m W clock recovery circuit, *IEEE J. Solid-State Circuits*, vol. 31, no. 4, pp. 472-480, Apr. 1996.
26. R. J. Baumert, P. C. Metz, M. E. Pedersen, R. L. Pritchett, and J. A. Young, A monolithic 50-200MHz CMOS clock recovery and retiming circuit, *IEEE Custom Integrated Circuits Conference*, pp. 14.5.5-14.5.4, 1989.

27. B. Lai and R. C. Walker, A monolithic 622Mb/s clock extraction data retiming circuit, *IEEE Inter. Solid-State Circuits Conference,* pp. 144-145, 1991.

28. B. Kim, D. M. Helman, and P. R. Gray, A 30MHz hybrid analog/digital clock recovery circuit in 2-μm CMOS, *IEEE J. Solid-State Circuits,* vol. 25, no. 6, pp. 1385-1394, Dec. 1990.

11

Continuous-Time Filters

John M. Khoury
Lucent Technologies

11.1 Introduction

Modern Very Large Scale Integrated (VLSI) circuits realize complex mixed analog-digital systems on a monolithic semiconductor chip. These systems generally incorporate signal processing operations that can be performed either in the digital domain using digital signal processing (DSP) techniques or in the analog domain with analog signal processing (ASP) circuits. ASP techniques fall into two basic categories: continuous-time or sampled-data. The selection of DSP, continuous-time ASP, or sampled-data ASP approaches is highly dependent on the system requirements; however, continuous-time filters are generally preferable in applications that require low power, high-frequency operation, and moderate dynamic range.

Fully integrated continuous-time filters have found wide application in many VLSI systems that include modems, telephone circuits, disk drive read channels, video processing circuits, and others. The applications usually fall into one of the three basic configurations shown in Fig. 11.1. In the top two views, the continuous-time filter provides anti-aliasing and smoothing functions for sample-data signal processing operations that are either performed with switched-capacitor (SC), switched-current (SI), or DSP filters. Generally, for these applications, the precise signal processing functions are kept in the sampled-data domain. The continuous-time filter can then have non-stringent frequency response specifications provided that the ratio of half the sampling rate to the band edge is large for the sampled-data filter. Often for lower power operation or extremely high frequency operation, the entire signal processing is performed with the continuous-time ASP as shown in the bottom of Fig. 11.1.

When designing a system, the natural question arises as to which is the best approach for performing the core signal processing operations: DSP or ASP. In general, DSP is usually the obvious choice if the application has the following attributes: (1) frequency response specifications that must be repeatable to within fractions of decibels (dB) over all manufacturing processes, (2) band edges are in the 100-kHz range and below, (3) dynamic range requirements exceed 80 dB, and (4) a high degree of programmability or coefficient adaptation is needed.

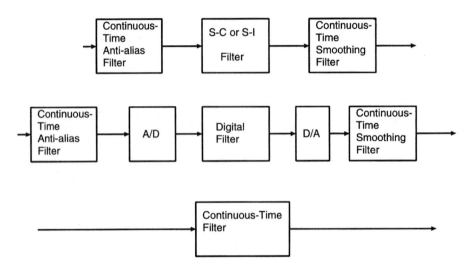

FIGURE 11.1   General uses of continuous-time filter.

ASP typically has a clear advantage when the critical frequencies in the filter exceed several hundred kHz. In today's 0.35-μm CMOS technologies, SC filters are generally limited to sampling rates below 100 MHz; hence, filter passbands will typically be below 25 MHz. SI filters have similar limitations. Continuous-time filters are then the only viable alternative for passbands in the 10's of MHz and higher.

Continuous-time filters may be designed for the entire frequency range from audio to above 100 MHz. When used in audio and above audio applications, continuous-time filters can achieve the lowest possible power of all the filtering techniques; dynamic ranges above 90 dB can be obtained and linearity above 80 dB is possible with certain continuous-time design approaches.[1-4] Higher frequency continuous-time filters can also be designed to achieve excellent linearity.[5] In the 10- to 150-MHz range, continuous-time filters usually will achieve linearity and dynamic range performance in the 60-dB range and cutoff frequency accuracy in the range of a few percent.[6-11] Integrated continuous-time filters to date have not achieved the performance required in the high-frequency, high-Q, and high-linearity wireless application. The fundamental limitations of thermal noise present in all integrated continuous-time, SC, and SI filters severely limits the achievable dynamic range under high-Q conditions.[12]

Integrated continuous-time filters differ from discrete active filters and SC or SI filters in that a frequency tuning circuit is almost always required to obtain accurate frequency response characteristics.[1,7,9,10,13-15] In VLSI chips, the capacitor values can vary by ±15% for linear capacitors such as double-poly or metal-metal capacitors. Similarly, the resistance element, whether it is a diffused resistor, poly-silicon resistor, or transistor will vary widely with processing and temperature. The combined effect often results in RC products that vary by as much as ±50%. In continuous-time filter applications such as anti-aliasing or reconstruction functions, such variation is often acceptable. However, to achieve corner frequency accuracy of a few percent, a tuning circuit is required. In addition to frequency tuning/scaling the filter, circuits to tune the quality factor of the filter are sometimes employed.[6,8,11,16-18]

The following sections in this chapter cover the state-variable implementation of continuous-time filters, the design of VLSI integrators, and the design of highly linear continuous-time filters. The chapter concludes with the design of tuning circuits.

11.2 State-Variable Synthesis Techniques

In the 1960s, considerable research was performed in the area of active filter design. At that time, the focus was on discrete circuit implementations that operated with single-ended circuitry. Although many creative and theoretically appealing approaches were invented and used commercially for discrete designs, only a few of the circuit topologies are well suited to VLSI implementations. An extensive discussion of active filter realizations can be found in Ref. 19.

Of all the possible active filter topologies possible, the *state-variable filter* is the most general in form and most widely used in VLSI continuous-time filters today. The key advantage of state-variable filters is that they require only two basic building blocks: (1) integrators and (2) weighted summers. In VLSI solutions, the integrators are realized with on-chip capacitors, an active element such as an operational amplifier (op-amp), and a resistive element or transconductance amplifier. Signal summation is performed in the voltage or current domains, depending on the technique used.

The topology of state-variable filters can take on many varied forms. In the most general case, a linear system with N state variables would consist of N integrators with signal coupling between any and all integrators. In practice, coupling between integrators is limited to make the design realizable. In the biquadratic (biquad) filter structure, the N-th order filter is realized as a cascade of second-order circuits, followed by a first-order circuit, if N is odd. The biquad approach is widely used for its simplicity, ease of design, and ease of debugging.[19]

An alternate form of state-variable filters, called the "leapfrog" topology, is realized by simulating the equations that govern RLC ladder filters.[19] In the leapfrog topology, only the state variables (i.e., integrators) that are adjacent to one another are coupled. Leapfrog filters are more difficult to design, but they generally offer improved passband magnitude response accuracy and better dynamic range performance than the cascade of biquads.

Biquadratic Filters

The biquad structure realizes the filtering function as a cascade of second-order filters. The structure decouples the poles of the system and can ease the overall design approach.

The general equation governing the biquadratic filter is

$$V_{out}(s) = K \frac{s^2 + \dfrac{\omega_{oz}}{Q_z} s + \omega_{oz}^2}{s^2 + \dfrac{\omega_{op}}{Q_p} s + \omega_{op}^2} V_{in}(s) \tag{11.1}$$

where $V_{out}(s)$ and $V_{in}(s)$ are the output and input signals of the biquad, respectively, K is a gain constant, ω_p and ω_z are the frequencies of the poles and zeros, and Q_p and Q_z are the quality factors of the poles and zeroes, respectively. Although many methods of realizing this transfer function are possible, the state-variable approach uses a loop of two integrators connected with negative feedback to realize the poles. Damping around one (or both) integrator(s) makes the corresponding integrator lossy and implements the pole quality factor, Q_p. The zeroes of the biquad can be achieved by (1) creating an output signal $V_{out}(s)$ that is the weighted summation of the two integrator outputs, as well as the input signal, $V_{in}(s)$, or (2) by summing scaled values of the input signal into both integrators as well as directly to the output. The block diagram of the generalized biquad, shown in Fig. 11.2, places the zeroes and adjusts the overall gain of the filter with the K_1, K_2, and K_3 constants. The block diagram of Fig. 11.2 can be easily converted to an integrated VLSI filtering technique with a one-for-one substitution of the integrators and weighted summers with the corresponding VLSI circuits. Integrated implementations will be discussed in the sections on G_m-C, G_m-OTA-C, and *MOSFET-C* filters.

If the high-order transfer function is of the form:

$$H(s) = \frac{N(s)}{D(s)} \tag{11.2}$$

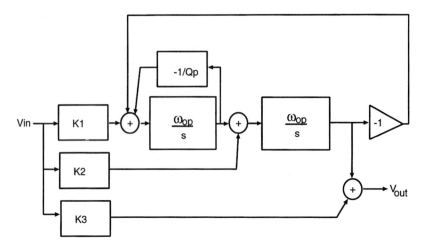

FIGURE 11.2   Biquad block diagram.

then $H(s)$ can be factored into second-order sections where the numerators and denominators of these biquads are at most second order, as in Eq. 11.1. Issues of how to arrange the cascade of second-order functions and how to pair the poles and zeroes can greatly affect the filter's performance in terms of signal swing, dynamic range, and dc offset accumulation. A few simple rules of how to realize a cascaded filter are enumerated here.

1. **Factor into Biquadratic Terms:**   Split the numerator, $N(s)$, and the denominator, $D(s)$, into products of second-order functions. If either $N(s)$ or $D(s)$ is odd-order, a first-order term will be necessary. The transfer function is then in the following form:

$$H(s) = \frac{N_1(s)N_2(s)N_3(s)\cdots}{D_1(s)D_2(s)D_3(s)\cdots} \qquad (11.3)$$

2. **Pair Poles and Zeroes:**   Convert Eq. 11.3 into a product of second-order transfer functions, $H_A(s)$ $H_B(s)$ $H_C(s)$ ..., by pairing each $N_i(s)$ with a $D_j(s)$ in such a way that $|H_A(j\omega)|$, $|H_B(j\omega)|$, $|H_C(j\omega)|$, etc., has as flat a magnitude response over the passband as possible. In this way, the signal at the various points in the cascade of the filter will be large and hence less susceptible to interference. Interference could be due to the thermal noise of the active and passive circuits, power supply noise, and crosstalk from digital circuits on-chip.

   To make $|H_A(j\omega)|$, $|H_B(j\omega)|$, $|H_C(j\omega)|$, etc., as flat as possible over the passband, pair the zeroes of $N_i(s)$ as close in frequency and Q as the poles of $D_j(s)$. This method minimizes the variation caused by $|N_i(j\omega)|/|D_j(j\omega)|$ because the effects of the pole and zero pairs tend to partially cancel.

3. **Choose Cascade Order:**   The next decision is to order the biquads (and maybe a first-order term). Many practical factors influence the optimum ordering. A few examples are:
   a. Order the cascade to equalize signal swing as much as possible throughout the filter to maximize dynamic range.
   b. Choose the first biquad to be lowpass or bandpass to reject high-frequency noise, eliminating overload in the remaining stages.
   c. If the offset at the filter output is critical, the last stage should be a highpass or bandpass to block the dc of previous stages.
   d. Avoid high-Q biquads at the last stage because these biquads have higher fundamental noise and worse sensitivity to power supply noise than low-Q stages.[12]

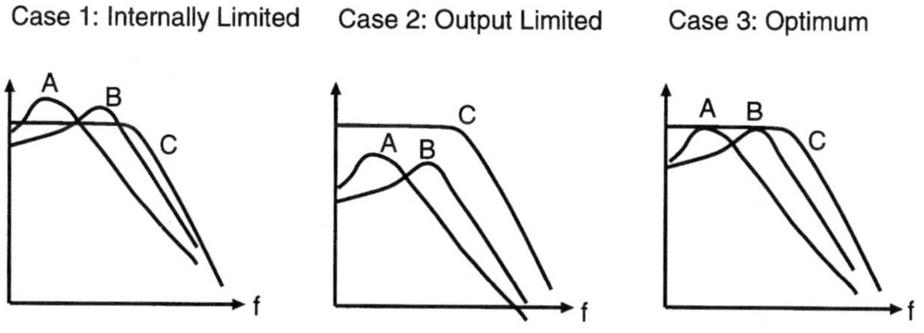

FIGURE 11.3 Dynamic range scaling.

    e. In general, do not place allpass stages at the end of the cascade because these have wideband noise. It is usually best to place allpass stages near the beginning of the filter.

    f. If several highpass or bandpass stages are available, one can place them at the beginning, middle, and end of the filter. This will prevent input dc offset from overloading the filter, will prevent internal offsets of the filter itself from accumulating (and hence decreasing available signal swing), and will provide a filter output that has low dc offset.

    g. The effect of thermal noise at the filter output varies with ordering; therefore, several decibels (dB) of SNR can often be gained with biquad reordering.

  4. **Dynamic Range Optimization:** Dynamic range optimization is simply the scaling of gains within the filter to make sure that the overload levels of the integrators (or summers) are equalized so that all elements will saturate at the same signal level.

If the frequency spectrum of the input signal is known, then dynamic range scaling of the filters should be performed with this signal. The maximum amplitude input signal should be provided and the gains scaled until all integrator and summer outputs are at their maximum level. Note that gain scaling should be performed so as not to modify any loop gains in the filter; otherwise, the transfer function would be altered.

If the frequency spectrum of the input signal is unknown, the typical approach is to assume the input signal is a single sinusoid. The filter is then dynamic range scaled so that for the maximum amplitude input sinusoid, all integrator and summer outputs have the same maximum value for any possible sinusoidal frequency. Usually, the frequency of the input sinusoid is swept over the filter's passband and the maximum levels are then gain scaled. Pictorially this can be seen in Fig. 11.3. Here, the filter consists of a cascade of three biquads: A, B, C. The frequency response to each biquad output is shown. In case 1, the signal will clip at the output of biquads A and B first; whereas in case 2, the output, C, will saturate first. It is only in case 3 — where the maximum gains have been equalized — that clipping occurs in all three biquads at the same level.

Dynamic range scaling must be performed not only at the output of each biquad, but also at the output of the internal integrator. As an example, consider the classical Tow-Thomas biquad in Fig. 11.4. The derivation of this biquad from the block diagram in Fig. 11.2 should be self-evident. The frequency response shows that the internal node V_x will clip at a lower input amplitude level than the output. The signal amplitude at node V_x must be reduced by a factor F. The reduction in gain can be achieved by lowering the impedance in the feedback loop of the first integrator by F. However, to maintain constant loop gain around the two integrator loop, the input resistor of the second integrator must become R/F. The result of this dynamic range scaling is shown in Fig. 11.5.

Leapfrog Filters

The leapfrog filter topology uses active integrators and weighted summers to simulate all the equations governing RLC ladder filters.[19] The question naturally arises as to why passive *ladder* filters should be chosen. First, a wealth of knowledge and design tables exist for these filters. Designers can easily use

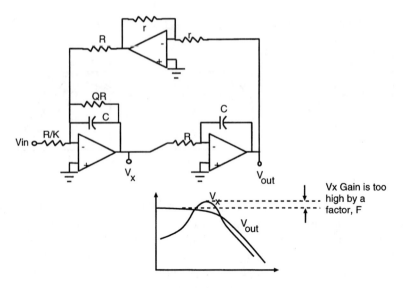

FIGURE 11.4   Tow-Thomas biquad prior to dynamic range scaling.

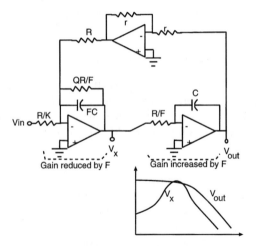

FIGURE 11.5   Tow-Thomas biquad after dynamic range scaling.

tabulated data to design classical ladder filters that implement Butterworth, Chebychev, Bessel, etc., responses. With a few simple steps, these ladders can be transformed into an active leapfrog topology with element values.[19] The second and more important reason to simulate ladder filters is that *in the passband*, the sensitivity of the filter's magnitude response to element value variation is extremely low. This low sensitivity is not true in the stopband, nor is it true for the phase response of the filter. Since leapfrog filters simulate all the equations governing the ladder filter, these sensitivity advantages carry over to the active realization. Finally, filters that are relatively insensitive to component errors *usually* have lower thermal noise. In most applications, leapfrog filters will have superior performance relative to biquadratic filters in terms of noise and passband magnitude response accuracy. Snelgrove and Sedra[20] analyzed biquad filters, leapfrog filters, and filters optimized for noise and magnitude response sensitivity. The leapfrog filters achieved performance close to the optimized design, but the biquad approach showed significantly degraded performance.

The design of leapfrog filters can be found in Ref. 19. Here, we will show by example the design of these filters. Consider the third-order lowpass doubly terminated ladder shown in Fig. 11.6. Since the

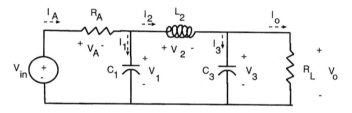

FIGURE 11.6 Third-order doubly terminated lowpass LC ladder filter.

active filter must simulate all equations governing the ladder, the first step is to write all the equations. The two-terminal branch relationships are:

$$I_A = \frac{V_A}{R_A}, \quad V_1 = \frac{I_1}{sC_1}, \quad I_2 = \frac{V_2}{sL_2}, \quad V_3 = \frac{I_3}{sC_3}, \quad I_O = \frac{V_O}{R_L} \tag{11.4}$$

The KVL equations are:

$$V_A = V_{in} - V_1, \quad V_2 = V_1 - V_3, \quad V_O = V_3 \tag{11.5}$$

The KCL equations are:

$$I_1 = I_A - I_2, \quad I_3 = I_2 - I_O \tag{11.6}$$

As can be seen in Eqs. 11.4 to 11.6, some variables are currents while others are voltages. In the implementations, usually all signal variables are either voltages or currents. Here, voltage signals will be assumed. To convert the current signals in Eqs. 11.4 to 11.6 to voltages, all currents can be scaled by a resistance r of arbitrary value (e.g., 1 Ω). After the scaling by r, Eqs. 11.4 to 11.6 become:

$$rI_A = \frac{rV_A}{R_A}, \quad V_1 = \frac{rI_1}{srC_1}, \quad rI_2 = \frac{V_2}{sL_2/r}, \quad V_3 = \frac{rI_3}{srC_3}, \quad rI_O = \frac{rV_o}{R_L},$$

$$V_A = V_{in} - V_1, \quad V_2 = V_1 - V_3, \quad V_O = V_3, \quad rI_1 = rI_A - rI_2, \quad rI_3 = rI_2 - rI_O \tag{11.7}$$

Using Eq. 11.7, the signal flow graph (SFG) shown in Fig. 11.7 can be obtained. Arrows flowing into a circle represent summation, and the values next to the arrows indicate a scaling operation. In the SFG, the KVL equations are implemented with the top two summation circles, while the KCL equations are

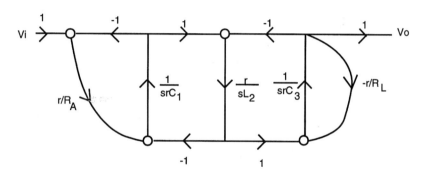

FIGURE 11.7 Signal flow graph representation of LC ladder filter.

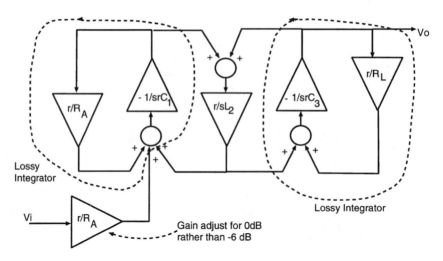

FIGURE 11.8 Leapfrog filter with gain scaling extracted.

on the bottom side. If one were to implement this SFG directly, the gain of the filter would be less than 0 dB (e.g., –6 dB for an equally terminated ladder). In fact, the gain would be the same as the original RLC ladder. By replicating the gain block, r/R_A, as shown in Fig. 11.8 an additional degree of freedom is obtained to implement arbitrary filter gains. Dotted lines are used to indicate that the integrators on the end of the filter are damped, while the inner integrator is lossless. In the realization of high-order ladder filters, the internal integrators will always be lossless, while the outside ones will be lossy due to the ladder terminations.

Highpass and bandpass leapfrog filters can be realized directly from the lowpass LC ladder with the use of the classical lowpass-to-highpass or lowpass-to-bandpass transformations.[19] For illustrative purposes, the bandpass case is considered here. Starting from a lowpass prototype with frequency domain variable s, a bandpass filter with bandwidth BW and center frequency ω_o can be realized in the frequency domain variable p with the following transformation:

$$s = \frac{p^2 + \omega_o^2}{pBW} \tag{11.8}$$

Applying the transformation element by element to a third-order lowpass ladder, one obtains the bandpass ladder shown in Fig. 11.9. An SFG can be generated directly from the bandpass ladder and the active filter realized. Alternatively, the lowpass-to-bandpass transformation can be applied directly to the lowpass active filter of Fig. 11.8, resulting in the bandpass active filter in Fig. 11.10. Notice that each integrator has been replaced with a bandpass biquad and that the biquads corresponding to the terminations are damped.

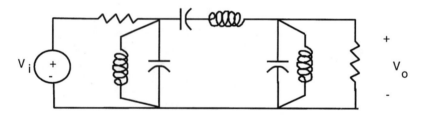

FIGURE 11.9 Bandpass ladder filter.

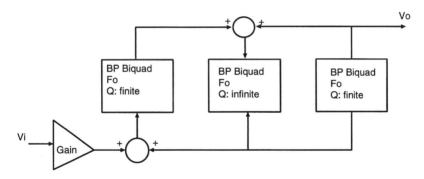

FIGURE 11.10   Bandpass leapfrog filter realization.

Designers proficient in the use of SFGs can readily transform the active leapfrog realization to include zeroes in the transfer function.

11.3 Realization of VLSI Integrators

Once the state-variable topology has been created, the VLSI filter realization is determined by the approach used for the integrator. This section describes the most common types of VLSI integrators and their corresponding summing circuits. The three most common types of implementations are the G_m-C, G_m-OTA-C and *MOSFET-C* filters. G_m-C filters are generally recognized to offer the highest possible frequency operation at the lowest power; however, the structures are sensitive to parasitic capacitances and generally have higher noise and offset than other techniques. G_m-OTA-C filters are far less parasitic sensitive than G_m-C designs, but at the cost of higher power. Finally, *MOSFET-C* filters generally are the most parasitic insensitive, and have the least noise and offset; however, the frequency of operation is *usually* the lowest of the three approaches. In BiCMOS technology where extremely wideband op-amps can be made, *MOSFET-C* techniques possess bandwidth capabilities approaching that of G_m-C and G_m-OTA-C filters.

G_m-C Integrators and Filters

G_m-C filters implement integrators with a transconductance amplifier loaded by a capacitor. As shown in Fig. 11.11 a differential transconductance amplifier (also called a transconductor) takes an input voltage, V_{ind}, and produces at its output a current $I_{out} = G_m V_{ind}$. This output current is integrated by the capacitor to produce the output voltage signal, V_{out}. The transfer function of the G_m-C integrator is

$$H(s) = \frac{G_m}{sC} = \frac{\omega_o}{s} \qquad (11.9)$$

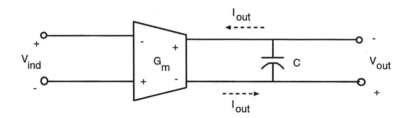

FIGURE 11.11 G_m-C integrator.

where ω_o is the unity-gain frequency of the integrator. The ideal integrator has infinite dc gain, a unity-gain frequency of ω_o, and a phase shift of $-90°$ for all frequencies. Capacitors in VLSI technology are usually high quality, so all stringent integrator requirements fall on the transconductor design. Since the transconductor is a voltage-to-current ($V \rightarrow I$) converter, it should have: (1) high input impedance to accurately sense the input voltage signal, (2) high output impedance so the output signal appears as a current source, (3) high dc gain, (4) wide bandwidth so as not to create phase and magnitude errors in the integrator response, (5) large signal handling capability at the input and output for good dynamic range, and (6) a well-defined and tunable $V \rightarrow I$ mechanism to be used for frequency scaling the filter to remove process and temperature variations. In CMOS or BiCMOS technology, achieving high input impedance is simple due to the gate terminal of the MOSFET. Designing for high output impedance can be achieved with cascoding and with the use of regulated cascodes[21]; however, there is a tradeoff between high bandwidth and high output impedance. The most difficult aspect of G_m cell design is making the $V \rightarrow I$ mechanism tunable simultaneously achieving good linearity in the presence of large input signal swings.

Building state-variable filters with G_m-C filters follows directly from the block diagrams or SFGs. Signal summation is performed in the current domain by placing transconductor outputs in parallel. Consider the SFG in Fig. 11.12. The bandpass G_m-C filter is readily implemented as in Fig. 11.13. The loop of two

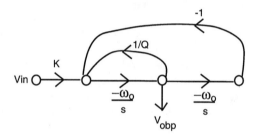

FIGURE 11.12   Signal flow graph representation of a state-variable biquad.

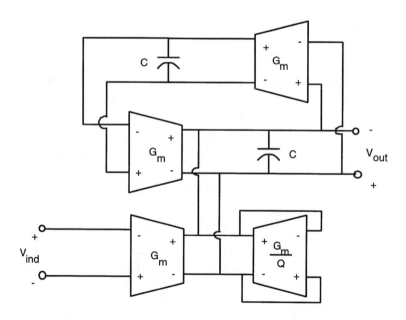

FIGURE 11.13 G_m-C realization of a bandpass biquad.

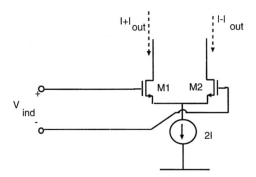

FIGURE 11.14   An MOS differential pair used as a $V \to I$ converter.

integrators is on the top of the figure, biquad damping is performed with the G_m/Q transconductor, and input signal scaling and summation are achieved with the remaining G_m cell. The G_m/Q transconductor connected in the negative feedback configuration on itself implements a resistor of value Q/G_m.

The key aspect in the design of G_m-C filters is the transconductor design and more specifically, the $V \to I$ converter. In the simplest case, the $V \to I$ converter can be a simple MOS differential pair, as shown in Fig. 11.14. The large signal differential output current, I_{outd}, is given by:[22]

$$I_{outd} = 2I_{out} = \mu_n \frac{C_{ox}W}{2L} V_{ind} \sqrt{\left(\frac{4I}{\mu_n C_{ox} W/(2L)} \right) - \left(V_{ind}\right)^2} \qquad (11.10)$$

I_{outd} is a non-linear function of the input V_{ind}. The transconductance of the differential pair, $G_m = dI_{outd}/dV_{ind}$, is maximum at $V_{ind} = 0$ and falls off for increased signal swing. The maximum G_m is:

$$G_m = \sqrt{2I\mu_n C_{ox} W/L} = g_{m1} = g_{m2} \qquad (11.11)$$

The transconductance can be tuned with the tail current $2I$ to frequency scale the G_m-C filter; however, the tuning range is small since G_m only varies as the square root of the current. In general, if the targeted filter application requires no programmability and the critical frequencies are nominally fixed, then roughly a 2:1 tuning range is needed to accommodate process and temperature variations. The tail current would then require a 4:1 variation, greatly impacting power dissipation. The more significant disadvantage of this $V \to I$ converter is its small linear input range. It can be shown that the linear differential input voltage range is much smaller than $\pm\sqrt{2}(V_{gs1,2-bias} - V_T)$, where $V_{gs1,2-bias}$ is the bias level of M1 and M2 for $V_{ind} = 0$. To maximize the linear input range $V_{gs1,2-bias}$ must be kept large by using small W/L ratios. Even with use of small W/L ratios, the input range is typically limited to less than ±200 mV for linearity of 40 to 60 dB.

Many linearization techniques have been invented using MOSFETs in the saturation and triode regions, as well as BiCMOS solutions. A few approaches are discussed here. In the basic MOS differential pair, the output current, I_{outd}, increases with V_{ind}; however, the rate of increase drops off at higher input amplitude levels. One solution is to have another source of current that is added to the transconductor's output. If that current is zero for low levels of V_{ind}, but increases with V_{ind}, the net effect is to linearize the overall G_m cell. Rather than adding a current, we can instead subtract a current from the G_m stages output. The amount to be subtracted would be maximum for $V_{ind} = 0$ and would drop to zero for large differential input signals. This technique uses an additional differential pair, as shown in Fig. 11.15.[10] Transistors M3 and M4 operate at lower current than M1 and M2 and are biased such that ($V_{gs3,4-bias} - V_T$) $\ll$ ($V_{gs3,4-bias} - V_T$). Detailed design equations for the sizing of M1-M4, I1, and I2 can be found in Ref. 10. The current through M3 and M4 will saturate at lower input voltages than the drain currents of

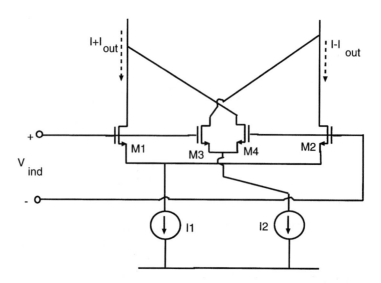

FIGURE 11.15   A cross-coupled linearized MOS $V \rightarrow I$ converter.

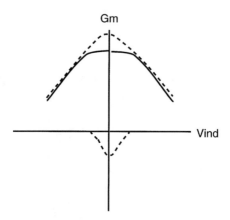

FIGURE 11.16   Individual differential pair G_m (dotted curves) and combined G_m (solid curve).

M1 and M2. The concept is more clearly understood with Fig. 11.16. The dotted lines show the transconductance of the individual differential pairs versus input differential signal. The dotted lines for positive G_m correspond to the transistors M1 and M2, while the negative G_m refers to M3 and M4. Adding the G_m curves, the solid curve is obtained. Notice now that the transconductance curve is flat for small V_{ind}. It is possible to add the outputs of multiple differential pairs with slightly different non-linearities to broaden the region over which G_m is constant. A related approach is given in Ref. 23.

A classical linearization method is to use a differential pair with source degeneration. However, in most technologies, the resistor used for degeneration would vary with temperature and processing and be nominally fixed in value. To afford tunability, the degeneration resistor can be replaced with a MOSFET, M3, operating in the triode region, as shown in the G_m cell of Fig. 11.17.[24] M1 and M2 act as source followers and the transconductor's signal current, I_{out}, is ideally the drain current of M3. The drain current of M3 can be expanded in a Taylor series for the case of zero drain-to-source voltage and obtains the following from the "3/2 Power" model[25] of the MOSFET:

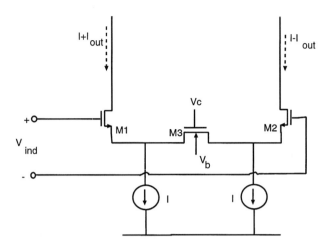

FIGURE 11.17 Source degenerated MOS $V \rightarrow I$ converter.

$$I = (W/L)\mu C_{ox}\left[(V_C - V_Q - V_T)(V_D - V_S) + \sum_{i=2}^{\infty} a_i(V_D^i - V_S^i)\right] \qquad (11.12)$$

where V_Q is the bias level of the source and drain with respect to the body, V_S and V_D are the voltages on the source and drain terminals, respectively, and the a_i are constants. Notice that if the source and drain voltages are balanced around a common-mode voltage V_Q, such that $V_D = V_Q + V_{ind}/2$ and $V_S = V_Q - V_{ind}/2$, then as can be seen from Eq. 11.12,

$$I = (W/L)\mu C_{ox}\left[(V_C - V_Q - V_T)V_{ind} + \sum_{i>1,odd}^{\infty} a_i V_{ind}^i\right] \approx (W/L)\mu C_{ox}(V_C - V_Q - V_T)V_{ind} \qquad (11.13)$$

For many applications, the remaining odd-order non-linearity is low enough (e.g., –65 dB) to be inconsequential. For applications requiring superior linearity, cross-coupled triode degenerated differential pairs can be used to theoretically cancel all high-order non-linearities.[24,26] Based on Eq. 11.13 the transconductance of the G_m cell is:

$$G_m = (W/L)\mu C_{ox}(V_C - V_Q - V_T) \qquad (11.14)$$

assuming M1 and M2 are ideal source followers. As desired, the G_m is tunable with the control voltage, V_C, connected to the gate of M3. The linear input range can in practice be on the order of ±1.0 V, provided that M3 remains in the triode region. The maximum input signal is thus equal to

$$V_{ind-max} = 2(V_C - V_Q - V_T) \qquad (11.15)$$

The maximum input signal swing is a function of the tuning control voltage, V_C; consequently, the dynamic range is tightly coupled to the tuning range. In some situations where a programmable filter is required, multiple transistors can replace M3 and the triode devices can either be connected to V_C, or switched off to implement ranges in the filter.[7] In contrast to the transconductors in Figs. 11.14 and 11.15, power dissipation is unaffected by tuning. Finally, source followers M1 and M2 must be extremely

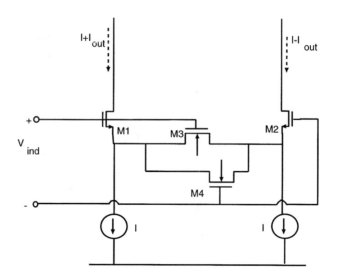

FIGURE 11.18 G_m cell operating in triode and saturation.

low impedance (i.e., have high g_m) to drive the resistance of M3. Either large W/L devices can be used for M1 and M2, or negative feedback can be used around M1 and M2 to reduce their impedance at their source terminals.[5,27]

As an alternative to requiring low impedance source followers, the G_m cell shown in Fig. 11.18 can be used.[9] For small input signals, M3 and M4 operate in the triode region as in the previous design; however, the effective control voltage to tune the devices is set to the gate-to-source bias level of M1 and M2. For large positive input differential signals, more current flows in M1, increasing the V_{GS} of M1. If M3 and M4 had fixed gate voltages, the current through them would drop off, resulting in lower G_m as in the design of Fig. 11.17. However, the gate voltage of M3 increases under this input condition and helps to maintain a constant G_m. By scaling M1 and M2 to M3 and M4 (e.g., to a ratio of about 7:1) the linear range can be expanded. The linear range is also larger than that of Fig. 11.17 because M3 and M4 can operate in both the triode and saturation regions.

All transconductor designs discussed to this point have achieved an expanded linear range as a function of device matching or use of balanced input differential signals. Since matching and signal balancing can never be perfect, the achievable linearity is a strong function of layout and processing. In contrast, the transconductor of Fig. 11.19 achieves high linearity by maintaining constant drain-to-source voltage across triode devices M1 and M2.[8,28,29] Using the basic triode equation for a MOSFET,

$$I = \frac{1}{2}\mu_n C_{ox} \frac{W}{L}\left[2\left(V_{gs} - V_t\right)V_{DS} - V_{DS}^2\right]$$ (11.16)

one can see that if the drain-to-source voltage is held constant, the relationship between V_{GS} and I is linear, except for an offset. Cascode devices Q1 and Q2 in Fig. 11.19 are used to hold $V_{DS1} = V_{DS2} = I_d R_d$, resulting in a linear transconductance of

$$g_m = \mu_n C_{ox} \frac{W}{L} R_d I_d$$ (11.17)

The G_m cell is easily tuned with the collector current of Q3, I_d. Q1-Q3 could be replaced with MOSFETs; however, since the transconductance of bipolar devices is higher than MOSPETs, the BiCMOS solution provides superior cascoding to hold the drain-to-source voltages of M1 and M2 constant.

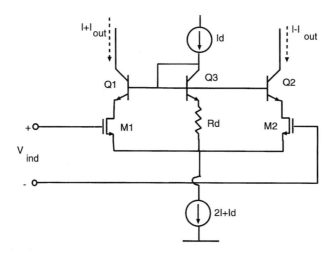

FIGURE 11.19 $V \rightarrow I$ converter with differential pair operating in triode with constant drain-to-source voltage.

Many alternative linearization schemes exist for the $V \rightarrow I$ converters in CMOS, BiCMOS, and bipolar technology.[5,17,23,30-32] Invariably, nearly all the techniques require matching of transistors and/or balanced signals to achieve optimal linearity performance. Also, many of the techniques, particularly the MOSFET-based transconductors, rely on simplified large signal models (e.g., square law) of the transistors to model and cancel the non-linearity. In reality, more complex transistor equations, as found in Ref. 33, are needed to better predict performance. Ultimately, only experimental results over many process lots must be used for guaranteeing a specified linearity.

Once the $V \rightarrow I$ converter design has been determined, the entire G_m amplifier or integrator can be assembled using known op-amp structures. The design in Fig. 11.20 uses the $V \rightarrow I$ converter of Fig. 11.17 with a folded-cascode output stage. The cascoding raises the output impedance of the amplifier and optionally A1 through A4 create regulated cascodes to raise the output impedance further.[21] The nominally equal capacitors C_{lp} and C_{ln} serve two functions. For differential-mode output signals, the capacitors integrate the output current. For common-mode signals, they provide high frequency common-mode

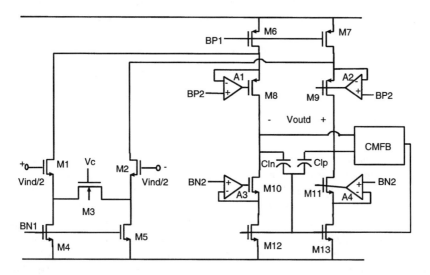

FIGURE 11.20 Folded cascode MOS G_m-C integrator.

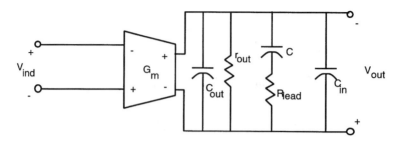

FIGURE 11.21   Non-idealities modeled in the G_m-C integrator.

feedback (CMFB) via the gates of M12 and M13, while low frequency CMFB is performed with standard techniques.[34] Folding the cascode structure permits larger output voltage swings and equal input and output common-mode voltage levels. Use of folding as opposed to an unfolded cascode (i.e., telescopic structure) will generally result in increased input-referred noise and offset due to the addition of transistors M6 and M7.

G_m-C Integrator Frequency Response Errors

The G_m-C integrator will have magnitude and phase errors due to parasitic capacitances and resistances. Consider the non-ideal integrator shown in Fig. 11.21. C is the integrating capacitance, c_{in} is the parasitic capacitance due to wire routing and the input of the next stage, and c_{out} and r_{out} are the parasitic output capacitance and resistance of the transconductor, respectively. R_{lead} is a series resistance that is sometimes added to provide phase lead for correction of parasitic effects. Assuming that the G_m amplifier has a dc level of G_{mo}, a parasitic pole at ω_p and a parasitic zero at ω_z, the transfer function of the non-ideal integrator can be derived as:

$$H_a(s) = \frac{G_{mo}\left(1+s/\omega_z\right)}{\left(1+s/\omega_p\right)} \frac{r_{out}}{1+sr_{out}\left(C+c_{out}+c_{in}\right)} \tag{11.18}$$

The ideal integrator has infinite dc gain, a rolloff of 6 dB/octave, a unity-gain frequency of $\omega_o = G_{mo}/C$ and a phase of 90° for all frequencies. Parasitics generally have a much stronger effect on the integrator's phase response than the magnitude response. The integrator phase errors in turn are the largest source of filter magnitude response errors. In general, the integrator phase accuracy at ω_o is most critical. The G_m-C integrator's phase error as a function of frequency is:

$$\phi_{I-error}(\omega) \approx \pi/2 + \left(\omega/\omega_z\right) - \left(\omega/\omega_p\right) - \arctan\left[G_{mo}r_{out}\left(\omega/\omega_{ox}\right)\right] \tag{11.19}$$

where ω_{ox} is the actual as opposed to the ideal unity-gain frequency. As an example, consider a G_m-C integrator with a unity-gain frequency of 20 MHz, $C = 1$ pF, $G_{mo} = 125.7$ μS, and $r_{out} = 1$ MΩ. If the transconductor has a parasitic pole at 300 MHz, but no parasitic zero, the resulting phase error at 20 MHz is –3.4°. Depending on the application, such an error may be acceptable. Two methods exist for correcting the phase error. The simplest approach is to add a zero to the transfer function to create phase lead of +3.4° at 20 MHz, resulting in zero net phase error at ω_o. The small value resistor, R_{lead}, in Fig. 11.21 is used for this purpose. Phase lead can also be created within the G_m amplifier with known feedforward techniques. If the accuracy of the phase is critical, tunable phase lead or lag can be performed.[6,8,11,16]

The natural question arises as to what integrator phase accuracy is required. Although the requirement is dependent on the filter topology, and transfer function, a good estimate can be determined by considering the damped loop of two integrators shown in Fig. 11.22. This loop of integrators is found in

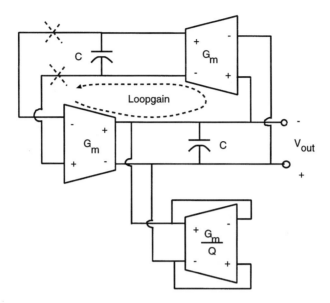

FIGURE 11.22 Integrator phase error impact on biquad filter Q.

cascade of biquad filters and leapfrog filters and is therefore quite general. The system poles are determined by the loop gain transfer function, $T(s)$,

$$T(s) = \left[\frac{G_m}{sC}\right]\left[\frac{G_m}{sC + G_m/Q}\right] = \left[\frac{1}{s/\omega_o}\right]\left[\frac{1}{s/\omega_o + 1/Q}\right] \tag{11.20}$$

Under ideal conditions and assuming a high value of Q, the loop gain phase shift is

$$\phi(\omega) \approx \frac{\omega_o}{\omega Q} \tag{11.21}$$

Notice that as Q increases, the net phase shift around the loop approaches zero; thus making any integrator phase errors a large source of damping errors. Consider two examples. First, the continuous-time filter used in hard disk drive read channels often has a Bessel response where all the poles are low Q. Assuming $Q \approx 2$, the nominal phase shift around the integrator loop, $\phi(\omega_o) \approx 26.6°$, is quite large. If the total integrator phase error is to be kept <0.1ϕ, then each integrator phase error must be <1°. In contrast, for a bandpass response that might be found in wireless systems with $Q = 100$, the nominal loop phase is $\phi(\omega_o) \approx 0.57°$, and each integrator phase error must be kept <0.03°. The low-Q integrator requirements can be met with fixed or tunable phase lead networks, whereas the high-Q applications will require phase tuning (Q tuning) circuits. Although this discussion has centered around G_m-C filters, the results are equally applicable to G_m-OTA-C and *MOSFET-C* filters.

G_m-OTA-C Filters

The G_m-C integrator of Fig. 11.11 achieves high frequency operation because the circuit configuration is open loop; however, the requirement for high impedance at the output nodes of the G_m amplifier is generally difficult to meet. Use of cascodes in the output stage is required, but the signal swing and resulting linearity often suffer. The second difficulty is that any parasitic capacitance due to routing or the input impedance of the next stage will lower the unity-gain frequency of the integrator. A portion

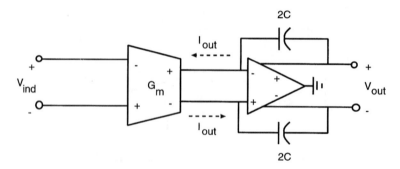

FIGURE 11.23 Fully balanced G_m-OTA-C integrator.

of the parasitic capacitance is the non-linear diode capacitance of the drain diffusions. As the signal frequency increases, more current flows into these non-linear capacitors, reducing the overall linearity of the integrator.

The G_m-OTA-C integrator of Fig. 11.23 adds an operational transconductance amplifier (*OTA*) connected with negative feedback via the integration capacitors at the output of the G_m amplifier. For this configuration, the G_m amplifier drives into a virtual short-circuit and hence requires no signal swing. Parasitic capacitances at the output of the G_m amplifier are held at virtual ground and the capacitances at the output of the *OTA* are driven by a low impedance, so in both cases, their impact on the integrator's frequency response and linearity is negligible. Although the parasitic capacitances do not shift the ω_o of the integrator, high-frequency parasitic poles are created that can cause phase errors near ω_o. For this reason, the G_m-OTA-C approach operates at a lower frequency than the G_m filtering technique.

The dc gain of the G_m-OTA-C integrator is

$$\frac{V_{out}(s)}{V_{ind}(s)} = G_{mo}r_{out}A_o \tag{11.22}$$

where G_{mo} is the dc transconductance of the transconductor, r_{out} is its output resistance, and A_o is the dc voltage gain of the *OTA*. The G_m-OTA-C integrator has two stages of gain, and the increase in gain relative to the G_m-C approach by the factor A_o results in a more ideal integrator characteristic at low frequency. Since two gain stages are used, cascoding is often eliminated or simplified so the improvement in the G_m-OTA-C integrator's dc gain over that of the G_m-C integrator may be less dramatic. Elimination of cascoding simplifies the circuit design, often results in reduced power dissipation, and enables lower voltage operation.

Signal summation in G_m-OTA-C filters is performed in the current domain at the summing node of the *OTA*. Instead of one G_m amplifier as in Fig. 11.23, the outputs of multiple transconductors with different values of G_m can be paralleled and drive the *OTA* summing node to obtain weighted summation and integration.

Many G_m-OTA-C integrator designs exist, but the one in Fig. 11.24 is a good example.[8] The transconductor consists of M1-M4, Q1, and Q2. The $V \rightarrow I$ mechanism is realized with triode operated devices, M1 and M2, that have constant drain-to-source voltages. Varying the base voltage of transistors Q1 and Q2 tunes the $V \rightarrow I$ converter. Unlike the $V \rightarrow I$ converter in Fig. 11.19, devices M1 and M2 are connected to ground rather than to a current source to improve the headroom in the G_m amplifier. The input common-mode voltage level at the gates of M1 and M2 must be well controlled so the drain current bias levels are fixed. M3 and M4 serve as simple current sources and their gate voltages are varied by a CMFB circuit (not shown) to control the output common-mode voltage of the G_m stage. The second stage consists of the *OTA*, integrating capacitors, and phase lead transistors, M5 and M6. M5 and M6 operate in triode with a zero drain-to-source bias voltage and act as resistors. Their role is to provide phase lead

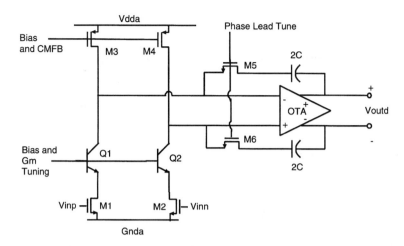

FIGURE 11.24   A BiCMOS G_m-*OTA-C* integrator with tunable phase lead.

in much the same way as the R_{lead} resistor does in the G_m-*C* design of Fig. 11.21. Here, the use of transistors instead of a fixed resistor permits the amount of phase lead to be adjusted for process and temperature variations. The phase lead tuning circuit in Ref. 8 is a dc control loop and does not use the more complex Q tuning techniques given in Refs. 6, 11, 16, and 18.

The *OTA* design can take the form of standard op-amp topologies. In Ref. 8 a simple one-stage BiCMOS design with cascoding was used, as shown in Fig. 11.25. CMFB control via the gates of M2 and M4 is required to control the average output voltages of V_{outp} and V_{outn}. Notice that although the G_m-*OTA-C* integrator requires two basic amplifiers (e.g., a G_m and an *OTA* amplifier), the amplifiers can often be simpler than the single G_m amplifier required in G_m-*C* filters. The power and bandwidth penalties of the G_m-*OTA-C* technique may therefore be less negative than at first view.

MOSFET-C Filters

MOSFET-C and G_m-*OTA-C* filters are similar, except that passive $V \rightarrow I$ devices are used to produce currents for summing into the *OTA* or op-amp in the *MOSFET-C* case, rather than an active $V \rightarrow I$

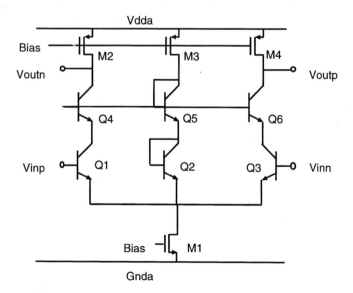

FIGURE 11.25   *OTA for G_m-OTA-C integrator of Fig. 11.24.*

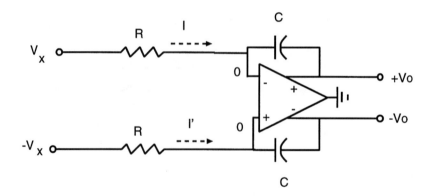

FIGURE 11.26   Balanced active RC integrator.

converter as in the G_m-*OTA-C* approach. Using two equal resistors for $V \to I$ conversion results in the fully balanced classical active-RC integrator of Fig. 11.26. For balanced inputs signals, V_x and $-V_x$, that have a common-mode level of zero, the input terminals of the op-amp are fixed at zero volts. The op-amp provides balanced outputs, $V_o(t)$ and $-V_o(t)$, that are derived to be:

$$V_o(t) - -V_o(t) = \frac{-1}{C} \int_{-\infty}^{t} \left[I(\tau) - I'(\tau) \right] d\tau \tag{11.23}$$

Therefore,

$$V_o(t) = \frac{-1}{RC} \int_{-\infty}^{t} V_x(\tau) d\tau \tag{11.24}$$

Obviously, using fixed resistors does not permit frequency scaling the integrator to remove temperature and process variations. If one replaces the resistors with MOSFETs operating in the triode region with zero drain-to-source bias, the *MOSFET-C* integrator shown in Fig. 11.27 results. Although each MOSFET is non-linear, operating with the fully balanced structure and assuming perfect matching of devices, the even-order non-linearities will cancel. Odd-order non-linearities are not canceled; how-

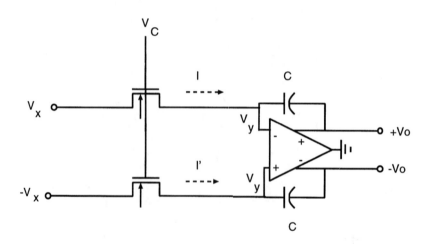

FIGURE 11.27   The *MOSFET-C* integrator.

ever, they are low enough for many applications. Using Eq. 11.12 the output of the *MOSFET-C* integrator can be derived[35] as:

$$V_o(t) \approx \frac{-(W/L)\mu C_{ox}(V_C - V_Q - V_T)}{C} \int_{-\infty}^{t} V_x(\tau)d\tau \qquad (11.25)$$

V_Q is the input common-mode bias level of the signals V_x and $-V_x$. The approximation results from the assumption that the odd-order non-linearities can be neglected. Notice that the op-amp input terminals are at voltage V_y, as opposed to zero. V_y will have a bias level of V_Q, but V_y will vary with the input signal V_x with a square-law characteristic.[36] For practical circuits, V_y will vary by a couple of hundred millivolts, resulting in modest input common-mode range requirements for the amplifier.

The derivation of the *MOSFET-C* integrator linear input/output characteristic assumes that both transistors remain in the triode region. Such a requirement results in a coupling of the minimum permissible tuning voltage, V_{C-min} and the maximum input signal swing, V_{x-max}.

$$V_{x-max} = V_{C-min} - V_Q - V_T \qquad (11.26)$$

Non-linearity cancellation in the *MOSFET–C* integrator depends on matched devices as well as fully balanced input signals. Since in a filter the signals that drive the integrator input come from a similar integrator stage, the op-amp used must have well-balanced outputs. If the signal balancing is imperfect in magnitude or if the signals are not exactly 180° out of phase, the even-order non-linearities will no longer be suppressed.[37] Magnitude errors of 1% and phase errors of 1° can be problematic.[37] The op-amp design must use a robust CMFB circuit consisting of an error amplifier and linear detector (i.e., two resistors) for the output common-mode detection, as described in Ref. 34. Many CMFB circuits that are adequate for differential switched-capacitor circuits do not provide adequate balancing for *MOSFET-C* filters.

Variations of the basic two-transistor fully balanced *MOSFET-C* integrator are possible.[26,38] In Ref. 38, the design of *MOSFET-C* filters with only single-ended output op-amps is described. In Ref. 26, the four-transistor *MOSFET-C* integrator shown in Fig. 11.28 in theory cancels even- and odd-order non-linearities. Assuming perfectly matched transistors and constant mobility of carriers in the transistors,[33] the integrator output can be derived with Eq. 11.12 as:

$$V_o(t) = \frac{-(W/L)\mu C_{ox}(V_{C1} - V_{C2})}{C} \int_{-\infty}^{t} V_x(\tau)d\tau \qquad (11.27)$$

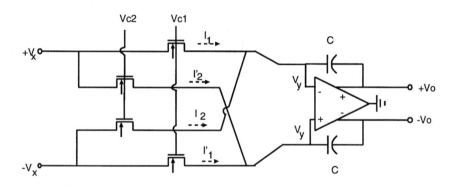

FIGURE 11.28 Four transistor *MOSFET-C* integrator.

Here, integrator tuning only depends on $V_{C1} - V_{C2}$, and is independent of the transistor threshold voltage, V_T, resulting in several advantages. First, in the basic two-transistor *MOSFET-C* integrator, body noise, V_B, can modulate the threshold voltage and couple into the integrator. The threshold voltage using the signal convention in the *MOSFET-C* integrator, is:[33]

$$V_T = V_{T0} + \gamma \left(\sqrt{V_Q - V_B + \phi_B} - \sqrt{\phi_B} \right) \tag{11.28}$$

where V_{T0} is the threshold voltage at zero source-to-body bias, and γ and ϕ_B are constants. Since the integrator's gain in the four-transistor *MOSFET-C* design is independent of V_T, immunity to body noise is achieved. The second advantage of the four-transistor *MOSFET-C* integrator is that very low integrator gains can be realized without loss of signal swing by making $V_{C1} - V_{C2}$ small, while simultaneously maximizing $V_{C1} + V_{C2}$. In contrast, to realize low gain, the basic *MOSFET-C* integrator requires $V_C - V_Q - V_T$ to be small, directly limiting the signal swing, as shown in Eq. 11.26. The primary disadvantage of the four-transistor *MOSFET-C* integrator is increased thermal noise and device sensitivity in comparison to the standard *MOSFET-C* circuit.[37] For most applications, the four-transistor integrator results in increased dynamic range (e.g., a few dB) and a larger tuning range.

Realizing *MOSFET-C* filters from SFGs, block diagrams, or classical active RC filters is straightforward. The *MOSFET-C* equivalent of the Tow-Thomas biquad (Fig. 11.4) is constructed as in Fig. 11.29. Each resistor in the active RC filter is replaced with a MOSFET, and the single-ended topology is converted to a balanced structure by mirroring all devices around the line of symmetry going between the op-amp input terminals and its output. Mirroring and use of a fully balanced structure is mandatory to obtain cancellation of even-order non-linearities in the triode-operated MOSFETs. Signal inversion is obtained by taking the opposite output terminal of the op-amp, so the inverting amplifier in the classical single-ended design is not required in the *MOSFET-C* approach. Weighted signal summation is achieved by connecting multiple transistors to the op-amp summing node and scaling the relative W/L ratios of the devices. The gates of all MOSFETs are connected to a single control voltage, V_C, for frequency tuning the filter.

The frequency response of *MOSFET-C* filters can become distorted due to integrator gain and phase errors, as in the case of G_m-C and G_m-OTA-C filters. As before, phase errors are more problematical. Integrator phase errors in the *MOSFET-C* approach are due to the finite op-amp gain–bandwidth product as well as the distributed capacitance of the MOSFET. The triode-operated MOSFET acts as a linear

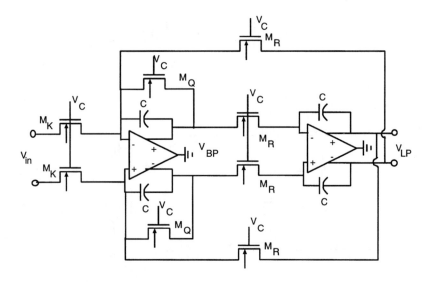

FIGURE 11.29   *MOSFET-C* version of Tow-Thomas biquad.

tunable resistor for small signals; however, parasitic capacitance between the channel-to-gate terminal and from the channel-to-body terminal makes the transistor act as a uniformly distributed RC transmission line.[33,37] The transistor is then a lowpass filter between the integrator input and the op-amp summing node, creating phase lag in the integrator's response.

The channel resistance R_T and capacitance C_T of the MOSFET are given by

$$R_T = \frac{1}{\mu C_{ox}\left(W/L\right)\left[V_C - V_Q - V_T\right]} \tag{11.29}$$

$$C_T = WLC_{ox}\left(b+1\right) \tag{11.30}$$

where b is the backgate effect and is approximately 0.1. As discussed in Ref. 37, for most applications, the MOSFET can be modeled as a first-order lumped circuit resulting in the balanced small-signal *MOSFET-C* integrator model of Fig. 11.30. The transfer function of the integrator is then

$$\frac{V_o}{V_i} = \frac{-\omega_o}{s}\left[\frac{1}{1+\dfrac{s}{\omega_\tau}}\right] \tag{11.31}$$

where $\omega_o = 1/R_T C$ and $\omega_\tau = 1/R_T C_T$. Clearly, distributed capacitance adds extra phase lag. Note that ω_o is proportional to $1/L$, but ω_τ is proportional to $1/L^2$, where L is the channel length. Therefore, *MOSFET-C* filters will in general have worse distributed effects for *low* frequency filters because long channel lengths are used.

MOSFET-C integrator phase errors due to distributed effects and op-amp finite gain-bandwidth product can be compensated by introducing a high frequency zero by either adding a small series resistor to each integrating capacitor or by adding capacitor C_c in parallel with the MOSFETs, as shown in Fig. 11.31. Assuming an op-amp frequency response model of $A(s) = \omega_T/s$, C_c should be selected as follows to null the integrator phase error at the unity-gain frequency:[37]

$$C_c \approx \frac{C_T}{6} + \frac{\omega_o}{\omega_T}C \tag{11.32}$$

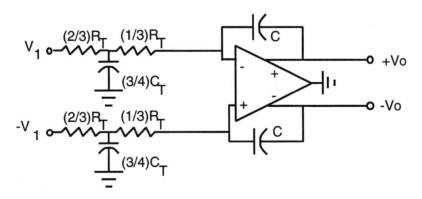

FIGURE 11.30 Small-signal model of *MOSFET-C* integrator with MOSFET channel capacitance.

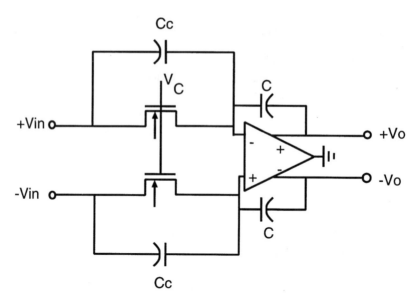

FIGURE 11.31   Phase compensated *MOSFET-C* integrator.

The use of a fixed phase compensation capacitor, such as C_c, or a fixed series resistor is often appropriate for low-quality factor filters, but not for high-Q filters (e.g., Q > 10).

Alternate Continuous-Time Filter Techniques

Although the vast majority of integrated continuous-time filters fall into the G_m-C, G_m-OTA-C, or *MOSFET-C* approaches, other techniques have been developed for applications requiring high linearity. The approaches described so far permit continuous frequency scaling of the filter and hence require MOS or bipolar transistors as variable resistors or transconductors. Since transistors are inherently non-linear, use of nominally matched devices, balanced signals, etc., must be employed to achieve overall linear operation. Practical issues generally limit the linearity to 60 dB for signal swings on the order of 1 V. To achieve linearity in the 80-dB range and higher, use of linear passive devices (e.g., resistors) has helped.[2-4,39,40]

The linearity of the *MOSFET-C* integrator can be vastly improved if less drain-to-source voltage is dropped across the MOSFETs for maximum input signals. Adding resistors in series with the input signals will improve linearity, but decrease the tuning range. With the addition of two transistors connected to a second control voltage V_{c2}, as in Fig. 11.32, the tuning range can be restored.[4] The majority of the input voltage is dropped across the resistors for excellent linearity, and the four transistors are used to simply steer signal current to ground or to the op-amp summing node. As V_{c2} is increased, more current is

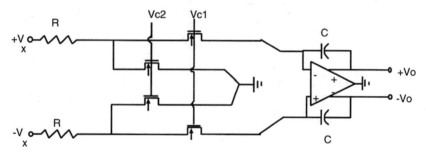

FIGURE 11.32   *R-MOSFET-C* integrator.

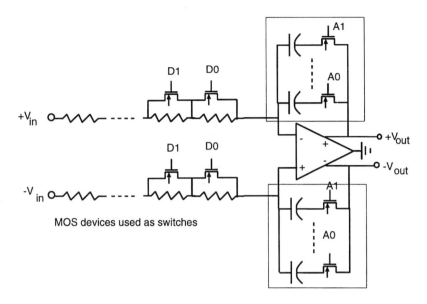

FIGURE 11.33 Programmable active RC integrator.

diverted to ground, leaving less for the summing node, so the integrator gain is reduced. This configuration has been used to achieve in excess of 90-dB linearity for digital audio applications.[4] In this technique, the integrator loop gain is low, restricting the signal bandwidth to levels significantly below standard *MOSFET-C* filters.

If continuous frequency scaling can be replaced with discretized tuning, then switchable active RC filters can be used to maximize linearity.[2,3,39,40] The basic concept, as shown in Fig. 11.33, is to build the integrator with linear resistors and capacitors. MOS switches are then used to program the values of capacitors and/or resistors. In the figure, a series connection of resistors is shown; however, parallel combinations are also feasible.[2] This technique has no inherent bandwidth restrictions, but in practice the MOS switches add parasitic capacitance that cause phase lag, disturbing the integrator's frequency response. A design tradeoff is then tuning resolution versus switch parasitics. To provide adequate tuning range to cover process and temperature variations (e.g., ±50%) and a minimization of switch parasitics, typically results in tuning resolution of ±2 to ±5%. Filter tuning can use microprocessor control[3] or a replica master integrator or delay circuit with up/down counters.[2,39,40] Finally, since the components are inherently linear, use of balanced structures is not mandatory and a larger class of classical active RC filters can be implemented on-chip. Single-amplifier biquad structures[19,40] then become attractive.

11.4 Filter Tuning Circuits

Integrated continuous-time filters require on-chip tuning circuits to control the corner or center frequencies in the face of process and temperature variations. Only in some non-critical applications, such as anti-alias or reconstruct filters, can the ±50% variation in filter time constants be tolerated. Filter tuning, or frequency scaling of the filter, can be accomplished with either direct or indirect tuning methods.

Direct Tuning

Direct tuning, illustrated conceptually in Fig. 11.34, is described in Ref. 41. The filter is periodically removed from the signal path and, after measurements are made with respect to a reference signal, the required adjustments are applied to the filter and held with analog or digital storage means. The filter is then returned to the signal path. The advantage of this approach is that the filter is measured directly

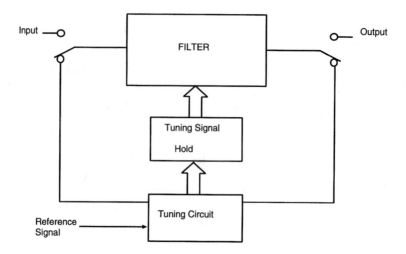

FIGURE 11.34 A direct-tuned filter.

and excellent accuracy is possible. Since many applications cannot tolerate interruption of the signal path, it is conceptually possible to use two filters[41]: one processes the signal while the other is tuned. However, the complexity and chip area overhead cause this alternative to be rarely used. Direct tuning is seldom used, but a production example does exist.[42]

Master-Slave Tuning

By far the most prevalent tuning method is the indirect or master-slave tuning technique. In this approach, a master circuit that is made of the same components as the main filter is tuned with respect to a reference signal or component. The tuning signal that adjusts the master is also used to tune the slave or main filter. Since components in the master and slave circuits are on the same chip, they will match accurately. The time constants of the slave will then track the master over all process and temperature variations. If the master and slave components are placed closely in the layout, and good layout techniques are used, then matching of a few percent is possible.

One simple dc master-slave frequency tuning technique, shown in Fig. 11.35, uses a precision off-chip resistor. A small voltage V is applied to the MOSFET and $-V$ is applied to the resistor. The circuit converges when $I_1 = I_2$, so that $r_{ds} = 1/R_s$. The process and temperature variations of the MOSFET resistance are then removed; however, capacitor values in the filter will need trimming during manufacture. An ingenious variation[43] replaces the external precision resistor with an on-chip switched-capacitor equivalent resistance as shown in Fig. 11.36. If switched-capacitor C_s is clocked at a rate f_s, the charge transfer from

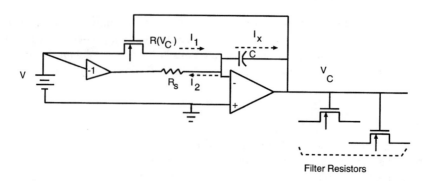

FIGURE 11.35 Reference resistor-based master-slave tuning circuit.

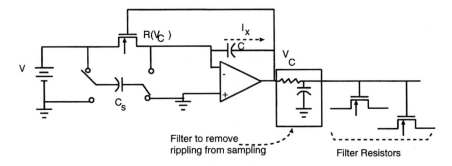

FIGURE 11.36 Switched-capacitor-based master-slave tuning circuit.

the input voltage V to the summing node is equivalent to that which would be obtained with a resistor of value

$$R_{sc} = \frac{1}{f_s C_s} \tag{11.33}$$

The average current flowing through the switched-capacitor will be forced equal to the drain current in the MOSFET due to the negative feedback resulting in $I_x = 0$. The voltage V_C that tunes the MOSFET in the master cell will then reach an average dc value with some ripple due to the switched-capacitor circuit operation. After additional filtering of V_C to remove the ripple, the tuning signal is routed to the slave filter. The slave filter will possess many time constants and critical frequencies. A critical frequency, f_x, will be determined by a MOSFET channel resistance r_{ds-x} and a capacitance, C_x. The tuning circuit then ensures that

$$f_x = f_s \left(\frac{r_{ds}}{r_{ds-x}} \right) \left(\frac{C_s}{C_x} \right) \tag{11.34}$$

The filter's critical frequency, f_x, is then accurate since it is determined by the clock reference, f_s, the ratio of the channel resistance of transistors, and the ratio of capacitors. Clearly, the switched-capacitor tuning technique has the benefit that it tunes time constants, not just resistance values. The master-slave dc tuning methods shown here are equally applicable to G_m-C and G_m-OTA-C filters.

Often, the master-slave tuning approach uses a phase-lock loop (PLL) and a master cell that is either a voltage-controlled filter (VCF) or a voltage-controlled oscillator (VCO). First, consider the case of a master VCF.[6,10] Many options exist for the design of the VCF, but a reasonable design uses a lowpass biquadratic filter. The transfer function of a lowpass biquad is given by

$$V_{o-lp}(s) = \frac{K\omega_o^2}{s^2 + \dfrac{\omega_o}{Q}s + \omega_o^2} V_{in}(s) \tag{11.35}$$

where K is a gain constant. If a sinewave at frequency ω_o is applied to the filter's input, the phase of the output will be $-90°$ with respect to the input, provided that the filter is accurately tuned. The PLL tuning circuit will adjust the RC products of the VCF until the phase shift is $-90°$. One embodiment of the VCF master-slave tuning method is shown in Fig. 11.37. The exclusive-OR (XOR) gate phase detector is preceded by two slicers that convert the filter's sinewave input and output signals to logic levels. The loop filter, $H(s)$, has a first-order lowpass response and the feedback loop forces the tuning voltage (or current)

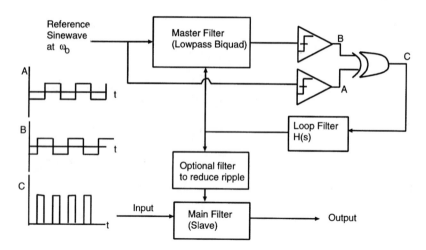

FIGURE 11.37   A VCF-based master-slave tuning system.

at the output of the loop filter to converge to a dc value, which simultaneously frequency tunes the master biquad and slave filter. Since the loop filter output will contain ripple, extra lowpass filtering can be applied outside the feedback loop so as not to compromise loop stability while eliminating the ripple on the tuning signal in the slave filter.

The tuning accuracy of this approach is directly linked to the loop gain of the PLL, which is derived as

$$\text{Loop gain} = K_D H(s) K_{VCF} = K_D H(s)\left(\frac{\partial \omega_o}{\partial V_C}\right)\left(\frac{-2Q}{\omega_o}\right) \tag{11.36}$$

K_D is the phase detector gain in volts/radian, K_{VCF} is the master filter's gain in radians/volt, V_C is the filter's tuning signal, and Q and ω_o are the quality factor and corner frequency of the master biquad, respectively. To maximize the tuning accuracy, the loop gain should be maximized. The gain is kept high by using a relatively high-Q master filter (e.g., $Q > 5$) and an integrator for the loop filter. Frequency tuning errors will occur if there are static phase errors in the PLL. Static phase errors will not cause PLL convergence difficulty; however, the relative phase of the biquad's input and output signals may differ from $-90°$. Static phase errors can occur due to delay mismatches in the slicers or rise and fall time asymmetries in the XOR phase detector.[6]

The PLL with VCF master-slave tuning approach is appealing because the master circuit is simply a biquad that closely resembles a portion of the slave filter. The only drawback is that this tuning method requires a sinusoidal reference signal that is not usually available in most system applications. Unfortunately, use of a commonly available square wave reference signal is unacceptable because the large harmonic content shifts the zero crossings at the master biquad's output, creating a systematic error in phase measurements, resulting in a fixed frequency tuning error. Triangular input signals work better since the harmonic content is lower.[6]

Since sinusoidal reference signals are rarely available, the more common PLL tuning approach uses a VCO in a classical PLL circuit.[44] Use of a master VCO with PLL has been widely applied to master-slave continuous-time filter tuning.[1,7,9,13,14,27] This master-slave tuning technique, depicted in Fig. 11.38, uses a master oscillator that is constructed with the same type of capacitors and resistors (or transconductors) used in the slave filter. The VCO produces a sinusoidal output, so a slicer converts the signal to a square wave for application to the digital phase-frequency detector. Some implementations use an XOR gate, but the phase-frequency detector enhances the capture range of the PLL and prevents harmonic locking.[44] In Fig. 11.38 the loop filter is implemented with a charge pump that has fixed source and sink current levels coupled to a resistor and capacitor in series to ground. The capacitor and charge pump form an

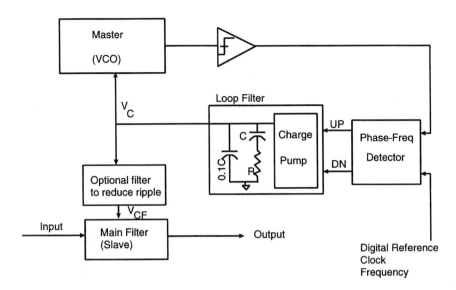

FIGURE 11.38 A VCO-based master-slave tuning system.

integrator and the series resistor introduces a zero to stabilize the loop. As in the previous case, the output of the loop filter tunes both the master oscillator and the slave filter. However, unlike the PLL with VCF tuning method, the PLL will lock the VCO to *exactly* the same frequency as the input reference signal even if static phase errors occur. In this application, static phase errors between the input reference and the VCO output are inconsequential.

The primary drawback with the PLL and VCO master-slave tuning method is that the VCO design is difficult. Virtually any poorly designed VCO can be placed in the PLL and lock to the reference frequency; however, the issue is whether the control voltage, V_C, that is developed will correctly tune the slave filter. In other words, the voltage-to-frequency conversion curve for the VCO and the slave filter must be identical or at least they must match to better than about ±1% over the full range of tuning levels. Designing a slave filter that has a well-defined tuning voltage-to-corner frequency curve is relatively easy. In contrast, unless extreme care is taken in the VCO design, the amplitude of oscillation and hence VCO linearity can be coupled to the tuning signal level. The amplitude of oscillations in the VCO must be well controlled in the linear region so that the MOSFETs or transconductors in the master operate at the same levels as in the slave so that the voltage-to-frequency characteristics are nominally the same. This issue has been addressed in Ref. 7, but more work needs to be done.

Q Tuning Loops

Frequency tuning circuits hold the filter's time constants fixed over process and temperature variations or, equivalently, hold constant an integrator's unity-gain frequency. In contrast, Q tuning loops adjust an integrator's phase response to achieve −90° over all process and temperature conditions. Rather than directly tune the phase response of an integrator, the Q of a bandpass biquad can be adjusted instead[6,11,16] as shown in Fig. 11.39. If the integrators have phase lag, the Q will be enhanced; if they have phase lead, the Q will be reduced. The input-output relationship of the bandpass biquad is

$$V_{o-bp}(s) = \frac{K\omega_o s}{s^2 + \dfrac{\omega_o}{Q}s + \omega_o^2} V_{in}(s) \qquad (11.37)$$

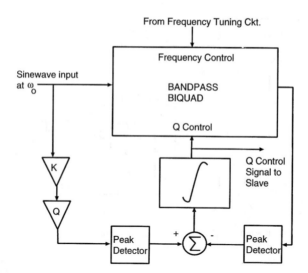

FIGURE 11.39 A master-slave Q tuning loop.

If a sinusoid of frequency ω_o is applied to the biquad input and the biquad is properly frequency scaled by a tuning loop (not shown), then the sinusoid at the biquad output will have an amplitude KQ. Since K will be defined by well-matched components, the output amplitude error can be directly attributed to errors in Q. By subtracting the amplitude of the biquad's input and output and integrating, the Q error can be driven to zero with the negative feedback shown. Amplitude measurements can use r.m.s.-type detectors or peak amplitude detecting circuits, as shown in the figure. Many variations of Q tuning loops exist.[6,11,16,18]

Master-slave Q tuning loops should be used with caution for two reasons. First, if the tuning loop incorrectly adjusts the integrator's phase response, even momentarily during a transient, the slave filter can break into oscillation. Second, integrator phase errors are in large part due to parasitic capacitances. For the master-slave Q tuning to be successful, parasitics in the master and slave must match and track, suggesting that meticulous attention to layout detail is required. Additionally, excellent modeling and extraction of the parasitics from the layout are necessary to precisely design these circuits.

11.5 Conclusion

Integrated continuous-time filters are now found in a variety VLSI chips spanning the frequency range from a few kHz for ultra-low power hearing aid applications to over 150 MHz for the hard disk drive read channel application. The continuous-time filtering techniques described in this chapter have become integral parts of production quality chips and now represent well-established analog filter design practices.

References

1. M. Banu, and Y. Tsividis, An elliptic continuous-time CMOS filter with on-chip automatic tuning, *IEEE J. Solid-State Circuits*, vol. SC-20, no. 6, pp. 1114-1121, Dec. 1985.
2. A. Durham, J. Hughes, and W. Redman-White, Circuit architectures for high linearity monolithic continuous-time filtering, *IEEE Trans. Circ. Syst.*, pp. 651-657, Sept. 1992.
3. H. Khorramabadi, M. Tarsia, and N. Woo, Baseband filters for IS-95 CDMA receiver applications featuring digital automatic frequency tuning, *1996 International Solid State Circuits Conference*, pp. 172-173.
4. U.-K. Moon and B.-S. Song, Design of a low-distortion 22-kHz fifth-order bessel filter, *IEEE J. Solid-State Circuits*, vol. 28, no. 12, pp. 1254-1264, Dec. 1993.

5. S. Willingham, K. Martin, and A. Ganesan, A BiCMOS low distortion 8-MHz low-pass filter, *IEEE J. Solid-State Circuits*, vol. 28, no. 12, pp. 1234-1245, Dec. 1993.

6. V. Gopinathan, Y. Tsividis, K-S Tan, R. Hester, Design considerations for high-frequency continuous-time filters and implementation of an antialiasing filter for digital video, *IEEE J. Solid-State Circuits*, vol. SC-25, no. 6, pp. 1368-1378, Dec. 1990.

7. J. Koury, Design of a 15 MHz continuous-time filter with on-chip tuning, *IEEE J. Solid-State Circuits,*, vol. 26, no. 12, pp. 1988-1997, Dec. 1991.

8. C. Laber and P. Gray, A 20MHz 6th order BiCMOS parasitic insensitive continuous-time filter and second order equalizer optimized for disk-drive read channels, *IEEE J. Solid-State Circuits*, vol. 28, pp. 462-470, Apr. 1993.

9. F. Krummenacher and N. Joehl, A 4 MHz CMOS continuous-time filter with on-chip automatic tuning, *IEEE J. Solid-State Circuits*, vol. 23, no. 3, pp. 750-758, June 1988.

10. H. Khorramabadi and P. R. Gray, High-frequency CMOS continuous-time filters, *IEEE J. Solid-State Circuits*, vol. SC-19, no. 6, pp. 939-948, Dec. 1984.

11. C. Chiou and R. Schaumann, Design and performance of a fully integrated bipolar 10.7 MHz analog bandpass filter, *IEEE Trans. Circ. Syst.*, vol. CAS-33, no. 2, pp. 116-124, Feb. 1986.

12. B-S Song, and P. R. Gray, Switched-capacitor high-Q bandpass filters for IF applications, *IEEE J. Solid-State Circuits*, vol. SC-21, no. 6, pp. 924-933, Dec. 1986.

13. K. W. Moulding, J. R. Quartly, P. J. Rankin, R. S. Thompson, and G. A. Wilson, Gyrator video filter IC with automatic tuning, *IEEE J. Solid-State Circuits*, vol. SC-15, no. 6 pp. 963-968, Dec. 1980.

14. K. S. Tan and P. R. Gray, Fully integrated analog filters using bipolar FET technology, *IEEE J. Solid-State Circuits*, vol. SC-13, no. 6, pp. 814-821, Dec. 1978.

15. K. R. Rao, V. Sethuraman, and P. K. Neelakantan, A novel 'follow the master' filter, *Proc. IEEE*, vol. 65, pp. 1725-1726, 1977.

16. D. Senderowicz, D. Hodges, and P. Gray, An NMOS integrated vector-locked loop, *Proc. IEEE Int. Symp. Circ. Syst.*, pp. 1164-1167, May 1982.

17. J. Silva-Martinez, M. Steyaert, and W. Sansen, A 10.7-MHz 68-dB SNR CMOS continuous-time filter with on-chip automatic tuning, *IEEE J. Solid-State Circuits*, vol. 27, no. 12, pp. 1843-1853, Dec. 1992.

18. S. Pavan and Y. Tsividis, An analytical solution for a class of oscillators, and its application to filter tuning, *IEEE Trans. Circ. Syst., I*, vol. 45, no. 5, May 1998.

19. A. Sedra, and P. Brackett, *Filter Theory and Design: Active and Passive*, Matrix Publishers, Inc., Beaverton, OR, 1978.

20. W. Snelgrove and A. Sedra, Synthesis and analysis of state-space active filters using intermediate transfer functions, *IEEE Trans. Circ. Syst.*, vol. CAS-33, no. 3, pp. 287-301, March 1986.

21. K. Bult and G. Geelen, A fast-settling CMOS op amp with 90 dB DC-gain and 116 MHz unity-gain frequency, *International Solid State Circuits Conference*, pp. 108-109, 1990.

22. P. Gray and R. Meyer, *Analysis and Design of Analog Integrated Circuits*, John Wiley & Sons, second edition, 1984.

23. A. Nedungadi and T. R. Viswanathan, Design of linear CMOS transconductance elements, *IEEE Trans. Circ. Syst.*, vol. CAS31, pp. 891-894, October 1984.

24. Y. Tsividis, Z. Czarnul, and S. C. Fang, MOS transconductors and integrators with high linearity, *Electronics Letters*, vol. 22, pp. 245-246, Feb. 1986.

25. M. Banu and Y. Tsividis, Detailed analysis of nonidealities in MOS fully integrated active RC filter based on balanced networks, *Proceedings of the Institute of Elect. Eng.*, vol. 131, Pt. G, no. 5, pp. 190-196, Oct. 1984.

26. Z. Czarnul, Modification of the banu-tsividis continuous-time integrator structure, *IEEE Trans. Circ. Syst.*, vol. CAS-33, no. 7, pp. 714-716, July 1986.

27. D. Welland, S. Phillip et al., A digital read/write channel with EEPR4 detection, *IEEE Intl. Solid-State Circuits Conference*, San Francisco, pp. 276-277, 352, 1994.

28. J. Pennock, CMOS triode transconductor for continuous-time active integrated filters, *IEE Electronic Letters*, vol. 21, no. 18, pp. 817-818, Aug. 1985.

29. R. Alini, A. Baschirotto, and R. Castello, Tunable BiCMOS continuous-time filter for high-frequency applications, *IEEE J. Solid-State Circuits*, vol. 27, no. 12, pp. 1905-1915, Dec. 1992.

30. Y. Tsividis and J. Voorman, *Integrated Continuous-Time Filters: Principles, Design and Applications*, IEEE Press, New York, 1993.

31. D. Calder, Audio frequency gyrator filters for an integrated paging receiver, *IEEE Conference, Mobile Radio Systems and Techniques*, no. 238, 1984.

32. J. Voorman, W. Bruls, and P. Barth, Integration of analog filters in a bipolar process, *IEEE J. Solid-State Circuits*, vol. SC-17, no. 4, pp. 713-722, Aug. 1982.

33. Y. Tsividis, *Operation and Modeling of the MOS Transistor*, McGraw-Hill, New York, 1987.

34. M. Bani, J. Khoury, and Y. Tsividis, Fully differential operational amplifiers with accurate output balancing, *IEEE J. Solid-State Circuits*, vol. SC-23, no. 6, pp. 1410-1414, Dec. 1988.

35. M. Banu and Y. Tsividis, Fully integrated active RC filters in MOS technology, *IEEE J. Solid-State Circuits*, vol. SC-18, pp. 644-651, Dec. 1983.

36. Y. Tsividis, M. Banu, and J. Khoury, Continuous-time MOSFET-C filters in VLSI, *IEEE J. Solid-State Circuits*, vol. SC-21, no. 1, Feb. 1986, pp. 15-30; and *IEEE Trans. Circ. Syst.*, vol. CAS-33, no. 2, pp. 125-140, Feb. 1986.

37. J. Khoury, and Y. Tsividis, Analysis and compensation of high frequency effects in integrated MOSFET-C continuous-time filters, *IEEE Trans. Circ. Syst.*, vol. CAS-34, no. 8, pp. 862-875, Aug. 1987.

38. M. Ismail, A new MOSFET capacitor integrator, *IEEE Trans. Circ. Syst.*, vol. CAS-32, no. 11, pp. 1194-1196, Nov. 1985.

39. A. Durham and W. Redman-White, Integrated continuous-time balanced filters for 16-b DSP interfaces, *IEEE J. Solid-State Circuits*, vol. 28, no. 7, pp. 835-839, July 1993.

40. R. Shariatdoust, K. Nagaraj, J. Khoury, S. Daubert, and D. Fasen, An integrating servo demodulator for hard disk drives, *IEEE 1993 Custom Integrated Circuits Conference*, pp. 10.6.1-10.6.5, 1993.

41. Y. Tsividis, Self-tuned filters, *Electronics Letters*, vol. 17, no. 12, pp. 406-407, June 1981.

42. G. Smolka, U. Riedle, U. Grehl, B. Jahn, F. Parzefall, W. Veit, and H. Werker, A low-noise trunk interface circuit with continuous-time filters and on-chip tuning, in *Integrated Continuous-Time Filters: Principles, Design and Applications*, Edited by Y. Tsividis and J. Voorman, IEEE Press, New York, 1993.

43. T. R. Viswanathan, S. Murtuza, V. Syed, J. Berry, and M. Staszel, Switched-capacitor frequency control loop, *IEEE J. Solid-State Circuits*, vol. SC-17, no. 4, pp. 775-778, Aug. 1982.

44. F. Gardner, *Phase-Lock Techniques*, John Wiley & Sons, New York, 1966.

12

Switched-Capacitor Filters

Andrea Baschirotto
Università di Pavia

12.1 Introduction

The accuracy of the absolute value of integrated passive devices (R and C) is very poor. As a consequence, the frequency response accuracy of integrated active-RC filters is poor and they are not feasible when high-accuracy performance is needed. A possible solution for the implementation of analog filters with accurate frequency response was given by the switched-capacitor (SC) technique in the late 1970s.[1,2] Their popularity has increased further since they can be realized with the same standard CMOS technology used for digital circuits. In this way, fully integrated low-cost high-flexibility mixed-mode systems have become possible. The main reasons for the large popularity of SC networks can be summarized as follows:

1. The basic requirements of SC filters fit the popular MOS technology features. In fact, the infinite input impedance of the operational amplifier (op-amp) is obtained using a MOS input device;

MOS transconductance amplifiers can be used since only capacitive load is present; precise switches are realized with a MOS transistor; and capacitors are available in the MOS process.

2. SC filter performance accuracy is based on the matching of integrated capacitors (and not on their absolute values). In a standard CMOS process, the capacitor matching error can be less than 0.2%. As a consequence, SC systems guarantee very accurate frequency response without component trimming. For the same reason, temperature and aging coefficients track reducing performance sensitivity to temperature and aging variations.

3. It is possible to realize SC filters with long time constants without using large capacitors and resistors. This means a chip area saving, with respect to active-RC filter implementations.

4. SC systems operate with closed-loop structures; this allows one to process large swing signals and to achieve large dynamic range.

On the other hand, the major drawbacks of the SC technique can be summarized in the following points:

1. To process a fully analog signal, an SC filter has to be preceded by an anti-aliasing filter and followed by a smoothing filter, which complicate the overall system and increase power and die size.

2. The op-amps embedded in an SC filter have to perform a large dc-gain and a large unity-gain bandwidth, much larger than the bandwidth of the signal to be processed. This limits the maximum signal bandwidth.

3. The power of noise of all the sources in the SC filter is folded in the band [0–Fs/2]. Thus, their noise power density is increased by the factor (Fs/2)/Fb, where Fs is the sampling frequency and Fb is the noise bandwidth at the source.

Since its first proposal, the SC technique has been deeply developed. Many different circuits solutions have been realized with the SC technique not only in analog filtering, but also in analog equalizers, analog-to-digital and digital-to-analog conversion (including in particular the oversampled SD converters), Sample&Hold, Track&Hold, etc.

In this chapter, an overview of the main aspects of the SC technique is given, leaving to the reader to study the large literature for details more related to his or her necessity. A few advanced solutions feasible for future SC systems are given in the last section.

12.2 Sampled-Data Analog Filters

An SC filter is a continuous-amplitude, sampled-data system. This means that the amplitude of the signals can assume any value within the possible range in a continuous manner. On the other hand, these values are assumed at certain time instants and then they are held for the entire sampling period. Thus, the resulting waveforms are not continuous in time but look like a staircase.

In an SC filter, the continuous-time input signal is firstly sampled at sampling frequency Fs and then processed through the SC network. This sampling operation results in a particular feature of the frequency response of the SC system. In the following, the aspects relative to the sampling action are illustrated from an intuitive point of view, while a more rigorous description can be found in Ref. 3.

The sampling operation extracts from the continuous-time waveform the values of the input signal at the instant k·Ts (k = 1,2,3, ...), where Ts is the sampling period (Ts = 1/Fs). This is shown in Fig. 12.1 for a single-input sinewave at $fo = 0.16 \cdot Fs$ (i.e., with $fo < Fs/2$).

If the input sinewave is at Fs + fo, the input sequence of analog samples is exactly equal to that previously obtained with fo as input frequency (see Fig. 12.2). Both sequences should then be processed in exactly the same way by the SC network, and the overall filter output sequence should then be again identical. The two input sinewaves are then indistinguishable after the sampling action. This effect is called *aliasing*. It can be demonstrated that a sinewave at frequency fo in the range [0–Fs/2] is aliased by the components at frequency f_{al} given by:

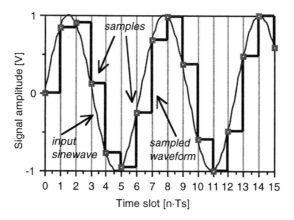

FIGURE 12.1   Sampling of the input signal.

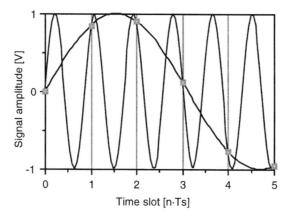

FIGURE 12.2   Aliasing between fo and FS + fo.

$$f_{al} = k \cdot Fs \pm fo \quad \left(k = 1, 2, 3, \ldots\right) \tag{12.1}$$

As a consequence, in order to avoid frequency aliasing (which means signal corruption), the input signal band of a sample data system must be limited to the [0–Fs/2] range. The range [0–Fs/2] is called *baseband* and the top limitation is an expression of the Nyquist theorem.

After the sampling, the SC network processes the sequence of samples, independently of how they have been produced. Since all the frequencies given in Eq. 12.1 produce the same sequence of samples, the gain for all of them is the same. This concept results in the fact that the transfer function of a sampled-data system is periodical with period equal to Fs, and it is symmetrical in its period. For instance, in Fig. 12.3 the frequency response amplitude for a lowpass filter is shown for frequencies higher than Fs.

As stated above, in order to avoid the aliasing effect corrupting the signal, it is necessary to limit the input signal bandwidth. This function is performed by the anti-aliasing (AA) filter, which is placed in front of the SC filter and operates in the continuous-time domain. From a practical point of view, the poles of the SC filter are typically much smaller than Fs/2, and only in the passband is the frequency response required to be accurate. On the other hand, the AA filter transfer function is not required to be accurate. Thus, the AA filter is usually implemented with active-RC filters (see Fig. 12.4).

At the output of the SC network, a staircase signal is produced. If a continuous-time output waveform is needed, a continuous-time smoothing filter must be added. The overall SC filter processing chain then results as shown in Fig. 12.5. Of course, in some cases, the input signal spectrum is already limited to

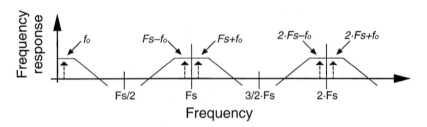

FIGURE 12.3   Periodicity of the sampled-data system transfer function.

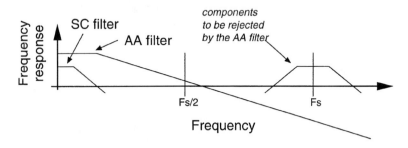

FIGURE 12.4   Transfer functions of switched-capacitor and anti-aliasing filters.

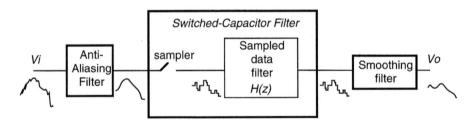

FIGURE 12.5   Overall SC filtering structure.

Fs/2 and then the AAF filter is not necessary; while in other cases, the final smoothing filter is no longer necessary, like when the SC filter is used in front of a sampled-data system (like an ADC, for instance).

12.3 The Principle of the SC Technique

The principle of the SC technique consists in simulating the resistor behavior with a switched-capacitor structure. In the structure of Fig. 12.6, where an ideal op-amp is used, the resistor Req is connected between Vi and a zero-impedance zero-voltage node (as a virtual ground is). This means that a continuous-time current I flows from Vi, through Req, into the virtual ground. This current is equal to:

$$I = Vi/Req \qquad (12.2)$$

The alternative SC structure is shown in Fig. 12.7(a). It is composed of an input sampling capacitor CS (connected through four switches to the input signal VI, to the op-amp input node, and to two ground nodes), an op-amp, and a

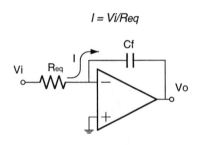

FIGURE 12.6   Basic RC integrator.

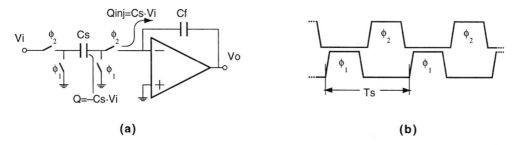

FIGURE 12.7 Basic SC integrator.

feedback (integrating) capacitor Cf. The clock phases driving the switches are shown in Fig. 12.7(b). A switch is closed (conductive) when its driving phase is high. It is necessary that the two clock phases are non-overlapping, in order to connect each capacitor plate to only one low-impedance node for each time slot.

During phase φ1, capacitor Cs is discharged. During phase φ2, Cs is connected between Vi and the virtual ground. So, a charge $Q = -Cs \cdot Vi$ is collected on its right-hand plate. Due to the charge conservation law applied at the virtual ground node, this charge collection corresponds to an injection in virtual ground of the same amount of charge but with the opposite sign, given by:

$$Qinj = Cs \cdot Vi \tag{12.3}$$

Notice that this charge injection is independent of the component in the op-amp feedback path. This charge injection occurs every clock period. Observing this effect for a long time slot T, the total charge injection Qtot is given by:

$$Qtot = Cs \cdot Vi \cdot \frac{T}{Ts} \tag{12.4}$$

This corresponds to a mean current (Imean) equal to:

$$Imean = \frac{Qtot}{T} = Cs \cdot \frac{Vi}{Ts} \tag{12.5}$$

Equating Eq. 12.2 with Eq. 12.5, the following relationship holds:

$$Req = \frac{Ts}{Cs} \tag{12.6}$$

This means that the structure composed of Cs and the four switches operated at Fs is equivalent to a resistor Req. This approximation is valid for Vi equal to a dc-value, as is the case in the proposed example, and it is still valid for Vi *slowly variable with respect to the clock period*; otherwise, the quantitative average operation of Eq. 12.5 is no longer valid. The limits of the approximation between a resistor and an SC structure as expressed in Eq. 12.6 implies fundamental differences in the exact design of SC filters when derived from active-RC filters in a one-to-one correspondence.

The synthesis of active filters is based on the use of some elementary blocks interconnected in different ways, depending on the type of adopted design philosophy. The different strategies and approaches for designing analog filters are well known from the continuous-time domain and they are also used in the case of SC filters, although the sampled-data nature of the SC filters can be profitably used either for simplifying or improving the design itself. In any case, basic building blocks are used to compose high-order filters.

In the following, the main basic blocks are described. They implement first-order (active integrators, undamped and damped, summers) and second-order (biquads) transfer functions in the z-domain (z is the state variable in the sampled data domain).

12.4 First-Order SC Stages

The Active SC Integrators

In Fig. 12.8(a-c), the standard integrators normally used in SC designs are shown. For each integrator, the transfer function in the z-domain is reported, assuming that the input signal is sampled during phase $\phi 1$ and is held to this value until the end of phase $\phi 2$, while the output is read during phase $\phi 2$.

The third integrator is called bilinear since it implements the bilinear mapping of the s-to-z transformation (s is the state variable in the continuous-time domain).

In all the above transfer functions, only capacitor ratios appear. For this reason, the SC filter transfer functions are sensitive only to the capacitor ratios (i.e., to the capacitor matching) and are independent of absolute capacitor value. This is a remarkable advantage of all SC networks.

An important feature of all these integrators is their insensitivity to parasitic capacitance. This can be verified by observing that any parasitic capacitance connected to the capacitor left-hand plate is not connected to the virtual ground and therefore does not contribute to the amount of injected charge. On the other hand, the stray capacitance connected to the capacitor right-hand armature could contribute to the charge transfer, but this capacitance is switched between two nodes (ground and virtual ground) at the same potential and thus no charge injection results.

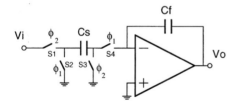

$$H_a(z) = \frac{V_o(z)}{V_i(z)} = \frac{Cs}{Cf} \cdot \frac{z^{-1}}{1-z^{-1}} \qquad (12.7a)$$

(a) - non-inverting

$$H_b(z) = \frac{V_o(z)}{V_i(z)} = -\frac{Cs}{Cf} \cdot \frac{1}{1-z^{-1}} \qquad (12.7b)$$

(b) - inverting

$$H_c(z) = \frac{V_o(z)}{V_i(z)} = 2 \cdot \frac{Cs}{Cf} \cdot \frac{1+z^{-1}}{1-z^{-1}} \qquad (12.7c)$$

(c) - bilinear

FIGURE 12.8   SC integrators.

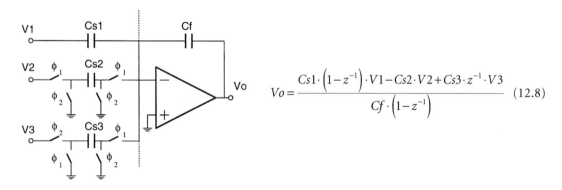

FIGURE 12.9 The SC summing integrator.

$$Vo = \frac{Cs1 \cdot \left(1 - z^{-1}\right) \cdot V1 - Cs2 \cdot V2 + Cs3 \cdot z^{-1} \cdot V3}{Cf \cdot \left(1 - z^{-1}\right)} \qquad (12.8)$$

The Summing Integrator

The SC operation is based on charge transfer. It is therefore easy to make weighted sum of multiple inputs by connecting different input branches to the same virtual ground. This concept is shown in the summing integrator of Fig. 12.9. The transfer function from the three input signals to the output is given in Eq. 12.8.

If the integrating feedback capacitor (Cf) is replaced by a feedback switched-capacitor (Csw), the structure does not maintain memory of its past evolution and a simple summing amplifier is obtained. The resulting structure is shown in Fig. 12.10, with the corresponding transfer function given in Eq. 12.9. This is the basic building block for the construction of SC filters implementing FIR frequency response.[4]

The Active Damped SC Integrator

A damped integrator can be realized by connecting a damping switched capacitor (Cd) in parallel to the integrating capacitor (Cf), as shown in Fig. 12.11. Both inverting and non-inverting circuits are possible, depending on the type of input sampling structure. Equation 12.10a is valid for the clock phases out of parentheses, while Eq. 12.10b is valid for the clock phases within parentheses.

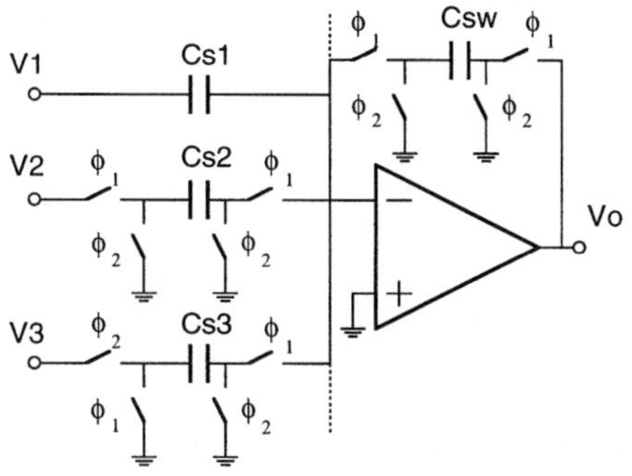

$$Vo = \frac{1}{Csw} \cdot \left(Cs1 \cdot \left(1 - z^{-1}\right) \cdot V1 - Cs2 \cdot V2 + Cs3 \cdot z^{-1} \cdot V3\right)$$
$$(12.9)$$

FIGURE 12.10 The SC summing amplifier.

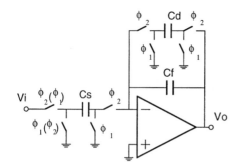

$$H_1\!\left(z\right) = \frac{Cd \cdot z^{-1}}{\left(Cd + Cf\right) - Cf \cdot z^{-1}} \tag{12.10a}$$

$$H_2\!\left(z\right) = -\frac{Cd}{\left(Cd + Cf\right) - Cf \cdot z^{-1}} \tag{12.10b}$$

FIGURE 12.11 Damped SC integrator.

A Design Example

As an example, the design of a damped SC integrator (Fig. 12.11) is given. A possible design approach is to derive the capacitor values of the SC structure from the R and C values in the equivalent continuous-time prototype, which is shown in Fig. 12.12, by the relationship of Eq. 12.6. It results that: $Cs = Ts/Rs$, and $Cd = Ts/Rd$. For instance, to have the pole frequency at 10 kHz with unitary dc-gain, a possible solution is: $Rs = Rd = 159.15$ kΩ, and $Cf = 10$ pF. Using $Fs = 1$ MHz, it is obtained that $Cs = Cd = 0.628$ pF.

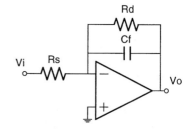

FIGURE 12.12 Continuous-time damped RC integrator.

The frequency response of the continuous-time and the SC damped integrator are shown in Fig. 12.13(a). Line I refers to the damped integrator of Fig. 12.11. Line II refers to a damped integrator with bilinear input branch (see Fig. 12.8(c)), while line III refers to the active-RC integrator of Fig. 12.12. In the passband, the frequency responses track very well. By increasing the input frequency, a difference becomes evident (as stated by the fact that Eq. 12.6 is valid for slowly variant signals). This is more pronounced if the frequency response is plotted up to $2 \cdot Fs$ (see Fig. 12.13(b)), where the periodic behavior of the sampled-data system frequency response is evident. Moreover, the key point of a sampled-data filter is the fact that the frequency response fixes the ratio between sampling-frequency and pole-frequency, i.e., with the above capacitor values, the pole frequency (fp) is 10 kHz for $Fs = 1$ MHz, while it decreases to 1 kHz if $Fs = 100$ kHz is used (i.e., the ratio fp/Fs remains constant). For this reason, the frequency response is plotted as a function of the normalized frequency f/Fs.

A limited stopband attenuation results for line I. This attenuation is improved using the bilinear input branch which implements a zero at Fs/2. This does not affect the frequency response in the passband, while a larger attenuation is obtained in the stopband (line II).

For the SC networks, the frequency response depends on capacitor ratios. Thus, the above extracted capacitor values can be normalized to the lowest one (which will be the unit capacitance). For this first-order cell example, the normalized capacitor values are: $Cf = 15.92$, and $Cs = Cd = 1$. The chosen value of the unit capacitance will not change the transfer function, while it will affect other filter features like die size, power consumption, and output noise.

From a general point of view, using Cf much larger than Cd corresponds to having a small damping impedance, as is the case in high-Q filters. This results in a large capacitor spread. On the other hand, to have a small time constant requires Cs much smaller than Cf; and also in this case, a large capacitor spread occurs. Since the unit capacitance cannot be smaller than a minimum technological value (to achieve a certain matching accuracy), a large capacitor spread means having to drive a large capacitor and thus requires a high power level to operate the op-amp. In addition, a large chip area is needed. Thus, in order to avoid large capacitor spread, possible solutions will be proposed in the following.

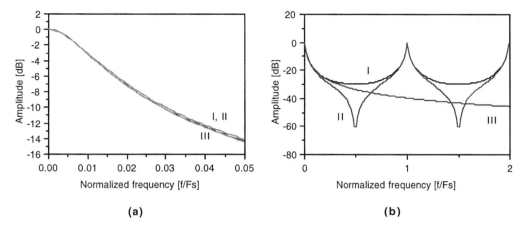

FIGURE 12.13 Frequency response comparison for different integrators.

12.5 Second-Order SC Circuit

The general expression of a second-order (biquadratic) z-transfer function can be written in the form:

$$H(z) = \frac{\gamma + \varepsilon \cdot z^{-1} + \delta \cdot z^{-2}}{1 + \alpha \cdot z^{-1} + \beta \cdot z^{-2}} \qquad (12.11)$$

The denominator coefficients (α, β) fix the pole frequency and quality factor, while the numerator coefficients $(\gamma, \varepsilon, \delta)$ define the types of filter frequency response. Several SC biquadratic cells have been proposed in literature to implement the above transfer function. In the following, two of them are presented: the Fleischer & Laker one and another one useful for high sampling frequency.

The Fleischer & Laker Biquad

A popular biquadratic cell, proposed by Fleischer & Laker,[5] is shown in Fig. 12.14 in its most general form. The cell is composed by:

- an input branch (capacitor G, H, I, and J) which allows selection of the zero positions, and therefore the kind of frequency response;
- a resonating loop (capacitor A, B, C, and D in conjunction with the damping capacitors E, and F) which sets the pole frequency and the pole quality factor. The two damping capacitors are not usually adopted together. In general E-capacitor is used for high-Q filters, while F-capacitor is preferred for low-Q circuits.

The transfer function of the Fleischer & Laker biquad cell can be written as:

$$H(z) = \frac{Vo}{Vi} = -\frac{D \cdot I + \left(A \cdot G - D \cdot (I+J)\right) z^{-1} + \left(J \cdot D - A \cdot H\right) z^{-2}}{D \cdot (B+F) - \left(D \cdot (2 \cdot B + F) - A \cdot (C+E)\right) z^{-1} + \left(D \cdot B - A \cdot E\right) z^{-2}} \qquad (12.12)$$

This cell allows one to synthesize any kind of transfer function by using parasitic insensitive structures, which ensures performance accuracy. The key observation related to this biquad cell is that the first op-amp operates in cascade to the second one; therefore, its settling is longer than the second one is.

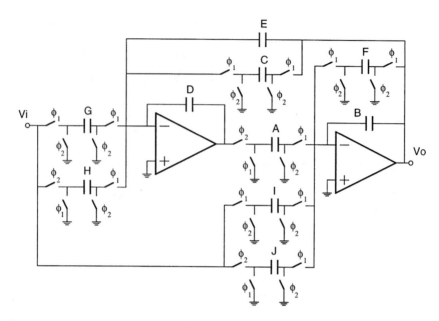

FIGURE 12.14   Fleischer & Laker biquadratic cell.

Design Methodology

At the first order, the SC circuits can be derived from continuous-time circuits implementing the desired frequency response, by a proper substitution of each resistor with the equivalent SC structures, as shown for the first-order cell. An alternative approach is to optimize the transfer function in the z-domain, and to numerically fit the desired transfer function with the transfer function implemented by the second-order cell. A third possibility is to adapt the s-domain transfer function to the z-domain signal processing of the SC structures. This procedure will be used in the following.

Consider the case of a given transfer function in s-domain, for instance, when an approximation table (Butterworth, Chebychev, Bessel, etc.) is used. The s-domain transfer function to be implemented is written as:

$$H(s) = \frac{a_2 \cdot s^2 + a_1 \cdot s + a_0}{s^2 + b_1 \cdot s + b_0} \tag{12.13}$$

This s-domain transfer function is transformed into a z-domain transfer function through the use of the bilinear s-to-z transformation:

$$s = \frac{2}{Ts} \cdot \frac{1 - z^{-1}}{1 + z^{-1}} \tag{12.14}$$

This transformation produces a warping of the frequency of interest. To avoid this error, a characteristic frequency for the given design (i.e., the frequencies ω_i of interest for the final filter mask) should be 'prewarped' according to the relationship between the angular frequencies in the s-domain (Ω_i) and in the z-domain (ω_i):

$$\Omega_i = \frac{2}{Ts} \cdot \tan\left(\frac{\omega_i Ts}{2}\right) \tag{12.15}$$

as it is indicated in Fig. 12.15 for a bandpass-type response.

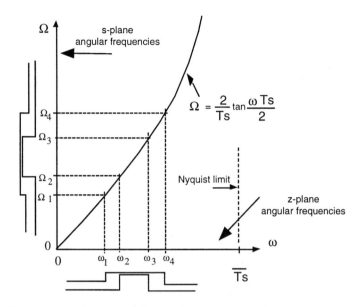

FIGURE 12.15 Bilinear mapping between continuous-time (Ω) and sampled-data (ω) frequency.

The characteristic frequency can be the −3dB frequency for a low-pass filter. The H′(s) that will satisfy the "prewarped" filter mask will be automatically transformed by Eq. 12.14 into a z-domain transfer function whose frequency response satisfies the desired filter mask in the ω-domain. Obviously, if $\omega_i \cdot T_s \ll 1$, no predistortion is needed, being $\Omega_i \approx \omega_i$. Assuming that $\omega_{i \cdot T_s} \ll 1$, H′(s) ≈H(s) and Eq. 12.14 is substituted directly into Eq. 12.13; the resulting coefficients of terms z^{-i} in the denominator are then equated to the corresponding ones in Eq. 12.12. Assuming A = B = D = 1, the capacitor values for the E-type and F-type can be extracted as follows:

$$F = 0 \quad E = \frac{b_1 T s}{1 + \dfrac{b_1 T s}{2} + \dfrac{b_0 T s^2}{4}} \quad C = \frac{b_0 T s^2}{1 + \dfrac{b_1 T s}{2} + \dfrac{b_0 T s^2}{4}} \tag{12.16a}$$

$$E = 0 \quad E = \frac{b_1 T s}{1 - \dfrac{b_1 T s}{2} + \dfrac{b_0 T s^2}{4}} \quad C = \frac{b_0 T s^2}{1 - \dfrac{b_1 T s}{2} + \dfrac{b_0 T s^2}{4}} \tag{12.16b}$$

When the bilinear transform is used, a second-order numerator is obtained, which can be written as given in Eq. 12.17, where simple solutions for the input capacitors are also given.

$$\text{Lowpass} \quad K\left(1 + z^{-1}\right)^2 \qquad I = J = IKI \quad G = 4IKI; \ H = 0 \tag{12.17a}$$

$$\text{Bandpass} \quad K\left(1 + z^{-1}\right) \cdot \left(1 - z^{-1}\right) \quad I = G = H = IKI; \quad J = 0 \tag{12.17b}$$

Highpass $\qquad K\left(1-z^{-1}\right)^2 \qquad\qquad I = J = IKI \qquad G = H = 0 \qquad$ (12.17c)

Design Example

As design example for the second-order cell, a bandpass response with $Q = 5$, $fo = 20$ kHz, and $Fs = 1$ MHz is considered (i.e., $fo/Fs = 0.02$). The transfer functions in the s-domain and in the z-domain (using bilinear transformation) are the following:

$$H(s) = \frac{s}{s^2 + \dfrac{2\cdot\pi\cdot fo}{Q}\cdot s + \left(2\cdot\pi\cdot fo\right)^2} = \frac{2.5133\cdot 10^4 \cdot s}{s^2 + 2.5133\cdot 10^4 \cdot s + 1.5791\cdot 10^{10}} \qquad (12.18)$$

$$H(z) = \frac{z^{-2} - 1}{z^{-2} - 1.9719\cdot z^{-1} + 0.98755} \qquad (12.19)$$

No frequency has been prewarped, since, applying Eq. 12.15 to fo, the prewarped pole frequency should be 19.765 kHz, with a negligible deviation of about 0.1%.

For the bandpass response, using the bilinear s-to-z mapping, the zero positions are at $\{z = 1, z = -1\}$. The frequency response is shown in Fig. 12.16 with line I. The normalized capacitance value, obtained equating the transfer function of Eq. 12.19 with the transfer function of the biquadratic cell of Eq. 12.12, are given in Table 12.1, for the E-type and F-type structures, in column I. A very large capacitor spread (>78) is needed. This results in large die area, and large power consumption. The capacitor spread could be reduced with a slight modification of the transfer function to the one given in the following:

$$H(z) = \frac{z^{-1} - 1}{z^{-2} - 1.9719\cdot z^{-1} + 0.98755} \qquad (12.20)$$

With respect to the bilinear transformation of Eq. 12.9, in this case, the zero at dc is maintained, while the zero at Nyquist frequency (at Fs/2, i.e., at $z = -1$) is eliminated. The normalized capacitor values are indicated again in Table 12.1, in Column II. It can be seen that a large reduction of the capacitor spread is obtained (from 80 to 8, for the E-type). The obtained frequency response is reported in Fig. 12.16 with line II. In the passband no significant changes occur; on the other hand in the stopband, the maximum signal attenuation is about –35 dB. In some applications, this solution is acceptable, also in consideration of the considerable capacitor spread reduction. For this reason, if not strictly necessary, the zero at Fs/2

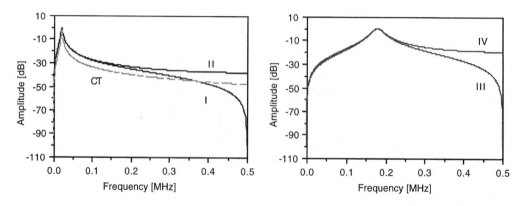

FIGURE 12.16   Frequency responses for different designs.

can be eliminated. However, reducing the factor fo/Fs results in reducing the stopband attenuation. For instance, for $fo = 200$ kHz (i.e., $fo/Fs = 0.2$), the frequency response with and without the Nyquist zero are reported in Fig. 12.16 with line III and IV, respectively. In this case, the stopband attenuation is reduced to −22 dB and therefore the Nyquist zero could be strongly needed. The relative normalized capacitor values are indicated in Table 12.1 in Column III (with zeros at $\{z = 1, z = -1\}$), and in Column IV (with zeros at $z = 1$).

A Biquadratic Cell for High Sampling Frequency

In the previous biquadratic cell, the two op-amps operate in cascade during the same clock phase. This requires that the second op-amp in cascade wait for the complete settling of the first op-amp to complete its settling. This, of course, reduces the maximum achievable sampling frequency, or, alternatively for a given sampling frequency, increases the required power consumption since op-amps with larger bandwidth are needed. In order to avoid this aspect, the biquad shown in Fig. 12.17 can be used. In this scheme, the two op-amps settle in different clock phase and thus they have the full clock phase time slot to settle.

The transfer function of the biquadratic cell is given in Eq. 12.21. As it can be seen a limitation occurs in the possible transfer function, since the term in z^{-2} is not present.

$$H(z) = -\frac{C3 \cdot C5 + C1 \cdot C4 - \left(C3 \cdot C5 - C2 \cdot C4\right) z^{-1}}{C3 \cdot \left(C8 + C6\right) + \left(C4 \cdot C9 - C3 \cdot C8 - 2 \cdot C3 \cdot C6\right) \cdot z^{-1} + C3 \cdot C6 \cdot z^{-2}} \qquad (12.21)$$

High-Order Filters

The previous first- and second-order cells can be used to build up high-order filters. The main architectures are taken from the theory for the active RC filters. Some of the most significant ones are: ladder[6] (with good amplitude response robustness with respect to component spread), cascade of first- and

TABLE 12.1 Capacitor Values for Different Designs

E-type	I	II	III	IV
A	10.131	1	12.366	1.0690
B	80.885	7.963	12.094	1.0000
C	1.2565	1	12.561	7.4235
D	10.131	8.0838	12.366	7.6405
E	1.9998	1.5915	1.9991	1.1815
F	0	0	0	0
G	1	1.5885	1	1
H	1	1.5885	1	1
I	1	0	1	0
J	0	0	0	0

F-type	I	II	III	IV
A	10.006	5.1148	11.305	5.9919
B	78.885	39.446	10.095	5.0497
C	1.2565	1	12.561	7.4235
D	10.006	8.1404	11.305	7.0793
E	0	0	0	0
F	1.9998	1	1.9991	1
G	1	1.5885	1	1
H	1	1.5885	1	1
I	1	0	1	0
J	0	0	0	0

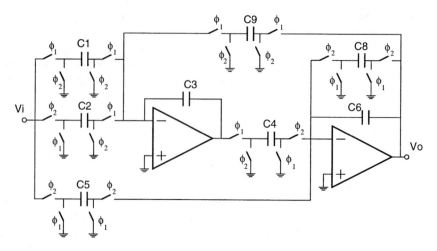

FIGURE 12.17   High-frequency biquadratic cell.

second-order cells (with good phase response robustness with respect to component spread), follow-the-leader feedback (for low-noise systems, like reconstruction filters in oversampled DAC).

12.6 Implementation Aspects

The arguments presented so far have to be implemented in actual integrated circuits. Such implementations have to minimize the effects of the non-idealities of the actual blocks, which are capacitors, switches, and op-amps. The capacitor behavior is quite stable, apart from capacitance non-linearities which affect the circuit performance only as a second-order effect.

On the other hand, switches and op-amps must be properly designed to operate in the SC system. The switches must guarantee a minimum conductance to ensure a complete charge transfer within the available timeslot. For the same reason, the op-amps must ensure large-dc gain, large unity-gain bandwidth, and large slew rate. For instance, in Fig. 12.18, the ideal output waveform of an SC network is

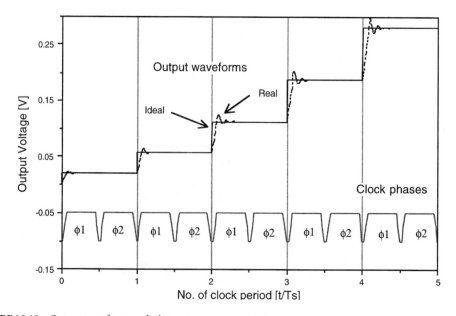

FIGURE 12.18   Output waveform evolution.

shown with a solid line, while the more realistic actual waveform is illustrated with the dotted line. The output sample is updated during phase $\phi1$, while it is held (at the value achieved at the end of phase $\phi1$) during phase $\phi2$. In phase $\phi1$, the output value moves from its initial to its final value. The slowness of this movement is affected by switch conductance, op-amp slew rate, and op-amp bandwidth.

The transient response of the system can be studied using the linear model of Fig. 12.19, where the conductive switches are replaced by their on-resistance Ron and the impulsive charge injection is replaced by a voltage step. The assumption of a complete linear system should allow one to study the system evolution exactly. In this case, the circuit time-constants depend on input branch ($\tau_{in} = 2 \cdot Ron \cdot Cs$), op-amp frequency response, and feedback factor.

Non-linear analysis is, however, necessary when op-amp slew rate occurs. This analysis is difficult to carry out and optimum performance can be achieved using computer simulations. Usually, for typical device models, 10% of the available timeslot (i.e., Ts/2) is used for slew rate, while 40% is used for linear settling.

Integrated Capacitors

Integrated capacitors in CMOS technology for SC circuits are mainly realized using poly1-poly2 structure, whose cross-section is shown in Fig. 12.20. This capacitor implementation guarantees linear behavior over a large signal swing. The main drawbacks of integrated capacitors are related to their absolute and relative inaccuracy, and to their associated parasitic capacitance.

The absolute value of integrated capacitors can change ±30% from their nominal values. However, the matching between equal capacitors can be on the order of 0.2%, provided that proper layout solutions are adopted (in close proximity, with guard rings, with common centroid structure). The matching of two capacitors of different value C can be expressed with the standard deviation of their ratio σ_C, which is correlated with the standard deviation of the ratio between two identical capacitors σ_{C1} by Eq. 12.22.[7]

$$\sigma_C = \frac{\sigma_{C1}}{\sqrt{C/C1}} \tag{12.22}$$

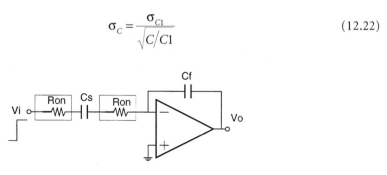

FIGURE 12.19 Linear model for transient analysis.

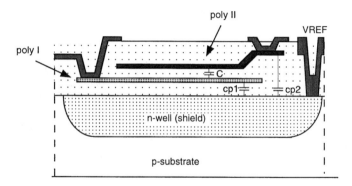

FIGURE 12.20 Poly1-poly2 capacitor cross-section.

This model can be used to evaluate the robustness of the SC system performance with respect to random capacitor variations using a Monte Carlo analysis.

The plates of a poly1-poly2 capacitor of value C present a parasitic capacitance toward the substrate, as shown in Fig. 12.18. Typically, this capacitance is about 10% of C for the bottom plate (cp1 = C/10), and it is 1% of C for the top plate (cp2 = C/100). In order to reduce the effect of these parasitic capacitances in the transfer function of the SC systems, it is useful to connect the top plate to the op-amp input node, and the bottom plate to low impedance nodes (op-amp output nodes or voltage sources). In addition, in Fig. 12.20, an n-well, biased with a clean voltage VREF; is placed under the poly1-poly2 capacitor in order to reduce noise coupling from the substrate, through parasitic capacitance.

MOS Switches

The typical situation during sampling operation is shown in Fig. 12.21(a) (this is the input branch of the integrator of Fig. 12.7(a)). The input signal Vi is sampled on the sampling capacitor Cs in order to have Vc = Vi.

In Fig. 12.21(b), the switches are replaced by a single-nMOS device which operates in the triode region with an approximately zero voltage drop between drain and source. The switch on-resistance Ron can be expressed as:

$$Ron = \frac{1}{\mu n \cdot Cox \cdot \frac{W}{L} \cdot \left(V_{gs} - V_{th}\right)} = \frac{1}{\mu n \cdot Cox \cdot \frac{W}{L} \cdot \left(V_G - Vi - V_{th}\right)} \tag{12.23}$$

where V_G is the amplitude of the clock driving phase, μn is the electron mobility, Cox is the oxide capacitance, and W and L are the width and length of the MOS device. Using V_{DD} = 5 V (i.e., V_G = 5 V), the dependence of Ron on the input voltage is plotted in Fig. 12.22(a). This means that if Ron is required by the capacitor value to be lower than a given value (to implement a low Ron·Cs time constant), a limitation in the possible input swing is given. For instance, if the maximum possible Ron is 2.5 kΩ, the maximum input signal swing is [0 V–3.5 V].

To avoid this limitation, a complementary switch can be used. It consists of an NMOS and a PMOS device in parallel, as shown in Fig. 12.23. The PMOS switch presents a Ron behavior complementary to that of the NMOS, as plotted in Fig. 12.22(b). The complete switch Ron is then given by the parallel of the two contributions which is sufficiently low for all the signal swing.

Using this solution requires one to distribute double clock lines controlling the NMOS and the PMOS. This could be critical for SC filters operating at high-sampling frequency, also in consideration of the synchronization of the two phases and of the digital noise from the distributed clocks which could reduce the dynamic range.

Once a minimum conductance is guaranteed, the structure can be studied using the linear model for the MOS devices S1 and S2 which operate in the triode region, resulting in the circuit of Fig. 12.24. In

(a)   **(b)**

FIGURE 12.21   (a) Ideal sampling structure, (b) sampling structure with NMOS switches.

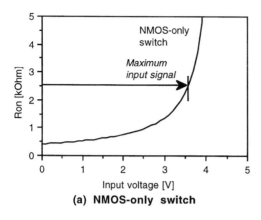

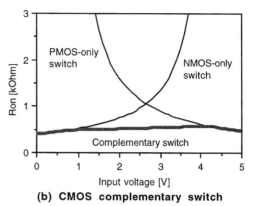

FIGURE 12.22 Switch on-resistance.

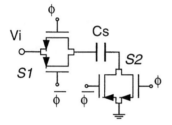

FIGURE 12.23 Sampling structure with complementary switches.

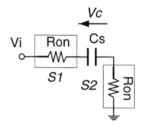

FIGURE 12.24 Sampling operation linear model.

this case, Vc follows Vi, through an exponential law with a time constant $\tau_{in} = Cs \cdot 2 \cdot Ron$. Typically, at least $6 \cdot \tau_{in}$ must be guaranteed in the sampling timeslot to ensure sufficient accuracy. For a given sampling capacitance value, this is achieved using switches with sufficiently low on-resistance and no voltage drop across its nodes. Large on-resistance results in a long time constant and incomplete settling, while a voltage drop results in an incorrect final value. MOS technology allows the implementation of analog switches satisfying both the previous requirements.

Transconductance Amplifier

The SC technique appears the natural application of available CMOS technology design features. This is true also for the case of the op-amp design. In fact, SC circuits require an infinite input op-amp impedance, as in the case of op-amp using a MOS input device. On the other hand, CMOS op-amps are particularly efficient when the load impedance is not resistive and low, but only capacitive, as in the case of SC circuits. In addition, SC circuits allow one to process a full swing (rail-to-rail) signal and this is possible for CMOS op-amps. The main requirements to be satisfied by the op-amp remain the bandwidth, the slew rate, and the dc-gain.

The bandwidth and the slew rate must be sufficiently large to guarantee accurate settling for all the signal steps. The op-amp gain must be sufficiently large to ensure a complete charge transfer. A tradeoff between large dc-gain (achieved with low-current and/or multistage structure) and large bandwidth (obtained at high-current and/or simple structure) must be optimized. For this case, the use of mixed technology (like BiCMOS) could help the proper design optimization.

12.7 Performance Limitations

The arguments described thus far are valid assuming an ideal behavior of the devices in the SC network (i.e., op-amp, switches, and capacitor). However, in actual realization, each of them presents non-idealities which reduce the performance accuracy of the complete SC circuit. The main limitations and their effects are described in the following. Finally, considerations about noise in SC systems conclude this section.

Limitation Due to the Switches

As described before, CMOS switches satisfy both low on-resistance and zero voltage-drop requirements. However, they introduce some performance limitations due to their intrinsic CMOS realization. The cross-section of an NMOS switch in its on-state is shown in Fig. 12.25. The connection between its nodes N1 and N2 is guaranteed by the presence of the channel, made up of the charge Qch. The amount of charge Qch can be written as:

$$Qch = \left(W \cdot L\right) \cdot Cox \cdot \left(V_G - Vi - V_{TH}\right) \tag{12.24}$$

where Vi is the channel (input) voltage. Both nodes N1 and N2 are at voltage Vi (no voltage drop between the switch nodes). In addition, the gate oxide which guarantees infinite MOS input impedance constitutes a capacitive connection between gate and both source and drain. This situation results in two non-ideal effects: charge injection and clock feedthrough.

Charge Injection

At the switch turn-off, the charge Qch given in Eq. 12.24 is removed from the channel and it is shared between the two nodes connected to the switch, with a partition depending on the node impedance level.

The charge k·Qch is injected in N2 and collected on a capacitor Cc. A voltage variation nVc across the capacitor arises, which is given by:

$$\Delta Vc = k \cdot \frac{Qch}{Cc} \tag{12.25}$$

For all the switches of a typical SC integrator, as shown in Fig. 12.26, this effect is important. For instance, for the switch S4 connected to the op-amp virtual ground, the charge injection into the virtual ground is collected in the feedback capacitor and it is processed as an input signal. The amount of this charge injection depends on different parameters (see Eq. 12.24 and Eq. 12.25). Charge Qch depends on switch size W, which however cannot be reduced beyond a certain level; otherwise, the switch on-resistance should increase. Thus, a tradeoff between charge injection and on-resistance is present. In addition, charge Qch depends on the voltage Vi which the switch is connected to. For the switches S2, S3, and S4, the injected charge is proportional to (V_G – Vgnd) and is always fixed; as a consequence, it can be considered like an offset. On the other hand, for the switch S1 connected to the signal swing, the

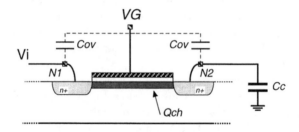

FIGURE 12.25   Switch charge profile of an NMOS switch in the on-state.

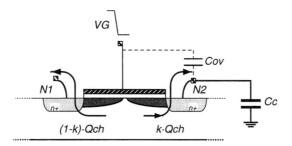

FIGURE 12.26   Charge displacement during turn-off.

channel charge Qch is dependent on (V_G – Vi), i.e., on the signal amplitude and thus also the charge injection is signal dependent. This creates an additional signal distortion.

Possible solutions for the reduction of the charge injection are: use of dummy switches, use of slowly variable clock phase, use of differential structures, use of delayed clock phases,[8] and use of signal-dependent charge pump.[9]

Dummy switches operate with complementary phases in order to sink the charge rejected by the original switches. The use of differential structures reduces the offset charge injection to the mismatch of the two differential paths. For the signal-dependent charge injection, the delayed phases of Fig. 12.26 are applied to the integrator of Fig. 12.26. This clock phasing is based on the concept that at the turn-off, S3 is open before S1. In such a way, when S1 opens, the impedance toward Cs is infinite and no signal-dependent charge injection occurs into Cs.

Clock Feedthrough

The clock feedthrough is the amount of signal that is injected in the sampling capacitor Cc from the clock phase through the MOS overlap capacitor (Cov) path, shown in Fig. 12.27, which is then proportional to the area of the switches. Using large switches, to reduce on-resistance, results in large charge injection and large clock feedthrough. This error is typically constant (it depends on capacitance partition) and therefore it can be greatly reduced by using differential structures. The voltage error ΔVc across a capacitance Cc due to the feedthrough of the clock amplitude ($V_{DD} - V_{SS}$) can be written as:

$$\Delta Vc = \left(V_{DD} - V_{SS}\right) \cdot \frac{Cov}{Cov + Cc} \tag{12.26}$$

Limitation Due to the Op-Amp

The operation of SC networks is based on the availability of a "good" virtual ground which ensures a complete charge transfer from the sampling capacitors to the feedback capacitor. Whenever this charge transfer is incomplete, the SC network performance derives from its nominal behavior. The main non-ideality causes from the op-amp are: finite dc-gain, finite bandwidth, finite slew-rate, and gain non-linearity.

Finite Op-Amp dc-Gain Effects[10,11]

The op-amp finite gain results in a deviation of output voltage at the end of the sampling period from the ideal one, as shown in Fig. 12.28(a). This output sample deviation can be translated in the SC system performance deviation. For the finite gain effect, an analysis which correlates the op-amp gain Ao with SC network performance deviation, can be carried out under the hypothesis that the op-amp bandwidth is sufficiently large to settle within the available timeslot.

For the case of the summing amplifier of Fig. 12.10, it can be demonstrated that the effect of the finite op-amp dc-gain (AO) is only in an overall gain error. For this reason SC FIR filters (based on this scheme) exhibit a low sensitivity to op-amp finite dc-gain. On the other hand, for the case of SC integrators, the

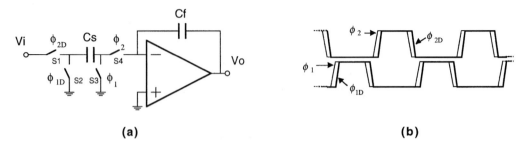

FIGURE 12.27   Clocking scheme for signal-dependent charge injection reduction.

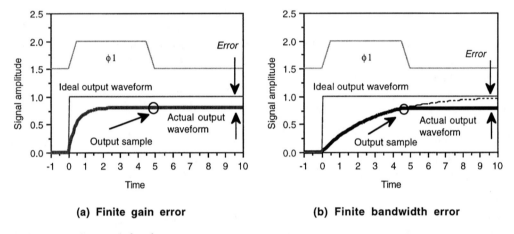

(a) Finite gain error   **(b) Finite bandwidth error**

FIGURE 12.28   Op-amp induced errors.

finite gain effect results in pole and gain deviation. For instance, the transfer function of the integrator of Fig. 12.8(a) becomes:

$$H_a(z) = \frac{V_o(z)}{V_i(z)} = \frac{Cs}{Cf} \cdot \frac{z^{-1}}{1 + \dfrac{1}{Ao}\left(1 + \dfrac{Cs}{Cf}\right) - \left(1 + \dfrac{1}{Ao}\right) \cdot z^{-1}} \qquad (12.27)$$

For a biquadratic cell, the op-amp finite gain results in pole frequency, pole quality factor deviations. The actual frequency and quality factor of the pole (f_oA and Q_A) are correlated to their nominal values (fo and Q) by the relationship:

$$f_{oA} = \frac{Ao}{1 + Ao} \cdot f_o \qquad Q_A = \frac{1}{\dfrac{1}{Q} + \dfrac{2}{Ao}} \approx \left(1 - \frac{2 \cdot Q}{Ao}\right) \cdot Q \qquad (12.28)$$

Finite Bandwidth and Slew-Rate[10,11]

Also, op-amp finite bandwidth and slew-rate result in incomplete charge transfer, which still corresponds with deviation of the output sample with respect to its nominal value. For the case of only finite bandwidth, the effect is shown in Fig. 12.28(b). An analysis similar to that of the finite gain for the finite bandwidth and slew-rate effect is not easily extracted. This is due to the fact that incomplete settling is caused by the correlation of a linear effect (e.g., the finite bandwidth) and a non-linear effect (e.g., the slew rate). In addition, this case is worsened by the fact that in some structures, several op-amps are connected in cascade and then each op-amp (a part of the first one) has to wait for the operation conclusion of the preceding one.

Op-Amp Gain Non-Linearity

Since the SC structure allows one to process large swing signals, for this signal swing, the op-amp has to perform constant gain. When the op-amp gain is not constant for all the necessary output swing, distortion arises. An analysis can be carried out for the case of the integrator of Fig. 12.8(a).[12] Assuming an op-amp input (vi)-to-output (vo) relationship expressed in the form:

$$vo = a_1 \cdot vi + a_2 \cdot vi^2 + a_3 \cdot vi^3 + \dots \tag{12.29}$$

The resulting harmonic components (for $\omega o \cdot Ts \ll 1$) are given by:

$$HD2 = \frac{a_2}{2 \cdot a_1^3 \beta} \; V_o \sqrt{1 + \left(\frac{Vo}{2 \cdot Vi} \right)^2} \tag{12.30a}$$

$$HD3 = \frac{a_3}{2 \cdot a_1^4 \beta} \; V_o^2 \sqrt{1 + \left(\frac{Vo}{3 \cdot Vi} \right)} \tag{12.30b}$$

The distortion can then be reduced by making constant low-gain (i.e., reducing a_2 and a_3) or by using a very large op-amp gain (i.e., increasing a_1). This second case is usually the adopted strategy.

Noise in SC Systems[13,14]

In SC circuits, the main noise sources are in the switches (thermal noise) and in the op-amp (thermal noise and $1/f$ noise). These noise sources are processed by the SC structure as an input signal, i.e., they are sampled (with consequent folding) and transferred to the output with a given transfer function. As explained for the signal, the output frequency range of an SC filter is limited in the band [0–Fs/2]. This means that for any noise source its sampled noise band, independent of how large it is at the source before sampling, is limited in the [0–Fs/2] range. On the other hand, the total power of noise remains constant after sampling; this means that the power density of sampled noise is increased by the factor Fb/(Fs/2),

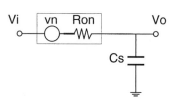

FIGURE 12.29 Sampling the noise on a capacitor.

where Fb is the noise band at its source. This can be seen for the switch noise in the following simple example. Consider the structure of Fig. 12.29, where the resistance represents the switch on-resistance. Its associated noise spectral density is given by $v_n^2 = 4kT \cdot Ron$ (where k is the Boltzmann's constant and T is the absolute temperature). The transfer function to the output node (the voltage over the capacitor) is H(s) = $\frac{1}{1 + s \cdot Ron \cdot Cs}$. The total output noise can be calculated from the expression:

$$n_o^2 = \int_0^\infty v_n^2 \cdot \left| H(s) \right|^2 \cdot df = \frac{kT}{Cs} \tag{12.31}$$

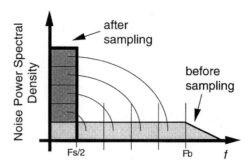

FIGURE 12.30   Folding of the noise power spectral density.

The total sampled noise is then given by kT/Cs, and presents a bandwidth of $Fs/2$. This means that the output noise power density is $kT/Cs·2/Fs$. (See Fig. 12.30.)

The same folding concept can be applied to the op-amp noise. For the op-amp $1/f$ noise, the corner frequency is usually lower than $Fs/2$. Therefore, the $1/f$ noise is not modified by the sampling. On the other hand, the white noise presents a bandwidth Fb larger than $Fs/2$. This means that this noise component is modified in a noise source of bandwidth $Fs/2$ and noise power density multiplied by the factor $Fb/(Fs/2)$. When the noise sources are evaluated in this way, the output noise of an SC cell can be evaluated by summing the different components properly weighted by their transfer functions from their source position to the output node.

A few considerations follow for the noise performance of an SC cell. Switch noise is independent of Ron, since its dependence is cancelled in the bandwidth dependence. Thus, this noise source is dependent only on Cs. Noise reduction is achieved by increasing Cs. This, however, trades with the power increase necessary to drive the enlarged capacitance. Of course, even if Ron does not appear in the noise expression, as the capacitor is enlarged, the Ron must be adequately decreased in order to guarantee a proper sampling accuracy.

For the op-amp noise, the noise band is usually correlated with the signal bandwidth. Therefore, a good op-amp settling (achieved with a large signal bandwidth) is in contrast to low-noise performance (achieved with reduced noise bandwidth). Therefore, in low-noise systems, the bandwidth of the op-amp is designed to be the minimum that guarantees proper settling.

12.8 Compensation Technique (Performance Improvements)

SC systems usually operate with a two-phase clock in which the op-amp is "really" active only during one phase, and during the other phase is "sleeping." Provided that the op-amp output node is not read during the second phase, this non-active phase could be used to improve the performance of the SC system, as shown in the following.[15] 1/f noise and offset can be reduced with Correlated Double Sampling (CDS) or the chopper technique. Similar structures are also able to compensate for the error due to a finite gain of the operational amplifier. On the other hand, proper structures are able to reduce the capacitor spread occurring in particular situations (high-Q or large time constant filters). Finally, the double-sampled-technique can be used to increase, by a factor of two, the sampling frequency of the SC system.

CDS Offset-Compensated SC Integrator

The extra phase available in a two-phase SC system can be used to reduce op-amp offset and 1/f noise effects at the output. A possible scheme is shown in Fig. 12.31,[16] and operates as follows. Capacitor Cof is used to sample, during $\phi1$, the offset voltage Voff as it appears in the inverting node of the op-amp with close to unitary feedback. During $\phi2$, the inverting node is still at a voltage very close to Voff, since

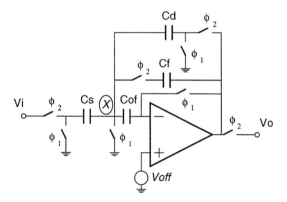

FIGURE 12.31   Offset-compensated SC integrator.

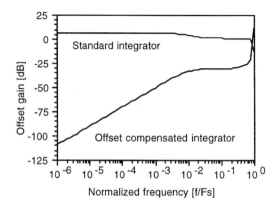

FIGURE 12.32   Offset-compensated SC integrator performance.

the bandwidth of Voff is assumed to be very small with respect to the sampling frequency. Capacitor Cof maintains the charge on its armatures and acts like a battery. Thus, node X is a good virtual ground, independent of the op-amp offset. In the same way, the output signal, read only during $\phi 2$, is offset independent. The effect of this technique can be simulated using the value of the first-order cell of the previous example (i.e., Cf = 15.92, and Cs = Cd = 1, and Cof = 1). The transfer function Vo/Voff is shown in Fig. 12.32. At low frequency, the Voff is highly rejected, while this is not the case of the standard (uncompensated) integrator. The main problem with this solution is due to the unity feedback operation of the structure during phase $\phi 1$. This requires the stability of the op-amp, which could require a very high power consumption.

Chopper Technique

An alternative solution to reduce offset 1/f noise at the output is given by the chopper technique. It consists of placing one SC mixer for frequency Fs/2 at the op-amp input and a similar one at the op-amp output. This action does not affect white noise. On the other hand, offset and 1/f noise are shifted to around Fs/2, not affecting the frequencies around dc, where the signal to be processed is supposed to be.

This concept is shown for a fully differential op-amp in Fig. 12.33. In Fig. 12.34, the input-referred noise power spectral density (PSD) without and with chopper modulation is shown. The white noise level (wnl) is not affected by the chopper operation and remains constant. It will be modified by the folding of the high-frequency noise, as previously described.

This technique is particularly advantageous for SC systems since the mixer can be efficiently implemented with the SC technique as shown in Fig. 12.35.

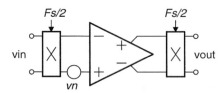

FIGURE 12.33   Op-amp with chopper.

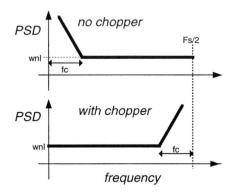

FIGURE 12.34   Input-referred noise power spectral density (PSD) without and with chopper technique.

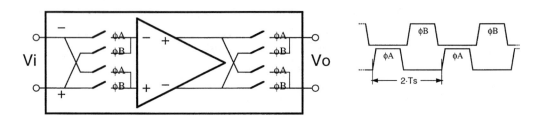

FIGURE 12.35   Op-amp with SC chopper configuration.

Finite-Gain Compensated SC Integrator

In the op-amp design, a tradeoff between op-amp dc-gain and bandwidth exists. Therefore, when a large bandwidth is needed, a finite dc-gain necessarily occurs, reducing SC filter performance accuracy. To avoid this, the available extra phase can be used to self-calibrate the structure with respect to the error due to the op-amp finite gain. In the literature, several techniques have been proposed. The majority of them are based on the concept of using a preview of the future output samples to pre-charge a capacitor placed in series to the op-amp inverting input node in order to create a "good" virtual ground (as for offset cancellation). The various approaches differ on how they get the preview and how they calibrate the new virtual ground. For the different cases, they can be effective for a large bandwidth,[17,18] for a small bandwidth,[19,20] or for a passband bandwidth.[21] As an example of this kind of compensation, one of the earliest proposed schemes is depicted in Fig. 12.36.

The op-amp finite gain makes the op-amp inverting input node different from the virtual ground ideal behavior and assume the value $-Vo/Ao$, where Vo is the output value and Ao is the op-amp dc-gain. In the scheme of Fig. 12.36, the future output sample is assumed to be close to the previous sample, sampled of Cg1. This limits the effectiveness of this scheme to signal frequencies f for which this assumption is valid (i.e., for $f/Fs \ll 1$). The circuit operates as follows. During $\phi 1$, auxiliary capacitor Cg1 samples the output; while during $\phi 2$, Cg1 is used to precharge Cg2 to $-Vo/Ao$, generating a good virtual ground at node X.

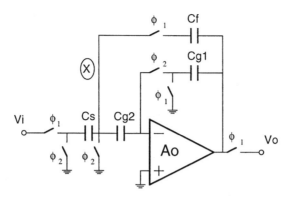

FIGURE 12.36   Finite-gain-compensated SC integrator.

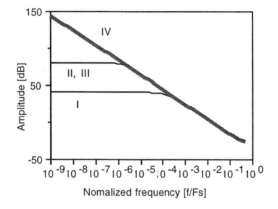

FIGURE 12.37   Finite-gain-compensated SC integrator performance.

In Fig. 12.37, the frequency responses of different integrators are compared. Line I refers to an uncompensated integrator with Ao = 100; line II refers to the uncompensated integrator with Ao = 10,000. This line matches with line III, which corresponds to the compensated integrator with Ao = 100. Finally, line IV is the frequency response of the ideal integrator. From this comparison, the compensation effect is to achieve an op-amp gain Ao performance similar to those achieved with an op-amp gain Ao^2.

Alternative solutions to the op-amp gain compensation are based on the use of a replica amplifier matched with the main one. Also in this way the effectiveness of the solution is to achieve performance accuracy relative to an op-amp dc-gain of Ao^2.

The Very-Long Time-Constant Integrator

In the design of Very-Long Time-Constant integrators using the scheme of Fig. 12.8(a), typical key points to be considered are:

- The capacitor spread: if the pole frequency fp is very low with respect to the sampling frequency Fs, then the capacitor spread S = Cf/Cs of a standard integrator (Fig. 12.8(a)) will be very large. This results in a large die area and reduced performance accuracy for poor matching.
- The sensitivity to the parasitic capacitances: proper structure can reduce capacitor spread. They, however, suffer from the presence of parasitic capacitance. Parasitic-insensitive or at least parasitic-compensated designs should then be considered.
- The offset of the operational amplifier: offset-compensated op-amps are needed when the op-amp offset contribution cannot be tolerated.

In the literature, several SC solutions have been proposed, more oriented toward reducing the capacitor spread than toward compensating either the parasitics or the op-amp offset.

A first solution is based on the use of a capacitive T-network in a standard SC integrator, as shown in Fig. 12.38.[22] The operation of the sampling T-structure is to realize a passive charge partition with the capacitors Cs1, and Cs2+Cs3. The final result is that only the charge on Cs3 is injected into the virtual ground. Therefore, the effect of this scheme is that Cs is replaced with the Cs_equiv, given by the expression:

$$Cs\_equiv = Cs3 \cdot \frac{Cs1}{Cs1 + Cs2 + Cs3} \tag{12.32}$$

The net gain of this approach is that, using $Cs2 = \sqrt{S} \cdot Cs1 = \sqrt{S} \cdot Cs3$, the capacitor spread is reduced to $\sqrt{S}$. For example, an integrator with Cs = 1 and Cf = 40, can be realized with Cs1 = 1, Cs2 = 6, Cs3 = 1, and Cf = 5, i.e., with the capacitor spread reduced to 6.

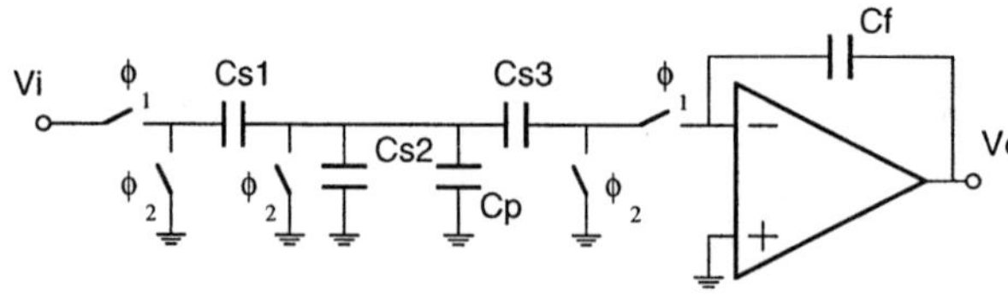

$$\frac{Vo}{Vi} = \frac{Cs1}{Cs1 + Cs2 + Cs3} \cdot \frac{Cs3}{Cf} \cdot \frac{1}{1 - z^{-1}} \tag{12.33}$$

FIGURE 12.38 A T-network long-time-constant SC integrator.

The major problem of the circuit of Fig. 12.38 is due to the fact that the T-network is sensitive to the parasitic capacitance Cp (due to Cs1, Cs2, and Cs3) in the middle node of the T-network, which is added to Cs2, reducing frequency response accuracy.

A parasitic-insensitive circuit is the one proposed by Nararaj[23] and shown in Fig. 12.39. In this case, the transfer function is given by Eq. 12.34. Also, in this case, $Cs = Cx = \sqrt{S} \cdot Cf$ are usually adopted to reduce the standard spread from S to $\sqrt{S}$. However, for the Nagaraj integrator, the op-amp is used on both phases, disabling the possibility of using double-sampled structure.

It is also possible to combine a long-time-constant scheme with an offset-compensated scheme to obtain a long-time-constant offset-compensated SC integrator.[15]

Double-Sampling Technique

If the output value of the SC integrator of Fig. 12.8(b) is read only at the end of $\phi2$, the requirement for the op-amp to settle can be relaxed. For the integrator of Fig. 12.8(b), the time available for the op-amp to settle is $T_s/2$. The equivalent double-sampled structure is shown in Fig. 12.40. The capacitor values

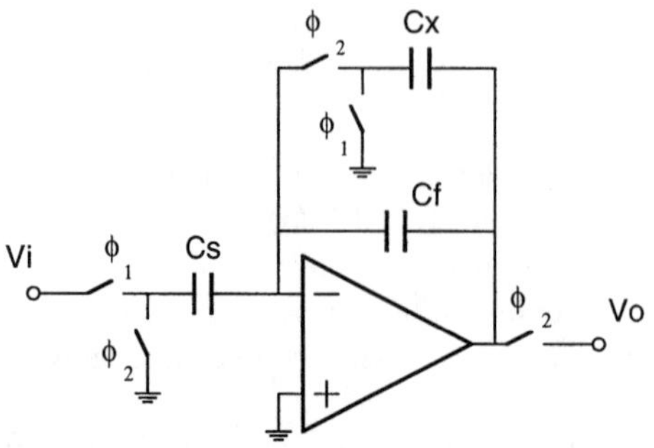

$$\frac{Vo}{Vi} = \frac{Cf}{Cs} \cdot \frac{Cx}{Cf + Cx} \cdot \frac{1}{1 - z^{-1}} \tag{12.34}$$

FIGURE 12.39 The Nagaraj's long-time-constant SC integrator.

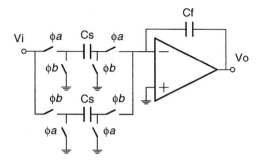

FIGURE 12.40   Double-sampled SC integrator.

for the two structures are the same, and thus they implement the same transfer function. The time evolution for the two structures are compared in Fig. 12.41. For the double-sampled SC integrator, the time available for the op-amp to settle is doubled.

This advantage can be used in two ways. First, when a high sampling frequency is required, if the op-amp cannot settle in Ts/2, the extra time allows it to reach the speed requirement (i.e., the double-sampling technique is used to increase the sampling frequency). Second, at low sampling frequency when the power consumption must be strongly reduced, a smaller bandwidth guaranteed by the op-amp reduces its power consumption.

The cost of the double-sampled structure is the doubling of all the switched capacitors. In addition, in the case of a small mismatch between the two parallel paths, mismatch energy could be present around Fs/4.[24]

12.9 Advanced SC Filter Solutions

In this section, alternative solutions able to overcome some basic limitations to SC system performance improvement are proposed. They deal with the tradeoff between bandwidth-vs.-gain in the op-amp design for high-frequency SC filter, and the implementation of low-voltage SC filters.

Precise Op-Amp Gain (POG) for High-Speed SC Structures[25,26]

Standard design of SC networks assumes operation with infinite gain and infinite bandwidth op-amps. However, in the op-amp design, a tradeoff exists between the speed and the gain. As a consequence with a high sampling frequency, the needed large bandwidth limits the op-amp gain to low values, thus limiting the achievable accuracy. Thus, standard design is less feasible for high-sampling frequency. A possible

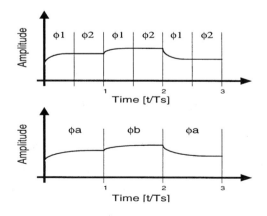

FIGURE 12.41   Double-sampled operation.

solution to this limitation is the Precise Op-amp Gain (POG) design approach, which consists of designing high-frequency SC networks, taking into account the precise gain value of the op-amps as a parameter in the capacitor design. The standard design op-amp tradeoff between *speed-and-gain* is then changed into the POG design tradeoff between *speed-and-gain precision*, which is more affordable in high-frequency op-amps.

If the op-amp dc-gain is Ao, the transfer function of the first-order cell shown in Fig. 12.42 is given by:

$$H_{POG}(z) = -\frac{Cs_{POG}}{\left(Cf_{POG} + Cd_{POG} + \dfrac{Cs_{POG} + Cf_{POG} + Cd_{POG}}{Ao}\right) - Cf_{POG}\left(1 + \dfrac{1}{Ao}\right)z^{-1}} \quad (12.35)$$

This expression for Ao equal to infinite becomes the standard case t.f H_{ST}, which is given by:

$$H_{ST}(z) = -\frac{Cs}{\left(Cf + Cd\right) - Cf\ z^{-1}} \quad (12.36)$$

To obtain the same t.f., the POG capacitor values are obtained from the standard values as given in the following:

$$Cs_{POG} = Cs \qquad Cf_{POG} = Cf \cdot \left(1 + \frac{1}{Ao}\right) \qquad Cd_{POG} = Cd\left(1 + \frac{1}{Ao}\right) + \frac{Cs}{Ao} \quad (12.37)$$

The concept here applied to the SC integrator can be applied to higher-order SC filters. It can be demonstrated that using the POG approach with an op-amp nominal gain Ao and an actual gain in the range [Ao(1 − ε), Ao(1 + ε)] achieves the same response accuracy as the standard approach with an infinite op-amp nominal gain with an actual gain given by:

$$A_{eff} = \frac{\left(Ao + 1\right)\left(1 \pm \varepsilon\right)}{\varepsilon} \approx \frac{Ao}{\varepsilon} \quad (12.38)$$

The value A_{eff} can then be defined as the effective gain of the POG approach. For example, for the op-amp with Ao = 100 and ε = 0.08, the same performance accuracy is achieved as when using standard design with op-amp gain A = 1250. This value of A_{eff} can then be used in Eq. 12.27 to evaluate filter performance accuracy.

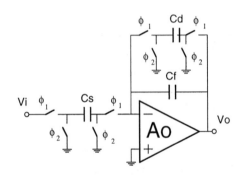

FIGURE 12.42   Dumped SC integrator.

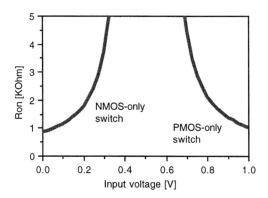

FIGURE 12.43   Switch Ron with V_{DD} = 1 V.

Low-Voltage Switched-Capacitor Solutions

In the last few years, the interest in low-power, low-voltage integrated systems has consistently grown due to the increasing importance of portable equipment and to the reduction of the supply voltage of modern standard CMOS scaled-down technology ICs. For the design of SC filters operating at reduced supply voltages,[27,28] capacitor properties are quite stable. On the other hand, at low supply, it is difficult to properly operate the MOS switches and the op-amps. With the supply voltage reduction, the MOS switches' overdrive voltage is lowered, inhibiting proper operation of classical transmission gate (complementary switches). The switch conductance for different input voltages changes, depending on the supply voltage V_{DD}. In Fig. 12.22, the case for V_{DD} = 5 V was shown. In Fig. 12.43, the case for V_{DD} = 1 V is reported for comparison. In this case, there is a critical voltage region centered around $V_{DD}/2$ for which both switches are not conducting. In SC circuits, to achieve rail-to-rail swing, the output of the op-amp must necessarily cross this critical region, where the switches connected to the op-amp output node will not properly operate at V_{DD} = 1 V.

On the other hand, op-amp operation can be achieved with proper design using a supply voltage as low as $V_{TH}+2\cdot V_{OV}$, with some modifications at system level. Switches and op-amp sections may use different supply voltages. In fact, using voltage multiplier (a possible scheme is shown in Fig. 12.44[29]), it is possible to generate on-chip a voltage higher than the supply voltage. This "multiplied" voltage can then be used to power the entire SC circuit (op-amp and switches) or to drive only the switches.

If the higher supply is used to bias the op-amp and switches, standard design solutions can be implemented. In addition, the op-amp powered with a higher supply voltage can manage a larger signal swing, with a consequential larger dynamic range. However, in a scaled-down technology, the maximum acceptable electric field between gate and channel (for gate oxide breakdown) and between drain and source (for hot electron damage) must be reduced. This puts an absolute limit on the value of the multiplied supply voltage. In addition, the need to supply a dc-current to the op-amp from the multiplied supply forces one must use an external capacitor, which is an additional cost.

An alternative approach consists of using the multiplied supply to drive only the switches. In this case, the voltage multiplier does not supply any dc-current, thus avoiding any external capacitor. This solution, like the previous

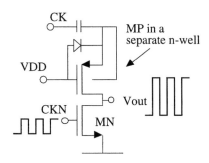

FIGURE 12.44   Charge-pump for on-chip voltage multiplication.

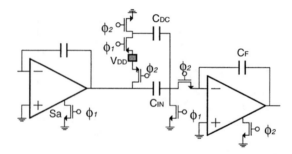

FIGURE 12.45 Switched-op-amp SC integrator.

one, must not exceed the limit of the technology associated with the gate oxide breakdown. Nonetheless, this approach is largely used because it allows the filter to operate at high sampling frequency.

In order to avoid any kind of voltage multiplier, the Switched-OpAmp (SOA) approach was proposed[30,31] and is based on the following considerations:

1. The best condition for the switches driven with a low supply voltage is to be connected either to ground or to V_{DD}. Thus, to properly operate, S2, S3, and S4 in Fig. 12.8(a) have to be referred to ground or to V_{DD} (i.e., the op-amp input dc-voltage has to be either ground or V_{DD}). This allows one to minimize the required op-amp supply voltage. On the other hand, the op-amp dc output voltage has to be at $V_{DD}/2$ in order to have rail-to-rail output swing.
2. The switch S1 connected to the signal swing cannot operate properly for the full signal swing, as explained before.

The resulting SOA SC integrator is shown in Fig. 12.45. The SOA approach uses an op-amp, which can operate in a tri-state mode. In this way, the critical output switch S1 is no longer necessary and it can be eliminated, moving the critical problem to the op-amp design. The function of the eliminated critical switch S1 is implemented by turning the op-amp on and off through Sa, which is connected to ground.

The input dc-voltage is set to ground. Therefore, all the switches are connected to ground (and realized with a single NMOS device) or to V_{DD} (and realized with a PMOS device), and are driven with the maximum overdrive, given by $V_{DD} - V_{TH}$. Capacitor C_{DC} in Fig. 12.45 gives a fixed charge injection into virtual ground, producing a voltage level shift between input (V_{in_DC}) and output (V_{out_DC}) op-amp dc-voltage. Since V_{in_DC} is set to ground, using $C_{DC} = C_{IN}/2$ sets $V_{out\_DC} = V_{DD}/2$. C_{DC} has no effect on the signal transfer function given by:

$$H(z) = \frac{C_{IN} \cdot z^{-1/2}}{C_F \cdot \left(1 - z^{-1}\right)} \tag{12.39}$$

A fully differential architecture of the scheme of Fig. 12.45 provides both signal polarities at each node, useful to build up high-order structures without any extra elements (e.g., inverting stage). In addition, any disturbance (offset or noise) injected by C_{DC} results in a common-mode signal, which is largely rejected by the fully differential operation.

The SOA approach suffers from the following problems:

1. SOA structure operates with an op-amp, which is turned on and off. Its turn-on time becomes the main limitation in the possible maximum sampling frequency.
2. In an SOA structure, the output signal is available only during one clock phase; while during the other clock phase, the output is set to zero (return-to-zero), as shown in Fig. 12.46. If the output signal is read as a continuous-time waveform, the return to zero has two effects: a loss of 6 dB in

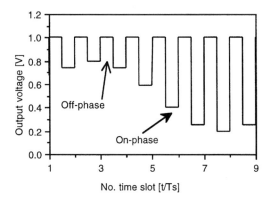

FIGURE 12.46   Switched-op-amp output waveform.

TABLE 12.2   Comparison Between Different Low-Voltage SC Designs

	Supply multiplier	Clock multiplier	Switched-op-amp
$V_{DDswith}$	V_{DDmult}	V_{DDmult}	V_{DD}
$V_{DDop-amp}$	V_{DDmult}	V_{DD}	V_{DD}
New op-amp design	No	No/Yes	Yes
New switch design	No	No	Yes
Output swing	+	–	–
Gate Break-down (V_{GS} limit)	–	–	+
Hot electron (V_{DS} limit)	–	+	+
Sampling frequency limitation	+	+	–
Power consumption	–	–	+
External component	Yes	No	No
Continuous waveform:Gain loss	1	1	1/2
Return-to-zero distortion	+	+	–

the transfer function, and an increased distortion due to the large output steps. On the other hand, when the SOA integrator is used in front of a sampled-data system (like an ADC), the output signal is sampled only when it is valid and both the above problems are cancelled.

A comparison between the three different low-voltage SC filter design approaches is given in Table 12.2.

References

The literature about SC filters is so wide that any list of referred publications should be considered incomplete. In the following, just a few papers related to the discussed topics are indicated.

1. B. J. Hosticka, R. W. Brodersen, and P. R. Gray, MOS sampled-data recursive filters using switched-capacitor integrators, *IEEE J. Solid-State Circuits,* vol. SC-12, pp. 600-608, Dec. 1997.
2. J. T. Caves, M. A. Copeland, C. F. Rahim, and S. D. Rosenbaum, Sampled analog filtering using switched capacitors as resistor equivalents, *IEEE J. Solid-State Circuits,* vol. SC-12, pp. 592-599, Dec. 1977.
3. R. Gregorian and G. C. Temes, *Analog MOS Integrated Circuits for Signal Processing,* John Wiley & Sons, 1986.
4. G. T. Uehara, and P. R. Gray, A 100MHz output rate analog-to-digital interface for PRML magnetic disk read channels in 1.2μ CMOS, *IEEE Intl. Solid-State Circuits Conference 1994, Digest of Tech. Papers,* pp. 280-281, 1994.

5. P. E. Fleischer and K. R. Laker, A family of active switched-capacitor biquad building blocks, *Bell Syst. Tech. J.*, vol. 58, pp. 2235-2269, 1979.

6. G. M. Jacobs, D. J. Allstot, R. W. Brodersen, and P. R. Gray, Design techniques for MOS switched-capacitor ladder filters, *IEEE Trans.Circ. Syst.*, vol. CAS-25, pp. 1014-1021, Dec. 1978.

7. J. B. Shyu, G. C. Temes, and F. Krummenacher, Random error effects in matched MOS capacitors and current sources, *IEEE J. Solid-State Circuits*, vol. SC-19, pp. 948-955, 1984.

8. D. G. Haigh and B. Singh, A switching scheme for switched-capacitor filters which reduces the effects of parasitic capacitances associated with switch control terminals, *IEEE Int. Symp. Circ. Syst.*, ISCAS 1993.

9. 9.T. Brooks, D. H. Robertson, D. F. Kelly, A. DelMuro, and S. W. Harston, A cascaded sigma-delta pipeline A/D converter with 1.25MHz signal bandwidth and 89dB SNR, *IEEE J. Solid-State Circuits*, vol. SC-32, pp. 1896-1906, Dec. 1997.

10. 10.G. C. Temes, Finite amplifier gain and bandwidth effects in switched capacitor filters, *IEEE J. Solid-State Circuits*, vol. SC-15, pp. 358-361, June 1980.

11. K. Martin and A. S. Sedra, Effects of the op amp finite gain and bandwidth on the performance of switched-capacitor filters, *IEEE Trans. Circ. Syst.*, vol. CAS-28, no. 8, pp. 822-829, Aug. 1981.

12. K. Lee and R. G. Meyer, Low-distortion switched-capacitor filter design techniques, *IEEE J. Solid-State Circuits*, Dec. 1985.

13. C. A. Gobet and A. Knob, Noise analysis of switched-capacitor networks, *IEEE Trans. Circ. Syst.*, vol. CAS-30, pp. 37-43, Jan. 1983.

14. J. H. Fischer, Noise sources and calculation techniques for switched-capacitor filters, *IEEE J. Solid-State Circuits*, vol. SC-17, pp. 742-752, Aug. 1982.

15. C. Enz and G. C. Temes, Circuit techniques for reducing the effects of opamp imperfections: autozeroing, correlated double sampling, and chopper stabilization, *Proc. IEEE*, vol. 84, no. 11, pp. 1584-1614, Nov. 1996.

16. K. K. K. Lam and M. A. Copeland, Noise-cancelling switched-capacitor (SC) filtering technique, *Electronics Letters*, vol. 19, pp. 810-811, Sept. 1983.

17. K. Nagaraj, T. R. Viswanathan, K. Singhal, and J. Vlach, Switched-capacitor circuits with reduced sensitivity to amplifier gain, *IEEE Trans. Circ. Syst.*, vol. CAS-34, pp. 571-574, May 1987.

18. L. E. Larson and G. C. Temes, Switched-capacitor building-blocks with reduced sensitivity to finite amplifier gain, bandwidth, and offset voltage, *IEEE Int. Symp. Circ. Syst. (ISCAS '87)*, pp. 334-338, 1987.

19. K. Haug, F. Maloberti, and G. C. Temes, Switched-capacitor integrators with low finite-gain sensitivity, *Electronics Letters*, vol. 21, pp. 1156-1157, Nov. 1985.

20. K. Nagaraj, J. Vlach, T. R. Viswanathan, and K. Singhal, Switched-capacitor integrator with reduced sensitivity to amplifier gain, *Electronics Letters*, vol. 22, pp. 1103-1105, Oct. 1986.

21. A. Baschirotto, R. Castello, and F. Montecchi, Finite gain compensated double-sampled switched-capacitor integrator for high-Q bandpass filters, *IEEE Trans. Circ. Syst.*, vol. CAS-I 39, no. 6, June 1992.

22. T. Huo and D. J. Allstot, MOS SC highpass/notch ladder filter, *Proc. IEEE Int. Symp. Circ. Syst.*, pp. 309-312, May 1980.

23. K. Nagaraj, A parasitic insensitive area efficient approach to realizing very large time constant in switched-capacitor circuits, *IEEE Trans. Circ. Syst.*, vol. CAS-36, pp. 1210-1216, Sept. 1989.

24. J. J. F. Rijns and H. Wallinga, Spectral analysis of double-sampling switched-capacitor filters, *IEEE Trans. Circ. Syst.*, vol. 38, no. 11, pp. 1269-1279, Nov. 1991.

25. A. Baschirotto, F. Montecchi, and R. Castello, A 15MHz 20mW BiCMOS switched-capacitor biquad operating with 150Ms/s sampling frequency, *IEEE J. Solid-State Circuits*, pp. 1357-1366, Dec. 1995.

26. A. Baschirotto, Considerations for the design of switched-capacitor circuits using precise-gain operational amplifiers, *IEEE Trans. Circ. Syst., II.*, vol. 43, no. 12, pp. 827-832, Dec. 1996.

27. R. Castello, F. Montecchi, F. Rezzi, and A. Baschirotto, Low-voltage analog filter, *IEEE Trans. Circ. Syst., II.*, pp. 827-840, Nov. 1995.

28. A. Baschirotto and R. Castello, 1V switched-capacitor filters, *Workshop on Advances in Analog Circuit Design,* Copenhagen, 28-30 Apr. 1998.

29. J. F. Dickson, On-chip high-voltage generation in MNOS integrated circuits using an improved voltage multiplier technique, *IEEE J. Solid-State Circuits,* vol. SC-11, no. 3, pp. 374-378, June 1976.

30. O. J. Crols and M. Steyaert, Switched-opamp: an approach to realize full CMOS switched-capacitor circuits at very low power supply voltages, *IEEE J. Solid-State Circuits,* vol. SC-29, no. 8, pp. 936-942, Aug. 1994.

31. A. Baschirotto and R. Castello, A 1V 1.8MHz CMOS Switched-opamp SC filter with rail-to-rail output swing, *IEEE J. Solid-State Circuits,* pp. 1979-1986, Dec. 1997.

13
Materials

Stephen I. Long
*University of California
at Santa Barbara*

13.1 Introduction

Very-high-speed digital integrated circuit design is a multidisciplinary challenge. First, there are several IC technologies available for very-high-speed applications. Each of these claims to offer unique benefits to the user. In order to choose the most appropriate or cost-effective technology for a particular application or system, the designer must understand the materials, the devices, the limitations imposed by process on yields, and the thermal limitations due to power dissipation.

Second, very-high-speed digital ICs present design challenges if the inherent performance of the devices is to be retained. At the upper limits of speed, there are no digital circuits, only analog. Circuit design techniques formerly thought to be exclusively in the domain of analog IC design are effective in optimizing digital IC designs for highest performance.

Finally, system integration when using the highest-speed technologies presents an additional challenge. Interconnections, clock and power distribution both on-chip and off-chip require much care and often restrict the achievable performance of an IC in a system.

The entire scope of very-high-speed digital design is much too vast to present in a single tutorial chapter. Therefore, we must focus the coverage in order to provide some useful tools for the designer. We will focus primarily on compound semiconductor technologies in order to restrict the scope. Silicon IC design tutorials can be found in other chapters in this handbook. This chapter gives a brief introduction to compound semiconductor materials in order to justify the use of non-silicon materials for the highest-speed applications. The transport properties of several materials are compared. Second, a technology-independent description of device operation for high-speed or high-frequency applications will be given in Chapter 14. The charge control methodology provides insight and connects the basic material properties and device geometry with performance. Chapter 15 describes the design basics of very-high-speed ICs. Static design methods are illustrated with compound semiconductor circuit examples, but are based on generic principles such as noise margin. The transient design methods emphasize analog circuit techniques and can be applied to any technology.

Finally, Chapter 16 describes typical circuit design approaches using FET and bipolar device technologies and presents applications of current interest.

13.2 Compound Semiconductor Materials

The compound semiconductor family is composed of the group III and group V elements shown in Table 13.1. Each semiconductor is formed from at least one group III and one group V element. Group IV elements such as C, Si, and Ge are used as dopants, as are several group II and VI elements such as Be or Mg for p-type and Te and Se for n-type. Binary semiconductors such as GaAs and InP can be grown in large single-crystal ingot form using the liquid-encapsulated Czochralski method[1] and are the materials of choice for substrates. At the present time, GaAs wafers with a diameter of 100 and 150 mm are most widely used. InP is still limited to 75 mm diameter.

TABLE 13.1 Column III, IV, and V Elements Associated with Compound Semiconductors

B	C	N
Al	Si	P
Ga	Ge	As
In	Sn	Sb

Three or four elements are often mixed together when grown as thin *epitaxial* films on top of the binary substrates. The alloys thus formed allow electronic and structural properties such as bandgap and lattice constant to be varied as needed for device purposes. Junctions between different semiconductors can be used to further control charge transport as discussed in Section 13.4.

13.3 Why III-V Semiconductors?

The main motivation for using the III-V compound semiconductors for device applications is found in their electronic properties when compared with those of the dominant semiconductor material, silicon. Figure 13.1 is a plot of steady-state *electron velocity* of several n-type semiconductors versus electric field. From this graph, we see that at low electric fields the slope of the curves (*mobility*) is higher than that of silicon. High mobility means that the resistivity will be less for III-V n-type materials, and it may be easier to achieve lower access resistance. *Access resistance* is the series resistance between the device contacts and the internal active region. An example would be the base resistance of a bipolar transistor. Lower resistance will reduce some of the fundamental device time constants to be described in Chapter 14 that often dominate device high-frequency performance. Figure 13.1 also shows that the peak electron velocity is higher for III-V materials, and the peak velocity can be achieved at much lower electric fields. High velocity reduces *transit time*, the time required for a charge carrier to travel from its source to its destination, and improves device high-frequency performance, also discussed in Chapter 14. Achieving this high velocity at lower electric fields means that the devices will reach their peak performance at lower voltages, which is useful for low-power, high-speed applications. Mobilities and peak velocities of several semiconductors are compared in Table 13.2.

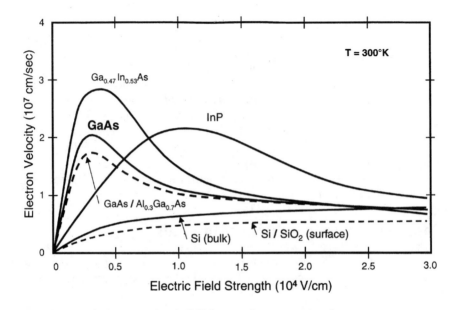

FIGURE 13.1 Electron velocity versus electric field for several n-type semiconductors.

TABLE 13.2 Comparison of Mobilities and Peak Velocities of Several n- and p-type Semiconductors

Semiconductor	E_G (eV)	ε_r	Electron Mobility (cm²/V-s)	Hole Mobility (cm²/V-s)	Peak Electron Velocity (cm/s)
Si (bulk)	1.12	11.7	1350	450	N.A.
Ge	0.66	15.8	3900	1900	N.A.
InP	1.35 D	12.4	4600	150	2.1×10^7
GaAs	1.42 D	13.1	8500	400	2×10^7
$Ga_{0.47}In_{0.53}As$	0.78 D	13.9	11,000	200	2.7×10^7
InAs	0.35 D	13.6	22,600	460	4×10^7
$Al_{0.3}Ga_{0.7}As$	1.80 D	12.2	1000	100	—
AlAs	2.17	10.1	280	—	—
$Al_{0.48}In_{0.52}As$	1.92 D	12.3	800	100	—

Note: In bandgap energy column, the symbol "D" indicates direct bandgap; otherwise, it is indirect bandgap. T = 300 K and "weak doping" limit.

On the other hand, as also shown in Table 13.2, p-type III-V semiconductors have rather poor hole mobility when compared with elemental semiconductor materials such as silicon or germanium. Holes also reach their peak velocities at much higher electric fields than electrons. Therefore, p-type III-V materials needed for the base of a bipolar transistor, for example, are used, but their thickness must be extremely small to avoid degradation in transit time. Lateral distances must also be small to avoid excessive series resistance. CMOS-like complementary FET technologies have also been developed,[2] but their performance has been limited by the poorer speed of the p-channel devices.

13.4 Heterojunctions

In the past, most semiconductor devices were composed of a single semiconductor element, such as silicon or gallium arsenide, and employed n- and p-type doping to control charge transport. Figure 13.2(a) illustrates an energy band diagram of a semiconductor with uniform composition that is in an applied electric field. Electrons will drift downhill and holes will drift uphill in the applied electric field. The electrons and/or holes could be produced by doping or by ionization due to light. In a *heterogeneous* semiconductor as shown in Fig. 13.2(b), the bandgap can be graded from wide bandgap on the left to narrow on the right by varying the composition. In this case, even without an applied electric field, a built-in quasi-electric field is produced by the bandgap variation that will transport both holes and electrons in the *same* direction.

The abrupt *heterojunction* formed by an atomically abrupt transition between AlGaAs and GaAs, shown in the energy band diagram of Fig. 13.3, creates discontinuities in the valence and conduction bands. The conduction band energy discontinuity is labeled ΔE_C and the valence band discontinuity, ΔE_V. Their sum equals the energy bandgap difference between the two materials. The potential energy steps caused by these discontinuities are used as barriers to electrons or holes. The relative sizes of these potential barriers depend on the composition of the semiconductor materials on each side of the heterojunction. In this example, an electron barrier in the conduction band is used to confine carriers into a narrow potential energy well

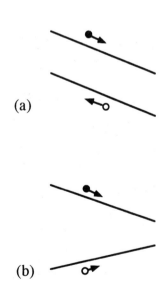

FIGURE 13.2 (a) Homogeneous semiconductor in uniform electric field, and (b) Heterogeneous semiconductor with graded energy gap. No applied electric field.

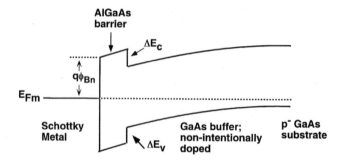

FIGURE 13.3   Energy band diagram of an abrupt heterojunction.

with triangular shape. Quantum well structures such as these are used to improve device performance through two-dimensional charge transport channels, similar to the role played by the inversion layer in MOS devices. The structure and operation of heterojunctions in FETs and BJTs will be described in Chapter 14.

The overall principle of the use of heterojunctions is summarized in a *Central Design Principle*:

> "Heterostructures use energy gap variations in addition to electric fields as forces acting on holes and electrons to control their distribution and flow."[3,4]

The energy barriers can control motion of charge both across the heterojunction and in the plane of the heterojunction. In addition, heterojunctions are most widely used in light-emitting devices, since the compositional differences also lead to either stepped or graded index of refraction, which can be used to confine, refract, and reflect light. The barriers also control the transport of holes and electrons in the light-generating regions.

Figure 13.4 shows a plot of bandgap versus lattice constant for many of the III-V semiconductors.[3] Consider GaAs as an example. GaAs and AlAs have the same lattice constant (approximately 0.56 nm) but different bandgaps (1.4 and 2.2 eV, respectively). An alloy semiconductor, AlGaAs, can be grown epitaxially on a GaAs substrate wafer using standard growth techniques. The composition can be selected by the Al-to-Ga ratio, giving a bandgap that can be chosen across the entire range from GaAs to AlAs.

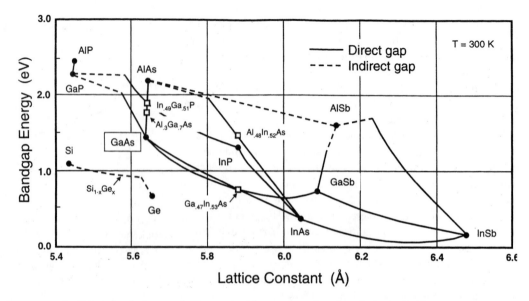

FIGURE 13.4   Energy bandgap versus lattice constant for compound semiconductor materials.

Since both lattice constants are essentially the same, very low lattice mismatch can be achieved for any composition of $Al_xGa_{1-x}As$. Lattice matching permits low defect density, high-quality materials to be grown that have good electronic and optical properties. It quickly becomes apparent from Fig. 13.4, however, that a requirement for lattice matching to the substrate greatly restricts the combinations of materials available to the device designer. For electron devices, the low mismatch GaAs/AlAs alloys, GaSb/AlSb alloys, $Al_{.48}In_{.52}As/InP/Ga_{.47}In_{.53}As$, and $GaAs/In_{.49}Ga_{.51}As$ combinations alone are available. Efforts to utilize combinations such as GaP on Si or GaAs on Ge that lattice match have been generally unsuccessful because of problems with interface structure, polarization, and autodoping.

For several years, lattice matching was considered to be a necessary condition if mobility-damaging defects were to be avoided. This barrier was later broken when it was discovered that high-quality semiconductor materials could still be obtained although lattice-mismatched if the thickness of the mismatched layer is sufficiently small.[5,6] This technique, called *pseudomorphic* growth, opened another dimension in III-V device technology, and allowed device structures to be optimized over a wider range of bandgap for better electron or hole dynamics and optical properties.

Two of the pseudomorphic systems that have been very successful in high-performance millimeter-wave FETs are the InAlAs/InGaAs/GaAs and InAlAs/InGaAs/InP systems. The $In_xGa_{1-x}As$ layer is responsible for the high electron mobility and velocity which both improve as the In concentration x is increased. Up to $x = 0.25$ for GaAs substrates and $x = 0.80$ for InP substrates have been demonstrated and result in great performance enhancements when compared with lattice-matched combinations.

References

1. Ware, R., Higgins, W., O'Hearn, K., and Tiernan, M., Growth and Properties of Very Large Crystals of Semi-Insulating Gallium Arsenide, presented at *18th IEEE GaAs IC Symp.*, Orlando, FL, 54, 1996.
2. Abrokwah, J. K., Huang, J. H., Ooms, W., Shurboff, C., Hallmark, J. A. et al., A Manufacturable Complementary GaAs Process, presented at *IEEE GaAs IC Symposium,* San Jose, CA, 127, 1993.
3. Kroemer, H., Heterostructures for Everything: Device Principles of the 1980s?, *Japanese J. Appl. Phys.*, 20, 9, 1981.
4. Kroemer, H., Heterostructure Bipolar Transistors and Integrated Circuits, *Proc. IEEE*, 70, 13, 1982.
5. Matthews, J. W. and Blakeslee, A. E., Defects in Epitaxial Multilayers. III. Preparation of Almost Perfect Layers, *J. Crystal Growth*, 32, 265, 1976.
6. Matthews, J. W. and Blakeslee, A. E., Coherent Strain in Epitaxially Grown Films, *J. Crystal Growth*, 27, 118, 1974.

14

Compound Semiconductor Devices for Digital Circuits

Donald B. Estreich
Hewlett-Packard Company

14.1 Introduction

An *active device* is an electron device, such as a transistor, capable of delivering power amplification by converting dc bias power into time-varying signal power. It delivers a greater energy to its load than if the device were absent. The *charge control* framework[1-3] discussed below presents a unified understanding of the operation of all electron devices and simplifies the comparison of the several active devices used in digital integrated circuits.

14.2 Unifying Principle for Active Devices: Charge Control Principle

Consider a generic electron device as represented in Fig. 14.1. It consists of three electrodes encompassing a charge *transport region*. The transport region is capable of supporting charge flow (electrons as shown in the figure) between an *emitting electrode* and a *collecting electrode*. A third electrode, called the *control electrode*, is used to establish the electron concentration within the transport region. Placing a *control charge*, Q_C, on the control electrode establishes a *controlled charge*, denoted as $-Q$, in the transport region. The operation of active devices depends on the *charge control principle:*[1]

Each charge placed upon the control electrode can at most introduce an equal and opposite charge in the transport region between the emitting and collecting electrode.

At most, we have the relationship, $|-Q| = |Q_C|$. Any parasitic coupling of the control charge to charge on the other electrodes, or remote parts of the device, will decrease the controlled charge in the transport region, that is $|-Q| < |Q_C|$ more generally. For example, charge coupling between the control

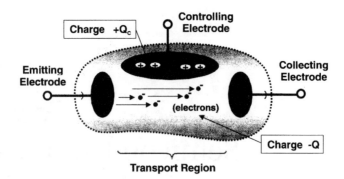

FIGURE 14.1   Generic charge control device consisting of three electrodes embedded around a charge transport region.

electrode and the collecting electrode forms a feedback or output capacitance, say C_o. Time variation of Q_C leads to the modulation of the current flow between emitting and collecting electrodes.

The generic structure in Fig. 14.1 could represent any one of a number of active devices (e.g., vacuum tubes, unipolar transistors, bipolar transistors, photoconductors, etc.). Hence, charge control analysis is very broad in scope, since it applies to all electronic transistors.

Starting from the charge control principle, we associate two characteristic time constants with an active device, thereby leading to a first-order description of its behavior. Application of a potential difference between the emitting and collecting electrodes, say V_{CC}, establishes an electric field in the transport region. Electrons in the transport region respond to the electric field and move across this region with a *transit time* τ_r. The transit time[*] is the first of the two important characteristic times used in charge control modeling. With charge $-Q$ in the transit region, the static (dc) current I_o between emitting and collecting electrodes is

$$I_o = -Q/\tau_r = Q_c/\tau_r \qquad (14.1)$$

A simple interpretation of τ_r is as follows: τ_r is equal to the length l of the transport region, divided by the average velocity of transit (i.e., $\tau_r = l/\langle v \rangle$). From this perspective, a charge of $-Q$ (coulombs) is swept out of the collecting electrode every τ_r seconds.

Now consider Fig. 14.2, showing the common-emitting electrode connection of the active device of Fig. 14.1 connected to input and output (i.e., load) resistances, say R_{in} and R_L, respectively. The second characteristic time of importance can now be defined: It is the *lifetime time constant*, and we denote it by τ. It is a measure of how long a charge placed on the control electrode will remain on the control terminal. The lifetime time constant is established in one of several ways, depending on the physics of the active device and/or its connection. The controlling charge may "leak away" by (1) discharging through the external resistor R_{in} as typically happens with FET devices, (2) recombining with intermixed oppositely charged carriers within the device (e.g., base recombination in a bipolar transistor), or (3) discharging through an internal shunt leakage path within the device. The dc current flowing to replenish the lost control charge is given by

$$I_{in} = -Q/\tau = Q_c/\tau \qquad (14.2)$$

The *static (dc) current gain* G_I of a device is defined as the current delivered to the output, divided by the current replenishing the control charge during the same time period. Where in τ seconds charge $-Q$ is both lost and replenished, charge Q_c times the ratio τ/τ_r has been supplied to the output resistor R_L. In symbols, the static current gain is

---

[*]The transit time τ_r is best interpreted as an average transit time per carrier (electron). We note that $1/\tau_r$ is common to all devices — it is related to a device's ultimate capability to process information.

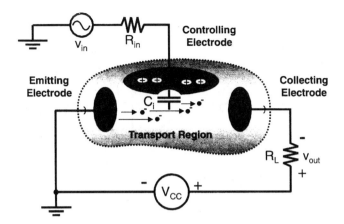

FIGURE 14.2   Generic charge control device of Fig. 14.1 connected to input and output resistors, R_{in} and R_L, respectively, with bias voltage and input signal applied.

$$G_I = I_o/I_{in} = \tau/\tau_r \tag{14.3}$$

provided $|-Q| = |Q_C|$ holds.

In the dynamic case, the process of *small-signal amplification* consists of an incremental variation of the control charge Q_c directly resulting in an incremental change in the controlled charge, $-Q$. The resulting variation in output current flowing in the load resistor translates into a time-varying voltage v_o. The charge control formalism holds just as well for large-signal situations. In the large-signal case, the changes in control charge are no longer small incremental changes. Charge control analysis under large charge variations is less accurate due to the simplicity of the model, but still very useful for approximate switching calculations in digital circuits.

An important dynamic parameter is the *input capacitance* C_i of the active device. Capacitance C_i is a measure of the work required to introduce a charge carrier in the transport region. Capacitance C_i is given by the change in charge Q for a corresponding change in input voltage v_{in}. It is desirable to maximize C_i in an active device. The *transconductance* g_m is calculated from

$$g_m = \left(\frac{\partial I_o}{\partial v_{in}}\right)_{v_o} = \left(\frac{\partial I_o}{\partial Q}\right) \cdot \left(\frac{\partial Q}{\partial v_{in}}\right) \tag{14.4}$$

The first partial derivative on the right-hand side of Eq. 14.4 is simply $(1/\tau_r)$, and the second partial derivative is C_i. Hence, the transconductance g_m is the ratio

$$g_m = \frac{C_i}{\tau_i} \tag{14.5}$$

A physical interpretation of g_m is the ratio of the work required to introduce a charge carrier to the average transit time of a charge carrier in the transport region. The transconductance is one of the most commonly used device parameters in circuit design and analysis.

In addition to C_i, another capacitance, say C_o, is introduced and associated with the collecting electrode. Capacitance C_o accounts for charge on the collecting electrode coupled to either static charge in the transport region or charge on the control electrode. A non-zero C_o indicates that the coupling between the controlling electrode and the charge in transit is less than unity (i.e., $|-Q| < |Q_C|$).

For small-signal analysis the capacitance parameters are usually taken at fixed numbers evaluated about the device's bias state. When using charge control in the large-signal case, the capacitance parameters

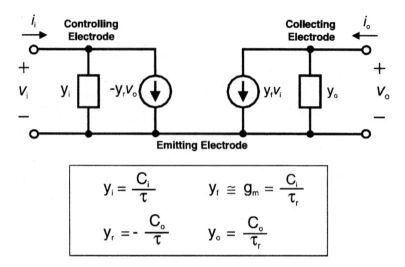

FIGURE 14.3   Two-port, small-signal, admittance charge control model with the emitting electrode selected as the common terminal to both input and output.

must include the voltage dependencies. For example, the input capacitance C_i can be strongly dependent upon the control electrode to emitting electrode and collecting electrode potentials. Hence, during the change in bias state within a device, the magnitude of the capacitance C_i is time varying. This variation can dramatically affect the switching speed of the active device. Parametric dependencies on the instantaneous bias state of the device are at the heart of accurate modeling of large-signal or switching behavior of active devices.

We introduce the *small-signal admittance charge control model* shown in Fig. 14.3. This model uses the emitting electrode as the common terminal in a two-port connection. The transconductance g_m is the magnitude of the real part of the *forward admittance* y_f and is represented as a voltage-controlled current source positioned from collecting-to-emitting electrode. The *input admittance*, denoted by y_i, is equivalent to (C_i/τ), where τ is the control charge lifetime time constant. Parameter y_i can be expressed in the form $(g_i + sC_i)$ where $s = j\omega$. An *output admittance*, similarly denoted by y_o, is given by (C_o/τ_r) where τ_r is the transit time and, in general $y_o = (g_o + sC_o)$. Finally, the *output-to-input feedback admittance* y_r is included using a voltage-controlled current source at the input. Often, y_r is small enough to approximate as zero (the model is then said to be *unilateral*).

Consider the frequency dependence of the *dynamic (ac) current gain* G_i. The low-frequency current gain is interpreted as follows: an incremental charge q_c is introduced on the control electrode with lifetime τ. This produces a corresponding incremental charge $-q$ in the transport region. Charge $-q$ is swept across the transport region every transit time τ_r seconds. In time τ, charge $-q$ crosses the transit region τ/τ_r times, which is identically equal to the low-frequency current gain.

The lifetime τ associated with the control electrode arises from charge "leaking off" the controlling electrode. This is modeled as an RC time constant at the input of the equivalent circuit shown in Fig. 14.4(a) with τ equal to $R_{eq}C_i$. R_{eq} is the equivalent resistance presented to capacitor C_i. That is, R_{eq} is determined by the parallel combination of $1/g_i$ and any external resistance at the input. The *break frequency* ω_B associated with the control electrode is

$$\omega_B = \frac{1}{\tau} = \frac{1}{R_{eq}C_i} \tag{14.6}$$

When the charge on the control electrode varies at a rate ω less than ω_B, G_i is given by τ/τ_r because charge "leaks off" the controlling electrode faster than $1/\omega$. Alternatively, when ω is greater than ω_B, G_i

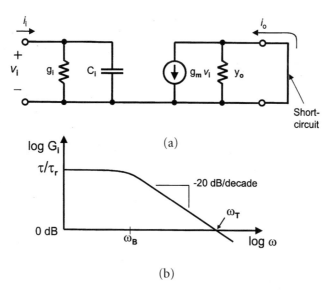

FIGURE 14.4 (a) Small-signal admittance model with output short-circuited, and (b) magnitude of the small-signal current gain G_i plotted as a function of frequency. The unity current gain crossover (i.e., $G_i = 1$) defines the parameter f_T (or ω_T).

decreases with increasing ω because the applied signal charge varies upon the control electrode more rapidly than $1/\tau$. In this case, G_i is inversely proportional to ω, that is,

$$G_i = \frac{1}{\omega \tau_r} = \frac{\omega_T}{\omega} \tag{14.7}$$

where ω_T is the common-emitter *unity current gain frequency*. At $\omega = \omega_T$ ($= 2\pi f_T$), the ac current gain equals unity, as illustrated in Fig. 14.4(b).

Consider the *current gain-bandwidth product* $G_i \Delta f$. A purely capacitive input impedance cannot define a bandwidth. However, a finite real impedance always appears at the input terminal in any practical application. Let R_i be the *effective input resistance* of the device (i.e., R_i will be equal to $(1/g_i)$ in parallel with the external resistance R_{in}). Since the input current is equal to q_c/τ and the output current is equal to q/τ_r, the current gain-bandwidth product becomes

$$G_i \cdot \Delta f = \frac{q/\tau_r}{q_c/\tau} \frac{\omega}{2\pi} \tag{14.8}$$

For $\omega \gg \omega_B$, at $\tau = 1/\omega$, and assuming $|q_c| = |-q|$,

$$G_i \cdot \Delta f = \frac{1}{2\pi \tau_r} = \frac{\omega_T}{2\pi} = f_T \tag{14.9}$$

f_T (or ω_T) is a widely quoted parameter used to compare or "benchmark" active devices. Sometimes, f_T (or ω_T) is interpreted as a measure of the maximum speed a device can drive a replica of itself. It is easy to compute and historically has been easy to measure with bridges and later using S-parameters. However, f_T does have interpretative limitations because it is defined as current into a short-circuit output. Hence, it ignores input resistance and output capacitance effects upon actual circuit performance.

Likewise, voltage and power gain expressions can be derived. It is necessary to define the output impedance before either can be quantified. Let R_o be the *effective output resistance* at the output terminal of the active device. Assuming both the input and output RC time constants to be identical (i.e., $R_iC_i = R_oC_o$), the *voltage gain* G_v can be expressed in terms of G_i as

$$G_v = G_i \frac{R_o}{R_i} = G_i \frac{C_i}{C_o} \qquad (14.10)$$

where R_o is the parallel equivalent output resistance from all resistances at the output node.

The *power gain* G_p is computed from the product of $G_i \cdot G_v$ along with the *power gain-bandwidth product*. These results are listed in Table 14.1 as summarized from Johnson and Rose.[1] These simple expressions are valid for all devices as interpreted from the charge control perspective. They provide for a first-order comparison, in terms of a few simple parameters, among the active devices commonly available. From an examination of Table 14.1, it is evident that maximizing C_i and minimizing τ_r leads to higher transconductance, higher parametric gains, and greater frequency response. This is an important observation in understanding how to improve upon the performance of any active device.

Whereas f_T has limitations, the frequency at which the maximum power gain extrapolates to unity, denoted by ω_{max}, is often a more useful indicator of device performance. The primary limitation of ω_{max} is that it is very difficult to calculate and is usually extrapolated from S-parameter measurements in which the extrapolation is approximate at best.

14.3 Comparing Unipolar and Bipolar Transistors

Unipolar transistors are active devices that operate using only a single charge carrier type, usually electrons, in their transport region. *Field-effect transistors* fall into the unipolar classification. In contrast, *bipolar transistors* depend on positive and negative charged carriers (i.e., both majority and minority carriers) within the transport region. A fundamental difference arises from the relative locations of the control

TABLE 14.1   Charge Control Relations for All Active Devices

Parameter	Symbol	Expression
Transconductance	g_m	$\dfrac{C_i}{\tau_r} \Leftrightarrow \omega_T C_i$
Current amplification	G_i	$\dfrac{1}{\omega\tau_r} \Leftrightarrow \dfrac{\omega_T}{\omega}$
Voltage amplification	G_v	$\dfrac{1}{\omega\tau_r}\dfrac{C_i}{C_o} \Leftrightarrow \dfrac{\omega_T}{\omega}\dfrac{C_i}{C_o}$
Power amplification	$G_p = G_i G_v$	$\dfrac{1}{\omega^2\tau_r^2}\dfrac{C_i}{C_o} \Leftrightarrow \dfrac{\omega_T^2}{\omega^2}\dfrac{C_i}{C_o}$
Current gain-bandwidth product	$G_i \cdot \Delta f$	$\dfrac{1}{\tau_r} \Leftrightarrow \omega_T$
Voltage gain-bandwidth product	$G_v \cdot \Delta f$	$\dfrac{1}{\tau_r}\dfrac{C_i}{C_o} \Leftrightarrow \omega_T \dfrac{C_i}{C_o}$
Power gain-bandwidth product	$G_p \cdot \Delta f^2$	$\dfrac{1}{\tau_r^2}\dfrac{C_i}{C_o} \Leftrightarrow \omega_T^2 \dfrac{C_i}{C_o}$

Note: Table assumes $R_iC_i = R_oC_o$. (After Johnson and Rose (Ref. 1), March 1959. © 1959 IEEE, reproduced with permission of IEEE.)

electrode and transport region — in unipolar devices, they are physically separated, whereas in bipolar devices, they are merged into the same physical region (i.e., base region). Before reviewing the physical operation of each, transport in semiconductors is briefly reviewed.

Charge Transport in Semiconductors[4-6]

Bulk semiconducting materials are useful because their conductivity can be controlled over many orders of magnitude by changing the doping level. Both electrons and holes[4] can conduct current in semiconductors. In integrated circuits metal, semiconductor, and insulator layers are used together in precisely positioned shapes and thicknesses to form useful device and circuit functions.

Fig. 14.1 illustrates the behavior of electron velocity as a function of local electric field strength for several important semiconducting materials. Two characteristic regions of behavior can be identified: a *linear* or *ohmic* region at low electric fields, and a *velocity-saturated* region at high fields. At low fields, current transport is proportional to the carrier's mobility. Mobility is a measure of how easily carriers move through a material.[4] At high fields, carriers saturate in velocity; hence, current levels will correspondingly saturate in active devices. The data in Fig. 13.1 assume low doping levels (i.e., $N_x < 10^{14}$ cm^{-3}). The dashed curve represents transport in a GaAs quantum well formed adjacent to an $Al_{0.3}Ga_{0.7}As$ layer — in this case, *interface scattering* lowers the mobility. A similar situation is found for transport in silicon at a semiconductor–oxide interface such as found in metal-oxide-semiconductor (MOS) devices.

Several general conclusions can be extracted from this data:

1. Compound semiconductors generally have higher electron mobilities than silicon.
2. At high fields (say E > 20,000 V/cm), saturated electron velocities tend to converge to values close to 1×10^7 cm/s.
3. Many compound semiconductors show a transition region between low and high electric field strengths with a *negative differential mobility* due to electron transfer from the Γ ($\mathbf{k} = 0$) valley to conduction band valleys with higher effective masses (this gives rise to the *Gunn Effect*[7]).

Hole mobilities are much lower than electron mobilities in all semiconductors. Saturated velocities of holes are also lower at higher electric fields. This is why n-channel field-effect transistors have higher performance than p-channel field-effect transistors, and why npn bipolar transistors have higher performance than pnp bipolar transistors. Table 14.2 compares electron and hole mobilities for several semiconducting materials.

Field-Effect (Unipolar) Transistor[8-10]

Fig. 14.5(a) shows a conceptual view of an n-channel field-effect transistor (FET). As shown, the n-type channel is a homogeneous semiconducting material of thickness b, with electrons supporting the drain-to-source current. A p-type channel would rely on mobile holes for current transport and all voltage polarities would be exactly reversed from those shown in Fig. 14.5(a). The control charge on the gate region (of length L and width W) establishes the number of conduction electrons per unit area in the channel by electrostatic attraction or exclusion. The cross-section on the FET channel in Fig. 14.5(b) shows a *depletion layer*, a region void of electrons, as an intermediary agent between the control charge and the controlled charge. This depletion region is present in *junction* FET and *Schottky barrier junction* (i.e., *metal-semiconductor junction*) MESFET structures.

In all FET structures, the gate is physically separated from the channel. By physically separating the control charge from the controlled charge, the gate-to-channel impedance can be very large at low frequencies. The gate impedance is predominantly capacitive and, typically, very low gate leakage currents are observed in high-quality FETs. This is a distinguishing feature of the FET — its high input impedance is desirable for many circuit applications.

The channel, positioned between the source and drain ohmic contacts, forms a resistor whose resistance is modulated by the applied gate-to-channel voltage. We know the gate potential controls the channel charge by the charge control relation. Time variation of the gate potential translates into a corresponding

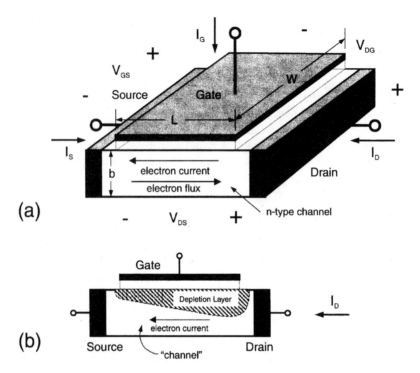

FIGURE 14.5   (a) Conceptual view of a field-effect transistor with the channel sandwiched between source and drain ohmic contacts and a gate control electrode in close proximity; and (b) cross-sectional view of the FET with a depletion layer shown such as would be present in a compound semiconductor MESFET.

time variation of the drain current (and the source current also). Therefore, transconductance g_m is the natural parameter to describe the FET from this viewpoint.

Fig. 14.6(a) shows the I_D-V_{DS} characteristic of the n-channel FET in the common-source connection with constant electron mobility and a long channel assumed. Two distinct operating regions appear in Fig. 14.6(a) — the *linear* (i.e., *non-saturated*) region, and the *saturated* region, separated by the dashed parabola. The origin of current saturation corresponds to the onset of *channel pinch-off* due to carrier exclusion at the drain end of the channel. Pinch-off occurs when the drain voltage is positive enough to

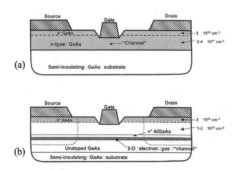

FIGURE 14.6   (a) Field-effect transistor drain current (I_D) versus drain-to-source voltage (V_{DS}) characteristic with the gate-to-source voltage (V_{GS}) as a stepped parameter; (b) I_D versus V_{GS} "transfer curve" for a constant V_{DS} in the saturated region of operation, revealing its "square-law" behavior; (c) transconductance g_m versus V_{GS} for a constant V_{DS} in saturated region of operation corresponding to the transfer curve in (b). These curves assume constant mobility, no velocity saturation, and the "long-channel FET approximation."

deplete the channel completely of electrons at the drain end; this corresponds to a gate-to-source voltage equal to the *pinch-off voltage*, denoted as $-V_p$ in Figs. 14.6(b) and (c). For constant V_{DS} in the saturated region, the I_D vs. V_{GS} *transfer curve* approximates "square law" behavior; that is,

$$I_D = I_{D,sat} = I_{DSS} \left[1 - \frac{V_{GS}}{\left(-V_P \right)} \right]^2 \quad \text{for } -V_p \leq V_{GS} \leq \varphi \tag{14.11}$$

where I_{DSS} is the drain current when $V_{GS} = 0$, and φ is a built-in potential associated with the gate-to-channel junction or interface (e.g., a metal-semiconductor Schottky barrier as in the MESFET). The symbol $I_{D,sat}$ denotes the drain current in the saturated region of operation. Transconductance g_m is linear with V_{GS} for the saturation transfer characteristic of Eq. (14.11) and is approximated by

$$g_m = \frac{\partial I_D}{\partial V_{GS}} \cong 2 \frac{D_{DSS}}{V_P} \left[1 - \frac{V_{GS}}{\left(-V_P \right)} \right] \quad \text{for } -V_p \leq V_{GS} \leq \varphi \tag{14.12}$$

Equations 14.11 and 14.12 are plotted in Figs. 14.6(b) and (c), respectively.

Bipolar Junction Transistors (Homojunction and Heterojunction)[7-11]

In the *bipolar junction transistor* (BJT), both the control charge and the controlled charge occupy the same region (i.e., the *base region*). A control charge is injected into the base region (i.e., this is the base current flowing in the base terminal), causing the emitter-to-base junction's potential barrier to be lowered. Barrier lowering results in majority carrier diffusion across the emitter-to-base junction. Electrons diffuse into the base and holes into the emitter in the npn BJT shown in Fig. 14.7. By controlling the emitter-to-base junction's physical structure, the dominant carrier diffusion across this n-p junction should be injection into the base region. For our npn transistor, the dominant carrier transport is electron diffusion into the base region where the electrons are minority carriers. They transit the base region, of base width W_b, by both diffusion and drift. When collected at the collector-to-base junction, they establish the collector current I_C. The base width must be short to minimize recombination in the base region (this is reflected in the current gain parameter commonly used with BJT and HBT devices).

In *homojunction* BJT devices, the emitter and base regions have the same bandgap energy. The respective carrier injection levels are set by the ratio of the emitter-to-base doping levels. For high emitter efficiency, that is, the number of carriers diffusing into the base being much greater than the number of carriers simultaneously diffusing into the emitter, the emitter must be much more heavily doped than the base region. This places a limit on the maximum doping level allowed in the base of the homojunction BJT, thereby leading to higher base resistance than the device designer would normally desire.[10] In contrast, the *heterojunction bipolar transistor* (HBT) uses different semiconducting materials in the base and emitter regions to achieve high emitter efficiency. A wider bandgap emitter material allows for high emitter efficiency while allowing for higher base doping levels which in turn lowers the parasitic base resistance. An example of a wider bandgap emitter transistor is shown in Fig. 14.8. In this example, the emitter is AlGaAs whereas the base and collector are formed with GaAs. Figure 14.8 shows the band diagram under normal operation with the emitter–base junction forward-biased and the collector–base junction reverse-biased. The discontinuity in the valence band edge at the emitter–base heterojunction is the origin of the reduced diffusion into the emitter region. The injection ratio determining the emitter efficiency depends exponentially on this discontinuity. If ΔE_g is the valence band discontinuity, the injection ratio is proportional to the exponential of ΔE_g normalized to the thermal energy kT:[10a]

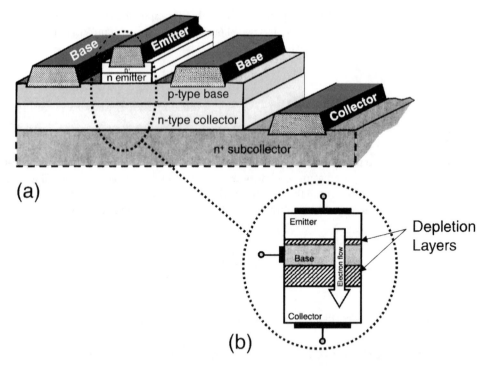

FIGURE 14.7   (a) Conceptual view of a bipolar junction transistor with the base region sandwiched between emitter and collector regions. Structure is representative of a compound semiconductor heterojunction bipolar transistor. (b) Simplified cross-sectional view of a vertically structured BJT device with primary electron flow represented by large arrow.

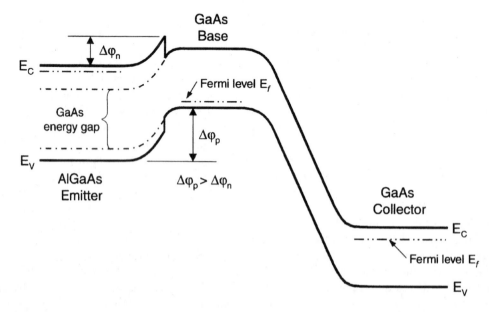

FIGURE 14.8   The bandgap diagram for an HBT AlGaAs/GaAs device with the wider bandgap for the AlGaAs emitter (solid line) compared with a homojunction GaAs BJT emitter (dot-dash line). The double dot-dashed line represents the Fermi level in each region.

$$\frac{J_n}{J_p} \propto \exp\left(-\Delta E_g / kT\right) \tag{14.13}$$

For example, ΔE_g equal to 8kT gives an exponential factor of approximately 8000, thereby leading to an emitter efficiency of nearly unity, as desired. The use of the emitter-base band discontinuity is a very efficient way to hold high emitter efficiencies.

In bipolar devices, the collector current I_C is given by the exponential of the *base-emitter forward voltage* V_{BE} normalized to the *thermal voltage* kT/q

$$I_C = I_S \exp\left(qV_{BE} / kT\right) \tag{14.14}$$

The saturation current I_S is given by a quantity that depends on the structure of the device; it is inversely proportional to the base doping charge Q_{BASE} and proportional to the device's area A, namely

$$I_S = \frac{qADn_i^2}{Q_{BASE}} \tag{14.15}$$

where the other symbols have their usual meanings (D is the minority carrier *diffusion constant* in the base, n_i is the *intrinsic carrier concentration* of the semiconductor, and q is the *electron's charge*).

A typical collector current versus collector-emitter voltage characteristic, for several increasing values of (forward-biased) emitter-base voltages, is shown in Fig. 14.9(a). Note the similarity to Fig. 14.6(a),

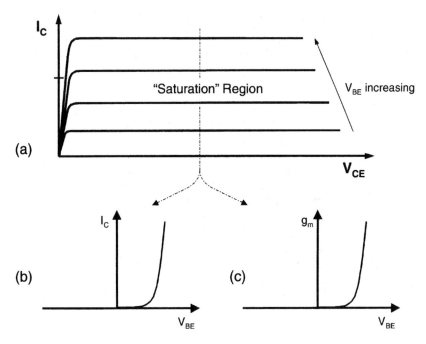

FIGURE 14.9 (a) Collector current (I_C) versus collector-to-emitter voltage (V_{CE}) characteristic curves with the base-to-emitter voltage (V_{BE}) as stepped parameter; (b) I_C versus V_{BE} "transfer curve" for a constant V_{CE} in saturated region of operation shows exponential behavior; and (c) transconductance g_m versus V_{BE} for a constant V_{CE} in the saturated region of operation corresponding to the transfer curve in (b).

with the BJT having a quicker turn-on for low V_{CE} values compared with the softer knee for the FET. The transconductance of the BJT and HBT is found by taking the derivative of Eq. 14.14, thus

$$g_m = \frac{\partial I_C}{\partial V_{BE}} = \frac{qI_S}{kT} \exp\left(qV_{BE}/kT\right) \tag{14.16}$$

Both I_C and g_m are of exponential form, as observed in Fig. 14.13; Eqs. 14.14 and 14.16 are plotted in Figs. 14.9(b) and (c), respectively. The transconductance of the BJT/HBT is generally much larger than that of the best FET devices (this can be verified by comparing Eq. (14.12) with Eq. (14.16) with typical parameter values inserted). This has significant circuit design advantages for the BJT/HBT devices over the FET devices because high transconductance is needed for high current drive to charge load capacitance in digital circuits. In general, higher g_m values allow a designer to use feedback to a greater extent in design and this provides for greater tolerance to process variations.

Comparing Parameters

Table 14.2 compares some of the more important features and parameters of the BJT/HBT device with the FET device. For reference, a common-source FET configuration is compared with a common-emitter BJT/HBT configuration. One of the most striking differences is the input impedance parameter. A FET has a high input impedance at low to mid-range frequencies because it essentially is a capacitor. As the frequency increases, the magnitude of the input impedance decreases as $1/\omega$ because a capacitive reactance varies as $\left|C_{gs}/\omega\right|$. The BJT/HBT emitter-base is a forward-biased pn junction, which is inherently a low impedance structure because of the lowered potential barrier to carriers. The BJT/HBT input is also capacitive (i.e., a large diffusion capacitance due to stored charge), but a large conductance (or small resistance) appears in parallel assuring a low input impedance even at low frequencies.

BJT/HBT devices are known for their higher transconductance g_m, which is proportional to collector current. An FET's g_m is proportional to the saturated velocity v_{sat} and its input capacitance C_{gs}. Thus, device structure and material parameters set the performance of the FET whereas thermodynamics play the key role in establishing the magnitude of g_m in a BJT/HBT.

TABLE 14.2 Comparing Electrical Parameters for BJT/HBT vs. FET

Parameter	BJT/HBT	FET
Input impedance Z	Low Z due to forward-biased junction; large diffusion capacitance C_{be}	High Z due to reverse biased junction or insulator; small depletion layer capacitance C_{gs}
Turn-on Voltage	Forward voltage V_{BE} highly repeatable; set by thermodynamics	Pinch-off voltage V_p not very repeatable; set by device design
Transconductance	High g_m [= $I_C/(kT/q)$]	Low g_m [$\cong v_{sat}C_{gs}$]
Current gain	β (or h_{FE}) = 50 to 140; β is important due to low input impedance	Not meaningful at low frequencies and falls as $1/\omega$ at high frequencies
Unity current gain cutoff frequency f_T	$f_T = g_m/2\pi C_{BE}$ is usually lower than for FETs	$f_T = g_m/2\pi C_{gs}$ (= $v_{sat}/2\pi L_g$) higher for FETs
Maximum frequency of oscillation f_{max}	$f_{max} = [f_T/(8\pi r_b C_{bc}]^{1/2}$	$f_{max} = f_T [r_{ds}/R_{in}]^{1/2}$
Feedback capacitance	C_{bc} large because of large collector junction	Usually C_{gd} is much smaller than C_{bc}
1/f Noise	Low in BJT/HBT	Very high 1/f noise corner frequency
Thermal behavior	Thermal runaway and second breakdown	No thermal runaway
Other		Backgating is problem in semi-insulating substrates

Thermodynamics also establishes the magnitude of the *turn-on voltage* (this follows simply from Eq. 14.14) in the BJT/HBT device. For digital circuits, turn-on voltage (or *threshold voltage*) is important in terms of repeatability and consistency for circuit robustness. The BJT/HBT is clearly superior to the FET in this regard because doping concentration and physical structure establish an FET's turn-on voltage. In general, these variables are less controllable. However, the forward turn-on voltage in the AlGaAs/GaAs HBT is higher (~1.4 V) because of the band discontinuity at the emitter–base heterojunction. For InP-based HBTs, the forward turn-on voltage is lower (~0.8 V) than that of the AlGaAs/GaAs HBT and comparable to the approximate 0.7 V found in silicon BJTs. This is important in digital circuits because reducing the signal swing allows for faster circuit speed and lowers power dissipation by allowing for reduced power supply voltages.

For BJT/HBT devices, *current gain* (often given the symbol of β or h_{FE}) is a meaningful and important parameter. Good BJT devices inject little current into the emitter and, hence, operate with low base current levels. The current gain is defined as the collector current divided by the base current and is therefore a measure of the quality of the device (i.e., traps and defects, both surface and bulk, degrade the current gain due to higher recombination currents). At low to mid-range frequencies, current gain is not especially meaningful for the high input impedance FET device because of the capacitive input.

The intrinsic gain of an HBT is higher because of its higher *Early voltage* V_A. The Early voltage is a measure of the intrinsic output conductance of a device. In the HBT, the change in the collector voltage has very little effect on the modulation of the collector current. This is true because the band discontinuity dominates the establishment of the current collected at the collector–base junction. A figure of merit is the *intrinsic voltage gain* of an active device, given by the product $g_m V_A$, and the HBT has the highest values compared to silicon BJTs and compound semiconductor FETs.

It is important to have a dynamic figure of merit or parameter to assess the usefulness of an active device for high-speed operation. Both the unity current gain cutoff frequency f_T and maximum frequency of oscillation f_{max} have been discussed in the charge control section above. Both of these figures of merit are used because they are simple and can generally be correlated to circuit speed. The higher the value of both parameters, the better the high-speed circuit performance. This is not the whole story because in digital circuits other factors such as output node-to-substrate capacitance, external load capacitances, and interconnect resistance also play an important role in determining the speed of a circuit.

Generally, $1/f$ noise is much higher in FET devices than in the BJT/HBT devices. This is usually of more importance in analog applications and oscillators however. Thermal behavior in high-speed devices is important as designers push circuit performance. Bipolar devices are more susceptible to thermal runaway than FETs because of the positive feedback associated with a forward-biased junction (i.e., a smaller forward voltage is required to maintain the same current at higher temperatures). This is not true in the FET; in fact, FETs generally have negative feedback under common biases used in digital circuits. Both GaAs and InP have poorer thermal conductivity than silicon, with GaAs being about one-third of silicon and InP being about one-half of silicon.

Finally, circuits built on GaAs or InP semi-insulating substrates are susceptible to *backgating*. Backgating is similar to the backgate-bias effects in MOS transistors, only it is not as predictable or repeatable as the well-known *backgate-bias effect* is in silicon MOSFETs on silicon lightly doped substrates. Interconnect traces with negatively applied voltages and located adjacent to devices can change their threshold voltage (or turn-on voltage). It turns out that HBT devices do not suffer from backgating, and this is one of their advantages. Of course, semi-insulating substrates are nearly ideal for microstrip transmission lines on top of the substrates because of their very low loss. Silicon substrates are much more lossy in comparison and this is a decided advantage in GaAs and InP substrates.

14.4 Typical Device Structures

In this section, a few typical device structures are described. We begin with FET structures and then follow with HBT structures. There are many variants on these devices and the reader is referred to the literature for more information.[10,11,13-16]

FET Structures

In the silicon VLSI world, the MOSFET (*metal-oxide-semiconductor field-effect transistor*) dominates. This device forms a channel at the oxide–semiconductor interface upon applying a voltage to the gate to attract carriers to this interface.[17] The thin layer of mobile carriers forms a two-dimensional sheet of carriers. One of the limitations with the MOSFET is that the oxide–semiconductor interface scatters the carriers in the channel and degrades the performance of the MOSFET. This is evident in Fig. 14.1 where the lower electron velocity at the Si-SiO$_2$ interface is compared with electron velocities in compound semiconductors. For many years, device physicists have looked for device structures and materials which increase electron velocity. FET structures using compound semiconductors have led to much faster devices such as the MESFET and the HEMT.

The MESFET (*metal-semiconductor FET*) uses a thin doped channel (almost always n-type because electrons are much more mobile in semiconductors) with a reverse-biased Schottky barrier for the gate control.[9] The cross-section of a typical MESFET is shown in Fig. 14.10(a). A recessed gate is used along with a highly doped n$^+$ layer at the surface to reduce the series resistance at both the source and drain connections. The gate length and electron velocity in the channel dominate in determining the high-speed performance of a MESFET. Much work has gone into developing processes that form shorter gate structures. For digital devices, lower breakdown voltages are permissible, and therefore shorter gate lengths and higher channel doping are more compatible with such devices. For a given semiconductor material, a device's breakdown voltage BV$_{GD}$ times its unity current gain cutoff frequency f_T is a constant. Therefore, it is possible to tradeoff BV$_{GD}$ for f_T in device design. A high f_T is required in high-speed digital circuits because devices with a high f_T over their logic swing will have a high g$_m$/C ratio for large-signal operation. A high g$_m$/C ratio translates into a device's ability to drive load capacitances.

It is also desirable to maximize the charge in the channel per unit gate area. This allows for higher currents per unit gate width and greater ability to drive large capacitive loads. The higher current per

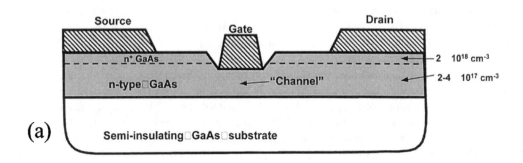

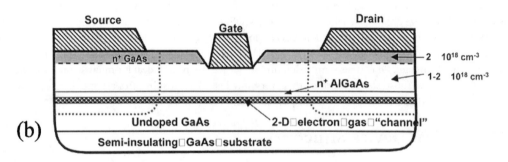

FIGURE 14.10 Typical FET cross-sections for (a) GaAs MESFET device with doped channel, and (b) AlGaAs/GaAs HEMT device with single quantum well containing and two-dimensional electron gas.

unit gate width also favors greater IC layout density. In the MESFET, the doping level of the channel sets this limit. MESFET channels are usually ion-implanted and the added lattice damage further reduces the electron mobility.

To achieve still higher currents per gate width and even higher figures of merit (such as f_T and f_{max}), the HEMT (*high electron mobility transistor*) structure has evolved.[10,16] The HEMT is similar to the MESFET except that the doped channel is replaced with a *two-dimensional quantum well* containing electrons (sometimes referred to as a *2-D electron gas*). The quantum well is formed by a discontinuity in conduction band edges between two different semiconductors (such as AlGaAs and GaAs in Fig. 14.3). From Fig. 14.4 we see that GaAs and $Al_{0.3}Ga_{0.7}As$ have nearly identical lattice constants but with somewhat different bandgaps. One compound semiconductor can be grown (i.e., using *molecular beam epitaxy* or *metalo-organic chemical vapor deposition* techniques) on a different compound semiconductor if the lattice constants are identical. Another example is $Ga_{0.47}In_{0.53}As$ and InP, where they are lattice matched. The difference in conduction band edge alignment leads to the formation of a quantum well. The greater the edge misalignment, the deeper the quantum well can be, and generally the greater the number of carriers the quantum well can hold. The charge per unit area that a quantum well can hold directly translates into greater current per unit gate width. Thus, the information in Fig. 14.4 can be used to *bandgap engineer* different materials that can be combined in lattice matched layers.

A major advantage of the quantum well comes from being able to use semiconductors that have higher electron velocity and mobility than the substrate material (e.g., GaAs) and also avoid charge impurity scattering in the quantum well by locating the donor atoms outside the quantum well itself. Figure 14.10(b) shows a HEMT cross-section where the dopant atoms are positioned in the wider bandgap AlGaAs layer. When these donors ionize, electrons spill into the quantum well because of its lower energy. Higher electron mobility is possible because the ionized donors are not located in the quantum well layer. A recessed gate is placed over the quantum well, usually on a semiconductor layer such as the AlGaAs layer in Fig. 14.10(b), allowing modulation of the charge in the quantum well.

There are only a few lattice-matched structures possible. However, semiconductor layers for which the lattice constants are not matched are possible if the layers are thin enough (of the order of a few nanometers). Molecular beam epitaxy and MOCVD make it possible to grow layers of a few atomic layers. Such structures are called *pseudomorphic HEMT* (PHEMT) devices.[10,16] This gives more flexibility in selecting quantum well layers which hold greater charge and have higher electron velocities and mobilities. The highest performance levels are achieved with pseudomorphic HEMT devices.

FET Performance

All currently used FET structures are n-channel because hole velocities are very low compared with electron velocities. Typical gate lengths range from 0.5 microns down to about 0.1 microns for the fastest devices. The most critical fabrication step in producing these structures is the gate recess width and depth.

The GaAs MESFET (ca. 1968) was the first compound semiconductor FET structure and is still used today because of its simplicity and low cost of manufacture. GaAs MESFET devices have f_T values in the 20 GHz to 50 GHz range corresponding to gate lengths of 0.5 microns down to 0.2 microns, and g_m values of the order of 200 to 400 mS/mm, respectively. These devices will typically have I_{DSS} values of 200 to 400 mA/mm, where parameter I_{DSS} is the common-source, drain current with zero gate voltage applied in a saturated state of operation.

In comparison, the first HEMT used an AlGaAs/GaAs material structure. These devices are higher performance than the GaAs MESFET (e.g., given an identical gate length, the AlGaAs/GaAs HEMT has an f_T about 50% to 100% higher, depending on the details of the device structure and quality of material). Correspondingly higher currents are achieved in the AlGaAs/GaAs HEMT devices.

Higher performance still is achieved using InP based HEMTs. For example, the $In_{0.53}Ga_{0.47}As/In_{0.52}Al_{0.48}As$ on InP lattice-matched HEMT have reported f_T numbers greater than 250 GHz with gate lengths of the order of 0.1 microns. Furthermore, such devices have I_{DSS} values approaching 1000 mA/mm and very high transconductances of greater than 1400 mS/mm.[16,18] These devices do have low breakdown voltages of the order of 1 or 2 V because of the small bandgap of InGaAs. Changing the stoichiometric ratios to $In_{0.14}Ga_{0.85}As/In_{0.14}Al_{0.30}As$ on a GaAs substrate produces a pseudomorphic HEMT structure. The $In_{0.14}Ga_{0.85}As$ is a strained layer when grown on GaAs. The use of strained layers gives the device designer more flexibility in accessing a wider variety of quantum wells depths and electronic properties.

Heterojunction Bipolar Structures

Practical heterojunction bipolar transistors (HBT) devices[13,15] are still evolving. Molecular beam epitaxy (MBE) is used to grow the doped layers making up the vertical semiconductor structure in the HBT. In fact, HBT structures were not really practical until the advent of MBE, although the idea behind the HBT goes back to around 1950 (Shockley). The vastly superior compositional control and layer thickness control with MBE is what made HEMTs and HBTs possible. The first HBT devices used an AlGaAs/GaAs junction with the wider bandgap AlGaAs layer forming the emitter region. Compound semiconductor HBT devices are typically mesa structures, as opposed to the more nearly planar structures used in silicon bipolar technology, because top surface contacts must be made to the collector, base, and emitter regions. Molecular beam epitaxy grows the stack of layers over the entire wafer, whereas, in silicon VLSI processes, selective implantations and oxide masking localize the doped regions. Hence, etching down to the respective layers allows for contact to the base and collector regions. An example of such a mesa HBT structure[14] is shown in Fig. 14.11. The HBT shown uses an InGaP emitter primarily for improved reliability over the AlGaAs emitter and a carbon-doped p^+ base GaAs layer.

Recently, InP-based HBT[15] have emerged as candidates for use in high-speed circuits. The two dominant heterojunctions are InP/InGaAs and AlInAs/InGaAs in InP devices. The small but significant bandgap difference between AlInAs directly on InP greatly limits its usefulness. InP-based HBT device structures are similar to those of GaAs-based devices and the reader is referred to Chapter 5 of Jalali and Pearton[10] for specific InP HBT devices. Generally, InP has advantages of lower surface recombination (higher current gain results), better electron transport, lower forward turn-on voltage, and higher substrate thermal conductivity.

HBT Performance

Typical current gain values in production-worthy HBT devices range from 50 at the low range to 140 at the high range. Cutoff frequency f_T values are usually quoted under the best (i.e., peak) bias conditions. For this reason f_T values must be carefully interpreted because in digital circuits, the bias state varies widely over the entire switching swing. For this reason, probably an averaged f_T value would be better, but it is difficult to determine. Typical f_T values for HBT processes in manufacturing (say 1998) are in

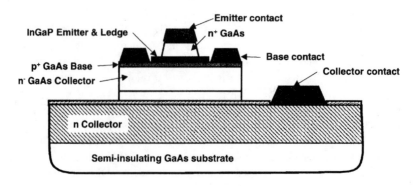

FIGURE 14.11 Cross-section of an HBT device with carbon-doped p^+ base and an InGaP emitter.[14] Note the commonly used mesa structure, where selective layer etching is required to form contacts to the base and collector regions.

the 50 to 140 GHz range. For example, for the HBT example in Fig. 14.11 with a 2 μm × 2 μm emitter f_T is approximately 65 GHz at a current density of 0.6 mA/μm^2 and its dc current gain is around 50. Of course, higher values for f_T have been reported for R&D or laboratory devices. In HBT devices, the parameter f_{max} is often lower than its f_T value (e.g., for the device in Fig. 14.11, f_{max} is about 75 GHz). Base resistance (refer to Table 14.2 for equation) is the dominant limiting factor in setting f_{max}. The best HBT devices have f_{max} values only slightly higher than their f_T values. In comparison, MESFET and HEMT devices typically have higher f_{max}/f_T ratios, although in digital circuits this may be of little importance.

Where the HBT really excels is in being able to generate much higher values of transconductance. This is a clear advantage in driving larger loading capacitances found in large integrated circuits. Biasing the HBT in the current range corresponding to the highest transconductance is essential to take advantage of the intrinsically higher transconductance.

References

1. Johnson, E. O. and Rose, A., Simple General Analysis of Amplifier Devices with Emitter, Control, and Collector Functions, *Proc. IRE*, 47, 407, 1959.
2. Cherry, E. M. and Hooper, D. E., *Amplifying Devices and Low-Pass Amplifier Design*, John Wiley & Sons, New York, 1968, Chap. 2 and 5.
3. Beaufoy, R. and Sparkes, J. J., The Junction Transistor as a Charge-Controlled Device, *ATE Journal*, 13, 310, 1957.
4. Shockley, W., *Electrons and Holes in Semiconductors*, Van Nostrand, New York, 1950.
5. Ferry, D. K., *Semiconductors*, Macmillan, New York, 1991.
6. Lundstrom, M., *Fundamentals of Carrier Transport*, Addison-Wesley, Reading, MA, 1990.
7. Sze, S. M., *Physics of Semiconductor Devices*, second ed., John Wiley & Sons, New York, 1981.
8. Yang, E. S., *Fundamentals of Semiconductor Devices*, McGraw-Hill, New York, 1978, Chap. 7.
9. Hollis, M. A. and Murphy, R. A., Homogeneous Field-Effect Transistors, *High-Speed Semiconductor Devices*, Sze, S. M., Ed., Wiley-Interscience, New York, 1990.
10. Pearton, S. J. and Shah, N. J., Heterostructure Field-Effect Transistors, *High-Speed Semiconductor Devices*, Sze, S. M., Ed., Wiley-Interscience, 1990, Chap. 5.
10a. Kroemer, H., Heterostructure Bipolar Transistors and Integrated Circuits, *Proc. IEEE*, 14, 13, 1982.
11. Muller, R. S. and Kamins, T. I., *Device Electronics for Integrated Circuits*, second ed., John Wiley & Sons, New York, 1986, Chap. 6 and 7.
12. Gray, P. E., Dewitt, D., Boothroyd, A. R., and Gibbons, J. F., *Physical Electronics and Circuit Models of Transistors*, John Wiley & Sons, New York, 1964, Chap. 7.
13. Asbeck, P. M., Bipolar Transistors, *High-Speed Semiconductor Devices*, S. M. Sze, Ed., John Wiley & Sons, New York, 1990, Chap. 6.
14. Low, T. S. et al., Migration from an AlGaAs to an InGaP Emitter HBT IC Process for Improved Reliability, presented at *IEEE GaAs IC Symposium Technical Digest*, Atlanta, GA, 143, 1998.
15. Jalali, B. and Pearson, S. J., *InP HBTs Growth, Processing and Applications*, Artech House, Boston, 1995.
16. Nguyen, L. G., Larson, L. E., and Mishra, U. K., Ultra-High-Speed Modulation-Doped Field-Effect Transistors: A Tutorial Review, *Proc. IEEE*, 80, 494, 1992.
17. Brews, J. R., The Submicron MOSFET, *High-Speed Semiconductor Devices*, Sze, S. M., Ed., Wiley-Interscience, New York, 1990, Chap. 3.
18. Nguyen, L. D., Brown, A. S., Thompson, M. A., and Jelloian, L. M., 50-nm Self-Aligned-Gate Pseudomorphic AlInAs/GaInAs High Electron Mobility Transistors, *IEEE Trans. Elect. Dev.*, 39, 2007, 1992.

15

Logic Design Principles and Examples

Stephen I. Long
*University of California
at Santa Barbara*

15.1 Introduction

The logic circuits used in high-speed compound semiconductor digital ICs must satisfy the same essential conditions for design robustness and performance as digital ICs fabricated in other technologies. The static or dc design of a logic cell must guarantee adequate voltage and/or current gain to restore the signal levels in a chain of similar cells. A minimum *noise margin* must be provided for tolerance against process variation, temperature, and induced noise from ground bounce, crosstalk, and EMI so that functional circuits and systems are produced with good electrical yield. Propagation delays must be determined as a function of loading and power dissipation.

Compound semiconductor designs emphasize speed, so logic voltage swings are generally low, τ_r low so that transconductances and f_T are high, and device access resistances are made as low as possible in order to minimize the lifetime time constant τ. This combination makes circuit performance very sensitive to parasitic R, L, and C, especially when the highest operation frequency is desired. The following sections will describe the techniques that can be used for static and dynamic design of high-speed logic.

15.2 Static Logic Design

A basic requirement for any logic family is that it must be capable of *restoring* the logic voltage or current swing. This means that the voltage or current gain with loading must exceed 1 over part of the transfer characteristic. Figure 15.1 shows a typical V_{out} vs. V_{in} *dc transfer characteristic* for a static ratioed logic inverter as is shown in the schematic diagram of Fig. 15.2. It can be seen that a chain of such inverters will restore steady-state logic voltage levels to V_{OL} and V_{OH} because the high-gain transition region around $V_{in} = V_{TH}$ will result in voltage amplification. Even if the voltage swing is very small, if centered on the *inverter threshold* voltage V_{TH}, defined as the intersection between the transfer characteristic and the $V_{in} = V_{out}$ line, the voltage will be amplified to the full swing again by each successive stage.

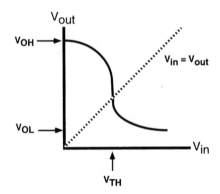

FIGURE 15.1   Typical voltage transfer characteristic for the logic inverter shown in Fig. 15.2.

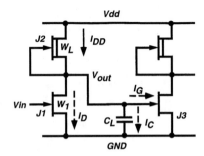

FIGURE 15.2   Schematic diagram of a direct-coupled FET logic (DCFL) inverter.

Ratioed logic implies that the *logic high and low voltages* V_{OH} and V_{OL} shown in Fig. 15.1 are a function of the widths W_1 and W_L of the FETs in the circuit shown in Fig. 15.2. In III-V technologies, this circuit is implemented with either MESFETs or HEMTs. The circuit in Fig. 15.2 is called *Direct Coupled FET Logic* or DCFL.

The logic levels of *non-ratioed* logic are independent of device widths. Non-ratioed logic typically occurs when the switching transistors do not conduct any static current. This is typical of logic families such as static CMOS or its GaAs equivalent *CGaAs*[2] which make use of complementary devices. Dynamic logic circuits such as precharged logic[1] and pass transistor logic[2,3] also do not require static current in pull-down chains. Such circuits have been used with GaAs FETs in order to reduce static power dissipation. They have not been used, however, for the highest speed applications.

Direct-Coupled FET Logic

DCFL is the most widely used logic family for the high-complexity, low-power applications that will be discussed in Chapter 16. The operation of DCFL shown in Fig. 15.2 is easily explained using a load line analysis. Currents are indicated by arrows in this figure. Solid arrows correspond to currents that are nearly constant. Dashed arrows represent currents that depend on the state of the output of the inverter. Figure 15.3 presents an I_D–V_{DS} characteristic of the enhancement mode (normally-off with threshold voltage $V_T > 0$) transistor J1. A family of characteristic curves is drawn representing several V_{GS} values. In this circuit, $V_{GS} = V_{in}$ and $V_{DS} = V_{out}$.

A load line representing the I–V characteristic of the active load ($V_{GS} = 0$) depletion-mode (normally-on with $V_T < 0$) transistor J2 is also superimposed on this drawing. Note that the logic low level V_{OL} is determined by the intersection of the two curves when $I_D = I_{DD}$. Load line 1 corresponds to a load device with narrow width; load line 2 for a wider device. It is evident that the narrow, weaker device will provide a lower V_{OL} value and thus will increase the logic swing. However, the weaker device will also have less

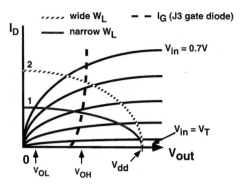

FIGURE 15.3 Drain current versus drain-source voltage characteristic of J1. The active load, J2, is also shown superimposed over the J1 characteristics as a load line. Two load lines corresponding to wide and narrow J2 widths are shown. In addition, the gate current I_G of J3 versus V_{out} limits the logic high voltage.

current available to drive any load capacitance, so the inverter with load line 1 will therefore be slower than the one with load line 2. There is therefore a tradeoff between speed and logic swing. So far, the analysis of this circuit is the same as that of an analogous nMOS E/D inverter.

In the case of DCFL logic inverters implemented with GaAs-based FETs, the Schottky barrier gate electrode of the next stage will limit the maximum value of V_{OH} to the forward voltage drop across the gate-source diode. This is shown by the gate diode I_G–V_{GS} characteristic also superimposed on Fig. 15.3. V_{OH} is given by the point of intersection between the load current I_{DD} and the gate current I_G, because a logic high output requires that the switch transistor J1 is off. V_{OH} will therefore also depend on the load transistor current. Effort must be made not to overdrive the gate since excess gate current will flow through the internal parasitic source resistance of the driven device J3, degrading V_{OL} of this next stage.

Source-Coupled FET Logic

A second widely used type of logic circuit — source-coupled FET Logic or SCFL is shown in Fig. 15.4. SCFL, or its bipolar counterpart, ECL, is widely used for very high-speed applications, which will be discussed in Chapter 16. The core of the circuit consists of a differential amplifier, J_1 and J_2, a current source J_3, and pull-up resistors R_L on the drains. The differential topology is beneficial for rejection of common-mode noise. The static design procedure can be illustrated again by a load-line analysis. A maximum current I_{CS} can flow through either J_1 or J_2.

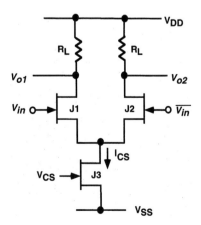

FIGURE 15.4 Schematic of differential pair J1,J2 used as a source-coupled FET logic (SCFL) cell.

Figure 15.5 shows the I_D–V_{DS} characteristic of J_1 for example. The maximum current I_{CS} is shown by a dotted line. The output voltage V_{O1} is either V_{DD} or $V_{DD} - I_{CS}R_L$; therefore, the maximum differential voltage swing, $\Delta V = 2\ I_{CS}R_L$, is determined by the choice of R_L. Next, the width of J_1 should be selected so that the change in V_{GS} needed to produce the voltage drop $I_{CS}R_L$ at the drain is less than $I_{CS}R_L$. This will ensure that the voltage gain is greater than 1 (needed to compensate for the source followers described below) and that the device is biased in its saturation region or cutoff at all times. The latter requirement is necessary if the maximum speed is to be obtained from the SCFL stage, since device capacitances are minimized in saturation and cutoff.

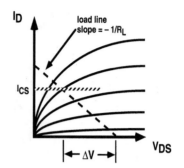

FIGURE 15.5 Load-line analysis of the SCFL inverter cell.

Source followers are frequently used on the output of an SCFL stage or at the inputs of the next stage. Figure 15.6(a) shows the schematic diagram of the follower circuit. The follower can serve two functions: level shifting and buffering of capacitive loads. When used as a level shifter, a negative or positive voltage offset can be obtained between input and output. The only requirement is that the V_{GS} of the source follower must be larger than the FET threshold voltage. If the source follower is at the output of an SCFL cell, it can be used as a buffer to reduce the sensitivity of delay to load capacitance or fanout.

The voltage gain of a source follower is always less than 1. This can be illustrated by another load-line analysis. Figure 15.6(b) presents the I_{D1}–V_{DS1} characteristic of the source follower FET, J1. A constant V_{GS1} is applied for every curve plotted in the figure. The load line (dashed line) of a depletion-mode, active current source J2 is also superimposed. In this circuit, the output voltage is $V_{out} = V_{DD} - V_{DS1}$. The V_{out} is determined by the intersection of the load line with the I_{D1} characteristic curves. The current of the pull-down current source is selected according to the amount of level shifting needed. A high current will result in a greater amount of level shift than a small load current. If the devices have high output resistance, and are accordingly very flat, very little change in V_{GS1} will be required to change V_{out} over the

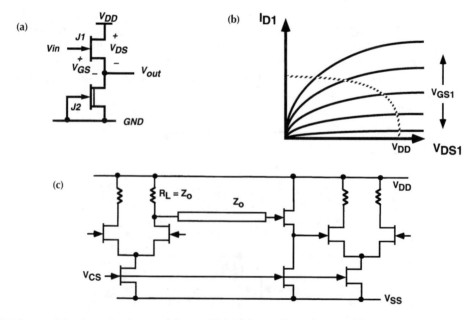

FIGURE 15.6 (a) Schematic of source follower, (b) load-line analysis of source follower, and (c) source follower buffer between SCFL stages.

full range from V_{OL} to V_{DD}. If V_{GS1} remains nearly constant, then V_{out} follows V_{in}, hence the name of the circuit. Since the input voltage to the source follower stage is $V_{in} = V_{GS1} + V_{out}$, a small change in V_{GS1} would produce an incremental voltage gain close to unity. If the output resistance is low, then the characteristic curves will slope upward and a larger range of V_{GS1} will be necessary to traverse the output voltage range. This condition would produce a low voltage gain. Small signal analysis shows that

$$A_v = \frac{1}{1 + 1 / \left[g_{m1} \left(r_{ds1} \| r_{ds2} \right) \right]} \tag{15.1}$$

The buffering effect of the source follower is accomplished by reducing the capacitive loading on the drain nodes of the differential amplifier because the input impedance of a source follower is high. Since the output tries to follow the input, the change in V_{GS} will be less than that required by a common source stage. Therefore, the input capacitance is dominated by C_{GD}, typically quite small for compound semiconductor FETs biased in saturation. The effective small-signal input capacitance is

$$C_G = C_{GD} + C_{GS} \left(1 - A_v \right) \tag{15.2}$$

where $A_v = dV_{out}/dV_{in}$ is the incremental voltage gain.

The source follower also provides a low output impedance, whose real part is approximately $1/g_m$ at low frequency. The current available to charge and discharge the load capacitance can be adjusted by the width ratio of J1 and J2. If the load is capacitive, V_{out} will be delayed from V_{in}. This will cause V_{GS1} to temporarily increase, providing excess current I_{D1} to charge the load capacitance. Ideally, for equal rise and fall times, the peak current available from J1 should equal the steady-state current of J2.

Source followers can also be used at the input of an SCFL stage to provide level shifting as shown in Fig. 15.6(c). In this case, the drain resistors, R_L, should be chosen to provide the proper termination resistance for the on-chip interconnect transmission line. These resistors provide a reverse termination to absorb signals that reflect from the high input impedance of the source follower. Alternatively, the drain resistors can be located at the gate of the source follower, thereby providing a shunt termination at the destination end of the interconnect. This practice results in good signal integrity, but because the practical values of characteristic impedance are less than 100 Ω, the current swing in the differential amplifier core must be large. This will increase power dissipation per stage.

SCFL logic structures generally employ more than one level of differential pairs so that complex logic functions (XOR, latch, and flip-flop) that require multiple gates to implement in logic families such as DCFL can be implemented in one stage. More details on SCFL gate structures and examples of their usage will be given in Chapter 16.

Static and Dynamic Noise Margin and Noise Sources

Noise margin is a measure of the ability of a logic circuit to provide proper functionality in the presence of noise.[4] There are many different definitions of noise margin, but a simple and intuitive one for the static or dc noise margin is illustrated by Fig. 15.8. Here, the transfer characteristic from Fig. 15.1 is plotted again. In order to evaluate the ability of a chain of such inverters to reject noise, a loop consisting of two identical inverters is considered. This might be representative of the positive feedback core of a bistable latch. Because the inverters are connected in a loop, an infinite chain of inverters is represented. The transfer characteristic of inverter 2 in Fig. 15.7 is plotted in gray lines. For inverter 2, $V_{out\,1} = V_{in2}$ and $V_{out2} = V_{in1}$. Therefore, the axes are reversed for the characteristic plotted for inverter 2. If a series noise source V_N were placed within the loop as shown, the maximum static noise voltage allowed will be represented by the *maximum width* of the loops formed by the transfer characteristic.[4] These widths, labeled V_{NL} and V_{NH}, respectively, for the low and high noise margins, are shown in the figure. If the

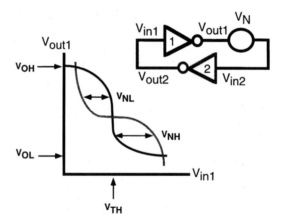

FIGURE 15.7   Voltage transfer characteristics of an inverter pair connected in a loop. Noise margins are shown as V_{NH} and V_{NL}.

voltage V_N exceeds V_{NL} or V_{NH}, the latch will be set into the opposite state and will remain there until reset. This would constitute a logic failure. Therefore, we must insist that any viable logic circuit provide noise margins well in excess of ambient noise levels in the circuit.

The static noise margin defined above utilized a dc voltage source V_N in series with logic inverters to represent static noise. This source might represent a static offset voltage caused by IR drop along IC power and ground distribution networks. The DCFL inverter, for example, would experience a shift in V_{TH} that is directly proportional to a ground voltage offset. This shift would skew the noise margins. The smallest noise margin would determine the circuit electrical yield. The layout of the power and ground distribution networks must consider this problem. The width of power and ground buses on-chip must be sufficient to guarantee a maximum IR drop that does not compromise circuit operation. It is important to note that this width is frequently much greater than what might be required by electromigration limits. It is essential that the designer consider IR drop in the layout. Some digital IC processes allow the topmost metal layer to form a continuous sheet, thereby minimizing voltage drops.

The static noise voltage source V_N might also represent static threshold voltage shifts on the active devices due to statistical process variation or backgating effects. Therefore, the noise margin must be several times greater than the variance in device threshold voltages provided by the fabrication process so that electrical yields will not be compromised.[5]

The above definition of maximum width noise margin has assumed a steady-state condition. It does not account for transient noise sources and the delayed response of the logic circuit to noise pulses. Unfortunately, pulses of noise are quite common in digital systems. For example, the ground potential can often be modified dynamically by simultaneous switching events on the IC chip.[6] Any ground distribution bus can be modeled as a transmission line with impedance Z_0 where

$$Z_0 = \sqrt{\frac{L_o}{C_o}} \tag{15.3}$$

Here, L_o is the equivalent series inductance per unit length and C_o the equivalent shunt capacitance per unit length. Since the interconnect exhibits a series inductance, there will be transient voltage noise ΔV induced on the line by current transients as predicted by

$$\Delta V = L\frac{dI}{dt} \tag{15.4}$$

This form of noise is often called *ground bounce*. The ground bounce ΔV is particularly severe when many devices are being switched synchronously, as would be the case in many applications involving flip-flops in shift registers or pipelined architectures. The high peak currents that result in such situations can generate large voltage spikes. For example, output drivers are well-known sources of noise pulses on power and ground buses unless they are carefully balanced with fully differential interconnections and are powered by power and ground pins separate from the central logic core of the IC.

Designing to minimize ground bounce requires minimization of inductance. Bakoglu[6] provides a good discussion of power distribution noise in high-speed circuits. There are several steps often used to reduce switching noise. First, it is standard practice to make extensive use of multiple ground pins on the chip to reduce bond-wire inductance and package trace inductance when conventional packaging is used. Bypass capacitance off-chip can be useful if it can be located inside the package and can have a high series resonant frequency. On-chip bypass capacitance is also helpful, especially if enough charge can be supplied from the capacitance to provide the current during clock transitions. The objective is to provide a low impedance between power and ground on-chip at the clock frequency and at odd multiples of the clock frequency. Finally, as mentioned above, high-current circuits such as clock drivers and output drivers should not share power and ground pins with other logic on-chip.

Crosstalk is another common source of noise pulses caused by electromagnetic coupling between adjacent interconnect lines. A signal propagating on a driven interconnect line can induce a crosstalk voltage and current in a coupled line. The duration of the pulse depends on the length of interconnect; the amplitude depends on the mutual inductance and capacitance between the coupled lines.[7]

In order to determine how much noise is acceptable in a logic circuit, the noise margin definition can be modified to accommodate the transient noise pulse situation. The logic circuit does not respond instantaneously to the noise pulse at the input. This delay in the response is attributed to the device and interconnect capacitances and the device current limitations which will be discussed extensively in Section 15.3. Consider the device input capacitance. Sufficient charge must be transferred during the noise pulse to the input capacitance to shift the control voltage either above or below threshold. In addition, this voltage must be maintained long enough for the output to respond if a logic upset is to occur. Therefore, a logic circuit can withstand a larger noise pulse amplitude than would be predicted from the static noise margin if the pulse width is much less than the propagation delay of the circuit. This increased noise margin for short pulses is called the dynamic noise margin (DNM). The DNM approaches the static NM if the pulse width is wide compared with the propagation delay because the circuit can charge up to the full voltage if not constrained by time.

The DNM can be predicted by simulation. Figure 15.8(a) shows the loop connection of the set-reset NOR latch similar to that which was used for the static NM definition in Fig.10.7. The inverter has been modified to become a NOR gate in this case. An input pulse train $V_1(t)$ of fixed duration but with gradually increasing amplitude can be applied to the set input. The latch was initialized by applying an initial condition to the reset input $V_2(t)$. The output response is observed for the input pulse train. At some input amplitude level, the output will be set into the opposite state. The latch will hold this state until it is reset again. The cross-coupled NOR latch thus becomes a logic upset detector, dynamically determining the maximum noise margin for a particular pulse width. The simulation can be repeated for other pulse widths, and a plot of the pulse amplitude that causes the latch to set for each pulse duration can be constructed, as shown in Fig. 15.8(b). Here, any amplitude or duration that falls on or above the curve will lead to a logic upset condition.

Power Dissipation

Power dissipation of a static logic circuit consists of a static and a dynamic component as shown below.

$$P_D = V_{DD}\bar{I}_{DD} + C_L \Delta V^2 \, f \eta \qquad (15.5)$$

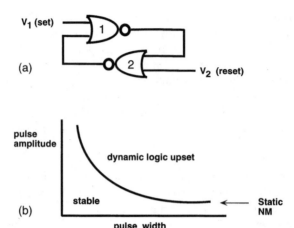

FIGURE 15.8   (a) Set-reset latch used to describe dynamic noise margin simulation, and (b) plot of the pulse amplitude applied to the set input in (a) that results in a logic upset.

In the case of DCFL, the current I_{DD} from the pull-up transistor J2 is relatively constant, flowing either in the pull-down (switch) device J1 or in the gate(s) of the subsequent stage(s). Taking its average value, the static power is $V_{DD}\overline{I_{DD}}$. The dynamic power $C_L\Delta V^2 f\eta$ depends on the frequency of operation f, the load capacitance C_L, and the duty factor η. η is application dependent. Since the voltage swing is rather small for the DCFL inverter under consideration (about 0.6 V for MESFETs), the dynamic power will not be significant unless the load capacitance is very large, such as in the case of clock distribution networks. The V_{DD} power supply voltage is traditionally 2 V because of compatibility with the bipolar ECL V_{TT} supply, but a V_{DD} as low as 1 V can be used for special low-power applications. Typical power dissipation per logic cell (inverter, NOR) depends on the choice of supply voltage and on I_{DD}. Power is typically determined based on speed and is usually in the range of 0.1 to 0.5 mW/gate. DCFL logic circuits are often used when the application requires high circuit density and very low power.

15.3 Transient Analysis and Design for Very-High-Speed Logic

Adequate attention must be given to static or dc design, as described in the previous section, in order to guarantee functionality under the worst-case situations. In addition, since the only reason to use the compound semiconductor devices for digital electronics at all is their speed, attention must be given to the dynamic performance as well. In this section, we will describe three methods for estimating the performance of high-speed digital logic circuit functional blocks. Each of these methods has its strengths and weaknesses.

The most effective methods for guiding the design are those that provide insight that helps to identify the dominant time constants that determine circuit performance. These are not necessarily the most accurate methods, but are highly useful because they allow the designer to determine what part of the circuit or device is limiting the speed. Circuit simulators are far more accurate (at least to the extent that the device models are valid), but do not provide much insight into performance limitations. Without simple analytical techniques to guide the design, performance optimization becomes a trial-and-error exercise.

Zero-Order Delay Estimate

The first technique, which uses the simple relationship between voltage and current in a capacitor,

$$I = C_L \frac{dV}{dt} \qquad (15.6)$$

is relevant when circuit performance is dominated by wiring or fan-out capacitance. This will be the case if the delay predicted by Eq. 15.6 due to the total loading capacitance, C_L, significantly exceeds the intrinsic delay of a basic inverter or logic gate. To apply this approach, determine the average current available from the driving logic circuit for charging (I_{LH}) and discharging (I_{HL}) the load capacitance. The logic swing ΔV is known, so low-to-high (t_{PLH}) and high-to-low (t_{PHL}) propagation delays can be determined from Eq. 15.6. These delays represent the time required to charge or discharge the circuit output to 50% of its final value. Thus, t_{PLH} is given by

$$t_{PLH} = \frac{C_L \Delta V}{2 I_{LH}} \qquad (15.7)$$

where I_{LH} is the average charging current during the output transition from V_{OL} to $V_{OL} + \Delta V/2$. The net propagation delay is given by

$$t_P = \frac{t_{PLH} + t_{PHL}}{2} \qquad (15.8)$$

At this limit, where speed is dominated by the ability to drive load capacitance, we see that increasing the currents will reduce t_P. In fact, the product of power (proportional to current) and delay (inversely proportional to current) is nearly constant under this situation. Increases in power lead to reduction of delay until the interconnect distributed RC delays or electromagnetic propagation delays become comparable to t_P.

The equation also shows that small voltage swing ΔV is good for speed if the noise margin and drive current are not compromised. This means that the devices must provide high transconductance.

For example, the DCFL inverter of Fig. 15.2 can be analyzed. Figure 15.9 shows equivalent circuits that represent the low-to-high and high-to-low transitions. The current available for the low-to-high transition, I_{PLH}, shown in Fig. 15.9(a), is equal to the average pullup current, I_{DD}. If we assume that $V_{OL} = 0.1$ V and $V_{OH} = 0.7$ V, then the ΔV of interest is 0.6 V. This brings the output up to 0.4 V at $V_{50\%}$. In this range of V_{out}, the active load transistor J2 is in saturation at all times for $V_{DD} > 1$ V, so I_{DD} will be relatively constant, and all of the current will be available to charge the capacitor.

The high-to-low transition is more difficult to model in this case. V_{out} will begin at 0.7 V and discharge to 0.4 V. The discharge current through the drain of J1 is going to vary with time because the device is below saturation over this range of V_{out}. Looking at the Vin = 0.7 V characteristic curve in Fig. 15.3, we see that its I_D–V_{DS} characteristic is resistive. Let's approximate the slope by $1/R_{on}$. Also, the discharge current I_{PHL} is the difference between I_{DD} and $I_{D1} = V_{out}/R_{on}$, as shown in Fig. 15.9(b). The average current available to discharge the capacitor can be estimated by

$$I_{HL} = \frac{V_{OH} + V_{50\%}}{2 R_{on}} - I_{DD} \qquad (15.9)$$

Then, t_{PHL} is estimated by

$$t_{PHL} = \frac{C_L \Delta V}{2 I_{HL}} \qquad (15.10)$$

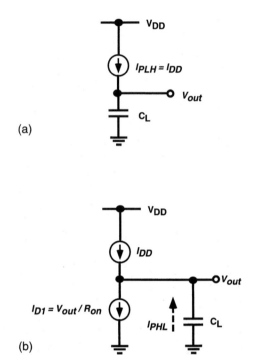

FIGURE 15.9   (a) Equivalent circuit for low-to-high transition; and (b) equivalent circuit for high-to-low transition.

Time Constant Delay Methods: Elmore Delay and Risetime

Time constant delay estimation methods are very useful when the wiring capacitance is quite small or the charging current is quite high. In this situation, typical of very-high-speed SSI and MSI circuits that push the limits of the device and process technology, the circuit delays are dominated by the devices themselves. Both methods to be described rely on a large-signal equivalent circuit model of the transistors, an approximation dubious at best. But, the objective of these techniques is not absolute accuracy. That is much less important than being able to identify the dominant contributors to the delay and risetime, since more accurate but less intuitive solutions are easily available through circuit simulation. The construction of the large signal equivalent circuit requires averaging of non-linear model elements such as transconductance and certain device capacitances over the appropriate voltage swing.

The propagation delay definition described above, the delay required to reach 50% of the logic swing, must be relaxed slightly to apply methods based on linear system analysis. It was first shown by Elmore in 1948[8] and apparently rediscovered by Ashar in 1964[9] that the delay time t_D between an impulse function $\delta(0)$ applied at $t = 0$ to the input of a network and the centroid or "center-of-mass" of the impulse response (output) is quite close to the 50% delay. This definition of delay t_D is illustrated in Fig. 15.10. Two conditions must be satisfied in order to use this approach. First, the step response of the network is monotonic. This implies that the impulse response is purely a positive function. Monotonic step response is valid only when the circuit poles are all negative and real, or the circuit is heavily damped. Due to feedback through device capacitances, this condition is seldom completely correct. Complex poles often exist. However, strongly underdamped circuits are seldom useful for reliable logic circuits because their transient response will exhibit ringing, so efforts to compensate or damp such oscillations are needed in these cases anyway. Then, the circuit becomes heavily damped or at least dominated by a single pole and fits the above requirement more precisely.

Second, the correspondence between t_D and t_{PLH} is improved if the impulse response is symmetric in shape, as in Fig. 15.10(b). It is shown in Ref. 9 that cascaded stages with similar time constants have a tendency to approach a Gaussian-shaped distribution as the number of stages becomes large. Most logic systems require several cascaded stages, so this condition is often true as well.

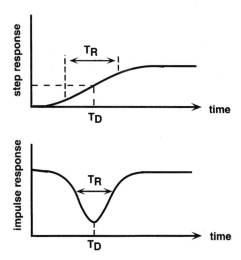

FIGURE 15.10 (a) Monotonic step response of a network; (b) corresponding impulse response. The delay t_D is defined as the centroid of the impulse response.

Assuming that these conditions are approximately satisfied, we can make use of the fact that the impulse response of a circuit in the frequency domain is given by its transfer (or network) function F(s) in the complex frequency $s = \sigma + j\omega$. Then, the propagation delay, t_D, can be determined by

$$
t_D = \frac{\int_0^\infty tf(t)dt}{\int_0^\infty f(t)dt} = \lim_{s \to 0} \frac{\int_0^\infty tf(t)e^{-st}dt}{\int_0^\infty f(t)e^{-st}dt} = \left[\frac{-\dfrac{d}{ds}F(s)}{F(s)}\right]_{s=0}
$$
(15.11)

Fortunately, the integration never needs to be performed. t_D can be obtained directly from the network function F(s) as shown. But, the network function must be calculated from the large-signal equivalent circuit of the device, including all important parasitics, driving impedances, and load impedances. This is notoriously difficult if the circuit includes a large number of capacitances or inductances.

Fortunately, in most cases, circuits of interest can be subdivided into smaller networks, cascaded, and the presumed linearity of the circuits can be employed to simplify the task. In addition, the evaluation of the function at $s = 0$ eliminates many terms in the equations that result. In particular, Tien[10] shows that two corollaries are particularly useful in cascading circuit blocks:

1. If the network function F(s) = A(s)/B(s), then

$$
t_D = \left[\frac{-\dfrac{d}{ds}A(s)}{A(s)}\right]_{s=0} + \left[\frac{\dfrac{d}{ds}B(s)}{B(s)}\right]_{s=0}
$$
(15.12)

2. If F(s) = A(s)B(s)C(s), then

$$
t_D = \left[\frac{-\dfrac{d}{ds}A(s)}{A(s)}\right]_{s=0} + \left[\frac{-\dfrac{d}{ds}B(s)}{B(s)}\right]_{s=0} + \left[\frac{-\dfrac{d}{ds}C(s)}{C(s)}\right]_{s=0}
$$
(15.13)

This shows that the total delay is just the sum of the individual delays of each circuit block. When computing the network functions, care must be taken to include the driving point impedance of the previous stage and to represent the previous stage as a Thevenin-equivalent open-circuit voltage source. A good description and illustration of the use of this approach in the analysis of bipolar ECL and CML circuits can be found in Ref. 10.

Risetime: the standard definition of risetime is the 10 to 90% time delay of the step response of a network. While convenient for measurement, this definition is analytically unpleasant to derive for anything except simple, first-order circuits. Elmore demonstrated that the standard deviation of the impulse response could be used to estimate the risetime of a network.[8] This definition provides estimates that are close to the standard definition. The standard deviation of the impulse response can be calculated using

$$T_R^2 = 2\pi \left[\int_0^\infty t^2 f(t) dt - t_D^2 \right] \tag{15.14}$$

Since the impulse response frequently resembles the Gaussian function, the integral is easily evaluated.

Once again, the integration need not be performed. Lee[11] has pointed out that the transform techniques can also be used to obtain the Elmore risetime directly from the network function F(s).

$$T_R^2 = 2\pi \left[\frac{\dfrac{d^2}{ds^2} F(s)}{F(s)} \right]_{s=0} - 2\pi \left[\frac{\dfrac{d}{ds} F(s)}{F(s)} \right]_{s=0}^2 \tag{15.15}$$

This result can also be used to show that the risetimes of cascaded networks add as the square of the individual risetimes. If two networks are characterized by risetimes T_{R1} and T_{R2}, the total risetime $T_{R,total}$ is given by the RMS sum of the individual risetimes

$$T_{R,total}^2 = T_{R1}^2 + T_{R2}^2 \tag{15.16}$$

Time Constant Methods: Open-Circuit Time Constants

The frequency domain/transform methods for finding delay and risetime are particularly valuable for design optimization because they identify dominant time constants. Once the time constants are found, the designer can make efforts to change biases, component values, or optimize the design of the transistors themselves to improve the performance through addressing the relevant bottleneck in performance. The drawback in the above technique is that a network function must be derived. This becomes tedious and time-consuming if the network is of even modest complexity. An alternate technique was developed[12,13] that also can provide reasonable estimates for delay, but with much less computational difficulty. The open-circuit time constant (OCTC) method is widely used for the analysis of the bandwidth of analog electronic circuits just for this reason. It is just as applicable for estimating the delay of very-high-speed digital circuits.

The basis for this technique again comes from the transfer or network function F(s) = Vo(s)/Vi(s). Considering transfer functions containing only poles, the function can be written as

$$F(s) = \frac{a_0}{b_n s^n + b_{n-1} s^{n-1} + \cdots + b_1 s + 1} \tag{15.17}$$

The denominator comes from the product of n factors of the form $(\tau_j s + 1)$, where τ_j is the time constant associated with the *j*-th pole in the transfer function. The b_1 coefficient can be shown to be equal to the sum

$$b_1 = \sum_{j=1}^{n} \tau_j \qquad (15.18)$$

of the time constants and b_2 the product of all the time constants. Often, the first-order term dominates the frequency response. In this case, the 3-dB bandwidth is then estimated by $\omega_{3dB} = 1/b_1$. The higher-order terms are neglected. The accuracy of this approach is good, especially when the circuit has a dominant pole. The worst error would occur when all poles have the same frequency. The error in this case is about 25%. Much worse errors can occur however if the poles are complex or if there are zeros in the transfer function as well. We will discuss this later.

Elmore has once again provided the connection we need to obtain delay and risetime estimates from the transfer function. The Elmore delay is given by

$$D = b_1 - a_1 \qquad (15.19)$$

where a_1 is the corresponding coefficient of the first-order zero (if any) in the numerator. The risetime is given by

$$T_R^2 = b_1^2 - a_1^2 + 2\left(a_2 - b_2\right) \qquad (15.20)$$

In Eq. 15.20, a_2 and b_2 correspond to the coefficients of the second-order zero and pole, respectively.

At this point, it would appear that we have gained nothing since finding that the time constants associated with the poles and zeros is well known to be difficult. Fortunately, it is possible to obtain the b_1 and b_2 coefficients directly by a much simpler method: open-circuit time constants. It has been shown that[11,12]

$$b_1 = \sum_{j=1}^{n} R_{jo} C_j = \sum_{j=1}^{n} \tau_{jo} \qquad (15.21)$$

that is, the sum of the time constants τ_{jo}, defined as the product of the effective open-circuit resistance R_{jo} across each capacitor C_j when all other capacitors are open-circuited, equals b_1. These time constants are very easy to calculate since open-circuiting all other capacitors greatly simplifies the network by decoupling many other components. Dependent sources must be considered in the calculation of the R_{jo} open-circuit resistances. Note that these open-circuit time constants are not equal to the pole time constants, but their sum gives the same result for b_1. It should also be noted that the individual OCTCs give the time constant of the network if the j-th capacitor were the only capacitor. Thus, each time constant provides information about the relative contribution of that part of the circuit to the bandwidth or the delay.[11] If one of these is much larger than the rest, this is the place to begin working on the circuit to improve its speed.

The b_2 coefficient can also be found by a similar process,[14] taking the sum of the product of time constants of all possible pairs of capacitors. For example, in a three-capacitor circuit, b_2 is given by

$$b_2 = R_{1o} C_1 R_{2s}^1 C_2 + R_{1o} C_1 R_{3s}^1 C_3 + R_{2o} C_2 R_{3s}^2 C_3 \qquad (15.22)$$

where the R_{js}^i resistance is the resistance across capacitor Cj calculated when capacitor C_i is *short*-circuited and all other capacitors are open-circuited. The superscript indicates which capacitor is to be shorted.

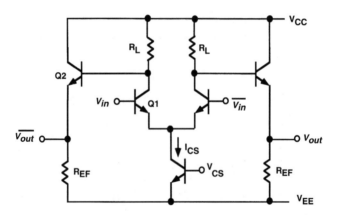

FIGURE 15.11   Schematic of basic ECL inverter.

So, R_{3s}^2 is the resistance across C_3 when C_2 is short-circuited and C_1 is open-circuited. Note that the first time constant in each product is an open-circuit time constant that has already been calculated. In addition, for any pair of capacitors in the network, we can find an OCTC for one and a SCTC for the other. The order of choice does not matter because

$$R_{io}C_iR_{js}^iC_j = R_{jo}C_jR_{is}^jC_i \qquad (15.23)$$

so we are free to choose whichever combination minimizes the computational effort.[14]

At this stage, it would be helpful to illustrate the techniques described above with an example. An ECL inverter whose schematic is shown in Fig. 15.11 is selected for this purpose. The analysis is based on work described in more detail in Ref. 15.

The first step is to construct the large-signal equivalent circuit. We will discuss how to evaluate the large-signal component values later. Figure 15.12 shows such a model applied to the ECL inverter, where

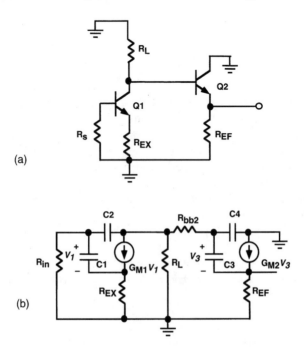

FIGURE 15.12   (a) Large-signal half-circuit model of ECL inverter; and (b) large-signal equivalent circuit of (a).

the half-circuit approximation has been used in Fig. 15.12(a) due to the inherent symmetry of differential circuits.[16] The hybrid-pi BJT model shown in Fig. 15.12(b) has been used with several simplifications. The dynamic input resistance, r_π, has been neglected because other circuit resistances are typically much smaller. The output resistance, r_o, has also been neglected for the same reason. The collector-to-substrate capacitance, C_{CS}, has been neglected because in III-V technologies, semi-insulating substrates are typically used. The capacitance to substrate is quite small compared to other device capacitances. Retained in the model are resistances R_{bb}, the extrinsic and intrinsic base resistance, and R_{EX}, the parasitic emitter resistance. Both of these are very critical for optimizing high-speed performance.

In the circuit itself, R_{IN} is the sum of the driving point resistance from the previous stage, probably an emitter follower output, and R_{bb1} of Q_1. R_L is the collector load resistor, whose value is determined by half of the output voltage swing and the dc emitter current, I_{CS}. $R_L = \Delta V/2I_{CS}$. The R_{EX} of the emitter follower is included in R_{EF}.

We must calculate open-circuit time constants for each of the four capacitors in the circuit. First consider C_1, the base-emitter diffusion and depletion capacitance of Q_1. C_2 is the collector-base depletion capacitance of Q_1. C_3 and C_4 are the corresponding base-emitter and base-collector capacitances of Q_2. Figure 15.13 represents the equivalent circuit schematic when $C_2 = C_3 = C_4 = 0$. A test source, V_1, is placed at the C_1 location. $R_{1o} = V_1/I_1$ is determined by circuit analysis to be

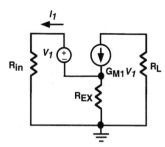

$$R_{1o} = \frac{R_{IN} + R_{EX}}{1 + G_{M1}R_{EX}} \quad (15.24)$$

FIGURE 15.13 Equivalent large-signal half-circuit model for calculation of R_{1o}.

Table 15.1 shows the result of similar calculations for R_{2o}, R_{3o}, and R_{4o}. The b_1 coefficient (first-order estimate of t_D) can now be found from the sum of the OCTCs:

$$b_1 = R_{1o}C_1 + R_{2o}C_2 + R_{3o}C_3 + R_{4o}C_4 \quad (15.25)$$

Considering the results in Table 15.1, one can see that there are many contributors to the time constants and that it will be possible to determine the dominant terms after evaluating the model and circuit parameters.

Next, estimates must be made of the non-linear device parameters, G_{Mi} and C_i. The large signal transconductances can be estimated from

$$G_M = \frac{\Delta I_C}{\Delta V_{BE}} \quad (15.26)$$

For the half-circuit model of the differential pair, the current ΔI_C is the full value of I_{CS} since the device switches between cutoff and I_{CS}. The ΔV_{BE} corresponds to the input voltage swing needed to switch the device between cutoff and I_{CS}. This is on the order of $3V_T$ (or 75 mV) for half of a differential input. So, $G_{M1} = I_{CS}/0.075$ is the large-signal estimate for transconductance of Q_1.

The emitter follower Q_2 is biased at I_{EF} when the output is at V_{OL}. Let us assume that an identical increase in current, I_{EF}, will provide the logic swing needed on the

TABLE 15.1 Effective Zero-Frequency Resistances for Open-Circuit Time-Constant Calculation for the Circuit of Fig. 15.12 $(G'_{Mi} = G_{Mi}/(1 + G_{Mi} R_{EX}))$

R_{1o}	$\dfrac{R_{IN} + R_{EX}}{1 + G_{M1}R_{EX}}$
R_{2o}	$R_{IN} + R_L + G'_{M1}R_{IN}R_L$
R_{3o}	$\dfrac{R_{bb} + R_L + R_{EF}}{1 + G_{M2}R_{EF}}$
R_{4o}	$R_{bb} + R_L$

output of the inverter to reach V_{OH}. Thus, $\Delta I_{C2} = I_{EF}$ and $R_{EF} = (V_{OH} - V_{OL})/I_{EF}$. The difference in V_{BE} at the input required to double the collector current can be calculated from

$$\Delta V_{BE} = V_T \ln(2) = 0.7V_T = 17.5\,\text{mV} \tag{15.27}$$

Thus, $G_{M2} = I_{EF}/0.0175$.

C_1 and C_3 consist of the parallel combination of the *depletion (space charge) layer capacitance*, C_{be}, and the *diffusion capacitance*, C_D. C_2 and C_4 are the base-collector depletion capacitances. Depletion capacitances are voltage varying according to

$$C(V) = C(0)\left(1 - \frac{V}{\phi}\right)^{-m} \tag{15.28}$$

where $C(0)$ is the capacitance at zero bias, ϕ is the built-in voltage, and m the grading coefficient. An equivalent large-signal capacitance can be calculated by

$$C = \frac{Q_2 - Q_1}{V_2 - V_1} \tag{15.29}$$

Q_i is the charge at the initial (1) or final (2) state corresponding to the voltages V_i. $Q_2 - Q_1 = \Delta Q$ and

$$\Delta Q = \int_{V_1}^{V_2} C(V)\,dV \tag{15.30}$$

The large-signal diffusion capacitance can be found from

$$C_D = G_M \tau_f \tag{15.31}$$

where τ_f is the forward transit delay (τ_r) as defined in Section 14.2.

Finally, the Elmore risetime estimate requires the calculation of b_2. Since there are four capacitors in the large-signal equivalent circuit, six terms will be necessary:

$$b_2 = R_{1o}C_1 R_{2s}^1 C_2 + R_{1o}C_1 R_{3s}^1 C_3 + R_{1o}C_1 R_{4s}^1 C_4 + R_{2o}C_2 R_{3s}^2 C_3 + R_{2o}C_2 R_{4s}^2 C_4 + R_{3o}C_3 R_{4s}^3 C_4 \tag{15.32}$$

R_{2s}^1 will be calculated to illustrate the procedure. The remaining short-circuit equivalent resistances are shown in Table 15.2. Referring to Fig. 15.14, the equivalent circuit for calculation of R_{2s}^1 is shown. This is the resistance seen across C_2 when C_1 is shorted. If C_1 is shorted, $V_1 = 0$ and the dependent current source is dead. It can be seen from inspection that

$$R_{2s}^1 = R_{IN} \| R_{EX} + R_L \tag{15.33}$$

TABLE 15.2   Effective Resistances for Short Circuit Time Constant Calculation for the Circuit of Fig. 15.12

R^1_{2s}	$R_{IN} \| R_{EX} + R_L$
R^1_{3s}	R_{3o}
R^1_{4s}	R_{4o}
R^2_{3s}	$\dfrac{\left(\dfrac{1}{G'_{M1}} \| R_{IN} \| R_L\right) + R_{bb} + R_{EF}}{1 + G_{M2} R_{EF}}$
R^2_{4s}	$\left(\dfrac{1}{G'_{M1}} \| R_{IN} \| R_L\right) + R_{bb}$
R^3_{4s}	$\left(R_L + R_{bb}\right) \| R_{EF}$

Time Constant Methods: Complications

As attractive as the time constant delay and risetime estimates are computationally, the user must beware of complications that will degrade the accuracy by a large margin. First, consider that both methods have depended on a restrictive assumption regarding monotonic risetime. In many cases, however, it is not unusual to experience complex poles. This can occur due to feedback which leads to inductive input or output impedances and emitter or source followers which also have inductive output impedance. When combined with a predominantly capacitive input impedance, complex poles will generally result unless the circuit is well damped. The time constant methods ignore the complex pole effects which can be quite significant if the poles are split and $\sigma \ll j\omega$. In this case, the circuit transient response will exhibit ringing, and time constant estimates of bandwidth, delay, and risetime will be in serious error. Of course, the ringing will show up in the circuit simulation, and if present, must be dealt with by adding damping resistances at appropriate locations.

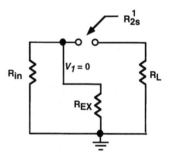

FIGURE 15.14   Equivalent circuit model for calculation of $R_{2s}{}^1$.

An additional caution must be given for circuits that include zeros. Although Elmore's equations can modify the estimates for t_D and T_R when there are zeros, the OCTC method provides no help in finding the time constants of these zeros. Zeros often occur in wideband amplifier circuits that have been modified through the addition of inductance for shunt peaking, for example. The addition of inductance, either intentionally or accidentally, can also produce complex pole pairs. Zeros are intentionally added for the optimization of speed in very-high-speed digital ICs as well; however, the large area required for the spiral inductors when compared with the area consumed by active devices tends to discourage the use of this method in all but the simplest (and fastest) designs.[11]

References

1. Yuan, J.-R. and Svensson, C., High-Speed CMOS Circuit Technique, *IEEE J. Solid-State Circuits*, 24, 62, 1989.
2. Weste, N. H. E. and Eshraghian, K., *Principles of CMOS VLSI Design — A Systems Perspective*, second ed., Addison-Wesley, Reading, MA, 1993.
3. Rabaey, J. M., *Digital Integrated Circuits: A Design Perspective*, Prentice-Hall, New York, 1996.
4. Hill, C. F., Noise Margin and Noise Immunity in Logic Circuits, *Microelectronics*, 1, 16, 1968.
5. Long, S. and Butner, S., *Gallium Arsenide Digital Integrated Circuit Design*, McGraw-Hill, New York, 1990, Chap. 3.
6. Bakoglu, H. B., *Circuits, Interconnections, and Packaging*, Addison-Wesley, Reading, MA, 1990, Chap. 7.
7. Long, S. and Butner, S., *Gallium Arsenide Digital Integrated Circuit Design*, McGraw-Hill, New York, 1990, Chap. 5.
8. Elmore, W. C., The Transient Response of Damped Linear Networks with Particular Regard to Wideband Amplifiers, *J. Appl Phys.*, 19, 55, 1948.
9. Ashar, K. G., The Method of Estimating Delay in Switching Circuits and the Fig. of Merit of a Switching Transistor, *IEEE Trans. Elect. Dev.*, ED-11, 497, 1964.
10. Tien, P. K., Propagation Delay in High Speed Silicon Bipolar and GaAs HBT Digital Circuits, *Int. J. High Speed Elect.*, 1, 101, 1990.
11. Lee, T. H., *The Design of CMOS Radio-Frequency Integrated Circuits*, Cambridge Univ. Press, Cambridge, U.K., 1998, Chap. 7.
12. Gray, P. E. and Searle, C. L., *Electronic Principles: Physics, Models, and Circuits*, John Wiley & Sons, New York, 1969, 531.
13. Gray, P. and Meyer, R., *Analysis and Design of Analog Integrated Circuits*, 3rd ed., John Wiley & Sons, New York, 1993, Chap. 7.
14. Millman, J. and Grabel, A., *Microelectronics*, second ed., McGraw-Hill, New York, 1987, 482.
15. Hurtz, G. M., Applications and Technology of the Transferred-Substrate Schottky-Collector Heterojunction Bipolar Transistor, M.S. Thesis, University of California, Santa Barbara, 1995.
16. Gray, P. and Meyer, R., *Analysis and Design of Analog Integrated Circuits*, 3rd ed., John Wiley & Sons, New York, 1993, Chap. 3.

16

Logic Design Examples

Charles E. Chang
Conexant Systems, Inc.

Meera Venkataraman
Troika Networks, Inc.

Stephen I. Long
*University of California
at Santa Barbara*

16.1 Design of MESFET and HEMT Logic Circuits

The basis of dc design, definition of logic levels, noise margin, and transfer characteristics were discussed in Chapter 16 using a DCFL and SCFL inverter as examples. In addition, methods for analysis of high-speed performance of logic circuits were presented. These techniques can be further applied to the design of GaAs MESFET, HEMT, or P-HEMT logic circuits with depletion-mode, enhancement-mode, or mixed E/D FETs. Several circuit topologies have been used for GaAs MESFETs, like direct-coupled FET logic (DCFL), source-coupled FET logic (SCFL), as well as dynamic logic families,[1] and have been extended for use with heterostructure FETs. Depending on the design requirements, whether it be high speed or low power, the designer can adjust the power-delay product by choosing the appropriate device technology and circuit topology, and making the correct design tradeoffs.

Direct-Coupled FET Logic (DCFL)

Among the numerous GaAs logic families, DCFL has emerged as the most popular logic family for high-complexity, low-power LSI/VLSI circuit applications. DCFL is a simple enhancement/depletion-mode GaAs logic family, and the circuit diagram of a DCFL inverter was shown in Fig. 16.2. DCFL is the only static ratioed GaAs logic family capable of VLSI densities due to its compactness and low power dissipation. An example demonstrating DCFL's density is Vitesse Semiconductor's 350K sea-of-gates array. The array uses a two-input DCFL NOR as the basic logic structure. The number of usable gates in the array is 175,000. A typical gate delay is specified at 95 ps with a power dissipation of 0.59 mW for a buffered two-input NOR gate with a fan-out of three, driving a wire load of 0.51 mm.[2] However, a drawback of DCFL is its low noise margin, the logic swing being approximately 600 mV. This makes the logic sensitive to changes in threshold voltage and ground bus voltage shifts.

DCFL NOR and NAND Gate

The DCFL inverter can easily be modified to perform the NOR function by placing additional enhancement-mode MESFETs in parallel as switch devices. A DCFL two-input NOR gate is shown in Fig. 16.1. If any input rises to V_{OH}, the output will drop to V_{OL}. If n inputs are high simultaneously, then V_{OL} will be decreased because the width ratio W_1/W_L in Fig. 16.2 has effectively increased by a factor of n. There is a limit to the number of devices that can be placed in parallel to form very wide NOR functions. The

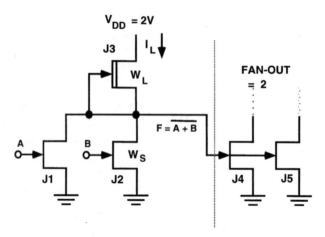

FIGURE 16.1   DCFL two-input NOR gate schematic.

drain capacitance will increase in proportion to the number of inputs, slowing down the risetime of the gate output. Also, the subthreshold current contribution from n parallel devices could become large enough to degrade V_{OH}, and therefore the noise margin. This must be evaluated at the highest operating temperature anticipated because the subthreshold current will increase exponentially with temperature according to:[3,4]

$$I_D = I_S \left[1 - \exp\left(\frac{cV_{DS}}{V_T} \right) \right] \left[\exp\left(\frac{bV_{DS}}{V_T} \right) \right] \left[\exp\left(\frac{aV_{GS}}{V_T} \right) \right] \tag{16.1}$$

The parameters a, b, and c are empirical fitting parameters. The first term arises from the diffusion component of the drain current which can be fit from the subthreshold I_D–V_{DS} characteristic at low drain voltage. The second and third terms represent thermionic emission of electrons over the channel barrier from source to drain. The parameters can be obtained by fitting the subthreshold I_D–V_{DS} and I_D–V_{GS} characteristics, respectively, measured in saturation.[5] For the reasons described above, the fan-in of the DCFL NOR is seldom greater than 4.

In addition to the subthreshold current loading, the forward voltage of the Schottky gate diode of the next stage drops with temperature at the rate of approximately –2mV/degree. Higher temperature operation will therefore reduce V_{OH} as well, due to this thermodynamic effect.

A NAND function can also be generated by placing enhancement-mode MESFETs in series rather than in parallel for the switch function. However, the low voltage swing inherent in DCFL greatly limits the application of the NAND function because V_{OL} will be increased by the second series transistor unless the widths of the series devices are increased substantially from the inverter prototype. Also, the switching threshold V_{TH} shown in Fig. 16.1 will be slightly different for each input even if width ratios are made different for the two inputs. The combination of these effects reduces the noise margin even further, making the DCFL NAND implementation generally unsuitable for VLSI applications.

Buffering DCFL Outputs

The output (drain) node of a DCFL gate sources and sinks the current required to charge and discharge the load capacitance due to wiring and fan-out. Excess propagation delay of the order of 5 ps per fan-out is typically observed for small DCFL gates. Sensitivity to wiring capacitance is even higher, such that unbuffered DCFL gates are never used to drive long interconnections unless speed is unimportant. Therefore, an output buffer is frequently used in such cases or when fan-out loading is unusually high.

The superbuffer shown in Fig. 16.2(a) is often used to improve the drive capability of DCFL. It consists of a source follower J_3 and pull-down J_4. The low-to-high transition begins when $V_{IN} = V_{OL}$. J_4 is cut off

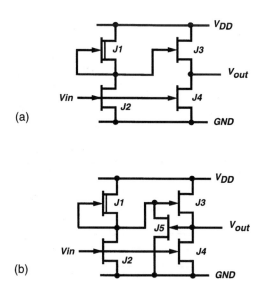

FIGURE 16.2 (a). Superbuffer schematic, and (b) modified superbuffer with clamp transistor. J5 will limit the output current when $V_{out} > 0.7$ V.

and J_3 becomes active, driving the output to V_{OH}. V_{OUT} follows the DCFL inverter output. For the output high-to-low transition, J_4 is driven into its linear region, and the output can be pulled to $V_{OL} = 0$ V in steady state. J_3 is cut off when the DCFL output (drain of J_1) switches from high to low. Since this occurs one propagation delay after the input switched from low-to-high, it is during this transition that the superbuffer can produce a current spike between V_{DD} and ground. J_4 attempts to discharge the load capacitance before the DCFL gate output has cut off J_3. Thus, superbuffers can become an on-chip noise source, so ground bus resistance and inductance must be controlled.

There is also a risk that the next stage might be overdriven with too much input current when driven by a superbuffer. This could happen because the source follower output is capable of delivering high currents when its V_{GS} is maximum. This occurs when $V_{out} = V_{OH} = 0.7$ V, limited by forward conduction of the gate diodes being driven. For a supply voltage of 2 V, a maximum $V_{GS} = 0.7$ V is easily obtained on J_3, leading to the possibility of excess static current flowing into the gates. This would degrade V_{OL} of the subsequent stage due to voltage drop across the internal source resistance. Figure 16.2(b) shows a modified superbuffer design that prevents this problem through the addition of a clamp transistor, J_5. J_5 limits the gate potential of J_3 when the output reaches V_{OH}, thus preventing the overdriving problem.

Source-Coupled FET Logic (SCFL)

SCFL is the preferred choice for very-high-speed applications. An SCFL inverter, a buffered version of the basic differential amplifier cell shown in Fig. 15.4, is shown in Fig. 16.3. The high-speed capability of SCFL stems from four properties of this logic family: small input capacitance, fast discharging time of the differential stage output nodes, good drive capability, and high F_t.

In addition to higher speed, SCFL is characterized by high functional equivalence and reduced sensitivity to threshold voltage variations.[5a] The current-mode approach used in SCFL ensures an almost constant current consumption from the power supplies and, therefore, the power supply noise is greatly reduced as compared to other logic families. The differential input signaling also improves the dc, ac, and transient characteristics of SCFL circuits.[6]

SCFL, however, has two drawbacks. First, SCFL is a low-density logic family due to the complex gate topology. Second, SCFL dissipates more power than DCFL, even with the high functional equivalence taken into account.

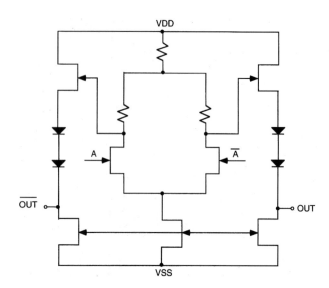

FIGURE 16.3   Schematic diagram of SCFL inverter with source follower output buffering.

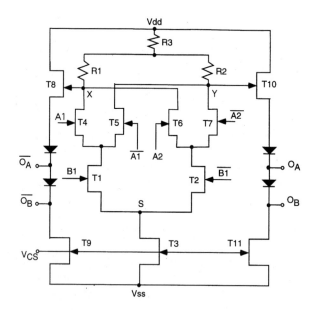

FIGURE 16.4   SCFL two-level series-gated circuit.

SCFL Two-Level Series-Gated Circuit

A circuit diagram of a two-level series-gated SCFL structure is shown in Fig. 16.4. 2-to-1 MUXs, XOR gates, and D-latches and flip-flops can be configured using this basic structure. If the A inputs are tied to the data signals and the B inputs are tied to the select signal, the resulting circuit is a 2-to-1 MUX. If the data are fed to the A1 input, the clock is connected to B and the A outputs $(O_A, \overline{O}_A)$ are fed back to the A2 inputs. The resulting circuit is a D-latch as seen in Fig. 16.5. Finally, an XOR gate is created by connecting $A_1 = \overline{A}_2$, forming a new input A_{IN} and $\overline{A}_1 = A_2$ to complementary new input $\overline{A}_{IN}$.

The inputs to the two levels require different dc offsets in order for the circuit to function correctly; thus, level-shifting networks using diodes or source followers are required. Series logic such as this also requires higher supply voltages in order to keep the devices in their saturation region. This will increase the power dissipation of SCFL.

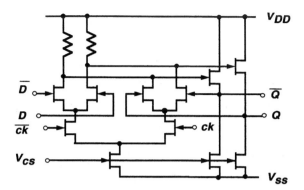

FIGURE 16.5 SCFL D latch schematic. Two cascaded latch cells with opposite clock phasing constitute a master-slave flip-flop.

The logic swing of the circuit shown in Fig. 16.4 is determined by the size of the current source T3 and the load resistors R1 and R2 (R1 = R2). Assuming T3 is in saturation, the logic swing on nodes X and Y is

$$\Delta V_{X,Y} = I_{ds3} \ R1 = I_{dss3} \ R1 \tag{16.2}$$

where I_{dss3} is the saturation current of T3 at Vgs = 0 V. The logic high and low levels on node X ($V_{X,H}$, $V_{X,L}$) are determined from the voltage drop across R3 and Eq. 16.2.

$$V_{X,H} = V_{dd} - (I_{dss3} \ R3) \tag{16.3}$$

$$V_{X,L} = V_{dd} - [I_{dss3} \ (R1 + R3)] \tag{16.4}$$

The noise margin is the difference between the minimum voltage swing required on the inputs to switch the current from one branch to the other (V_{SW}) and the logic swing $\Delta V_{X,Y}$. V_{SW} is set by the ratio between the sizes of the switch transistors (T1,T2, T4-T7) and T3. For symmetry reasons, the sizes of all the switch transistors are kept the same size. Assuming the saturation drain-source current of an FET can be described by the simplified square-law equation:

$$I_{ds} = \beta W (V_{gs} - V_T)^2 \tag{16.5}$$

where V_T is the threshold voltage, W is the FET width, and β is a process-dependent parameter. V_{SW} is calculated assuming all the current from T3 flows only through T2.

$$V_{SW} = |V_T| \sqrt{\left(W3 / W2 \right)} \tag{16.6}$$

For a fixed current source size (W3), the larger the size of the switch transistors, the smaller the voltage swing required to switch the current and, hence, a larger noise margin. Although a better noise margin is desirable, it needs to be noted that the larger switch transistors means increased input capacitance and decreased speed. Depending on the design specifications, noise margin and speed need to be traded off.

Since all FETs need to be kept in the saturation region for the correct operation of an SCFL gate, level-shifting is needed between nodes A and B and the input to the next gate, in order to keep T1, T2, and T4-T7 saturated. T3 is kept in saturation if the potential at node S is higher than $V_{SS} + V_{ds,sat}$. The potential at node S is determined by the input voltages to T1 and T2. V_S settles at a potential such that the drain-source current of the conducting transistor is exactly equal to the bias current, I_{dss3}, since no current flows through the other transistor. The minimum logic high level at the output node B ($V_{OB,H}$) is

$$V_{OB,H} \geq V_{ss} + V_{ds,sat} + V_{gs} = V_{ss} + V_{ds,sat} + V_{SW} + V_{th} \qquad (16.7)$$

To keep T9 and T11 in saturation, however, requires that

$$V_{OB,H} \geq V_{ss} + V_{ds,sat} + V_{SW} \qquad (16.8)$$

As with the voltage on node S, the drain voltages of T1 and T2 are determined by the voltage applied to the A inputs. The saturation condition for T1 and T2 is

$$V_{OA,H} - V_{SW} - V_{th} - V_S = V_{OA,H} - V_{OB,H} \geq V_{ds,sat} \qquad (16.9)$$

Equation 16.9 shows that the lower switch transistors are kept in saturation if the level-shifting difference between the A and B outputs is larger than the FET saturation voltage. Since diodes are used for level-shifting, the minimum difference between the two outputs is one diode voltage drop, V_D. If $V_{ds,sat} > V_D$, more diodes are required between the A and B outputs.

The saturation condition for the upper switch transistors, T4 to T7, is determined by the minimum voltage at nodes A and B and the drain voltage of T1 and T2.

$$V_{A,min} - (V_{OA,H} - V_{SW} - V_{th}) \geq V_{ds,sat} \qquad (16.10)$$

Substituting Eq. 16.4 into Eq. 16.10 yields

$$\left(V_{dd} - I_{dss3} * \left(R1 + R3\right)\right) - \left(V_{OA,H} - V_{SW} - V_{th}\right) \geq V_{ds,sat} \qquad (16.11)$$

Rewriting Eq. 16.11 using Eq. 16.8 gives the minimum power supply range

$$V_{dd} - V_{ss} \geq I_{dss3} * (R1 + R3) + 3V_{ds,sat} \qquad (16.12)$$

Equation 16.11 allows the determination of the minimum amount of level-shifting required between nodes A and B to the outputs

$$V_{A,H} - V_{OA,H} \geq V_{ds,sat} + I_{dss3} * R1 - V_{th} - V_{SW} \qquad (16.13)$$

Equations 16.8 to 16.13 can be used for designing the level shifters. The design parameters available in the level-shifters are the widths of the source followers (W8, W10), the current sources (W9, W11), and the diode (W_D). Assuming the current source width (W9) is fixed, the voltage drop across the diodes is partially determined by the ratio (W_D/W9). This ratio should not be made too small. Operating Schottky diodes at high current density will result in higher voltage drop, but this voltage will be partially due to the $I_D R_S$ drop across the parasitic series resistance. Since this resistance is often process dependent and difficult to reproduce, poor reproducibility of V_D will result in this case.

The ratio between the widths of the source follower and the current source (W8/W9) determines the gate-source voltage of the source follower (V_{gs8}). V_{gs8} should be kept below 0.5 V to prevent gate-source conduction.

The dc design of the two-level series-gated SCFL gate in Fig. 16.4 can be accomplished by applying Eqs. 16.2 to 16.13. Ratios between most device sizes can be determined by choosing the required noise margin and logic swing. Only W3 in the differential stage and W9 among the level-shifters are unknown at this stage. All other device sizes can be expressed in terms of these two transistor widths.

The relation between W3 and W9 can be determined only by considering transient behavior. For a given total power dissipation, the ratio between the power dissipated in the differential stage and the output buffers determines how fast the outputs are switched. If fast switching at the outputs is desired, more power needs to be allocated to the output buffers and, consequently, less power to the differential

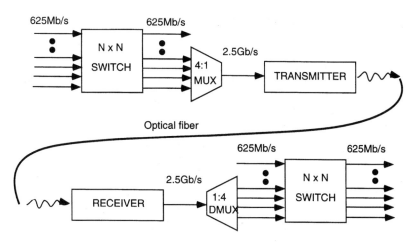

FIGURE 16.6   2.5-Gb/s optical communication system.

stage. While this allocation will ensure faster switching at the output, the switching speed of the differential stage is reduced because of the reduced current available to charge and discharge the large input capacitance of the output buffers.

Finally, it is useful to note that scaling devices to make a speed/power tradeoff is simple in SCFL. If twice as much power is allocated to a gate, all transistors and diodes are made twice as wide while all resistors are reduced by half.[6]

Advanced MESFET/HEMT Design Examples

High-Speed TDM Applications

The need for high bandwidth transmission systems continues to increase as the number of bandwidth-intensive applications in the areas of video imaging, multimedia, and data communication (such as database sharing and database warehousing) continues to grow. This has led to the development of optical communication systems with transmission bit rates, for example, of 2.5 Gb/s and 10 Gb/s. A simplified schematic of a 2.5 Gb/s communication system is shown in Fig. 16.6.

As seen in Fig. 16.6, MUXs, DMUXs, and switches capable of operating in the Gb/s range are crucial for the operation of these systems. GaAs MESFET technology has been employed extensively in the design of these high-speed circuits because of the excellent intrinsic speed performance of GaAs. SCFL is especially well suited for these circuits where high speed is of utmost importance and power dissipation is not a critical factor.

The design strategies employed in the previous subsection can now be further applied to a high-speed 4:1 MUX, as shown in Fig. 16.7. It was shown that the two-level series gated SCFL structure could be easily configured into a D-latch. The MSFF in the figure is simply a master-slave flip-flop containing two D-latches. The PSFF is a phase-shifting flip-flop that contains three D-latches and has a phase shift of 180° compared with an MSFF.

The 4:1 MUX is constructed using a tree-architecture in which two 2:1 MUXs merge two input lines each into one output operating at twice the input bit rate. The 2:1 MUX at the second stage takes the two outputs of the first stage and merges it into a single output at four times the primary input bit rate. The architecture is highly pipelined, ensuring good timing at all points in the circuit. The inherent propagation delay of the flip-flops ensures that the signals are passed through the selector only when they are stable.[6]

The interface between the two stages of 2:1 MUXs is timing-critical, and care needs to be taken to obtain the best possible phase margin at the input of the last flip-flop. To accomplish this, a delay is added between the CLK signal and the clock input to this flip-flop. The delay is usually implemented

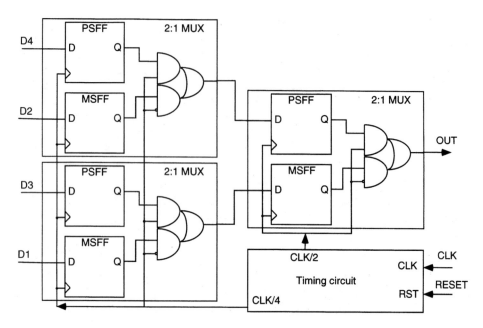

FIGURE 16.7   High-speed 4:1 multiplexer (MUX).

using logic gates because their delays are well characterized in a given process. Output jitter can be minimized if 50% duty-cycle clock signals are used. Otherwise, a retiming MSFF will be needed at the output of the 4:1 MUX.

The 4:1 MUX is a good example of an application of GaAs MESFETs with very-high-speed operation and low levels of integration. Vitesse Semiconductor has several standard products operating at the Gb/s range fabricated in GaAs using their own proprietary E/D MESFET process. For example, the 16 × 16 crosspoint switch, VSC880, has serial data rates of 2.0 Gb/s. The VS8004 4-bit MUX is a high-speed, parallel-to-serial data converter. The parallel inputs accept data at rates up to 625 Mb/s and the differential serial data output presents the data sequentially at 2.5 Gb/s, synchronous with the differential high-speed clock input.[2]

While the MESFET technologies have proven capable at 2.5 and 10 Gb/s data rates for optical fiber communication applications, higher speeds appear to require heterojunction technologies. The 40-Gb/s TDM application is the next step, but it is challenging for all present semiconductor device IC technologies. A complete 40-Gb/s system has been implemented in the laboratory with 0.1-μm InAlAs/InGaAs/InP HEMT ICs as reported in Refs. 7 and 8. Chips were fabricated that implemented multiplexers, photodiode preamplifiers, wideband dc 47-GHz amplifiers, decision circuits, demultiplexers, frequency dividers, and limiting amplifiers. The high-speed static dividers used the super-dynamic FF approach.[9]

A 0.2-μm AlGaAs/GaAs/AlGaAs HEMT quantum well device technology has also demonstrated 40-Gb/s TDM system components. A single chip has been reported that included clock recovery, data decision, and a 2:4 demultiplexer circuit.[10] The SCFL circuit approach was employed.

Very-High-Speed Dynamic Circuits

Conventional logic circuits using static DCFL or SCFL NOR gates such as those described above are limited in their maximum speed by loaded gate delays and serial propagation delays. For example, a typical DCFL NOR-implemented edge-triggered DFF has a maximum clock frequency of approximately $1/5\tau_D$ and the SCFL MSFF is faster, but it is still limited to $1/2\tau_D$ at best. Frequency divider applications that require clock frequencies above 40 GHz have occasionally employed alternative circuit approaches which are not limited in the same sense by gate delays and often use dynamic charge storage on gate nodes for temporarily holding a logic state. These approaches have been limited to relatively simple circuit functions such as divide-by-2 or -4.

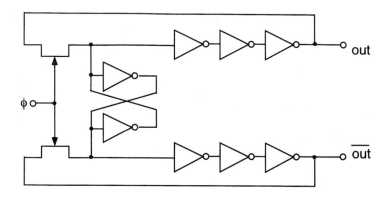

FIGURE 16.8   Dynamic frequency divider (DFD) divide-by-2 circuit. (Ref. 11, ©1989 IEEE, with permission.)

The dynamic frequency divider (DFD) technique is one of the well-known methods for increasing clock frequency closer to the limits of a device technology. For an example, Fig. 16.8 shows a DFD circuit using a single-phase clock, a cross-coupled inverter pair as a latch to reduce the minimum clock frequency, and pass transistors to gate a short chain of inverters.[11,12] These have generally used DCFL or DCFL superbuffers for the inverters. The cross-coupled inverter pair can be made small in width, since its serial delay is not in the datapath. But the series inverter chain must be designed to be very fast, generally requiring high power per inverter in order to push the power-delay product to its extreme high-speed end. Since fan-out is low, the intrinsic delays of an inverter in a given technology can be approached.

The maximum and minimum clock frequencies of this circuit can be calculated from the gate delays of the n series inverters as shown in Eqs. 16.14 and 16.15. An odd number n is required to force an inversion of the data so that the circuit will divide-by-2. Here, t_1 is the propagation delay of the pass transistor switches, J1 and J2, and t_D is the propagation delay of the DCFL inverters. The parameter "a" is the duty cycle of the clock. For a 50% clock duty cycle, the range of minimum to maximum clock frequency is about 2 to 1.

$$f_{\phi max} = \frac{1}{t_1 + nt_D} \tag{16.14}$$

$$f_{\phi min} = \frac{a}{t_1 + nt_D} \tag{16.15}$$

Clock frequencies as high as 51 GHz have been reported using this approach with a GaAs/AlGaAs P-HEMT technology.[12] The power dissipation was relatively high, 440 mW. Other DFD circuit approaches can also be found in the literature.[13-15]

A completely different approach, as shown in Fig. 16.9, utilizes an injection-locked push-pull oscillator (J1 and J2) whose free running frequency is a subharmonic of the desired input frequency.[16] FETs J3 and J4 are operating in their ohmic regions and act as variable resistors. The variation in V_{GS1} and V_{GS2} cause the oscillator to subharmonically injection-lock to the input source. Here, a divide-by-4 ratio was demonstrated with an input frequency of 75 GHz and a power dissipation of 160 mW using a 0.1-µm InP-based HEMT technology with $f_T = 140$ GHz and $f_{max} = 240$ GHz. This divider also operated in the 59–64 GHz range with only −10 dBm RF input power. The frequency range is limited by the tuning range of the oscillator. In this example, about 2 octaves of frequency variation was demonstrated.

Finally, efforts have also been made to beat the speed limitations of a technology by dynamic design methods while still maintaining minimum power dissipation. The quasi-dynamic FF[8,9] and quasi-differential FF[17] are examples of circuit designs emphasizing this objective. The latter has achieved 16-GHz clock frequency with approximately 2 mW of power per FF.

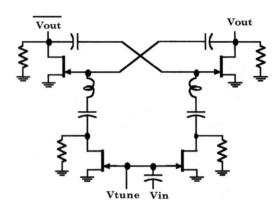

FIGURE 16.9   Injection-locked oscillator divide-by-4. (Ref. 16, ©1996 IEEE. With permission.)

16.2 HBT Logic Design Examples

From a circuit topology perspective, both III-V HBTs and silicon BJTs are interchangeable, with myriad similarities and a few essential differences. The traditional logic families developed for the Si BJTs serve as the starting point for high-speed logic with III-V HBTs. During the period of intense HBT development in the 1980s and early 1990s, HBTs have implemented ECL, CML, DTL, and, I²L logic topologies as well as novel logic families with advanced quantum devices (such as resonant tunneling diodes[17a]) in the hopes of achieving any combination of high-speed, low-power, and high-integration level. During that time, III-V HBTs demonstrated their potential integration limits with an I²L 32-bit microprocessor[18] and benchmarked its high-speed ability with an ECL 30-GHz static master/slave flip-flop based frequency divider. During the same time, advances in Si based technology, especially CMOS, have demonstrated that parallel circuit algorithms implemented in a technology with slower low-power devices capable of massive integration will dominate most applications. Consequently, III-V-based technologies such as HBTs and MESFET/HEMT have been relegated to smaller but lucrative niche markets.

As HBT technology evolved into a mature production technology in the mid-1990s, it was clear that III-V HBT technology had a clear advantage in high-speed digital circuits, microwave integrated circuits, and power amplifier markets. Today, in the high-speed digital arena, III-V HBTs have found success in telecom and datacom lightwave communication circuits for SONET/ATM-based links that operate from 2.5 to 40 Gb/s. HBTs also dominate the high-speed data conversion area with Nyquist-rate ADCs capable of gigabit/gigahertz sampling rates/bandwidths, sigma-delta ADCs with very high oversampling rates, and direct digital synthesizers with gigahertz clock frequencies and ultra-low spurious outputs. In these applications, the primary requirement is ultra-high-speed performance with LSI (10 K transistors) levels of integration. Today, the dominant logic type used in HBT designs is based on non-saturating emitter coupled pairs such as ECL and current-mode logic (CML), which is the focus of this chapter.

III-V HBT for Circuit Designers

III-V HBTs and Si BJTs are inherently bipolar in nature. Thus, from a circuit point of view, both share many striking similarities and some important differences. The key differences between III-V HBT technology and Si BJT technology, as discussed below, can be traced to three essential aspects: (1) heterojunction vs. homojunction, (2) III-V material properties, and (3) substrate properties.

First, the primary advantage of a base–emitter heterojunction is that the wide bandgap emitter allows the base to be doped higher than the emitter (typically 10 to 50X in GaAs/AlGaAs HBTs) without a reduction in current gain. This translates to lower base resistance for improved f_{max} and reduces base width modulation with V_{ce} for low output conductance. Alternatively, the base can be made thinner for lower base-transit time (τ_b) and higher f_t without having R_b too high. If the base composition is also graded from high bandgap to low, an electric field can be established to sweep electrons across the base for reduced τ_b and higher f_t. With a heterojunction B-E and a homojunction B-C, the junction turn-on voltage is higher in the B-E than it is in the B-C. This results in a common-emitter I–V curve offset from the off to saturation transition. This offset is approximately 200 mV in GaAs/AlGaAs HBTs. With a highly doped base, base punch-through is not typically observed in HBTs and does not limit the f_t-breakdown voltage product as in high-performance Si BJTs and SiGe HBT with thin bases. Furthermore, if a heterojunction is placed in the base–collector junction, a larger bandgap material in the collector can increase the breakdown voltage of the device and reduce the I–V offset.

Second, III-V semiconductors typically offer higher electron mobility than Si for overall lower τ_b and collector space charge layer transit times (τ_{cscl}). Furthermore, many III-V materials exhibit velocity overshoot in the carrier drift velocity. When HBTs are designed to exploit this effect, significant reductions in τ_{cscl} can result. With short collectors, the higher electron mobility can result in ultra-high f_t; however, this can also be used to form longer collectors with still acceptable τ_{cscl}, but significantly reduced C_{bc} for high f_{max}. The higher mobility in the collector can also lead to HBTs with lower turn on resistance in the common emitter I–V curves.

Since GaAs/AlGaAs and GaAs/InGaP have wider bandgaps than Si, the turn-on voltage of the B–E ($V_{be,on}$) junction is typically on the order of 1.4 V vs. 0.9 V for advanced high-speed Si BJT. InP-based HBTs can have $V_{be,on}$ on the order of 0.7 V; however, most mature production technologies capable of LSI integration levels are based on AlGaAs/GaAs or InGaP/GaAs. The base–collector turn-on voltage is typically on the order of 1 V in GaAs-based HBTs. This allows V_{ce} to be about 600 mV lower than V_{be} without placing the device in saturation. The wide bandgap material typically results in higher breakdown voltages, so III-V HBTs typically have a high Johnson figure of merit (f_t * breakdown voltage) compared with Si- and SiGe-based bipolar transistors.

The other key material differences between III-V vs. silicon materials are the lack of a native stable oxide in III-V, the extensive use of poly-Si in silicon-based processes, and the heavy use of implants and diffusion for doping silicon devices. III-V HBTs typically use epitaxial growth techniques, and interconnect step height coverage issues limit the practical structure to one device type, so PNP transistors are not typically included in an HBT process. These key factors contribute to the differences between HBTs and BJTs in terms of fabrication.

Third, the GaAs substrate used in III-V HBTs is semi-insulating, which minimizes parasitic capacitance to ground through the substrate, unlike the resistive silicon substrate. Therefore, the substrate contact as in Si BJTs is unnecessary with III-V HBTs. In fact, the RF performance of small III-V HBT devices can be measured directly on-wafer without significant de-embedding of the probe pads below 26 GHz. For interconnects, the line capacitance is typically dominated by parallel wire-to-wire capacitance, and the loss is not limited by the resistive substrate. This allows for the formation of high-Q inductors, low-loss transmission lines, and longer interconnects that can be operated in the 10's of GHz. Although BESOI and SIMOX Si wafers are insulating, the SiO_2 layer is typically thin resulting in reduced but still significant capacitive coupling across this thin layer.[19]

Most III-V substrates have a lower thermal conductivity than bulk Si, resulting in observed self-heating effects. For a GaAs/AlGaAs HBT, this results in observed negative output conductance in the common-emitter I–V curve measured with constant I_b. The thermal time constant for GaAs/AlGaAs HBTs is on the order of microseconds. Since thermal effects cannot track above this frequency, the output conductance of HBTs at RF (> 10 MHz) is low but positive. This effect does result in a small complication for HBT models based on the standard Gummel Poon BJT model.

Current-Mode Logic

The basic current-mode logic (CML) buffer/inverter cell is shown in Fig. 16.10. The CML buffer is a differential amplifier that is operated with its outputs clipped or in saturation. The differential inputs (V_{in} and V_{in}') are applied to the bases of Q_1 and Q_2. The difference in potential between V_{in} and V_{in}'

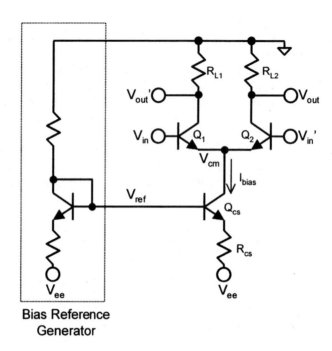

**Bias Reference
Generator**

FIGURE 16.10   Standard differential CML buffer with a simple reference generator.

determines which transistor I_{bias} is steered through, resulting in a voltage drop across either load resistance R_{L1} or R_{L2}. If $V_{in} = V_{OH}$ and $V_{in}' = V_{OL}$ ($V_{in,High} > V_{in,Low}$), Q_1 is on and Q_2 is off. Consequently, I_{bias} completely flows through R_{L1}, causing V_{out}' to drop for a logic low. With Q_2 off, V_{out} floats to ground for a logic high. If the terminal assignment of V_{out} and V_{out}' were reversed, this CML stage would be an inverter instead of a buffer.

The logic high V_{OH} of a CML gate is 0 V. The logic low output is determined by $V_{OL} = -R_{L1}I_{bias}$. With $R_{L1}/R_{L2} = 200$ Ωs, and $I_{bias} = 2$ mA, the traditional logic low of a CML gate is –400 mV. As CML gates are cascaded together, the outputs of one stage directly feed the inputs of another CML gate. As a result, the base-collector of the "on" transistor is slightly forward-biased (by 400 mV in this example). For high-speed operation, it is necessary to keep the switching transistors out of saturation. With a GaAs base-collector turn-on voltage near 1V, 500 to 600 mV forward-bias is typically tolerated without any saturation effects. In fact, this bias shortens the base-collector depletion region, resulting in the highest f_t vs. V_{ce} (f_{max} suffers due to increase in C_{bc}). As a result, maximum logic swing of a CML gate is constrained by the need to keep the transistors out of saturation. As the transistor is turned on, the logic high is actively pulled to a logic low; however, as the transistor is turned off, the logic low is pulled up by a RC time constant. With a large capacitive loading, it is possible that the risetime is slower than the falltime, and that may result in some complications with high-speed data.

A current mirror (Q_{cs} and R_{cs}) sets the bias (I_{bias}) of the differential pair. This is an essential parameter in determining the performance of CML logic. In HBTs, the f_t dependence on I_c is as follows:

$$1/2\pi f_t = \tau_{ec} = nkT/\left(qI_c\left(C_{bej} + C_{bc} + C_{bed}\right)\right) + R_c\left(C_{bej} + C_{bc}\right) + \tau_b + \tau_{cscl} \qquad (16.16)$$

where τ_{ec} is the total emitter-to-collector transit time, qI_c/nkT is the transconductance (g_m), C_{bc} is the base-collector capacitance, C_{bej} is the base–emitter junction capacitance, C_{bed} is the B–E diffusion capacitance, R_c is the collector resistance, τ_b is the base transit time, and τ_{cscl} is the collector space charge layer transit time. At low currents, the transit time is dominated by the device g_m and device capacitance. As the bias increases, τ_{ec} is eventually limited by τ_b and τ_{cscl}. As this limit approaches, Kirk effect typically starts to increase τ_b/τ_{cscl}, which decreases f_t. In some HBTs, the peak f_{max} occurs a bit after the peak f_t. With this in mind, optimal performance is typically achieved when I_{bias} is near $I_{c,maxft}$ or $I_{c,maxfmax}$. In some HBT technologies, the maximum bias may be constrained by thermal or reliability concerns. As a rule of thumb, the maximum current density of HBTs is typically on the order of 5×10^4 A/cm^2.

The bias of CML and ECL logic is typically set with a bias reference generator, where the simplest generator is shown in Fig. 16.10. Much effort has been invested in the design of the reference generator to maintain constant bias with power supply and temperature variation. In HBT, secondary effects of heterojunction design typically result in slightly varying ideality factor with bias. This makes the design of bandgap reference circuits quite difficult in most HBT technologies, which complicates the design of the reference generator. Nevertheless, the reference generators used today typically result in a 2% variation in bias current from −40 to 100 C with a 10% variation in power supply. In most applications, the voltage drop across R_{cs} is set to around 400 mV. With V_{ee} set at −5.2 V, V_{ref} is typically near −3.4 V. With constant bias, as V_{ee} moves by ±10%, then the voltage drop across R_{cs} remains constant, so V_{ref} moves by the change in power supply (about ±0.5 V). Since the logic levels are referenced to ground, the average value of V_{cm} (around −1.4 V) remains constant. This implies that changes in the power supply are absorbed by the base–collector junction of Q_{cs}, and it is important that this transistor is not deeply saturated.

Since the device goes from the cutoff mode to the forward active mode as it switches, the gate delay is difficult to predict analytically with small-signal analysis. Thus, large-signal models are typically used to numerically compute the delay in order to optimize the performance of a CML gate. Nevertheless, the small-signal model, frequency-domain approach described in Chapter 16.3 (Elmore) leads to the following approximation of a CML delay gate with unity fan-out:[19a]

$$\tau_{d,cml} = \left(1 + g_m R_L\right) R_b C_{bc} + R_b \left(C_{be} + C_d\right) + \left(2C_{bc} + 1/2C_{be} + 1/2C_d\right)/g_m \quad (16.17)$$

where C_d is the diffusion capacitance of $g_m(\tau_b + \tau_{cscl})$. Furthermore, by considering the difference in charge storage at logic high and logic low, divided by the logic swing, the effective CML gate capacitance can be expressed[19b] as

$$C_{cml} = C_{be}/2 + 2C_{bc} + C_s + \left(\tau_b + \tau_{cscl}\right)/R_L \quad (16.18)$$

where C_s is collector-substrate and interconnect capacitances. Both equations show that the load resistor and bias (which affects g_m and device capacitors) have a strong effect on performance.

For a rough estimate of the CML maximum speed without loading, one can assume that f_t is the gain bandwidth product. With the voltage gain set at $g_m R_L$, the maximum speed is $f_t/(g_m R_L)$. In the above example, at 1 mA average bias, $g_{m,int} = 1/26$ S at room temperature. Assuming the internal parasitic emitter resistance R_E is 10 ohms and using the fact that $g_{m,ext} = g_{m,int}/(1 + R_E g_{m,int})$, the effective extrinsic g_m is 1/36 mhos. With a 200-ohm load resistor, the voltage gain is approximately 5.5. With a 70-GHz f_t HBT process, the maximum switching rate is about 13 GHz. Although this estimate is quite rough, it does show that high-speed CML logic desires high device bias and low logic swing. In most differential circuits, only 3 to 4 kT/q is needed to switch the transistors and overcome the noise floor. With such low levels and limited gain, the output may not saturate to the logic extremes, resulting in decreasing noise margin. In practice, the differential logic level should not be allowed to drop below 225 mV. With a 225-mV swing vs. 400 mV, the maximum gate bandwidth improves to 23 GHz from 13 GHz.

Emitter-Coupled Logic

By adding emitter followers to the basic HBT CML buffer, the HBT ECL buffer is formed in Fig. 16.11. From a dc perspective, the emitter followers (Q_{ef1} and Q_{ef2}) shift the CML logic level down by V_{be}. With the outputs at V_{outA}/V_{outA}' and 400 mV swing from the differential pair, the first level ECL logic high is −1.4 V and the ECL logic low is −1.8 V. For some ECL logic gates, a second level is created through a

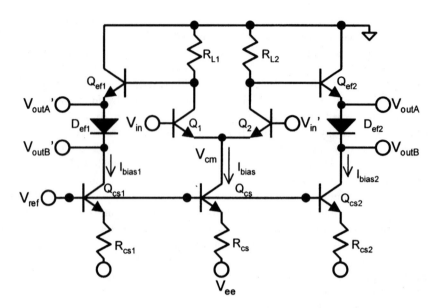

FIGURE 16.11 Standard differential ECL buffer with outputs taken at two different voltage levels.

Schottky diode voltage shift (D_{ef1}/D_{ef2}). The typical Schottky diode turn-on voltage for GaAs is 0.7 V, so the output at V_{outB}/V_{outB}' is −2.1 V for a logic high and −2.5 V for a logic low. In general, the HBT ECL levels differ quite a bit from the standard Si ECL levels. Although resistors can be used to bias Q_{ef1}/Q_{ef2}, current mirrors (Q_{cs1}/Q_{cs2} and R_{cs1}/R_{cs2}) are typically used. Current mirrors offer stable bias with logic level at the expense of higher capacitance, while resistors offer lower capacitance but the bias varies more and may be physically quite large.

From an ac point of view, emitter followers have high input impedance (approximately β times larger than an unbuffered input) and low output impedance (approx. $1/g_m$), which makes it an ideal buffer. In comparison with CML gates, since the differential pair now drives a higher load impedance, the effect of loading is reduced, yielding increased bandwidth, faster edge rates, and higher fan-out. The cost of this improvement is the increase in power due to the bias current of the emitter followers. For example, in a 50 GHz HBT process, a CML buffer (fan-out = 1, I_{bias} = 2mA, R_{L1}/R_{L2} = 150 Ω), the propagation delay (t_D) is 14.8 ps with a risetime [20 to 80%] (t_r) of 31 ps and a falltime [20 to 80%] (t_f) of 21 ps. In comparison, an ECL buffer with level 1 outputs (fan-out = 1, I_{bias} = 2 mA, I_{bias1}/I_{bias2} = 2 mA, R_{L1}/R_{L2} = 150 Ω) has t_d = 14 ps, t_f = 9 ps, and t_r = 16 ps. With a threefold increase in P_{diss}, the impedance transformation of the EF stage results in slightly reduced gate delays and significant improvements in the rise/falltimes. With the above ECL buffer modified for level 2 (level shifted) outputs, the performance is only slightly lower with t_D = 14.2 ps, t_f = 11 ps, and t_r = 22 ps.

In general, emitter followers tend to have bandwidths approaching the f_t of the device, which is significantly higher than the differential pair. Consequently, it is possible to obtain high-speed operation with the EF biased lower than would be necessary to obtain the maximum device f_t. With the ECL level 1 buffer, if the I_{bias1}/I_{bias2} is lowered to 1 mA from 2 mA, the performance is still quite high, with t_d = 15 ps, t_f = 13 ps, and t_r = 18 ps. Although t_D approaches the CML case, the t_f and t_r are still significantly better.

As the EF bias is increased, its driving ability is also increased; however, at some point with high bias, the output impedance of the EF becomes increasingly inductive. When combined with large load capacitance (as in the case of high fan-out or long interconnect), it may result in severe ringing in the output that can result in excessive jitter on data edges. The addition of a series resistor between the EF output and the next stage can help to dampen the ringing by increasing the real part of the load. This change, however, increases the RC time constant, which usually results in a significant reduction in performance.

In practice, changing the impedance of the EF bias source (high impedance current source or resistor bias) does not have a significant effect on the ringing. As a result, the primary method to control the ringing is through the EF bias, which places a very real constraint on bandwidth, fan-out, and jitter that needs to be considered in the topology of real designs. In some FET DCFL designs, several source followers are cascaded together to increase the input impedance and lower the output resistance between two differential pairs for high-bandwidth drive. Due to voltage headroom limits, it is very difficult to cascade two HBT emitter followers without causing the current source to enter deep saturation. In general, ECL gates are typically used for the high-speed sections due to significant improvement in rise/falltimes (bandwidth) and drive ability, although the power dissipation is higher.

ECL/CML Logic Examples

Typically, ECL and CML logic is mixed throughout high-speed GaAs/AlGaAs HBT designs. As a result, there are three available logic levels that can be used to interconnect various gates. The levels are CML (0/–400 mV), ECL1 (–1.4/–1/8 V), and ECL2 (–2.1/–2.5 V). To form more complex logic functions, typically two levels of transistors are used to steer the bias current. Figure 16.12 shows an example of an CML AND/NAND gate. For the ECL counterpart, it is only necessary to add the emitter followers. The top input is V_{inA}/V_{inA}'. The bottom input is V_{inB}/V_{inB}'. In general, the top can be driven with either the CML or the ECL1 inputs, and the bottom level can be driven by ECL1 and ECL2 levels. The choice of logic input levels is typically dictated by the design tradeoff between bandwidth, power dissipation, and fan-out. As seen in Fig. 16.12, only when V_{inA} and V_{inB} are high will I_{bias} current be steered into the load resistor that makes $V_{out} = V_{OH}$. All other combinations will make $V_{out} = V_{OL}$, as required by the AND function. Due to the differential nature, if the output terminal labels were reversed, this would be a NAND gate.

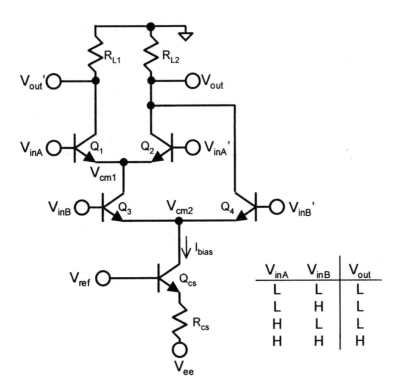

V_{inA}	V_{inB}	V_{out}
L	L	L
L	H	L
H	L	L
H	H	H

FIGURE 16.12   Two-level differential CML AND gate.

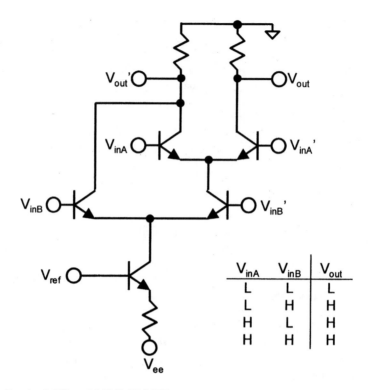

FIGURE 16.13 Two-level differential CML OR/NOR gate.

For the worst-case voltage headroom, V_{inA}/V_{inA}' is driven with ECL1 levels, resulting in V_{cm1} of -2.8 V. With an ECL2 high on V_{inB} (-2.1 V), the lower stage (Q_3/Q_4) has a B-C forward-bias of 0.7 V, which may result in a slight saturation. This also implies that V_{cm2} is around -3.5 V, which results in an acceptable nominal 100 mV forward-bias on the current source transistor (Q_{cs}). As V_{ee} becomes less negative, the change in V_{ee} is absorbed across Q_{cs}, which places Q_{cs} closer into saturation. In saturation, the current source holds I_e in Q_{cs} constant; so if I_b increases (due to saturation), then I_c decreases. For some current source reference designs that cannot source the increased I_b, the increased loading due to saturated I_b may lower V_{ref}, which would have a global effect on the circuit bias. If the current source reference can support the increase in I_b, then the bias of only the local saturated differential pair starts to decrease leading to the potential of lower speed and lower logic swing. For HBTs, the worst-case Q_{cs} saturation occurs at low temperature, and the worst-case saturation for Q_3/Q_4 occurs at high temperature since V_{be} changes by -1.4 mV/C and $V_{diode} = -1.1$ mV/C (for constant-current bias). It is possible to decrease the forward-bias of the lower stage by using the base-emitter diode as the level shift to generate the second ECL levels; however, the power supply voltage needs to increase from -5.2 to possibly -6 V.

With the two-level issues in mind, Fig. 16.13 illustrates the topology for a two-level OR/NOR gate. This design is similar to an AND gate except that $V_{out} = V_{OL}$ if both V_{inA} and V_{inB} are low. Otherwise, Vout = V_{OH}. By using the bottom differential pair to select one of the two top differential pairs, many other prime logic functions can be implemented. In Fig. 16.14, the top pairs are wired such that $V_{out} = V_{OL}$ if $V_{inA} = V_{inB}$, forming the XOR/XNOR block. If the top differential pairs are thought of as selectable buffers with a common output as shown in Fig. 16.15, then a basic 2:1 MUX cell is formed. Here, V_{inB}/V_{inB}' determines which input (V_{inA1}/V_{inA1}' or V_{inA2}/V_{inA2}') is selected to the output. This concept can be further extended to a 4:1 MUX if the top signals are CML and the control signals are ECL1 and ECL2, as shown in Fig. 16.16. Here, the MSB (ECL1) and LSB (ECL2) determine which of the four inputs are selected. With the 2:1 MUX in mind, if each top differential pair had separate output resistors with a common input, a 1:2 DEMUX is formed as shown in Fig. 16.17.

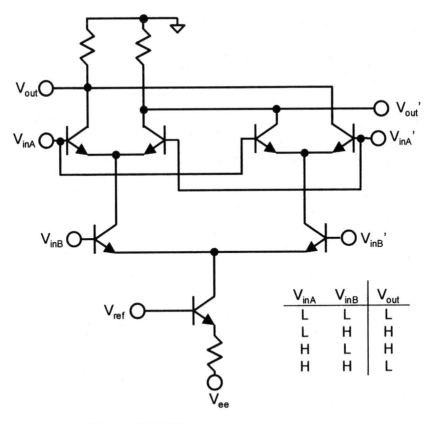

FIGURE 16.14 Two-level differential CML XOR gate.

V_{inA}	V_{inB}	V_{out}
L	L	L
L	H	H
H	L	H
H	H	L

The last primary cell of importance is the latch. This is shown in Fig. 16.18. Here, the first differential pair is configured as a buffer. The second pair is configured as a buffer with positive feedback. The positive feedback causes any voltage difference between the input transistors to be amplified to full logic swing and that state is held as long as the bias is applied. With this in mind, as the first buffer is selected ($V_{inB} = V_{OH}$), the output is transparent to the input. As $V_{inB} = V_{OL}$, the last value stored in the buffer is held, forming a latch, which, in this case, is triggered on the falling edge of the ECL2 level. When two of these blocks are connected together in series, it forms the basic master-slave flip-flop.

Advanced ECL/CML Logic Examples

With small signal amplifiers, the cascode configuration (common base on top of a common emitter stage) typically reduces the Miller capacitance for higher bandwidth. With the top-level transistor on, the top transistor forms a cascode stage with the lower differential amplifier. This can lead to higher bandwidths and reduced rise/falltimes. However, in large-signal logic, the bottom transistor must first turn on the top cascode stage before the output can change. This added delay results in a larger propagation delay for the bottom pair vs. the top switching pair. In the case of the OR/NOR, as in Fig. 16.13, if Q_1 or Q_2 switches with Q_3 on or if Q_4 switches, the propagation delay is short. If Q_3 switches, it must first turn on either Q_1/Q_2, leading to the longest propagation delay. In this case, the longest delay limits the usable bandwidth of the AND gate. Likewise, in the XOR case, the delay of the top input is shorter than the lower input, which results in asymmetric behavior and reduced bandwidth. For a 10-GHz flip-flop, this can result in as much as a 10-ps delay from the rising edge of the clock to the sample point of the data. This issue must be taken into account in determining the optimal input data phase for lowest bit errors when dealing with digital data.

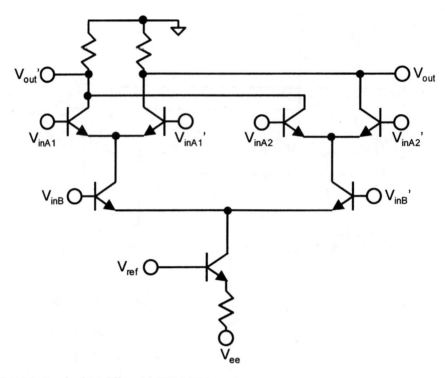

FIGURE 16.15   Two-level 2:1 differential CML MUX gate.

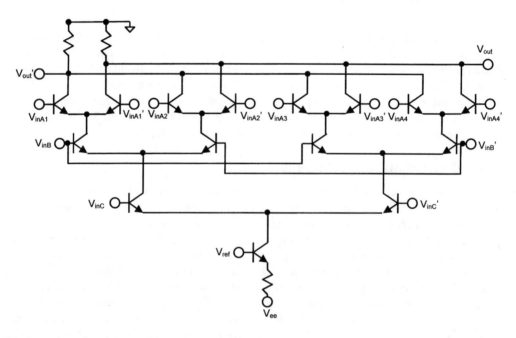

FIGURE 16.16   Three-level 4:1 differential CML MUX gate.

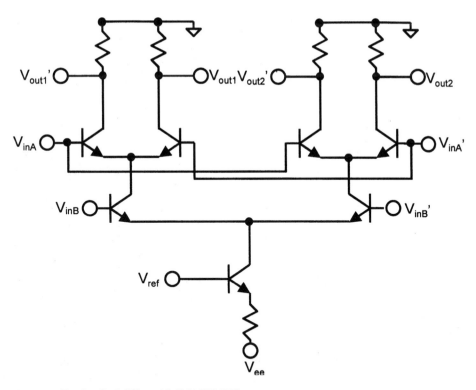

FIGURE 16.17 Two-level 1:2 differential CML DEMUX gate.

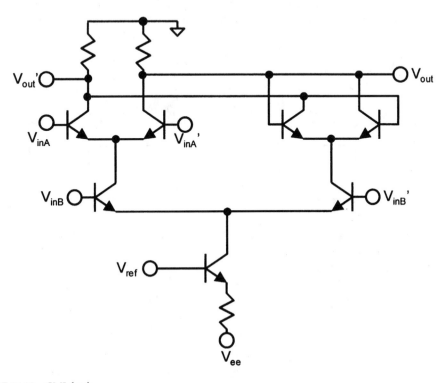

FIGURE 16.18 CML latch.

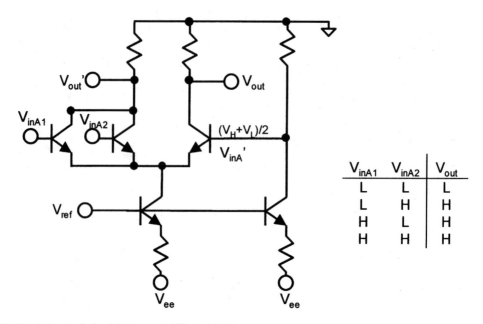

FIGURE 16.19 Single-level CML quasi-differential OR gate.

One solution to the delay issue is to use a quasi-differential signal. In Fig. 16.19, a single-ended single-level OR/NOR gate is shown. Here, a reference generator of $(V_H + V_L)/2$ is applied to V_{inA}'. If either of the V_{inA1} or V_{inA2} is high, then $V_{out} = V_{OH}$. This design has more bandwidth than the two-level topology shown in Fig. 16.12, but some of the noise margin may be sacrificed.

Figure 16.20 shows an example of a single-level XOR gate with a similar input level reference. In this case, $I_{bias1} = I_{bias2} = I_{bias3}$. The additional I_{bias3} is used to make the output symmetric. Ignoring I_{bias3}, when V_{inA} is not equal to V_{inB}, I_{bias1} and I_{bias2} are used to force $V_{out}' = V_{OL}$. When $V_{inA} = V_{inB}$, $V_{out} = V_{out}'$ since

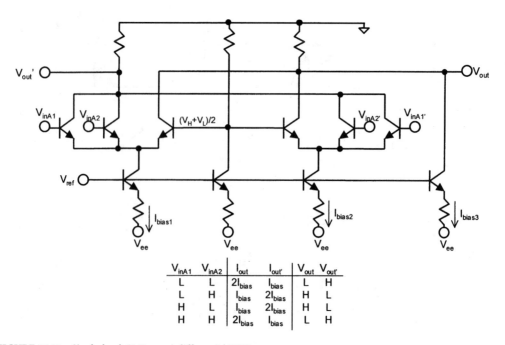

FIGURE 16.20 Single-level CML quasi-differential XOR gate.

both are lowered by I_{bias}, resulting in an indeterminate state. To remedy this, I_{bias3} is added to V_{out} to make the outputs symmetric. This design results in higher speed due to the single-level design; however, the noise margin is somewhat reduced due to the quasi-differential approach and the outputs have a common-mode voltage offset of $R_L I_{bias}$.

In a standard differential pair, the output load capacitance can be broken into three parts. The base-collector capacitance of the driving pair, the interconnect capacitance, and the input capacitance of the next stage. The interconnect capacitance is on the order of 5 to 25 fF for adjacent to nearby gates. The base-collector depletion capacitance is on the order of 25 fF. Assuming that the voltage gain is 5.5, the effective C_{bc} or Miller capacitance is about 140 fF. $C_{be,j}$, when the transistor is off, is typically less than 6 fF. The $C_{be,d}$ capacitance when the transistor is on is of the order of 50 to 200 fF. These rough numbers show that the Miller effect has a significant effect on the effective load capacitance. For the switching transistor, the Miller effect increases both the effective internal C_{bc} as well as the external load. In these situations, a cascode stage may result in higher bandwidth and sharper rise/falltimes with a slight increase in propagation delay. Figure 16.21 shows a CML gate with an added cascode stage. Due to the 400-mV logic swing, the cascode bases are connected to ground. For higher swings, the cascode bases can be biased to a more negative voltage to avoid saturation. The cascode requires that the input level be either ECL1 or ECL2 to account for the V_{be} drop of the cascode. Since the base of the cascode is held at ac ground, the Miller effect is not seen at the input of the common-base stage as the output voltage swings. From the common-emitter point of view, the collector swings only about 60 mV per decade change in I_c; thus, the Miller effect is greatly reduced. The reduction of the Miller effect through cascoding reduces the effect of both the internal transistor C_{bc} and load capacitance due to C_{bc} of the other stages, which results in the reduced rise/falltimes, especially at the logic transition region.

As both of the transistors in a cascode turn off, the increased charge stored in both transistors that has to discharge through an RC time constant may result in a slower edge rate near the logic high of a rising edge. In poorly designed cascode stages, the corner point between the fast-rising edge to the slower-rising edge may occur near the 20/80% point, canceling out some of the desired gains. Furthermore, with the emitter node of the off common-base stage floating in a high-impedance state, its actual voltage varies with time (large RC discharge compared to the switching time). This can result in some "memory"

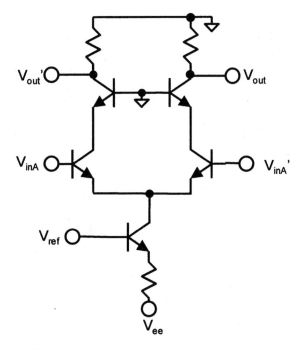

FIGURE 16.21 CML buffer with a cascode output stage.

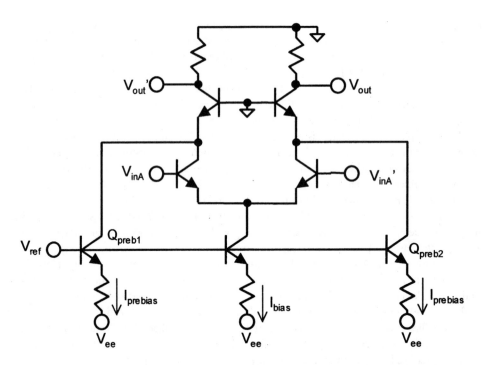

FIGURE 16.22   CML buffer with a cascode output stage and bleed current to keep the cascode "on."

effects where the actual node voltage depends on the previous bit patterns. In this case, as the transistor turns on, the initial voltage may vary, which can result in increased jitter with digital data. With these effects in mind, the cascoded CML design can be employed with performance advantages in carefully considered situations.

One way to remedy the off cascode issues is to use prebias circuits as shown in Fig. 16.22. Here, the current sources formed with Q_{preb1} and Q_{preb2} ($I_{prebias} \ll I_{bias}$) ensures that the cascode is always slightly on by bleeding a small bias current. This results in improvements in the overall rise- and falltimes, since the cascode does not completely turn off. This circuit does, however, introduce a common-mode offset in the output that may reduce the headroom in a two-level ECL gate that it must drive. Furthermore, a series resistor can be introduced between the bleed point and the current source to decouple the current source capacitance into the high-speed node. This design requires careful consideration to the design tradeoffs involving the ratio of $I_{bias}/I_{prebias}$ as well as the potential size of the cascode transistor vs. the switch transistors for optimal performance. When properly designed, the bleed cascode can lead to significant performance advantages.

In general, high-speed HBT circuits require careful consideration and design of each high-speed node with respect to the required level of performance, allowable power dissipation, and fan-out. The primary tools the designer has to work with are device bias, device size, ECL/CML gate topology, and logic level to optimize the design. Once the tradeoff is understood, CML/ECL HBT-based circuits have formed some of the faster circuits to date. The performance and capability of HBT technology in circuit applications are summarized below.

HBT Circuit Design Examples

A traditional method to benchmark the high-speed capability of a technology is to determine the maximum switching rate of a static frequency divider. This basic building block is employed in a variety of high-speed circuits, which include frequency synthesizers, demultiplexers, and ADCs. The basic static frequency divider consists of a master/slave flip-flop where the output data of the slave flip-flop is fed back to the input data of the master flip-flop. The clock of the master and the clock of the slave flip-flop are connected together, as shown in Fig. 16.23. Due to the low transistor count and importance in many larger high-speed circuits, the frequency divider has emerged as the primary circuit used to demonstrate the high-speed potential of new technologies. As HBT started to achieve SSI capability in 1984, a frequency divider with a toggle rate of 8.5 GHz[19c] was demonstrated. During the transition from research to pilot production in 1992, a research-based AlInAs/GaInAs HBT (f_t of 130 GHz and f_{max} of 90 GHz) was able to demonstrate a 39.5 GHz divide-by 4.[20] Recently, an advanced AlInAs/GaInAs HBT technology (f_t of 164 GHz and f_{max} of 800 GHz) demonstrated a static frequency divider operating at 60 GHz.[21] This HBT ECL-based design, to date, reports the fastest results for any semiconductor technology, which illustrates the potential of HBTs and ECL/CML circuit topology for high-speed circuits.

Besides high-speed operation, production GaAs HBTs have also achieved a high degree of integration for LSI circuits. For ADCs and DACs, the turn-on voltage of the transistor (V_{be}) is determined by material constants; thus, there is significantly less threshold variation when compared to FET-based technologies. This enables the design of high-speed and accurate comparators. Furthermore, the high linearity characteristics of HBTs enable the design of wide dynamic range and high linearity sample-and-hold circuits. These paramount characteristics result in the dominance of GaAs HBTs in the super-high performance/high-speed ADCs. An 8-bit 2 gigasamples/s ADC has been fabricated with 2500 transistors. The input bandwidth is from dc to 1.5 GHz, with a spur-free dynamic range of about 48 dB.[22]

Another lucrative area for digital HBTs is in the area of high-speed circuits that are employed in fiber-optic based telecommunications systems. The essential circuit blocks (such as a 40-Gb/s 4:1 multiplexers[23] and 26-GHz variable gain-limiting amplifiers[24]) have been demonstrated with HBTs in the research lab. In general, the system-level specifications (SONET) for telecommunication systems are typically very stringent compared with data communication applications at the same bit rate. The tighter specifications in telecom applications are due to the long-haul nature and the need to regenerate the data several times before the destination is reached. Today, there are many ICs that claim to be SONET-compliant at OC-48 (2.5 Gb/s) and some at OC-192 (10 Gb/s) bit rates. Since the SONET specifications apply on a system level, in truth, there are very few ICs having the performance margin over the SONET specification for use in real systems. Due to the integration level, high-speed performance, and reliability of HBTs, some of the first OC-48 (2.5 Gb/s) and OC-192 (10 Gb/s) chip sets (e.g., preamplifiers, limiting amplifiers,

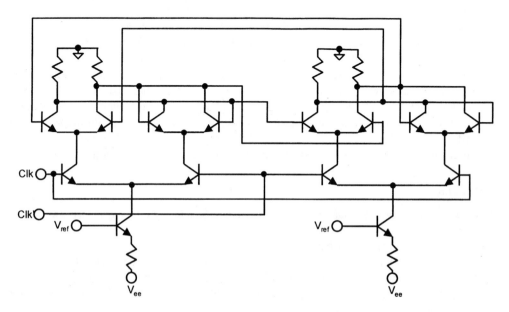

FIGURE 16.23 2:1 Frequency divider based on two CML latches (master/slave flip-flop).

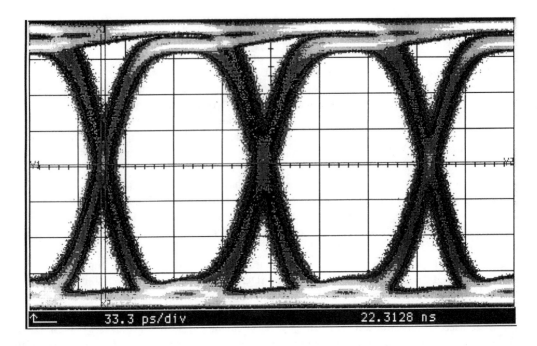

FIGURE 16.24   Typical 10 Gbps eye diagram for OC-192 crosspoint switch.

clock and data recovery circuits, multiplexers, demultiplexers, and laser/modulator drivers) deployed are based on GaAs HBTs. A 16×16 OC-192 crosspoint switch has been fabricated with a production 50 GHz f_t and f_{max} process.[25] The LSI capability of HBT technology is showcased with this 9000 transistor switch on a $6730 \times 6130 \ \mu m^2$ chip. The high-speed performance is illustrated with a 10 Gb/s eye diagram shown in Fig. 16.24. With less than 3.1 ps of RMS jitter (with four channels running), this is the lowest jitter 10-Gb/s switch to date. At this time, only two 16×16 OC-192 switches have been demonstrated[25,26] and both were achieved with HBTs. With a throughput of 160,000 Mb/s, these HBT parts have the largest amount of aggregate data running through it of any IC technology.

In summary, III-V HBT technology is a viable high-speed circuit technology with mature levels of integration and reliability for real-world applications. Repeatedly, research labs have demonstrated the world's fastest benchmark circuits with HBTs with ECL/CML-based circuit topology. The production line has shown that current HBTs can achieve both the integration and performance level required for high-performance analog, digital circuits, and hybrid circuits that operate in the high gigahertz range. Today, the commercial success of HBTs can be exemplified by that fact that HBT production lines ship several million HBT ICs every month and that several new HBT production lines are in the works. In the future, it is expected that advances in Si based technology will start to compete in the markets currently held by III-V technology; however, it is also expected that III-V technology will move on to address ever higher speed and performance issues to satisfy our insatiable demand for bandwidth.

References

1. Long, S. and Butner, S., *Gallium Arsenide Digital Integrated Circuit Design,* McGraw-Hill, New York, 1990, 210.
2. Vitesse Semiconductor, *1998 Product Selection Guide,* 164, 1998.
3. Troutman, R. R., Subthreshold Design Considerations for Insulated Gate Field-Effect Transistors, *IEEE J. Solid-State Ciruits,* SC-9, 55, 1974.
4. Lee, S. J. et al., Ultra-low Power, High-Speed GaAs 256 bit Static RAM, presented at *IEEE GaAs IC Symp.,* Phoenix, AZ, 1983, 74.

5. Long, S. and Butner, S., *Gallium Arsenide Digital Integrated Circuit Design*, McGraw-Hill, New York, 1990, Chap. 2.

5a. Long, S. and Butner, S., *Gallium Arsenide Digital Integrated Circuit Design*, McGraw-Hill, New York, 1990, Chap. 3.

6. Lassen, P. S., High-Speed GaAs Digital Integrated Circuits for Optical Communication Systems, Ph.D Dissertation, Tech. U. Denmark, Lyngby, Denmark, 1993.

7. Miyamoto, Y., Yoneyama, M., and Otsuji, T., 40-Gbit/s TDM Transmission Technologies Based on High-Speed ICs, presented at *IEEE GaAs IC Symp.*, Atlanta, GA, 51, 1998.

8. Otsuji, T. et al., 40 Gb/s IC's for Future Lightwave Communications Systems, *IEEE J. Solid-State Circuits*, 32, 1363, 1997.

9. Otsuji, T. et al., A Super-Dynamic Flip-Flop Circuit for Broadband Applications up to 24 Gb/s Utilizing Production-Level 0.2 μm GaAs MESFETs, *IEEE J. Solid-State Circuits*, 32, 1357, 1997.

10. Lang, M., Wang, Z. G., Thiede, A., Lienhart, H., Jakobus, T. et al., A Complete GaAs HEMT Single Chip Data Receiver for 40 Gbit/s Data Rates, presented at *IEEE GaAs IC Symposium*, Atlanta, GA, 55, 1998.

11. Ichioka, T., Tanaka, K., Saito, T., Nishi, S., and Akiyama, M., An Ultra-High Speed DCFL Dynamic Frequency Divider, presented at *IEEE 1989 Microwave and Millimeter-Wave Monolithic Circuits Symposium*, 61, 1989.

12. Thiede, A. et al., Digital Dynamic Frequency Dividers for Broad Band Application up to 60 GHz, presented at *IEEE GaAs IC Symposium*, San Jose, CA, 91, 1993.

13. Rocchi, M. and Gabillard, B., GaAs Digital Dynamic IC's for Applications up to 10 GHz, *IEEE J. Solid-State Circuits*, SC-18, 369, 1983.

14. Shikata, M., Tanaka, K., Inokuchi, K., Sano, Y., and Akiyama, M., An Ultra-High Speed GaAs DCFL Flip Flop – MCFF (Memory Cell type Flip Flop), presented at *IEEE GaAs IC Symp.*, Nashville, TN, 27, 1988.

15. Maeda, T., Numata, K. et al., A Novel High-Speed Low-Power Tri-state Driver Flip Flop (TD-FF) for Ultra-low Supply Voltage GaAs Heterojunction FET LSIs, presented at *IEEE GaAs IC Symp.*, San Jose, CA, 75, 1993.

16. Madden, C. J., Snook, D. R., Van Tuyl, R. L., Le, M. V., and Nguyen, L. D., A Novel 75 GHz InP HEMT Dynamic Divider, presented at *IEEE GaAs IC Symposium*, Orlando, FL, 137, 1996.

17. Maeda, T. et al., An Ultra-Low-Power Consumption High-Speed GaAs Quasi-Differential Switch Flip-Flop (QD-FF), *IEEE J. Solid-State Circuits*, 31, 1361, 1996.

17a. Ware, R., Higgins, W., O'Hearn, K., and Tiernan, M., Growth and Properties of Very Large Crystals of Semi-Insulating Gallium Arsenide, presented at *18th IEEE GaAs IC Symp.*, Orlando, FL, 54, 1996.

18. Yuan, H. T., Shih, H. D., Delaney, J., and Fuller, C., The Development of Heterojunction Integrated Injection Logic, *IEEE Trans. Elect. Dev.*, 36, 2083, 1989.

19. Johnson, R. A. et al., Comparison of Microwave Inductors Fabricated on Silicon-on-Sapphire and Bulk Silicon, *IEEE Microwave and Guided Wave Letters*, 6, 323, 1996.

19a. Ashar, K. G., The Method of Estimating Delay in Switching Circuits and the Fig. of Merit of a Switching Transistor, *IEEE Trans. Elect. Dev.*, ED-11, 497, 1964.

19b. Asbeck, P. M., Bipolar Transistors, *High-Speed Semiconductor Devices*, S. M. Sze, Ed., John Wiley & Sons, New York, 1990, Chap. 6.

19c. Matthews, J. W. and Blakeslee, A. E., Coherent Strain in Epitaxially Grown Films, *J. Crystal Growth*, 27, 118, 1974.

20. Jensen, J., Hafizi, M., Stanchina, W., Metzger, R., and Rensch, D., 39.5 GHz Static Frequency Divider Implemented in AlInAs/GaInAs HBT Technology, presented at *IEEE GaAs IC Symposium*, Miami, FL, 103, 1992.

21. Lee, Q., Mensa, D., Guthrie, J., Jaganathan, S., Mathew, T. et al., 60 GHz Static Frequency Divider in Transferred-substrate HBT Technology, presented at *IEEE International Microwave Symposium*, Anaheim, CA, 1999.

22. Nary, K. R., Nubling, R., Beccue, S., Colleran, W. T. et al., An 8-bit, 2 Gigasample per Second Analog to Digital Converter, presented at *17th Annual IEEE GaAs IC Symposium,* San Diego, CA, 303, 1995.

23. Runge, K., Pierson, R. L., Zampardi, P. J., Thomas, P. B., Yu, J. et al., 40 Gbit/s AlGaAs/GaAs HBT 4:1 Multiplexer IC, *Electronics Letters,* 31, 876, 1995.

24. Yu, R., Beccue, S., Zampardi, P., Pierson, R., Petersen, A. et al., A Packaged Broadband Monolithic Variable Gain Amplifier Implemented in AlGaAs/GaAs HBT Technology, presented at *17th Annual IEEE GaAs IC Symposium,* San Diego, CA, 197, 1995.

25. Metzger, A. G., Chang, C. E., Campana, A. D., Pedrotti, K. D., Price, A. et al., A 10 Gb/s High Isolation 16×16 Crosspoint Switch, Implemented with AlGaAs/GaAs HBT's, to be published, 1999.

26. Lowe, K., A GaAs HBT 16×16 10 Gb/s/channel Cross-Point Switch, *IEEE J. Solid-State Circuits,* 32, 1263, 1997.

Index